1 MONTH OF
FREE
READING

at
www.ForgottenBooks.com

By purchasing this book you are eligible for one month membership to ForgottenBooks.com, giving you unlimited access to our entire collection of over 1,000,000 titles via our web site and mobile apps.

To claim your free month visit: www.forgottenbooks.com/free470858

ISBN 978-0-428-61311-2
PIBN 10470858

This book is a reproduction of an important historical work. Forgotten Books uses state-of-the-art technology to digitally reconstruct the work, preserving the original format whilst repairing imperfections present in the aged copy. In rare cases, an imperfection in the original, such as a blemish or missing page, may be replicated in our edition. We do, however, repair the vast majority of imperfections successfully; any imperfections that remain are intentionally left to preserve the state of such historical works.

A

MANUAL OF THE MOLLUSCA;

OR,

RUDIMENTARY TREATISE

OF

RECENT AND FOSSIL SHELLS.

BY

S. P. WOODWARD, F.G.S.

ASSOCIATE OF THE LINNEAN SOCIETY;
ASSISTANT IN THE DEPARTMENT OF MINERALOGY AND GEOLOGY
IN THE BRITISH MUSEUM; AND
MEMBER OF THE COTTESWOLDE NATURALISTS' CLUB.

ILLUSTRATED WITH
NUMEROUS ENGRAVINGS AND WOODCUTS.

LONDON:
JOHN WEALE, 59, HIGH HOLBORN.
MDCCCLI-VI.

I. 1851
II. 1854
III. 1856

A

MANUAL OF THE MOLLUSCA;

OR,

RUDIMENTARY TREATISE

OF

RECENT AND FOSSIL SHELLS.

BY

S. P. WOODWARD, F.G.S.

ASSOCIATE OF THE LINNEAN SOCIETY;
ASSISTANT IN THE DEPARTMENT OF MINERALOGY AND GEOLOGY
IN THE BRITISH MUSEUM; AND
MEMBER OF THE COTTESWOLDE NATURALISTS' CLUB.

ILLUSTRATED WITH
NUMEROUS ENGRAVINGS AND WOODCUTS.

LONDON:
JOHN WEALE, 59, HIGH HOLBORN.
MDCCCLI-VI.

LONDON:
PRINTED BY WILLIAM OSTELL,
HART STREET, BLOOMSBURY.

A

MANUAL OF THE MOLLUSCA;

OR, A

RUDIMENTARY TREATISE

OF

RECENT AND FOSSIL SHELLS.

BY

S. P. WOODWARD,

ASSOCIATE OF THE LINNEAN SOCIETY;
ASSISTANT IN THE DEPARTMENT OF MINERALOGY AND GEOLOGY
IN THE BRITISH MUSEUM; AND
MEMBER OF THE COTTESWOLDE NATURALISTS' CLUB.

ILLUSTRATED BY

A. N. WATERHOUSE AND JOSEPH WILSON LOWRY.

LONDON:

JOHN WEALE, 59, HIGH HOLBORN.

MDCCCLI.

LONDON
PRINTED BY WILLIAM OSTELL,
MART STREET, BLOOMSBURY.

PREFACE.

THIS Manual, which for six years has occupied the writer's unceasing attention, was intended as a companion to Col. Portlock's Geology; and the desire to make it worthy of that association has led to an amount of labour and expense which only a very extended circulation will repay.

The plan and title have been taken from the "Manuel des Mollusques" of M. Sander Rang, (1829)—incomparably the best work of its kind—for an acquaintance with which the writer was indebted, sixteen years ago, to his friend and master, VILLIAM LONSDALE—the founder of the "Devonian System" n Geology.

On the subject of classification and nomenclature he has ollowed the advice and example of his former colleague in the Ɉeological Society, the late PROF. EDWARD FORBES; without vhose approval he has seldom added to, or deviated from, the ɔractice and plan of the "History of British Mollusca."

That he was right in taking this course, he has now the sanction of the highest authority in this country for believing;—since the same scheme has been employed by PROF. OWEN in the Hunterian Lectures and Catalogue. It has also been adopted by DR. E. BALFOUR in the Madras Museum; by the REV. PROF. HENSLOW, in his Report to the British Association on the Formation of Typical Collections; and by PROF. MORRIS in his Catalogue of British Fossils.

It was the writer's desire, by abstaining from the introduction of personal and peculiar views, and by adhering to whatever was well established and sanctioned by the best examples, to make the work suitable for the use of Natural History Classes in the Universities.*

To facilitate reference, and meet the most general requirements, the number of large groups and genera of shells has been restricted as much as possible, and those less important or less understood, have been treated as "sub-genera." A great many duplicate and unnecessary names have been mentioned only, as will be seen by a glance at the Index, where they are printed *in italics;* the writer's own wishes coincide with those of the distinguished botanist SIR J. E. SMITH, that "the system should not be encumbered with such names;" but they have been admitted in deference to custom, and general opinion. It has even been suggested that an additional list of synonymes might be given at the end, and some progress was made in preparing one; but it was found that it would occupy the whole of the "Third Part," and consisted of names chiefly obsolete, or "based on misconception of characters, and of the purpose of generic appellations." (Forbes and Hanley, IV. 265.)†

The rules of the British Association, intended to secure uniformity, have called into existence a few active opponents, seeking to distinguish themselves by the employment of pre-Linnean, and MS. names, on the pretence of carrying out the "law of priority," (p. 60.) But this folly has reached its height and will fall into contempt when it has lost its novelty.‡

---

* The former parts have been already adopted as a text book at Edinburgh, in the largest natural history class in the kingdom, under Prof. E. Forbes; and also by Profs. King and Melville, of Queen's College, Galway; Prof. Tennant, of King's, and Prof. Morris, of University Coll., London; and Prof. Sedgwick at Cambridge.

† All the blundering and bad spelling of English and French genus-makers will be found carefully recorded in the "Index Generum Malaco-zoorum," by the accurate and lamented DR. HERRMANNSEN,—a work indispensable to every writer on Conchology.

‡ One example will suffice. In an "Athenæum" report, by Prof. E. Forbes, the name "Lottia fulva" was *misprinted* "Jothia fulva:" but although immediately corrected, the *erratum* was formally installed as a "new genus," in the works of Gray, Philippi, Catlow, Adams, and other conchologists!

The investigation of dates is the most disheartening work upon which the time of an author can be employed; it is never safe to take them second-hand, and even reference to the original works is not always satisfactory.*

Two lists of *Errata* have been given, and it is earnestly recommended that these corrections be made *with pen and ink* at the places indicated. Small and self-evident typographical errors have not been enumerated; the difficulty of avoiding them, in a treatise of this kind, can only be appreciated by those who have had personal experience.

Those portions of the work have been treated in most detail which throw light on particular branches of anatomy and physiology; or on great natural-history problems, such as the value of species and genera, and the laws of geographical and geological distribution. It is in these departments that the affinity of natural science to the highest kinds of human knowledge is most distinctly seen; and in them the richest and noblest results are to be obtained. For to the thoughtful and earnest investigator, nature ever discloses indications of harmony and order, and reflects the attributes of the Maker.

The recreations of the young seldom fail to exercise a serious influence on after life; and the utility of their pursuits must greatly depend on the spirit in which they are followed. If wisely chosen and conscientiously prosecuted they may help to form habits of exact observation; they may train the eye and mind to seize upon characteristic facts, and to discern their real import; to discriminate between the essential and the accidental, and to detect the relations of phenomena, however widely separated and apparently unlike. In this way "la belle Science," (as Mr. Gaskoin calls Conchology!) may acquire the influence of pursuits more usually resorted to for mental development and discipline.

The writer desires again to acknowledge the assistance he

---

* The *dates* on the title pages of Journals and Transactions of Scientific Societies, are not usually *dates of publication*, but refer to the years *for which* they are issued to the subscribers. It is almost impossible, afterwards, to correct these false dates.

received in preparing particular portions of the work; and especially from Mr. T. Davidson, FGS. in the investigation of the Brachiopoda; Mr. J. W. Wilton of Gloucester, in reference to the lingual dentition of the Gasteropods; Mr. T. Huxley, FRS. for the revision of the chapter on the Tunicata; and to Mr. Albany Hancock, of Newcastle, for advice and information, often only to be obtained by new and careful investigations.

To Mr. H. Cuming he is indebted for the use of books and specimens; to the officers of the Museum, especially to Dr. Baird, and Mr. Waterhouse, for encouragement and sympathy; and to the Council of the Geological Society (1853-4) for the expression of their approval by the Wollaston award.

The wood-cuts have been principally executed by Miss A. N. Waterhouse of Marlborough House, from original drawings by the author; and although printed from stereotypes they have the advantage of accurately representing what was wished to be shewn.

The engravings of Mr. Wilson Lowry, speak for themselves; many of the figures are from the specimens in his cabinet; and the interest he has taken in the work will be seen in the care with which the technical characters of the shells are expressed.

BARNSBURY, March, 1856.

Directions to the Binder.

In binding the complete work, the Tables of Contents of the three Parts should be placed together at the beginning.

The Plates should be arranged in pairs, face to face, with the Explanation opposite to each.

CONTENTS.

CHAPTER IV.

CHAPTER V.

CHAPTER VI.

SYNOPSIS OF THE GENERA.

CONTENTS.

CONCLUSION.

CHAPTER I.

CHAPTER II.

CHAPTER III.

CHAPTER IV.

CHAPTER V.

SUPPLEMENTARY NOTES ON THE MOLLUSCA.

NOTICE.

THE second part of this Manual is now in preparation, and will be published early in the summer. It will contain an account of the remaining orders of shell-fish: a chapter on the Geographical Distribution of the Mollusca, with a Map of the Marine and Terrestrial Provinces ; a chapter on the distribution of Fossil Shells ; another on the methods of collecting and preserving Land, Fresh-water, and Sea-shells ; the Preface ; and an Index of the genera and technical terms.

The writer desires to acknowledge his obligations to Mr. Hugh Cumming, Professor Edward Forbes, and other gentlemen who have assisted him by advice, and the loan of specimens ; also to Mr. Van Voorst, for permission to copy some interesting figures from the " British Mollusca ;" and his thanks are most especially due to Mr. John Edward Gray, Keeper of the Zoological Department of the British Museum, for access to his library and cabinet, and the use of some of the best engravings which illustrate these pages.

Kingdom ANIMALIA.

Sub-kingdom I. VERTEBRATA.

Class I. Mammalia.
II. Aves.
III. Reptilia.
IV. Pisces.

Sub-kingdom II. MOLLUSCA.

Class I. Cephalopoda.
II. Gasteropoda.
III. Pteropoda.
IV. Brachiopoda.
V. Conchifera.
VI. Tunicata.

Sub-kingdom III. ARTICULATA.

Class I. Insecta.
II. Arachnida.
III. Crustacea.
IV. Cirripeda.
V. Anellata.
VI. Entozoa.

Sub-kingdom IV. RADIATA.

Class I. Acalepha.
II. Echinodermata.
III. Zoophyta.
IV. Foraminifera.
V. Infusoria.
VI. Amorphozoa.

A

MANUAL OF THE MOLLUSCA.

INTRODUCTION.

CHAPTER I.

ON THE POSITION OF THE MOLLUSCA IN THE ANIMAL KINGDOM.

ALL known animals are constructed upon four different types, and constitute as many natural divisions or sub-kingdoms.

1. The first of these primary groups is characterized by an internal skeleton, of which the essential, or ever-present part, is a backbone, composed of numerous joints, or *vertebræ*. These are the animals most familiar to us; beasts, birds, reptiles, and fishes, are four classes which agree in this one ·respect, and are hence collectively termed vertebrate animals, or the *vertebrata*.

2. Another type is exemplified in the common garden-snail, the nautilus, and the oyster; animals whose soft bodies are protected by an external shell, which is harder than bone, and equally unlike the skeleton of fishes, and the hard covering of the crab and lobster. These creatures form the subject of the present history, and are called *mollusca.**

---

* *Mollusca* soft (animals), from *mollis*. The Greeks termed them *Malakia*, whence the modern word *Malacology*, or the study of shell-fish.

3. The various tribes of insects, spiders, crabs, and worms, have no internal skeleton; but to compensate for it, their outer integument is sufficiently hard to serve at once the purposes of bones, and of a covering and defence. This external armature, like the bodies and limbs which it covers, is divided into segments or joints, which well distinguishes the members of this group from the others. The propriety of arranging worms with insects will be seen, if it be remembered, that even the butterfly and bee commence existence in a very worm-like form. This division of jointed animals bears the name of the *articulata*.

4. The fourth part of the animal kingdom consists of the coral-animals, star-fishes, sea-jellies, and those countless microscopic beings which swarm in all waters. Whilst other animals are bi-lateral, or have a right and left side, and organs arranged in pairs,—these have their organs placed in a circle around the mouth or axis of the body, and have hence obtained the appellation of *radiata*.

These groups illustrate successively the grand problems of animal economy. The lower divisions exhibit the perfectionizing of the functions of nutrition and reproduction; the higher groups present the most varied and complete development of the senses, locomotive powers, and instincts. We may also trace in them an ideal progression from the simplest to the most complicated structure and conditions. Commencing with the Infusorial monad, we may ascend in imagination by a succession of closely allied forms, to the sea-urchin and holothuria[*]; and thence by the lowest organized worms, upwards to the flying insect. Or, starting at the same point, we may pass from the polypes to the tunicaries; and from the higher kinds of shell-fish to the true fishes, and so on to those classes whose physical organization is most nearly identical with our own.

The *mollusca* are thus related to two of the other primary groups;—by the affinity of their simpler forms to the *zoophytes*,

[*] See the History of British Star-fishes, by Professor E. Forbes.

and of their highest class to the fishes;—to the cirripedes and other articulate animals, they present only superficial and illusive resemblance.

And further, we shall find that although it is customary to speak of shell-fish as "less perfect" animals, yet they really attain the perfection of their own type of structure; indeed it would seem to have been impossible to make any further advance, physical, or psychical, except by adopting a widely different *plan* from that on which the molluscous animals have been constructed.

The evidence afforded by geological researches at present tends to shew that the four leading types of animal structure have existed simultaneously from the very beginning of life upon the globe;* and though perpetually varying in the form under which they were manifested, they have never since entirely ceased to exist.

By adding to the living population of the world, those forms which peopled it in times long past, we may arrive at some dim conception of the great scheme of the animal kingdom. And if at present we see not the limits of the temple of nature, nor fully comprehend its design,—at least we can feel sure that there is a boundary to this present order of things; and that there has been a plan, such as we, from our mental constitution, are able to appreciate, and to study with ever-increasing admiration.

---

* Mr E. Logan, Geological Surveyor of the Canadas, has discovered *foot prints of a tortoise,* near Montreal, in the "Lingula Shale," or oldest fossibiferous rock at present known.

CHAPTER II.

CLASSES OF THE MOLLUSCA.

THE *mollusca* are animals with soft bodies, enveloped in a muscular skin, and usually protected by a univalve or bivalve shell. That part of their integument which contains the viscera and secretes the shell, is termed the *mantle ;* in the univalves it takes the form of a sac, with an opening in front, from which the head and locomotive organs project : in the bivalves it is divided into two lobes.

The univalve mollusca are *encephalous*, or furnished with a distinct head ; they have eyes and tentacula, and the mouth is armed with jaws. Cuvier has divided them into three classes, founded on the modifications of their feet, or principal locomotive organs.

1. The cuttle-fishes constitute the first-class, and are termed *cephalopoda*,* because their feet, or more properly *arms*, are attached to the head, forming a circle round the mouth. -

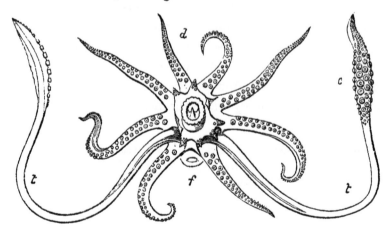

Fig. 1.†   *Oral aspect of a Cephalopod.*

* From *Cephale*, the head and *poda* feet. See the frontispiece and pl. I.

† Fig. 1. *Loligo vulgaris*, Lam. ¼. From a specimen taken off Tenby, by J. S. Bowerbank, Esq. The mandibles are seen in the centre, surrounded by the circular lip, the buccal membrane (with two rows of small cups on its lobes), the eight sessile arms, and the long pedunculated tentacles (t), with their enlarged extremities or clubs (c). The *dorsal* arms are lettered (d), the funnel (f).

2. In the *gasteropoda*,* or snails, the under side of the body forms a single muscular foot, on which the animals creep or glide.

Fig. 2.   *A Gasteropod.*†

3. ·The *pterpoda*‡ only inhabit the sea, and swim with a pair of fins, extending outwards from the sides of the head.

Fig. 3.   *A Pteropod.* §

The other mollusca are *acephalous*, or destitute of any distinct head; they are all aquatic, and most of them are attached, or have no means of moving from place to place. They are divided into three classes, characterized by modifications in their breathing-organ and shell.

4. The *brachiapoda*¶ are bivalves, having one shell placed on the back of the animal, and the other in front; they have no

* *Gaster*, the under side of the body.

† Fig. 2. *Helix desertorum.* Forskal. From a living specimen in the British Museum, March, 1850.

‡ *Pteron*, a wing.

§ Fig. 3. *Hyalœa tridentata*, Lam., from Quoy and Gaimard.

¶ *Brachion*, an arm; these organs were supposed to take the place of the feet in the preceding classes.

special breathing organ, but the mantle performs that office ; they take their name from two long ciliated arms, developed from the sides of the mouth, with which they create currents that bring them food.

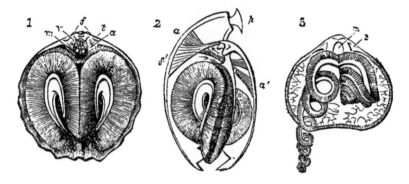

Fig. 4, 5, 6.   *Brachiopoda*.*

5. The *conchifera*,† or ordinary bivalves, (like the oyster), breathe by two pairs of gills, in the form of flat membraneous plates, attached to the mantle ; one valve is applied to the right, the other to the left side of the body.

6. The *tunicata* have no shell, but are protected by an elastic, gelatinous tunic, with two orifices ; the breathing-organ takes the form of an inner *tunic*, or of a riband stretched across the internal cavity.

Five of these modifications of the molluscan type of organization, were known to Linnæus, who referred the animals of all his genera of shell-fish to one or other of them ;‡ but unfortunately he did not himself adopt the truth which he was the first

* Fig. 4. (3). *Rhynchonella psittacea*, Chem, sp., dorsal valve, with the animal (after Owen). 5, 6, *Terebratula australis*, Quoy. From specimens collected by Mr. Jukes. (2). Ideal side view of both valves, (f, the retractor muscles, by which the valves are opened). (1). Dorsal valve. These woodcuts have been kindly lent by Mr. J. E. Gray.

† *Conchifera*, Shell-bearers.

‡ The Linnæan types were—Sepia, Limax, Clio, Anomia, Ascidia. *Terebratula* was included with Anomia, its organization being unknown.

to see; and here, as in his botany, employed an artificial, in preference to a natural method.

The systematic arrangement of natural objects ought not, however, to be guided by convenience, nor "framed merely for the purposes of easy remembrance and communication." The true method must be suggested by the objects themselves, by their qualities and relations;—it may not be easy to learn,—it may require perpetual modification and adjustment,—but inasmuch as it represents the existing state of knowledge it will aid in the UNDERSTANDING of the subject, whereas a "dead and arbitrary arrangement" is a perpetual bar to advancement, "containing in itself no principle of progression." (*Coleridge.*)

 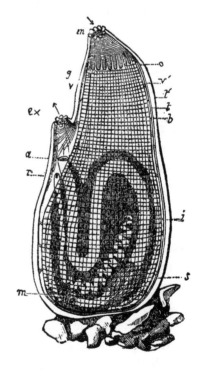

Fig. 7. *A Bivalve.**   Fig. 8. *A Tunicary.†*

Mya truncata, L. ½. From Forbes and Hanley.

† *Ascidia mentula*, Müll. Ideal representation; from a specimen dredged by Mr. Bowerbank, off Tenby.

Chapter III.

HABITS AND ECONOMY OF THE MOLLUSCA.

EVERY living creature has a history of its own; each has charac-
teristics by which it may be known from its relatives; each has
its own territory, its appropriate food, and its duties to perform
in the economy of nature. Our present purpose, however, is to
point out those circumstances and trace the progress of those
changes which are not peculiar to individuals or to species,
but have a wider application, and form the history of a great
class.

In their infancy the molluscous animals are more alike, both
in appearance and habits, than in after life; and the fry of the
acquatic races are almost as different from their parents as the
caterpillar from the butterfly. The analogy, however, is reversed
in one respect; for whereas the adult shell-fish are often sedentary, or walk with becoming gravity, the young are all swimmers,
and by means of their fins and the ocean-currents, they travel
to long distances, and thus diffuse their race as far as a suitable
climate and conditions are found. Myriads of these little
voyagers drift from the shores into the open sea and there perish;
their tiny and fragile shells become part of a deposit that is for
ever increasing over the bed of the deep sea,—at depths too
great for any living thing to inhabit. (*Forbes.*)

Some of these little creatures shelter themselves beneath the
shell of their parent for a time, and many can spin silken threads
with which they moor themselves, and avoid being drifted away.
They all have a protecting shell, and even the young bivalves
have eyes at this period of their lives, to aid them in choosing an
appropriate locality.

After a few days, or even less, of this sportive existence, the sedentary tribes settle in the place they intend to occupy during the remainder of their lives. The tunicary cements itself to rock or sea-weed; the shipworm adheres to timber, and the *pholas* and *lithodomus* to limestone rocks, in which they soon excavate a chamber which renders their first means of anchorage unnecessary. The *mya* and razor-fish burrow in sand or mud; the mussel and *pinna* spin a byssus; the oyster and *spondylus* attach themselves by spines or leafy expansions of their shell; the *brachiopoda* are all fixed by similar means, and even some of the gasteropods become voluntary prisoners, as the *hipponyx* and *vermetus*.

Other tribes retain the power of travelling at will, and shift their quarters periodically, or in search of food; the river-mussel drags itself slowly along by protruding and contracting its flexible foot; the cockle and *trigonia* have the foot bent, enabling them to make short leaps; the scallop (*pector opercularis*) swims rapidly by opening and shutting its tinted valves. Nearly all the gasteropods creep like the snail, though some are much more active than others; the pond-snails can glide along the surface of the water, shell-downwards; the nucleobranches and pteropods swim in the open sea. The cuttle-fishes have a strange mode of walking, head-downwards, on their outspread arms; they can also swim with their fins, or with their webbed arms, or by expelling the water forcibly from their branchial chamber; the calamary can even strike the surface of the sea with its tail, and dart into the air like the flying-fish. (*Owen.*)

By these means the mollusca have spread themselves over every part of the habitable globe; every region has its tribe; every situation its appropriate species; the land-snails frequent moist places, or woods, or sunny banks and rocks, climb trees, or burrow in the ground. The air-breathing limneids live in fresh-water, only coming occasionally to the surface; and the auriculas live on the sea-shore, or in salt-marshes. In the sea, each zone of depth has its molluscous fauna. The limpet and periwinkle live between tide-marks, where they are left dry twice a-day; the *trochi* and *purpuræ* are found at low water, amongst the sea-weed; the mussel affects muddy shores, the cockle rejoices in extensive sandy flats. Most of the finely-coloured shells of the tropics are found in shallow water, or amongst the breakers. Oyster-banks are usually in four or five fathom water; scallop-banks at twenty fathoms. Deepest of all, the *terebratulæ* are found, commonly at fifty fathoms, and sometimes at one hundred fathoms, even in Polar seas. The fairy-like *pteropoda*, the oceanic-snail, and multitudes of other floating molluscs, pass their lives on the open sea, for ever out of sight of land; whilst the *litiopa* and *scyllæa* follow the gulf-weed in its voyages, and feed upon the green delusive banks.

The food of the mollusca is either vegetable, infusorial, or animal. All the land-snails are vegetable-feeders, and their depredations are but too well known to the gardener and farmer; many a crop of winter corn and spring tares has been wasted by the ravages of the "small grey slug." They have their likings, too, for particular plants, most of the pea-tribe and cabbage-tribe are favourites, but they hold white mustard in abhorence, and fast or shift their quarters while that crop is on the ground.* Some, like the "cellar-snail," feed on cryptogamic vegetation, or on decaying leaves; and the slugs are attracted by *fungi*, or any odorous substances. The round-mouthed sea-snails are nearly all vegetarians, and consequently limited to the shore and the shallow waters in which sea-weeds grow. Beyond fifteen fathoms, almost the only vegetable production is the nullipore; but here corals and horny zoophytes take the place of *algæ* and afford a more nutritious diet.

The whole of the bivalves, and other head-less shell-fish, live on infusoria, or on microscropic vegetables, brought to them by the current which their ciliary apparatus perpetually excites; such, too, must be the sustenance of the *magilus*, sunk in its

---

* Dilute lime-water and very weak alkaline solutions are more fatal to snails than even salt.

coral bed, and of the *calyptræa*, fettered to its birth-place by its calcarious foot.

The carnivorous tribes prey chiefly on other shell-fish, or on zoophytes; since, with the exception of the cuttle-fishes, their organization scarcely adapts them for pursuing and destroying other classes of animals. One remarkable exception is formed by the *stylina*, which lives parasitically on the star-fish and sea-urchin; and another by the testacelle, which preys on the common earth-worm, following it in its burrow, and wearing a buckler, which protects it in the rear.

Most of the siphonated univalves are animal-feeders; the carrion-eating stromb and whelk consume the fishes and other creatures whose remains are always plentiful on rough and rocky coasts. Many wage war on their own relatives, and take them by assault; the bivalve may close, and the operculated nerite retire into his home, but the enemy, with rasp-like tongue, armed with silicious teeth, files a hole through the shell,—vain shield where instinct guides the attack! Of the myriads of small shells which the sea heaps up in every sheltered "ness," a large proportion will be found thus bored by the whelks and purples; and in fossil shell-beds, such as that in the Touraine, nearly half the bivalves and sea-snails are perforated,—the relics of antediluvian banquets.

This is on the shore, or on the bed of the sea; far away from land the *carinaria* and *firola* pursue the floating *acalephe;* and the argonaut, with his relative the *spirula*, both carnivorous, are found in the "high seas," in almost every quarter of the globe. The most active and rapacious of all are the calamaries and cuttles, who vindicate their high position in the naturalists' "system," by preying even on fishes.

As the shell-fish are great eaters, so in their turn they afford food to many other creatures; fulfilling the universal law of eating, and being eaten. Civilized man still swallows the oyster, although snails are no longer reckoned "a dainty dish;" mussels, cockles, and periwinkles are in great esteem with children and the other unsophisticated classes of society; and so are scallops and the *haliotis*, where they can be obtained. Two kinds of whelk are brought to the London market in great quantities; and the arms of the cuttle-fish are eaten by the Neapolitans, and also by the East Indians and Malays. In seasons of scarcity, vast quantities of shell-fish are consumed by the poor inhabitants of the Scotch and Irish coasts.* Still more are regularly collected for bait; the calamary is much used in the cod-fishery, off New-foundland, and the limpet and whelk on our own coasts.

Many wild animals feed on shell-fish; the rat and the racoon seek for them on the sea-shore when pressed by hunger; the South-American otter, and the crab-eating opossum constantly resort to salt-marshes, and the sea, and prey on the mollusca; the great whale lives habitually on the small floating pteropods; sea-fowl search for the litoral species at every ebbing tide; whilst, in their own element, the marine kinds are perpetually devoured by fishes. The haddock is a "great conchologist;" and some good northern sea-shells have been rescued, unbroken, from the stomach of the cod; whilst even the strong valves of the *cyprina* are not proof against the teeth of the cat-fish *(anarhicas)*.

They even fall a prey to animals much their inferiors in sagacity; the star-fish swallows the small bivalve entire, and dissolves the animal out of its shell; and the bubble-shell *(phyline)*, itself predacious, is eaten both by star-fish and sea-anemone *(actinia)*.

The land-snails afford food to many birds, especially to the thrush tribe; and to some insects, for the luminous larva of the glow-worm lives on them, and some of the large predacious beetles *(e. g. carabus violaceus* and *goerius olens)*, occasionally kill slugs.

The greatest enemies of the *mollusca*, however, are those of their own nation; scarcely one-half the shelly tribes graze peace-

* See Hugh Miller's " Scenes and Legends of the North of Scotland."

fully on sea-weed, or subsist on the nutrient particles which the sea itself brings to their mouths; the rest browse on living zoophytes, or prey upon the vegetable-feeders.

Yet in no class is the instinct of "self-preservation" stronger, nor the means of defence more adequate; their shells seem expressly given to compensate for the slowness of their movement, and the dimness of their senses. The cuttle-fish escapes from attack by swimming backwards and beclouding the water with an inky discharge; and the sea-hare *(aplysia)* pours out, when irritated, a copious purple fluid, formerly held to be poisonous. Others rely on passive resistance, or on concealment for their safety. It has been frequently remarked that molluscs resemble the hue and appearance of the situation they frequent; thus, the limpet is commonly overgrown with *balani* and sea-weed, and the ascidian with zoophytes, which form an effectual disguise; the *lima* and *modiola* spin together a screen of grotto-work. One ascidian *(a. cochligera)* coats itself with shell-sand, and the carrier-trochus cements shells and corals to the margin of its habitation, or so loads it with pebbles, that it looks but like a little heap of stones.

It must be confessed that the instincts of the shell-fish are of a low order, being almost limited to self-preservation, the escape from danger, and the choice of food. Their history offers none of those marvels which the entomologist loves to relate. An instance of something like social feeling has been observed in a Roman snail *(helix pomatia)* who, after escaping from a garden, returned to it in quest of his fellow-prisoner;—but the accomplished naturalist who witnessed the circumstance hesitated to record a thing so unexampled. The limpet, too, if we may trust the observations of Mr. Robert, of Lyme Regis, is fond of home, or at least possesses a knowledge of topography, and returns to the same roost after an excursion with each tide. Professor Forbes has immortalized the sagacity of the razor-fish, who submits to be salted in his hole, rather than expose himself to be caught, after finding that the enemy is lying in wait for him.

On the other hand, Mr. Bowerbank has a curious example of "instinct at fault," in the fossil spine of a sea-urchin, which appears to have been drilled by a carnivorous gasteropod!

We have spoken of shell-fish as articles of food, but they have other uses, even to man; they are the toys of children, who hear in them the roaring of the sea; they are the pride of "collectors" —whose wealth is in a cone or "wentle-trap;"* and they are the ornaments of barbarous tribes. The Friendly-islander wears the orange-cowry as a mark of chieftanship (*Stutchbury*), and the New Zealander polishes the *elenchus* into an ornament more brilliant than the "pearl ear-drop" of classical or modern times. (*Clarke.*) One of the most beautiful substances in nature is the shell-opal, formed of the remains of the ammonite. The forms and colours of shells (as of all other natural objects), answer some particular purpose, or obey some general law; but besides this, there is much that seems specially intended for our study, and calculated to call forth enlightened admiration. Thus the tints of many shells are concealed during life by a dull external coat, and the pearly halls of the nautilus are seen by no other eyes than ours. Or descending to mere "utility," how many tracts of coast are destitute of limestone, but abound in shell-banks which may be burned into lime; or in shell-sand, for the use of farmers.†

---

* The extravagant prices that have been given for rare shells, are less to be regretted, because they have induced voyagers to collect. Mere shell-collecting, however, is no more *scientific* than pigeon-fancying, or the study of old china. For *educational* purposes the best shells are the *types* of genera, or species which illustrate particular points of structure; and, fortunately for students, the prices are much diminished of late years. A *Carinaria* once "worth 100 guineas" (Sowerby) is now worth 1s. only; a Wentle-trap which fetched 40 guineas in 1701 (Rumphius) was worth only 20 guineas in 1753, and may now be had for 5s.! The *Conus gloria-maris* has fetched £50 more than once, and *Cypræa* umbilicata has been sold for £30 this year, 1850.

† Shell-sand is only beneficial on peaty soils, or heavy clay land. It sometimes hardens into limestone, as on the coast of Devon; and at Guadaloupe, where it contains litoral shells and human skeletons of recent date.

Not much is known respecting the individual duration of the shell-fish, though their length of life must be very variable. Many of the aquatic species are annuals, fulfilling the cycle of their existence in a single year; whole races are entombed in the wintry tide of mud that grows from year to year in the beds of rivers, and lakes, and seas; thus, in the Wealden clay we find layer above layer of small river-snails, alternating with thin strata of sediment, the index of immeasureably distant years. Dredgers find that whilst the adults of some shell-fish can be taken at all seasons, others can be obtained late in the autumn or winter only; those caught in spring and summer being young, or half-grown; and it is a common remark that *dead* shells (of some species) can be obtained of a larger size than any that we find alive, because they attain their full-growth at a season when our researches are suspended. Some species require part of two years for their full development; the young of the *doris* and *eolis* are born in the summer time, in the warm shallows near the shore; on the approach of winter they retire to deeper water, and in the following spring return to the tidal rocks, attain their full-growth early in the summer, and after spawning-time disappear.

The land-snails are mostly biennial; hatched in the summer and autumn, they are half-grown by the winter-time, and acquire their full-growth in the following spring or summer. In confinement, a garden-snail will live for six or eight years; but in their natural state it is probable that a great many die in their second winter, for clusters of empty shells may be found, adhering to one another, under ivied walls, and in other sheltered situations; the animals having perished in their hybernation. Some of the spiral sea-shells live a great many years, and tell their age in a very plain and interesting manner, by the number of fringes *(varices)* on their whirls; the contour of the *ranella* and *murex* depends on the regular recurrence of these ornaments, which occur after the same intervals in well-fed individuals, as in their less fortunate kindred. The Ammonites appear, by their *varices,* or periodic mouths (pl. III., fig. 3), to have lived and continued growing for many years.

Many of the bivalves, like the mussel and cockle, attain their full-growth in a year. The oyster continues enlarging his shell by annual "shoots," for four or five years, and then ceases to grow outwards; but very aged specimens may be found, especially in a fossil state, with shells an inch or two in thickness. The giant-clam *(tridacna)*, which attains so large a size that poets and sculptors have made it the cradle of the sea-goddess,— must enjoy an unusual longevity; living in the sheltered lagoons of coral-islands, and not discursive in its habits, the corals grow up around, until it is often nearly buried by them; but although there seems to be no certain limit to its life (though it may live a century for all that we know), yet the time will probably come when it will be overgrown by its neighbours, or choked with sediment.

The fresh-water molluscs of cold climates bury themselves during winter, in the mud of their ponds and rivers; and the land-snails hide themselves in the ground, or beneath moss and dead leaves. In warm climates they become torpid during the hottest and driest part of the year.

Those genera and species which are most subject to this "summer sleep," are remarkable for their tenacity of life; and numerous instances have been recorded of their importation from distant countries, in a living state. In June, 1850, a living pond-mussel was sent to Mr. Gray, from Australia, which had been more than a year out of water.*   The pond-snails *(ampullariæ)* have been found alive in logs of mahogany from Honduras (Mr. Pickering); and M. Caillaud carried some from Egypt to Paris, packed in saw-dust. Indeed, it is not easy to ascertain the limit of their endurance; for Mr. Laidlay having placed a number in a drawer for this purpose, found them alive after *five*

---

* "It was alive 498 days after it was taken from the pond; and in the interim had been only twice for a few hours in water, to see if it was alive."— *Rev. W. O. Newnham.*

years, although in the warm climate of Calcutta.   The *cyclosto-mas*, which are also *operculated*, are well known to survive impri-sonments of many months ; but in the ordinary land-snails such cases are more remarkable.   Some of the large tropical *bulimi*, brought by Lieutenant Graves from Valparaiso, revived after being packed, some for thirteen, others for twenty months.   In 1849, Mr. Pickering received from Mr. Wollaston a basket-full of Madeira snails (of twenty or thirty different species), three-fourths of which proved to be alive, after several months' confine-ment, including a sea-voyage.   Mr. Wollaston has himself told us that specimens of two Madeira snails (*helix papilio* and *tecti-formis*) survived a fast and imprisonment in pill-boxes, of two years and a half, and that a large number of the small *helix turricula*, brought to England at the same time, were *all* living after being inclosed in a dry bag for a year and a half.

But the most interesting example of resuscitation occurred to a specimen of the Desert snail, from Egypt, chronicled by Dr. Baird.*   This individual was fixed to a tablet in the British Museum, on the 25th of March, 1846 ; and on March 7th, 1850, it was observed that he must have come out of his shell in the interval (as the paper had been discoloured, apparently in his attempt to get away) ; but finding escape impossible, had again retired, closing his aperture with the usual glistening film ; this led to his immersion in tepid water, and marvellous recovery. He is now (March 13th, 1850) alive and flourishing, and has sat for his portrait.   (Fig. 2.)

The permanency of the shell-bearing races is effectually pro-vided for by their extreme fecundity ; and though exposed to a hundred dangers in their early life, enough survive to re-people the land and sea abundantly.   The spawn of a single *doris* may contain 600,000 eggs (Darwin) ;   a river-mussel has been estimated to produce 300,000 young in one season, and the oyster cannot be much less prolific.   The land-snails have fewer enemies, and, fortunately, lay fewer eggs.

* An. Nat. Hist. 1850.

Lastly, the *mollusca* exhibit the same instinctive care with insects and the higher animals, in placing their eggs in situations where they will be safe from injury, or open to the influences of air and heat, or surrounded by the food which the young will require. The tropical *bulimi* cement leaves together, to protect and conceal their large, bird-like, eggs; the slugs bury theirs in the ground; the oceanic-snail attaches them to a floating raft;

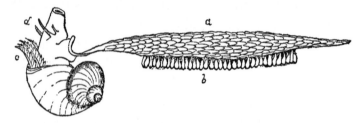

Fig. 9.   *Ianthina with its raft.*

and the argonaut carries them in her frail boat. The horny capsules of the whelk are clustered in groups, with spaces pervading the interior, for the free passage of sea-water; and the nidamental ribbon of the *doris* and *eolis* is attached to a rock, or some solid surface from which it will not be detached by the waves. The river-mussel and *cyclas* carry their parental care still further, and nurse their young in their own mantle, or in a special *marsupium*, designed, like that of the opossum, to protect them until they are strong enough to shift for themselves.

If any one imbued with the spirit of Paley or Chateaubriand, should study these phenomena, he might discover more than the "barren facts" which alone appear, without significance, to the unspiritual eye; he would see at every step fresh proofs of the wisdom and goodness of God, who thus manifests his greatness by displaying the same care for the maintenance of his feeblest creatures, as for the well-being of man, and the stability of the world.

CHAPTER IV.

STRUCTURE AND PHYSIOLOGY OF THE MOLLUSCA.

MOLLUSCOUS animals possess a distinct nervous system, instruments appropriated to the five senses, and muscles by which
they execute a variety of movements.   They have organs, by
which food is procured and digested,—a heart, with arteries
and veins, through which their colourless fluids circulate,—a
breathing-organ,—and in most instances, a protecting shell.
They produce eggs ; and the young generally pass through one
preparatory, or larval, stage.

The nervous system, upon which sensation and the exercise
of muscular motion depend, consists of a brain or principal
centre, and of various nerves possessing distinct properties : the
optic nerves are only sensible of light and colours ; the *auditory*
nerves convey impressions of sound ; the *olfactory*, of odours ;
the *gustatory*, of flavours ; whilst the nerves of touch or feeling
are widely·diffused, and indicate in a more general way the presence of external objects.   The nerves by which motion is produced, are distinct from these, but so accompany them as to appear like parts of the same cords.   Both kinds of nerves cease
to act when their connection with the centre is interrupted or
destroyed.   There is reason to believe, that most of the movements of the lower animals result from the reflection of external
stimulants (like the process of *breathing* in man), without the
intervention of the will.*

In the *mollusca*, the principal part of the nervous system is
a ring surrounding the throat *(œsophagus)*, and giving off nerves
to different parts of the body.   The points from which the nerves
radiate, are enlargements, termed centres *(ganglia)*, those on the

* See Müller's Elements of Philosophy, edited by Dr. Baly.

sides and upper part of the ring represent the brain, and suppl[
nerves to the eyes, tentacles, and mouth; other centres, con
nected with the lower side of the œsophageal ring, send nerve
to the foot, viscera, and respiratory organ. In the bivalves, th
branchial centre is the most conspicuous, and is situated on th
posterior adductor muscle. In the tunicaries, the correspondin[
nervous centre may be seen between the two orifices in th
muscular tunic. This scattered condition of the nervous centre
is eminently characteristic of the entire sub-kingdom.

Organs of special sense.—Sight. The eyes are two in num
ber, placed on the front or sides of the head; sometimes the
are *sessile,* in others stalked, or placed on long pedicels *(om
matophora).* The eyes of the cuttle-fishes resemble those (
fishes in their large size and complicated structure. Each con
sists of a strong fibrous globe *(slerotic),* transparent in fron
(cornea), with the opposite internal surface *(retina)* covered b
a dark pigment which receives the rays of light. This chambe
is occupied by an aqueous humour, a crystalline lens, and a v
treous humour, as in the human eye. In the *strombidœ,* th
eye is not less highly organised, but in most of the *gasteropod*
it has a more simple structure, and perhaps only possesses ser
sibility of light without the power of distinct vision. Th
larval bivalves have also a pair of eyes in the normal positic
(fig. 30) near the mouth; but their development is not con
tinued, and the adults are either eyeless, or possess merely ri
dimentary organs of vision, in the form of black dots *(ocell*
along the margin of the mantle.* These supposed eyes hav
been detected in a great many bivalves, but they are most con
spicuous in the scallop, which has received the name of *arg[*
from Poli, on this account (fig. 10).

In the tunicaries similar *ocelli* are placed between the te]
tacles which surround the orifices.

* "Each possesses a cornea, lens, choroid and nerve; they are, witho
doubt, organs of vision."—*Garner.*

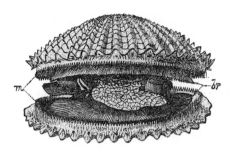

Fig. 10.   *Pecten varius.**

Sense of Hearing.   In the highest cephalopods, this organ consists of two cavities in the rudimentary cranium which protects the brain ; a small calcarious body or *otolithe* is suspended in each, as in the vestibular cavities of fishes.† Similar auditory capsules occur near the base of the tentacles in the *gasteropoda*, and they have been detected, by the vibration of the otolithes, in many bivalves and brachiopods.   With the exception of *tritonia* and *eolis*, none of mollusca have been observed to emit sounds.   (*Grant*).

Sense of Smell.   This faculty is evidently possessed by the cuttle-fishes and gasteropods ; snails discriminate their food by it, slugs are attracted by offensive odours, and many of the marine *zoophaga* may be taken with animal baits.   In the pearly nautilus, there is a hollow plicated process beneath each eye,

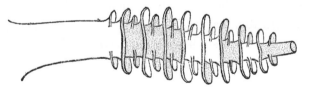

Fig. 11.   *Tentacle of a Nudibranch.*‡

* *Pecten varius*, L., from a specimen dredged by Mr. Bowerbank, off Tenby ; *m*, the pallial curtains ; *br*, the branchiœ.

† In the Octopods, there is a foramen near the eye, and in some of the Calamaries a plicated organ, which M. D'Orbigny regards as an external ear.

‡ Fig. 11. Tentacle of *Eolis coronata*, Forbes, from Alder and Hancock.

which M. Valenciennes regards as the organ of smell*. Messrs
Hancock and Embleton attribute the same function to the la
mellated tentacles of the nudibranchs, and compare them witl
the olfactory organs of fishes.

The labial tentacles of the bivalves are considered to be or
gans for discriminating food, but in what way is unknown (fig
18. *l. t.*) The *sense of taste*, is also indicated rather by th
habits of the animals, and their choice of food, than by th
structure of a special organ. The *acephala* appear to exercis
little discrimination in selecting food, and swallow anything tha
is small enough to enter their mouths, including living animal
cules, and even the sharp *spicula* of sponges. In some instances
however, the oral orifice is well guarded, as in *pecten* (fig. 10.
In the *Encephala*, the tongue is armed with spines, employec
in the comminution of the food, and cannot possess a very de·
licate sense. The more ordinary
and diffused *sense of touch* is possessed by all the mollusca ; it is
exercised by the skin, which is
everywhere soft and lubricous,
and in a higher degree by the
fringes of the bivalves (fig. 12),

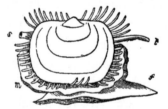

Fig. 12.   *Lepton Squamosum.*†

and by the filaments and tentacles (*vibracula*) of the gasteropods
the eye-pedicels of the snail are evidently endowed with grea
sensitiveness in this respect. That shell-fish are not very sensi
ble of pain, we may well believe, on account of their tenacity o
life, and the extent to which they have the power of reproducin¿
lost parts.

Muscular System. The muscles of the *mollusca* are prin
cipally connected with the skin, which is exceedingly contractil
in every part. The snail affords a remarkable, though familia

---

* Mr. Owen regards the membraneous *lamellæ* between the oral tentacl·
and in front of the mouth, as the seat of the olfactory sense. *See Fig.* 44.

† Fig. 12. *Lepton sqaumosum Mont.*, from a drawing by Mr. Alder, i
the British Mollusca ; copied by permission of Mr. Van Voorst.

nstance, when it draws in its eye-stalks, by a process like the nversion of a glove-finger; the branching gills of some of the ea-slugs, and the tentacles of the cuttle-fishes, are also emiently contractile.*

The inner tunic of the *ascidians* (fig. 8, *t.*) presents a beau-iful example of muscular tissue, the crossing fibres having much he appearance of basket-work; in the transparent *salpians,* hese fibres are grouped in flat bands, and arranged in characeristic patterns.   In this class (*tunicata*) they act only as *phincters* (or circular muscles), and by their sudden contraction xpel the water from the branchial cavity.   The muscular foot of he bivalves is extremely flexible, having layers of circular fibres or its protrusion, (fig. 18. *f*) and longitudinal bands for its reraction (fig. 30 *h*); its structure and mobility has been com-ared to that of the human tongue.   In the burrowing shell-fish such as *solen*), it is very large and powerful, and in the boring pecies, its surface is studded with silicious particles (*spicula*), which render it a very efficient instrument for the enlargement f their cells.   (*Hancock.*)   In the attached bivalves it is not

eveloped, or exists only in a rudimenary state, and is subsidiary to a gland which secretes the material of those threads with which the mussel and *pinna* attach hemselves. (Fig. 13.)   These threads re termed the *byssus;* the plug of the *nomia,* and the pedicel of *terebratula* re modifications of the *byssus.*

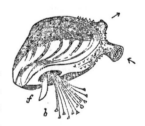

Fig. 13.   *Dreissena.*†

In the cuttle-fishes alone, we find muscles attached to inernal cartilages which represent the bones of *vertebrate* animals; he muscles of the arms are inserted in a cranial cartilage, and hose of the fins in the lateral cartilages, the equivalents of the ectoral fins of fishes.

* The muscular fibres of shell-fish do not exhibit the transverse stripes which characterize *voluntary* muscles in the higher animals.

† Fig. 13. *Dreissena polymorpha* (Pallas sp.) from the Surrey timber-ocks. *f*, foot. *b*, byssus.

which M. Valenciennes regards as the organ of smell*. Mes:
Hancock and Embleton attribute the same function to the
mellated tentacles of the nudibranchs, and compare them w
the olfactory organs of fishes.

The labial tentacles of the bivalves are considered to be
gans for discriminating food, but in what way is unknown (
18. *l. t.*) The *sense of taste*, is also indicated rather by
habits of the animals, and their choice of food, than by
structure of a special organ. The *acephala* appear to exer
little discrimination in selecting food, and swallow anything t
is small enough to enter their mouths, including living anin
cules, and even the sharp *spicula* of sponges. In some instan
however, the oral orifice is well guarded, as in *pecten* (fig.]
In the *Encephala*, the tongue is armed with spines, emplo
in the comminution of the food, and cannot possess a very

licate sense. The more ordinary
and diffused *sense of touch* is possessed by all the mollusca; it is
exercised by the skin, which is
everywhere soft and lubricous,
and in a higher degree by the
fringes of the bivalves (fig. 12),

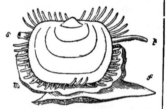

Fig. 12.   *Lepton Squamosum.*

and by the filaments and tentacles (*vibracula*) of the gasteropo
the eye-pedicels of the snail are evidently endowed with gr
sensitiveness in this respect. That shell-fish are not very ser
ble of pain, we may well believe, on account of their tenacit
life, and the extent to which they have the power of reproduc
lost parts.

Muscular System. The muscles of the *mollusca* are pr
cipally connected with the skin, which is exceedingly contrac
in every part. The snail affords a remarkable, though fami

* Mr. Owen regards the membraneous *lamellæ* between the oral tenta
and in front of the mouth, as the seat of the olfactory sense. *See Fig.* 4:

† Fig. 12. *Lepton sqaumosum Mont.*, from a drawing by Mr. Aldei
the British Mollusca; copied by permission of Mr. Van Voorst.

instance, when it draws in its eye-stalks, by a process like the inversion of a glove-finger; the branching gills of some of the sea-slugs, and the tentacles of the cuttle-fishes, are also eminently contractile.*

The inner tunic of the *ascidians* (fig. 8, *t.*) presents a beautiful example of muscular tissue, the crossing fibres having much the appearance of basket-work; in the transparent *salpians*, these fibres are grouped in flat bands, and arranged in characteristic patterns. In this class (*tunicata*) they act only as *sphincters* (or circular muscles), and by their sudden contraction expel the water from the branchial cavity. The muscular foot of the bivalves is extremely flexible, having layers of circular fibres for its protrusion, (fig. 18. *f*) and longitudinal bands for its retraction (fig. 30 *h*); its structure and mobility has been compared to that of the human tongue. In the burrowing shell-fish (such as *solen*), it is very large and powerful, and in the boring species, its surface is studded with silicious particles (*spicula*), which render it a very efficient instrument for the enlargement of their cells. (*Hancock.*) In the attached bivalves it is not

developed, or exists only in a rudimentary state, and is subsidiary to a gland which secretes the material of those threads with which the mussel and *pinna* attach themselves. (Fig. 13.) These threads are termed the *byssus;* the plug of the *anomia,* and the pedicel of *terebratula* are modifications of the *byssus.*

Fig. 13.   *Dreissena.*†

In the cuttle-fishes alone, we find muscles attached to internal cartilages which represent the bones of *vertebrate* animals; the muscles of the arms are inserted in a cranial cartilage, and those of the fins in the lateral cartilages, the equivalents of the pectoral fins of fishes.

---

* The muscular fibres of shell-fish do not exhibit the transverse stripes which characterize *voluntary* muscles in the higher animals.

† Fig. 13. *Dreissena polymorpha* (Pallas sp.) from the Surrey timber-docks. *f*, foot. *b*, byssus.

Muscles of a third kind are attached to the shell. The valve of the oyster (and other *mono-myaries*) are connected by a singl muscle; those of the *cytherea* (and other *di-myaries*), by two the contraction of which brings the valves together. They ai hence named *adductors ;* and the part of the shell to which the are attached is always indicated by scars. (Fig. 14, *a. a'*).

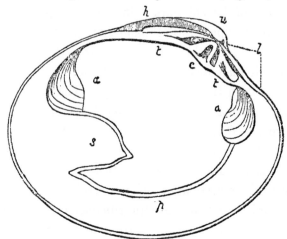

Fig. 14. *Left valve of Cytherea chione.**

The border of the mantle is also muscular, and the place its attachment is marked in the shell by a line called the *palli impression* (*p*); the presence of a bay, or *sinus* (*s*), in this lin shews that the animal had retractile siphons; the foot of tl animal is withdrawn by *retractor* muscles also attached to tl shell, and leaving small scars near those of the adducto (Fig. 30*).

The gasteropods withdraw into their shells when alarme by a shell-muscle, which passes into the foot, or is attached the *operculum ;* its impression is horse-shoe-shaped in the lin pet, as also in *navicella, concholepas*, and the nautilus; it b

* Fig. 14. *Cytherea chione*, L., coast of Devon, (original); *h*, the hin ligament; *u*, the umbo; *l*, the lunule; *c*, cardinal tooth; *t t'*, lateral teetl *a*, anterior adductor; *a'*, posterior adductor; *p*, pallial impression; *s*, sint occupied by retractor of the siphons.

comes deeper with age. In the spiral univalves, the scar is less conspicuous, being situated on the *columella*, and sometimes divided, forming two spots. It corresponds to the posterior *retractors* in the bivalves.

Digestive system. This part of the animal economy is all-important in the *radiate* classes, and scarcely of less consequence in the *mollusca*. In the ascidians (fig. 8, *i*), the alimentary canal is a convoluted tube, in part answering to the *œsophagus*, and in part to the intestine; the stomach is distinguished by longitudinal folds, which increase its extent of surface ; it receives the secretion of the liver by one or more apertures. In those bivalves, which have a large foot, the digestive organs are concealed in the upper part of that organ ; the mouth is unarmed, except by two pairs of soft membranous *palpi*, which look like accessory gills (fig. 18. *l. t.*) The ciliated arms of the brachipods, occupy a similar position (figs. 4, 5, 6), and are regarded as their equivalents. The encephalous mollusca are frequently armed with horny jaws, working vertically like the mandibles of a bird ; in the land-snails, the upper jaw is opposed only by the denticulated tongue, whilst the limneïds have two additional horny jaws, acting laterally. The tongue is muscular, and armed with recurved spines (or *lingual teeth*), arranged in a great variety of patterns, which are eminently characteristic of the genera.* Their teeth are amber-coloured, glossy, and translucent ; and being silicious (they are insoluble in acid), they can be used like a file, for the abrasion of very hard substances. With them the limpet rasps the stony nullipore, the whelk bores holes in other shells, and the cuttle-fish doubtless uses its tongue in the same manner as the cat. The tongue, or lingual ribbon, usually forms a triple band, of which the central part is called the *rachis*, and the lateral tracts *pleuræ*, the rachidian teeth

* The preparation of the lingual ribbon as a permanent microscopic object, requires some nicety of manipulation, but the arrangement of the teeth may be seen by merely compressing part of the animal between two pieces of glass.

sometimes form a single series, overlapping each other, or there
are lateral teeth on each side of a median series.   The teeth on
the pleuræ are termed *uncini;* they are extremely numerous in
the plant-eating gasteropods.   (Fig. 15. A.)*

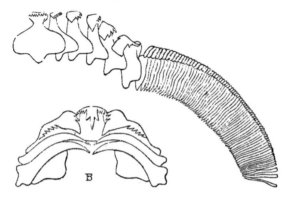

Fig. 15.   *Lingual Teeth of Mollusca.*

Sometimes the tongue forms a short semi-circular ridge,
contained between the jaws ; at others, it is extremely elongated,
and when withdrawn, its folds extend backwards to the stomach.
The lingual ribbon of the limpet is longer than the whole animal ; the tongue of the whelk has 100 rows of teeth ;   and
the great slug has 160 rows, with 180 teeth in each row.

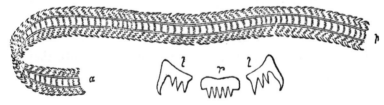

Fig. 16.   *Tongue of the Whelk.*†

The front of the tongue is frequently curved, or bent quite
over ; it is the part of the instrument in use, and its teeth are

* Fig. 15. A.   Lingual teeth of *trochus cinerarius* (after Lovén).   Only
the median tooth, and the (5) lateral teeth, and (90) *uncini* of one side of a
single row are represented.   B. One row of the lingual teeth of *cypræa
europæa;* consisting of a median tooth, and three *uncini* on each side of it.

† Fig. 16.   Lingual ribbon of *buccinum undatum* (original), from a preparation communicated by Wm. Thomson, Esq., of King's College.

often broken or blunted. The posterior part of the lingual ribbon usually has its margins rolled together, and united, forming a tube, which is presumed to open gradually. The new teeth are developed from behind forwards, and are brought successively into use, as in the sharks and rays amongst fishes. In the *bulladæ* the *rachis* of the tongue is unarmed, and the business of communicating the food is transferred to an organ which resembles the gizzard of a fowl, and is often paved with calcarious plates, so large and strong as to crush the small shell-fish which are swallowed entire. In the *aplysia*, which is a vegetable-feeder, the gizzard is armed with numerous small plates and spines. The stomach of some bivalves contains an instrument called the " crystalline stylet," which is conjectured to have a similar use.

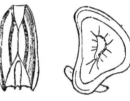

Fig. 17. *Gizzard of Bulla.*

In the cephalopods there is a crop in which the food may accumulate, as well as a gizzard for its trituration.

The *liver* is always large in the mollusca (fig. 10) ; its secretion is derived from arterial blood, and is poured either into the stomach, or the commencement of the intestine. In the nudibranchs, whose stomachs are often remarkably branched, the liver accompanies all the gastric ramifications, and even enters the respiratory papillæ on the backs of the eolids. The existence of a *renal* organ has been ascertained in most classes; in the bivalves it was detected by the presence of uric acid. The intestine is more convoluted in the herbivorous than in the carnivorous tribes : in the bivalves and in *haliotis* it passes through the ventricle of the heart ; its termination is always near the respiratory aperture (or excurrent orifice, when there are

* Fig. 17. Gizzard of *bulla lignaria* (original). Front and side view of a half-grown specimen, with the part nearest the head of the animal downwards ; in the front view the plates are in contact. The *cardiac* orifice is in the centre, in front ; the *pyloric* orifice is on the posterior dorsal side, near the small transverse plate.

two*), and the excrements are carried away by the water which has already passed over the gills.

Besides the organs already mentioned, the encephalous mollusks are always furnished with well-developed *salivary glands*, and some have a rudimentary *pancreas ;* many have also special glands for the secretion of coloured fluids, such as the purple of the *murex*, the violet liquid of *ianthina* and *aplysia*, the yellow of the *bulladæ*, the milky fluid of *eolis*, and the inky secretion of the cuttle-fishes. A few exhale peculiar odours, like the garlic-snail (*helix alliaria*) and *eledone moschata*. Many are phosphorescent, especially the floating tunicaries (*salpa* and *pyrosoma*), and bivalves which inhabit holes (*pholadidæ*). Some of the cuttle-fishes are slightly luminous ; and one land-slug, the *phosphorax*, takes its name from the same property.

Circulating system. The *mollusca* have no distinct absorbent system, but the product of digestion (*chyle*) passes into the general abdominal cavity, and thence into the larger veins, which are perforated with numerous round apertures. The circulating organs are the heart, arteries, and veins ; the blood is colourless, or pale bluish white. The *heart* consists of an *auricle* (sometimes divided into two), which receives the blood from the gills ; and a muscular *ventricle* which propels it into the arteries of the body. From the capillary extremities of the arteries it collects again into the veins, circulates a second time through the respiratory organ, and returns to the heart as arterial blood. Besides this *systemic* heart, the circulation is aided by two additional *branchial* hearts in the cuttle-fishes ; and by four in the brachiopoda. Mr. Alder has counted from 60 to 80 pulsations per minute in the nudibranchs, and 120 per minute in a *vitrina*. Both the arteries and veins form occasionally wide spaces, or

---

* In most of the gasteropods the intestine returns upon itself, and terminates on the right side, near the head. Occasionally it ends in a perforation more or less removed from the margin of the aperture, as in *trochotoma, fissurella, macrochisma*, and *dentalium*. In *chiton* the intestine is straight, and terminates posteriorly.

sinuses; in the cuttle-fishes the œsophagus is partly or entirely surrounded by a venous *sinus;* and in the *acephala* the viceral cavity itself forms part of the circulating system.

The circulation in the tunicaries presents a most remarkable exception to the general rule, for their blood ebbs and flows in the same vessels, as it was supposed to do in the human veins before the time of Harvey. In the transparent *salpæ* it may be seen passing from the heart into vessels connected with the *viscera* and tunics, and thence into the branchial vessels; but when this has continued for a time, the movement ceases, and recommences in the opposite direction, passing from the heart to the gill and thence to the system. (*Lister.*) In the compound tunicaries, there is a common circulation through the connecting medium, in addition to the individual currents.

Aquiferous canals. Sea-water is admitted to the visceral cavity of many of the mollusks (as it is also in radiate animals), by minute canals, opening externally in the form of pores. These *aquiferous pores* are situated either in the centre of the creeping disc, as in *cypræa, conus,* and *ancillaria;* or at its margin, as in *haliotis, doris,* and *aplysia.* In the cuttle-fishes, they are variously placed, on the sides of the head, or at the bases of the arms; some of them conduct to the large sub-orbital pouches, into which the tentacles are retracted.

Respiratory system. The respiratory process consists in the exposure of the blood to the influence of air, or water containing air; during which oxygen is absorbed and carbonic acid liberated. It is a process essential to animal life, and is never entirely suspended, even during hybernation. Those air-breathers that inhabit water are obliged to visit the surface frequently; and stale water is so inimical to the water-breathers, that they soon attempt to escape from the confinement of a glass or basin, unless the water is frequently renewed.* In general,

When aquatic plants are kept in the same glass with water-breathing snails, a balance is produced; which enables both to live without change of water.

fresh-water is immediately fatal to marine species, and salt-water to those which properly inhabit fresh; but there are some which affect brackish water, and many which endure it to a limited extent. The depth at which shell-fish live, is influenced by the quantity of oxygen which they require; the most active and energetic races live only in shallow water, or near the surface; those found in very deep water are the lowest in their instincts, and are specially organized for their situation. Some water-breathers require only moist sea-air, and a bi-diurnal visit from the tide,—like the periwinkle, limpet, and *kellia;* whilst many air-breathers live entirely in the water or in damp places by the water-side. In fact, the nature of the respiratory process is the same, whether it be aquatic or aërial, and it is essential in each case that the surface of the breathing-organ should be preserved moist. The process is more complete in proportion to the extent and minute sub-division of the vessels, in which the circulating fluid is exposed to the revivifying influence.

The land-snails (*pulmonifera*), have a lung, or air-chamber, formed by the folding of the mantle, over the interior of which the pulmonary vessels are distributed; this chamber has a round orifice, on the right side of the animal, which opens and closes at irregular intervals. The air in this cavity seems to renew itself with sufficient rapidity (by the law of diffusion), without any special mechanism.

In the aquatic shell-fish, respiration is performed by the mantle, or by a portion of it specialized, and forming a gill (*branchia*). It is effected by the mantle alone in one family of tunicaries (*pelonaiadæ*), in all the *brachiopoda*, and in one family of gasteropods (*actæonidæ*).

In most of the *tunicata*, the breathing organ forms a distinct *sac* lining the muscular tunic, or mantle (fig. 8. *b.*); this sac has only one external aperture, and conducts to the mouth, which is situated at its base. It is a sieve-like structure, and its inner surface is clothed with vibratile cilia* which create a perpetual

---

* From *cilium*, an eyelash; they are only visible under favourable circum-

current, setting in through the (branchial) orifice, escaping through the meshes of the net, and passing out by the anal orifice of the outer tunics.   The regularity of this current is interrupted only by spasmodic contractions of the mantle, occurring at irregular intervals, by which the creature spirts out water from *both* orifices, and thus clears its cavity of such accumulated particles as are rejected by the mouth ; and too large to escape through the branchial pores.   In the salpians, these contractions are *rythmical*, and have the effect of propelling them backwards.   In the ordinary bivalves, the gills form two membranous plates on each

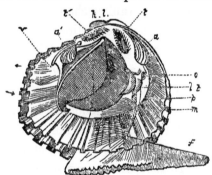

side of the body ; the muscular mantle is still sometimes united, forming a chamber with two orifices, into one of which the water flows, whilst it escapes from the other; there is a third opening in front, for the foot, but this in no wise influences the branchial circulation.   Some-

Fig. 18. *Trigonia pectinata.**

times the orifices are drawn out into long tubes, or *siphons*, especially in those shell-fish which burrow in sand.   (Figs. 19 and 7.)

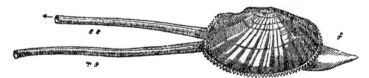

Fig. 19. *Bivalve with long siphons.*†

stances, with the aid of a microscope ; but the currents they cause are easily made perceptible by dropping fine sand into the water over them.

  * *Trigonia pectinata,* Lam. (original).   Brought from Australia by the late Captain Owen Stanley.   The gills are seen in the centre through the transparent mantle.   *o,* mouth; *l t,* labial tentacles; *f,* foot   *v,* vent.

  † Fig. 19. *Psammobia vespertina,* Chemn. after Poli, reduced one half. The arrows indicate the direction of the current.   *r s,* respiratory siphon. *e s,* excurrent siphon.

Those bivalves which have no siphons, and even those in which the mantle is divided into two lobes, are provided with valves or folds which render the respiratory channels just as complete in effect. These currents are not in any way connected with the opening and closing of the valves, which is only done in moving ; or in efforts to expel irritating particles.*

In some of the *gasteropoda* the respiratory organs form tufts, exposed on the back and sides (as in the *nudibranches*), or protected by a fold of the mantle (as in the *inferobranches* and *tectibranches* of Cuvier). But in most the mantle is inflected, and forms a vaulted chamber over the back of the neck, in which are contained the pectinated or plume-like gills (fig. 61). In the carnivorous gasteropods (*siphonostomata*) the water passes into this chamber through a *siphon*, formed by a prolongation of the upper margin of the mantle, and protected by the *canal* of the shell ; after traversing the length of the gill, it returns and escapes through a posterior siphon, generally less developed, but very long in *ovulum volva*, and forming a tubular spine in *typhis*.

In the plant-eating sea-snails (*holostomata*), there is no true siphon, but one of the " neck-lappets" is sometimes curled up and performs the same office, as in *paludina* and *ampullaria* (fig. 84). The in-coming and out-going currents in the branchial chamber, are kept apart by a valve-like fringe, continued from the neck-lappet. The out-current is still more effectually isolated in *fissurella*, *haliotis*, and *dentalium*, where it escapes by a hole in the shell, far removed from the point at which it entered. Near this outlet are the anal, renal, and generative orifices.

The cephalopods have two or four plume-like gills, symmetrically placed in a branchial chamber, situated on the under-side

---

* If a river-mussel be placed in a glass of water, and fine sand let fall gently over its respiratory orifices, the particles will be seen to rebound from the vicinity of the upper aperture, whilst they enter the lower one rapidly. But as this kind of food is not palatable, the creature will soon give a plunge with its foot, and closing its valves, spirt the water (and with it the sand) from both orifices ; the motion of the foot is, of course, intended to change its position.

of the body; the opening is in front, and occupied by a *funnel*, which, in the nautilus, closely resembles the siphon of the *paludina*, but has its edges united in the cuttle-fishes. The free edge of the mantle is so adapted that it allows the water to enter the branchial chamber on each side of the funnel; its muscular walls then contract and force the water through the funnel, an arrangement chiefly subservient to locomotion.* Mr. Bowerbank has observed, that the *eledone* makes twenty respirations per minute, when resting quietly in a basin of water.

In most instances, the water on the surface of the gills is changed by ciliary action alone; in the *cephalopods* and *salpians*, it is renewed by the alternate expansion and contraction of the respiratory chamber, as in the *vertebrate* animals.

The respiratory system is of the highest importance in the economy of the mollusca, and its modifications afford most valuable characters in classification. It will be observed that the Cuvierian *classes* are based on a variety of particulars, and are very unequal in importance; but the *orders* are characterized by their respiratory conditions, and are of much more nearly equal value.

	Orders.	*Classes.*
	Dibranchiata. Owen.	CEPHALOPODA.
	Tetrabranchiata. Owen.	
	Nucleobranchiata. Bl.	
ENCEPHALA	Prosobranchiata. M. Edw.	GASTEROPODA.
	Pulmonifera. Cuv.	
	Opisthobranchiata. M. Edw.	
	Aporobranchiata. Bl.	PTEROPODA.
	Palliobranchiata. Bl.	BRACHIOPODA.
ACEPHALA	Lamellibranchiata. Bl.	CONCHIFERA.
	Heterobranchiata. Bl.	TUNICATA.

The Shell. The relation of the shell to the breathing-organ is very intimate; indeed, it may be regarded as a *pneumo-skeleton*,

* A very efficient means of locomotion in the slender pointed calamaries, which dart backwards with the recoil, like rockets.

being essentially a calcified portion of the mantle, of which the breathing-organ is at most a specialised part.*

The shell is so characteristic of the mollusca that they have been commonly called "testacea" (from *testa* "a shell"), in scientific books; and the popular name of "shell-fish," though not quite accurate, cannot be replaced by any other epithet in common use. In one whole class, however, and in several families, there is nothing that would be popularly recognised as a shell.

Shells are said to be *external* when the animal is contained in them, and *internal* when they are concealed in the mantle; the latter, as well as the shell-less species, being called *naked* mollusks.

Three-fourths of the *mollusca* are *univalve*, or have but one shell; the others are mostly *bivalve*, or have two shells; the *pholads* have accessory plates, and the shell of *chiton* consists of eight pieces. Most of the *multivalves* of old authors were articulate animals (*cirripedes*), erroneously included with the *mollusca*, which they resemble only in outward appearance.

All, except the argonaut, acquire a rudimental shell before they are hatched, which becomes the *nucleus* of the adult shell; it is often differently shaped and coloured from the rest of the shell, and hence the *fry* are apt to be mistaken for distinct species from their parents.

In *cymba* (fig. 20) the nucleus is large and irregular; in *fusus antiquus* it is cylindrical; in the *pyramidellidæ* it is oblique; and it is spiral in *carinaria*, *atlanta*, and many limpets, which are symmetrical when adult.

The rudimentary shell of the *nudibranchs* is shed at an early

---

* In its most reduced form the shell is only a hollow cone, or plate, protecting the breathing organ and heart, as in *limax, testacella, carinaria*. Its peculiar features always relate to the condition of the breathing-organ; and in *terebratula* and *pelonaia* it becomes identified with the gill. In the nudibranchs the vascular mantle performs wholly or in part the respiratory office. In the cephalopods the shell becomes complicated by the addition of a distinct, internal, chambered portion (*phragmocone*), which is properly a *visceral* skeleton; in *spirula* the shell is reduced to this part.

age, and never replaced. In this respect the molluscan shell differs entirely from the shell of the crab and other articulate animals, which is periodically cast off and renewed.

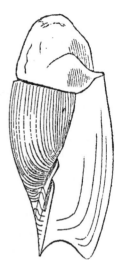

In the bivalves the embryonic shell forms the *umbo* of each valve; it is often very unlike the after-growth, as in *unio pictorum, cyclas henslowiana* and *pecten pusio*. In attached shells like the oyster and anomia the umbo frequently presents an exact imitation of the surface to which the young shell originally adhered.

Shells are composed of carbonate of lime, with a small proportion of animal matter. The source of this lime is to be looked for in their food.

Fig. 20. *Cymba.*

Modern inquiries into organic chemistry have shown that vegetables derive their elements from the mineral kingdom (air, water, and the soil), and animals theirs from the vegetable. The sea-weed filters the salt-water, and separates lime as well as organic elements; and lime is one of the most abundant mineral matters in land plants. From this source the *mollusca* obtain lime in abundance, and, indeed, we find frequent instances of shells becoming unnaturally thickened through the superabundance of this earth in their systems. On the other hand, instances occur of thin and delicate-shelled varieties, in still, deep water, or on clay bottoms; whilst in those districts which are wholly destitute of lime, like the lizard in Cornwall, and similar tracts of magnesian-silicate in Asia Minor, there are no mollusca. (*Forbes.*)

The texture of shells is various and characteristic. Some, when broken, present a dull lustre like marble or china, and are termed *porcellanous*; others are pearly or *nacreous*; some have a *fibrous* structure; some are *horny*, and others *glassy* and *translucent*.

* Fig. 20. *Cymba proboscidalis,* Lam., from a very young specimen in the cabinet of Hugh Cuming, Esq., from Western Africa.

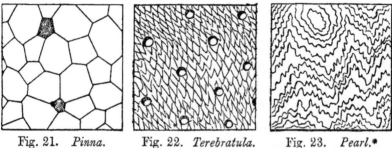

Fig. 21.   *Pinna.*   Fig. 22.   *Terebratula.*   Fig. 23.   *Pearl.**

The *nacreous* shells are formed by alternate layers of very thin membrane and carbonate of lime, but this alone does not give the pearly lustre which appears to depend on minute undulations of the layers, represented in fig. 23.   This lustre has been successfully imitated on engraved steel buttons.   Nacreous shells, when polished, form "mother of pearl;" when digested in weak acid, they leave a membraneous residue which retains the original form of the shell.   This is the most easily destructible of shell-textures, and in some geological formations we find only casts of the nacreous shells, whilst those of fibrous texture are completely preserved.

Pearls are produced by many bivalves, especially by the Oriental pearl-mussel (*avicula margaritifera*), and one of the British river-mussels (*unio margaritiferus*).   They are caused by particles of sand, or other foreign substances, getting between the animal and its shell; the irritation causes a deposit of nacre, forming a projection on the interior, and generally more brilliant than the rest of the shell.   Completely spherical pearls can only be formed loose in the muscles, or other soft parts of the animal. The Chinese obtain them artificially, by introducing into the living mussel foreign substances, such as pieces of mother-of-pearl fixed to wires, which thus become coated with a more brilliant material.

* Figs. 21, 22, 23. Magnified sections of shells, from Dr. Carpenter. Fragments of shell ground very thin, and cemented to glass slides with Canada balsam, are easily prepared, and form curious microscopic objects. A great variety of them may be procured of Mr. C. M. Topping, of Pentonville.

Similar prominences and concretions—pearls which are not *pearly*—are formed inside porcellanous shells; these are as variable in colour as the surfaces on which they are formed.*

The *fibrous* shells consist of successive layers of prismatic cells containing translucent carbonate of lime; and the cells of each successive layer correspond, so that the shell, especially when very thick (as in the fossil *inoceramus* and *trichites*), will break up vertically, into fragments, exhibiting on their edges a structure like arragonite, or satin-spar.   Horizontal sections exhibit a cellular net-work, with here and there a dark cell, which is empty. (fig. 21.)

The oyster has a *laminated* structure, owing to the irregular accumulation of the cells in its successive layers, and breaks up into horizontal plates.

In the boring-shells (*pholadidæ*) the carbonate of lime has an atomic arrangement like arragonite, which is considerably harder than calcarious spar; in other cases the difference in hardness depends on the proportion of animal matter, and the manner in which the layers are aggregated.†

In many bivalve shells there occurs a minute *tubular structure*, which is very conspicuous in some sections of pinna and oyster-shell.

The *brachiopoda* exhibit a characteristic structure by which the smallest fragment of their shells may be determined; it consists of elongated and curved cells, matted together, and often perforated by circular holes, arranged in quincunx order (fig. 22).

But the most complex shell-structure is presented by the *porcellanous* gasteropoda. These consist of three strata which readily separate in fossil shells, on account of the removal of their

---

* They are pink in *turbinellus* and strombus; white in *ostrea*; white or glassy, purple or black in mytilus; rose-coloured and translucent in *pinna*. (*Gray*.)

† The *specific gravity* of floating shells (such as *argonanta* and *ianthina*) is lower than that of any others. (*De la Beche.*)

animal cement. In fig. 24, *a* represents the outer, *b* the middle, and *c* the inner stratum; they may be seen, also, in fig. 25.

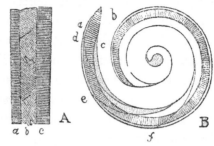

Each of these three strata is composed of very numerous vertical plates, like cards placed on edge; and the direction of the plates is sometimes transverse in the central stratum, and lengthwise in

Fig. 24. *Sections of a cone.**

the outer and inner (as in *cypræa, cassis, ampullaria,* and *bulimus*), or longitudinal in the middle layer, and transverse in the others (*e. g. conus, pyrula, oliva,* and *voluta*).

Each *plate,* too, is composed of a series of prismatic cells, arranged obliquely (45°), and their direction being changed in the successive plates, they cross each other at right angles. Tertiary fossils best exhibit this structure, either at their broken edge, or in polished sections.† *(Bowerbank).*

The argonaut-shell, and the bone of the cuttle-fish, have a peculiar structure; and the *Hippurite* is distinguished by a cancellated texture, unlike any other shell, except, perhaps, some of the *cardiaceæ* and *chamaceæ.*

Epidermis. All shells have an outer coat of animal matter called the "epidermis" (or *periostracum*), sometimes thin and transparent, at others thick and opaque. It is thick and olive-coloured in all fresh-water shells, and in many *arctic* sea-shells (*e. g. cyprina* and *astarte*); the colours of the land-shells often

---

 * Sections of *conus ponderosus,* Brug., from the Miocene of the Touraine. A, longitudinal section of a fragment, B, complete horizontal section; *a,* outer layer; *b,* middle; *c,* inner layer; *d, e, f,* lines of growth.

 † It is necessary to bear in mind that fossil shells are often *pseudomorphous,* or mere casts, in spar or chalcedony, of cavities once occupied by shells; such are the fossils found at Blackdown, and many of the London clay fossils at Barton. The Palæozoic fossils are often *metamorphic,* or have undergone a re-arrangement of their particles, like the rocks in which they occur.

depend on it ; sometimes it is silky as in *helix sericea*, or fringed with hairs, as in *trichotropis;* in the whelk and some species of *triton* and *conus* it is thick and rough like coarse cloth, and in some *modiolas* it is drawn out into long beard-like filaments.

In the cowry and other shell-fish with large mantle lobes, the epidermis is more or less covered up by an additional layer of shell deposited externally.

The *epidermis* has life, but not sensation, like the human scarf-skin ; and it protects the shell against the influence of the weather, and chemical agents ; it soon fades, or is destroyed, after the death of the animal, in situations where, whilst living, it would have undergone no change.   In the bivalves it is organically connected with the margin of the mantle.

It is most developed in shells which frequent damp situations, amongst decaying leaves, and in fresh-water shells.   All freshwaters are more or less saturated with carbonic-acid gas, and in limestone countries hold so much lime in solution as to deposit it in the form of *tufa* on the mussels and other shells.*   But in the absence of lime to neutralise the acid, the water acts on the shells, and would dissolve them entirely if it were not for their protecting epidermis.   As it is we can often recognise fresh-water shells by the erosion of those parts where the epidermis was thinnest, namely, the points of the spiral shells and the *umbones* of the bivalves, those being also the parts longest exposed.   Specimens of *melanopsis* and *bithinia* become truncated again and again in the course of their growth, until the adults are sometimes only half the length they should be, and the discoidal *planorbis* sometimes becomes perforated by the removal of its inner whirls ; in these cases the animal closes the break in its shell with new layers.   Some of the unios thicken their umbones enormously, and form a layer of animal matter with each new layer of shell, so that the river-action is arrested at a succession of steps.

---

* As at Tisbury, in Wiltshire, where remarkable specimens of *anodons* were obtained by the late Miss Benett.

FORMATION AND GROWTH OF THE SHELL.

The shell, as before stated, is formed by the *mantle* of the shell-fish, indeed, each layer of it was once a portion of the mantle, either in the form of a simple membrane, or as a layer of cells; and each layer was successively calcified (or hardened with carbonate of lime) and thrown off by the mantle to unite with those previously formed. Being extra-vascular it has no inherent power of repair. (*Carpenter.*)

The epidermis and cellular structures are formed by the margin (or *collar*) of the mantle; the membranous and nacreous layers, by the thin and transparent portion which contains the viscera; hence we find the pearly texture only as a lining inside the shell, as in the *nautilus,* and all the *aviculidæ* and *turbinidæ.*

If the margin of a shell is fractured during the life-time of the animal, the injury will be completely repaired by the reproduction both of the epidermis and of the outer layer of shell with its proper colour. But if the apex is destroyed, or a hole made at a distance from the aperture, it will merely be closed with the material secreted by the visceral mantle. Such inroads are often made by boring worms and shells, and even by a sponge (*cliona*) which completely mines the most solid shells. In Mr. Gray's cabinet is the section of a cone, in whose apex a colony of *lithodomi*

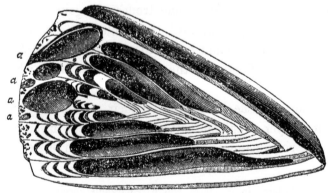

Fig. 25.   *Section of a cone perforated by lithodomi.*

had settled, compelling the animal to contract itself, faster even than it could form shell to fill up the void.

Lines of growth.   So long as the animal continues growing, each new layer of shell extends beyond the one formed before it; and, in consequence, the external surface becomes marked with *lines of growth.*   During winter, or the season of rest which corresponds to it, shells cease to grow; and these periodic resting-places are often indicated by interruptions of the otherwise regular lines of growth and colour, or by still more obvious signs.   It is probable that this pause, or cessation from growth, extends into the breeding season; otherwise there would be two periods of growth, and two of rest in each year.   In many shells the growth is uniform; but in others each stage is finished by the development of a fringe, or ridge (*varix*), or of a row of spines, as in *tridacna* and *murex.* (*Owen, Grant.*)

Adult characters.   The attainment of the full-growth proper to each species is usually marked by changes in the shell.

Some bivalves, like the oyster, and *gryphæa* (fig. 26), continue to increase in thickness long after they have ceased to grow outwards;

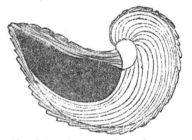

Fig. 26.   *Section of gryphæa.**

the greatest addition is made to the lower valve, especially near the umbo; and in the *spondylus* some parts of the mantle secrete more than others, so that cavities, filled with fluid, are left in the substance of the shell.

The adult *teredo* and *fistulana* close the end of their burrows; the *pholadidea* fills up the great *pedal* opening of its valves; and the *aspergillum* forms the porous disk from which it takes its name.   Sculptured shells, particularly *ammonites*, and species of *rostellaria* and *fusus*, often become plain in the last part of their

* Fig. 26. Section of *gryphæa incurva*, Sby. Lias, Dorset, (original; diminished one half), the upper valve is not much thickened; the interior is filled with lias.

growth. But the most characteristic change is
the thickening and contraction of the aperture
in the univalves. The young cowry (fig. 27)
has a thin, sharp lip, which becomes curled inwards, and enormously thickened and toothed in
the adult; the *pteroceras* (pl. 4, fig. 3) deve-
lopes its scorpion-like claws, only when full-
grown; and the land-snails form a thickened lip,
or narrow their aperture with projecting processes, so that it is a marvel how they pass in
and out, and how they can exclude their eggs,
(*e. g.* pl. 12, fig. 4, *anastoma;* and fig. 5, *helix
hirsuta*).

Fig. 27. *Young
Cowry.**

Yet at this time they would seem to require more space and
accommodation in their houses than before, and there are several
curious ways in which this is obtained. The *neritidæ* and *auriculidæ* dissolve all the internal spiral column† of their shells;
the cone (fig. 24, B,) removes all but a paper-like portion of its
inner whirls; the cowry goes still further, and continues removing
the *internal* layers of its shell-wall, and depositing new layers
externally with its overlapping mantle (fig. 76), until, in some
cases, all resemblance to the young shell is lost in the adult.

The power which mollusks possess of dissolving portions of
their own shells, is also exhibited by the *murices*, in removing
those spines from their whirls which interfere with their growth;
and by the *purpuræ* and others in wearing away the wall of their
aperture. The agency in these cases is supposed to be chemical.

Decollated shells. It frequently happens that as spiral shells
become adult they cease to occupy the upper part of their cavity;
the space thus vacated is sometimes filled with solid shell, as in
magilus; or it is partitioned off, as in *vermetus, euomphalus,*
turritella and *triton* (fig. 62). The deserted apex is sometimes
very thin, and becoming dead and brittle, it breaks away, leaving

---

* *Cypræa testudinaria*, L., young.

† This is sometimes done by the hermit-crab to the shells it occupies.

e shell truncated, or decollated. This happens constantly with
e *truncatellæ, cylindrellæ,* and *bulimus decollatus ;* amongst the
sh-water shells it depends upon local circumstances, but is very
mmon with *pirena* and *cerithidea.*

Forms of shells. These will be described particularly under
ch class ; enough has been said to show that in the molluscan
ell (as in the vertebrate skeleton) indications are afforded o
my of the leading affinities and structural peculiarities of the
imal. It may sometimes be difficult to determine the genus of
shell, especially when its form is very simple; but this results
ore from the imperfection of our technicalities and systems,
in from any want of co-ordination in the animal and its shell.

Monstrosities. The whirls of spiral shells are sometimes
arated by the interference of foreign substances, which adhere
them when young; the garden-snail has been found in this
idition, and less complete instances are common amongst sea-
lls. Discoidal shells occasionally become spiral (as in speci-
ns of *planorbis* found at Rochdale), or irregular in their
wth, owing to an unhealthy condition. The discoidal *ammo-
es* sometimes show a slight tendency to become spiral, and
re rarely become unsymmetrical, and have the keel on one
e, instead of in the middle.

All attached shells are liable to interference in their growth,
l malformations consequent on their situation in cavities, or
m coming in contact with rocks. The *dreissena polymorpha*
torts the other fresh-water mussels by fastening their valves
lh its *byssus ;* and *balani* sometimes produce strange protube-
ces on the back of the cowry, to which they have attached
mselves when young.*

In the miocene tertiaries of Asia Minor, Professor Forbes

---

* In the British Museum there is a *helix terrestris* (chemn.) with a small
k passing through it, and projecting from the apex and umbilicus. Mr.
ering has, in his collection, a *helix hortensis* which got entangled in a nut-
l when young, and growing too large to escape, had to endure the incubus
he end of its days.

growth. But the most characteristic change is
the thickening and contraction of the aperture
in the univalves. The young cowry (fig. 27)
has a thin, sharp lip, which becomes curled inwards, and enormously thickened and toothed in
the adult; the *pteroceras* (pl. 4, fig. 3) deve-
lopes its scorpion-like claws, only when full-
grown; and the land-snails form a thickened lip,
or narrow their aperture with projecting processes, so that it is a marvel how they pass in
and out, and how they can exclude their eggs,
(*e. g.* pl. 12, fig. 4, *anastoma;* and fig. 5, *helix
hirsuta*).

Fig. 27. *You
Cowry.**

Yet at this time they would seem to require more space ад
accommodation in their houses than before, and there are several
curious ways in which this is obtained. The *neritidæ* and *а·i-
culidæ* dissolve all the internal spiral column† of their shе;
the cone (fig. 24, B,) removes all but a paper-like portion of its
inner whirls; the cowry goes still further, and continues removing
the *internal* layers of its shell-wall, and depositing new laуrs
externally with its overlapping mantle (fig. 76), until, in sме
cases, all resemblance to the young shell is lost in the adult.

The power which mollusks possess of dissolving portion of
their own shells, is also exhibited by the *murices*, in removing
those spines from their whirls which interfere with their growi;
and by the *purpuræ* and others in wearing away the wall of tаir
aperture. The agency in these cases is supposed to be chem. l.

Decollated shells. It frequently happens that as spiral shеlls
become adult they cease to occupy the upper part of their cavr;
the space thus vacated is sometimes filled with solid shell, a in
magilus; or it is partitioned off, as in *vermetus, euomphaв,
turritella* and *triton* (fig. 62). The deserted apex is sometiеs
very thin, and becoming dead and brittle, it breaks away, leaviмg

---

* *Cypræa testudinaria*, L., young.

† This is sometimes done by the hermit-crab to the shells it occupies.

the shell truncated, or decollated.   This happens constantly with the *truncatellæ, cylindrellæ*, and *bulimus decollatus ;* amongst the fresh-water shells it depends upon local circumstances, but is very common with *pirena* and *cerithidea*.

Forms of shells.   These will be described particularly under each class ; enough has been said to show that in the molluscan shell (as in the vertebrate skeleton) indications are afforded o many of the leading affinities and structural peculiarities of the animal.   It may sometimes be difficult to determine the genus of a shell, especially when its form is very simple; but this results more from the imperfection of our technicalities and systems, than from any want of co-ordination in the animal and its shell.

Monstrosities.   The whirls of spiral shells are sometimes separated by the interference of foreign substances, which adhere to them when young; the garden-snail has been found in this condition, and less complete instances are common amongst sea-shells.   Discoidal shells occasionally become spiral (as in specimens of *planorbis* found at Rochdale), or irregular in their growth, owing to an unhealthy condition.   The discoidal *ammonites* sometimes show a slight tendency to become spiral, and more rarely become unsymmetrical, and have the keel on one side, instead of in the middle.

All attached shells are liable to interference in their growth, and malformations consequent on their situation in cavities, or from coming in contact with rocks.   The *dreissena polymorpha* distorts the other fresh-water mussels by fastening their valves with its *byssus ;* and *balani* sometimes produce strange protuberances on the back of the cowry, to which they have attached themselves when young.*

In the miocene tertiaries of Asia Minor, Professor Forbes

---

* In the British Museum there is a *helix terrestris* (chemn.) with a small stick passing through it, and projecting from the apex and umbilicus.   Mr. Pickering has, in his collection, a *helix hortensis* which got entangled in a nut-shell when young, and growing too large to escape, had to endure the incubus to the end of its days.

discovered whole races of *neritina, paludina,* and *melanopsis,* with whirls ribbed or keeled, as if through the unhealthy influence of brackish water. The fossil periwinkles of the Norwich Crag are similarly distorted, probably by the access of fresh-water; parallel cases occur at the present day in the Baltic.

Reversed shells. Left-handed, or reversed varieties of spiral shells have been met with in some of the very common species, like the whelk and garden-snail. *Bulimus citrinus* is as often sinistral as dextral; and a reversed variety of *fusus antiquus* was more common than the normal form in the *pliocene* sea. Other shells are constantly reversed, as *pyrula perversa,* many species of *pupa,* and the entire genera, *clausilia, cylindrella, physa,* and *triphoris.* Bivalves less distinctly exhibit variations of this kind; but the attached valve of *chama* has its *umbo* turned to the right or left indifferently; and of two specimens of *lucina childreni* in the British Museum, one has the right, the other the left valve flat.

The colours of shells are usually confined to the surface beneath the epidermis, and are secreted by the border of the mantle, which often exhibits similar tints and patterns (*e. g. voluta undulata,* fig. 73). Occasionally the inner strata of porcellanous shells are differently coloured from the exterior, and the makers of shell-cameos avail themselves of this difference to produce white or rose-coloured figures on a dark ground.*

The secretion of colour by the mantle depends greatly on the action of light; shallow-water shells are, as a class, warmer and brighter coloured than those from deep water; and bivalves which are habitually fixed or stationary (like *spondylus* and *pecten pleuronectes*) have the upper valve richly tinted, whilst the lower one is colourless. The backs of most spiral shells are darker

---

* Cameos in the British Museum, carved on the shell of *cassis cornuta,* are white on an orange ground; on *c. tuberosa,* and *madagascariensis,* white upon dark claret-colour; on *c. rufa,* pale salmon-colour on orange; and on *strombus gigas,* yellow on pink. By filing some of the olives (*e. g. oliva utriculus*) they may be made into very different coloured shells.

than the under sides; but in *ianthina* the base of the shell is habitually turned upwards, and is deeply dyed with violet. Some colours are more permanent than others; the red spots on the *naticas* and *nerites* are commonly preserved in tertiary and oolitic fossils, and even in one example (of n. subcostata schl.) from *Devonian* limestone. *Terebratula hastata*, and some *pectens* of the carboniferous period, retain their markings; the *orthoceras anguliferus* of the *Devonian* beds has zig-zag bands of colour; and a terebratula of the same age, from arctic North America,[*] is ornamented with several rows of dark red spots.

The operculum. Most spiral shells have an *operculum*, or lid, with which to close the aperture when they withdraw for shelter (see gasteropoda). It is developed on a particular lobe at the posterior part of the foot, and consists of horny layers, some times hardened with shelly matter (fig. 28).

Fig. 28.   *Trochus ziziphinus.*[†]

It has been considered by Adanson, and more recently by Mr. Gray, as the equivalent of the dextral valve of the conchifera; but however similar in appearance, its anatomical relations are altogether different. In position it represents the *byssus* of the bivalves (Lovén); and in function it is like the plug with which unattached specimens of *bysso-arca* close their aperture. (*Forbes.*)

Homologies of the shell.[‡] The shell is so simple a structure that its modifications present few points for comparison; but even these are not wholly understood, or free from doubt. The

---

[*] Presented to the British Museum by Sir John Richardson.

[†] *Trochus ziziphinus*, from the original, taken in Pegwell Bay abundantly. This species exhibits small tentacular processes, neck-lappets, side-lappets, tentacular filaments, and an operculigerous lobe.

[‡] Parts which correspond in their real nature—(their origin and develop.ment)—are termed *homologous;* those which agree merely in appearance, or office, are said to be *analogous.*

bivalve shell may be compared to the outer tunic of the *ascidian*, cut open and converted into separable valves.   In the *conchifera* this division of the mantle is vertical, and the valves are right and left.   In the *brachiopoda* the separation is horizontal, and the valves are dorsal and ventral.   The *monomyarian* bivalves lie habitually on one side (like the *pleuronectidæ* among fishes) ; and their shells, though *really* right and left, are termed " upper" and " lower" valves.   The univalve shell is the equivalent of *both* valves of the bivalve.   In the *pteropoda* it consists of dorsal and ventral plates, comparable with the valves of *terebratula*.   In the *gasteropoda* it is equivalent to both valves of the *conchifera* united above.*   The nautilus shell corresponds to that of the gasteropod ; but whilst its chambers are shadowed forth in many spiral shells, the *siphuncle* is something additional ; and the entire shell of the cuttle-fish and argonaut† have no known equivalent or parallel in the other molluscous classes.   The student might imagine a resemblance in the shell of the *orthoceras* to a *back-bone ;* but the true homologue of the vertebrate skeleton is found in the neural and muscular cartilages of the cephalopod ; whilst its *phragmocone* is but the representative of the calcarious axis (or *splanchno-skeleton*) of a coral, such as *amplexus* or *siphonophyllia*.

Temperature and hybernation.   Observations on the *temperature* of the *mollusca* are still wanted ; it is known, however, to vary with the medium in which they live, and to be sometimes a degree or two higher or lower than the external temperature; with snails (in cool weather), it is generally a degree or two higher.

The *mollusca* of temperate and cold climates are subject to *hybernation ;* during which state the heart ceases to beat, respira-

---

* Compare *fissurella* or *trochus* (fig. 28) with *lepton squamosum* (fig. 12). The disk of *hipponyx* is analogous to the ventral plate of hyalæa and terebratula.

† The argonaut shell is compared by Mr. Adams to the nidamental capsules of the *whelk ;* a better analogue would have been found in the *raft* of the *ianthina*, which is secreted by the *foot* of the animal, and serves to *float* the egg-capsules.

tion is nearly suspended, and injuries are not healed.   They also *æstivate*, or fall into a summer sleep when the heat is great; but in this the animal functions are much less interrupted. (*Muller.*)

Reproduction of lost parts.   It appears from the experiments of Spallanzani, that snails, whose ocular tentacles have been destroyed, reproduce them completely in a few weeks; others have repeated the trial with a like result.   But there is some doubt whether the renewal takes place if the brain of the animal be removed as well as its horns.   Madame Power has made similar observations upon various marine snails, and has found that portions of the foot, mantle, and tentacles, were renewed.   Mr. Hancock states that the species of *eolis* are apt to make a meal off each other's *branchiæ*, and that, if confined in stale water, they become sickly and lose those organs; in both cases they are quickly renewed under favourable circumstances.

Reproduction by gemmation.   The social and compound tunicaries resemble zoophytes, in the power they possess of budding out new individuals, and thus of multiplying their communities indefinitely, as the leaves on a tree.   This gemmation takes place only at particular points, so that the whole assemblages are aggregated in characteristic patterns.   The buds of the social tunicaries are supported at first by their parents, those of the compound families by the general circulation, until they are in a state to contribute to the common weal.

Viviparous reproduction.   This happens in a few species of gastropods, through the retention of the eggs in the oviduct, until the young have attained a considerable growth.   It also *appears* to take place in the acephalans, because their eggs generally remain within some part of the shell of the parent until hatched.

Alternate generation.   Amongst the tunicaries an example is found of regulated diversity in the mode of reproduction.   The salpians produce long chains of embryos, which, unless broken by accident, remain connected during life;—each individual of these compound specimens produces *solitary young*, often so un-

D

like the parent as to have been described and named by natural-ists as distinct species;—these solitary salpians again produce chains of embryos, like their grand-parents. (*Chamisso.*)

Oviparous reproduction. The sexes are distinct in the most highly organised (or *diœcious*) mollusca; they are united in the (*monœcious*) land-snails, pteropods, brachiopods, tunicaries, and in part of the conchifers. The prosobranchs pair; but in the diœcious acephalans and cuttle-fishes, the *spermatozoa* are merely discharged into the water, and are inhaled with the respiratory currents by the other sex. The monœcious land-snails require reciprocal union; the limneîds unite in succession, forming float-ing chains.

The *eggs* of the land-snails are separate, and protected by a shell, which is sometimes albuminous and flexible, at others cal-carious and brittle; those of the fresh-water species are soft, mucous, and transparent. The spawn of the sea-snails consists of large numbers of eggs, adhering together in masses, or spread out in the shape of a strap or ribbon, in which the eggs are ar-ranged in rows; this *nidamental ribbon* is sometimes coiled up spirally, like a watch-spring, and attached by one of its edges.

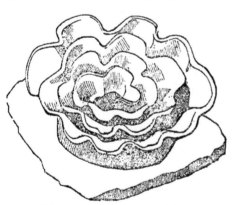

Fig. 29.   *Spawn of Doris.**

The eggs of the carnivo-rous gasteropods are in-closed in tough albuminous capsules, each containing numerous germs; these are deposited singly, or in rows, or agglutinated in groups, equalling the parent animal in bulk (fig. 70). The nidamental capsules of the cuttle-fish are clus-tered like grapes, each containing but one embryo; those of the calamary are grouped

* Nidamental ribbon of *Doris Johnstoni.* (*Alder and Hancock.*)

in radiating masses, each elongated capsule containing 30 or 40 ova. The material with which the eggs are thus cemented together, or enveloped, is secreted by the *nidamental gland*, an organ largely developed in the female gasteropods and cephalopods (fig. 43, n).

Development. The molluscan *ovum* consists of a coloured yolk (*vitellus*), surrounded by albumen. On one side of the yolk is a pellucid spot, termed the *germinal vesicle*, having a spot or *nucleus* on its surface. This germinal vesicle is a nucleated cell, capable of producing other cells like itself; it is the essential part of the egg, from which the *embryo* is formed; but it undergoes no change without the influence of the *spermatozoa*.[*] After impregnation, the germinal vesicle, which then subsides into the centre of the yolk, divides spontaneously into two; and these again divide and subdivide into smaller and still smaller globules, each with its pellucid centre or nucleus, until the whole presents a uniform granular appearance. The next step is the formation of a ciliated *epithelium* on the surface of the embryonic mass; movements in the albumen become perceptible in the vicinity of the *cilia*, and they increase in strength, until the embryo begins to revolve in the surrounding fluid.[†]

[*] No instance of "partheno-genesis" is known among the *mollusca*; the most "equivocal" case on record is that related by Mr. Gaskoin. A specimen of *helix lactea*, Müll., from the South of Europe, after being *two years* in his cabinet, was discovered to be still living; and on being removed to a plant-case it revived, and six weeks afterwards had produced twenty young ones!

[†] According to the observations of Professor Lovén (on certain bivalve mollusca), the ova are excluded immediately after the inhalation of the spermatozoa, and apparently from their influence; but impregnation does not take place within the ovary itself. The spermatozoa of *cardium pygmæum* were distinctly seen to penetrate, in succession, the outer envelopes of the ova, and arrive at the vitellus, when they disappeared. With respect to the "germinal vesicle;" according to Barry, it first approaches the inner surface of the vitelline membrane, in order to receive the influence of the spermatozoa; it then retires to the centre of the yolk, and undergoes a series of spontaneous subdivisions. In M. Lovén's account, it is said to "burst" and par-

Up to this point nearly the same appearances are presented by the eggs of all classes of animals,—they manifest, so far, a complete "unity of organization." In the next stage, the development of an organ, fringed with stronger *cilia*, and serving both for locomotion and respiration, shews that the embryo is a *molluscous animal;* and the changes which follow soon point out the particular *class* to which it belongs. The rudimentary *head* is early distinguishable, by the black eye-specks; and the *heart*, by its pulsations. The digestive and other organs are first " sketched out," then become more distinct, and are seen to be covered with a transparent shell. By this time the embryo is able to move by its own muscular contractions, and to swallow food; is is therefore " hatched," or escapes from the egg.

The embryo tunicary quits the egg in the cloacal cavity of its parent, and is at this time provided with a swimming instrument, like the tail of the tadpole, and with processes by which it attaches itself as soon as it finds a suitable situation.

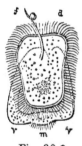

Fig. 30.*

The young bivalves also are hatched before they leave their parent, either in the gill cavity or in a special sac attached to the gills (as in *cyclas*), or in the interspaces of the external branchial laminæ (as in *unio*). At first they have a swimming disk, fringed with long *cilia*, and armed with a slender tentacular filament (*flagellum*). At a later period this disk disappears progressively, as the labial palpi are developed; and they acquire a foot, and with it the power of spinning a byssus. They now

tially dissolve, whilst the egg remains in the ovary, and before impregnation; it then passes to the centre of the yolk, and undergoes the changes described by Barry, along with the yolk, whilst the *nucleus* of the germinal vesicle, or some body exactly resembling it, is seen occupying a small prominence on the surface of the vitelline membrane, until the metamorphosis of the yolk is completed, when it disappears, in some unobserved manner, without fulfilling any recognized purpose.

* Fig. 30. Very young fry of *crenella marmorata*, Forbes, highly magnified; *d*, disk, bordered with cilia: *f*, flagellum; *v v*, valves; *m*, ciliated mantle.

have a pair of eyes, situated near the labial tentacles (fig. 30*, *e*), which are lost at a further stage, or replaced by numerous rudimentary organs placed more favourably for vision, on the border of the mantle.

Most of the aquatic *gasteropoda* are very minute when hatched, and they enter life under the same form,—that which

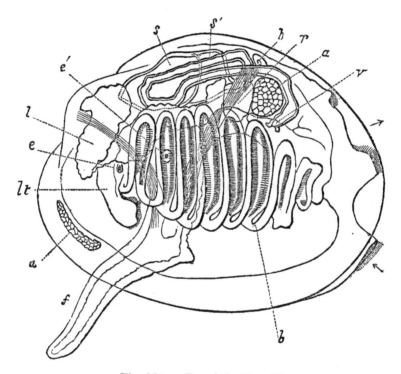

Fig. 30*. *Fry of the Mussel.**

has been already referred to as permanently characteristic of the *pteropoda*. (Fig. 60.)

The *Pulmonifera* and *Cephalopoda* produce large eggs, con-

---

* Fig. 30*. Fry of *mytilus edulis*, after Lovén. *e*, eye; *e′*, auditory capsule; *l t*, labial tentacles; *s s′*, the stomach; *b*, branchiæ; *h*, heart; *v*, vent; *l*, liver; *r*, renal organ; *a*, anterior adductor; *a′*, posterior adductor; *f*, foot. The arrows indicate the incurrent and excurrent openings; between which the margins of the mantle are united in the fry.

taining sufficient nutriment to support the *embryo* until it has

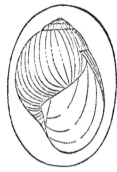

attained considerable size and development; thus, the newly-born cuttle-fish has a shell half an inch long, consisting of several layers, and the *bulimus ovatus* has a shell an inch in length when hatched. (Fig. 31.) These are said to undergo no transformation, because their larval stage is concealed in the egg. The embryonic development of the cuttle-fishes has not been observed; t is probable that they would reveal more curious changes than occur in any other class.

Fig. 31.*

The researches of *John Hunter*† into the embryonic condition of animals, led him to the conclusion that *each stage* in the development of the highest animals corresponded to the *permanent form* of some one of the inferior orders. This grand generalisation has since been more exactly defined and established by a larger induction of facts, some of which we have already described, and may now be stated thus:—

In the earliest period of existence all animals display one uniform condition; but after the first appearance of special development, uniformity is only met with amongst the members of the same primary division, and with each succeeding step it is more and more restricted. From that first step, the members of each primary group assume forms and pass through phases which have no parallels, except in the division to which each belongs. The mammal exhibits no likeness, at any period, to the *adult* mollusk, the insect, or the star-fish; but only to the

---

* Egg and young of *bulimus ovatus*, Mull. sp., Brazil, from specimens in the collection of Hugh Cuming, Esq.

† " In his printed works the finest elements of system seem evermore to flit before him, twice or thrice only to have been seized, and after a momentary detention to have been again suffered to escape. At length, in the astonishing preparations for his museum, he constructed it, for the scientific apprehension, out of the unspoken alphabet of nature." (*Coleridge*.)

ovarian stage of the invertebrata, and to more advanced stages of the classes formed upon its own type. And so also with the highest organized *mollusca;* after their first stage they resemble the simpler orders of their own sub-kingdom, but not those of any other group.

These are the views of Professor Owen—the successor of Hunter—by whom it has been most clearly shewn and steadfastly maintained, that the "unity of organization" manifested by the animal world results from the design of a Supreme Intelligence, and cannot be ascribed to the operation of a mechanical "law."

Chapter V.

CLASSIFICATION.

THE objects of classification are, *first*, the convenient and intelligible arrangement of the species;* and, *secondly*, to afford a summary, or condensed exposition, of all that is known respecting their structure and relations.

In studying the shell-fish, we find resemblances of two kinds. First, agreements of structure, form, and habits; and, secondly, resemblances of form and habits without agreement of structure. The first are termed relations of *affinity;* the second of *analogy.*

Affinities may be near, or remote. There is some amount of affinity common to all animals; but, like relationships amongst men, they are recognized only when tolerably close. Resemblances of structure which subsist from a very early age are presumed to imply original relationship; they have been termed

---

* At least 12,000 recent, and 15,000 fossil species of molluscous animals are known.

genetic (or *histological*), and are of the highest importance. Those which are superinduced at a later period, are of less consequence.

Analogies. Modifications relating only to peculiar habits are called *adaptive;* or *teleological,* from their relation to final causes.* A second class of analogical resemblances are purely external and illusive; they have been termed *mimetic* (*Strickland*), and, by their frequency, almost justify the notion that a certain set of forms and colours are repeated, or represented in every class and family. In all artificial arrangements, these mimetic resemblances have led to the association of widely different animals in the same groups.† Particular forms are also *represented* geographically‡ and geologically,§ as well as systematically.

In all attempts to characterise groups of animals, we find, that in advancing from the smaller to the larger combinations, many of the most obvious external features become of less avail, and we are compelled to seek for more constant and comprehensive signs in the phases of embryonic development, and the condition of the circulating, respiratory, and nervous systems.

Species. All the specimens, or individuals, which are so much alike that we may reasonably believe them to have descended from a common stock, constitute a *species.* It is a particular provision for preventing the blending of species, that *hybrids* are always barren; and it is certain, in the case of shells, that a great many kinds have not changed in form, from the tertiary

---

* For example, the paper nautilus, from its resemblance to *carinaria*, was long supposed to be the shell of a nucleobranche, parasitically occupied by the " *ocythoë.*"

† E. g. Aporrhaïs with strombus, and ancylus with patella.

‡ Monoceros imbricatum and buccinum antarcticum take the place, in South America, of our common whelk and purple, and solen gladiolus and solen americanus of our solen siliqua and ensis.

§ The frequent recurrence of similar species in successive strata may lead beginners to attribute too much to the influence of time and external circumstances; but such impressions disappear with further experience.

period to the present day,—a lapse of many thousand years,—
and through countless generations. When individuals of the
same brood differ in any respect, they are termed *varieties;* for
example, one may be more exposed to the light, and become
brighter coloured; or it may find more abundant food, and grow
larger than the rest. Should these peculiarities become permanent at any place, or period,—should all the specimens on a
particular island or mountain, or in one sea, or geological formation, differ from those found elsewhere,—such permanent variety
is termed a *race;* just as, in the human species, there are white
and coloured races. The species of some genera are less subject
to variation than others; the *nuculæ,* for example, although very
numerous, are always distinguishable by good characters. Other
genera, like *ammonites, terebratula,* and *tellina,* present a most
perplexing amount of variation, resulting from age, sex, supply
of food, variety of depth, and of saltness in the water. And
further, whilst in some genera every possible variety of form
seems to have been called into existence, in others only a few,
strikingly distinct forms, are known.

Genera are groups of species, related by community of structure in all essential respects. The genera of bivalves have been
characterised by the number and position of their hinge-teeth;
those of the spiral univalves, by the form of their apertures;
but these technical characters are only valuable so far as they
indicate differences in the animals themselves.

Families are groups of genera, which agree in some more
general characters than those which unite species into genera.
Those which we have employed are mostly modifications of the
artificial families framed by Lamarck, a plan which seemed more
desirable, in the present state of our knowledge, than a subdivision into very numerous families, without assignable characters.

The *orders and classes* of mollusca have already been referred
to; those now in use are all extremely natural.

It has been sometimes asserted that these groups are only
scientific contrivances, and do not *really* exist in nature; but

this is a false as well as a degrading view of the matter. The labours of the most eminent systematists have been directed to the discovery of the subordinate value of the characters derivable from every part of the animal organization; and, as far as their information enabled them, they have made their systems expressive " of all the highest facts, or generalisations, in natural history." (*Owen.*)

M. Milne Edwards has remarked, that the actual appearance of the animal kingdom is not like a well-regulated army, but like the starry heavens, over which constellations of various magnitude are scattered, with here and there a solitary star which cannot be included in any neighbouring group.

This is exceedingly true; we cannot expect our systematic groups to have equal numerical values,* but they ought to be of equal structural importance; and they will thus possess a *symmetry* of order, which is superior to mere numerical regularity.

All the most philosophic naturalists have entertained a belief that the development of animal forms has proceeded upon some regular plan, and have directed their researches to the discovery of that "reflection of the divine mind." Some have fancied that they have discovered it in a mystic number, and have accordingly converted all the groups into *fives*.† We do not undervalue these speculations, yet we think it better to describe things so far only as we know them.

Great difficulty has always been found in placing groups according to their affinities. This cannot be effected in—the way in which we are compelled to describe them—a single series; for each group is related to *all* the rest; and if we extend the representation of the affinities to very small groups, any arrange-

* The numerical development of groups is *inversely proportional to the bulk of the individuals composing them*. (*Waterhouse.*)

† The *quinarians* make out five molluscous classes, by excluding the *tunicata*; the same end would be attained in a more satisfactory manner by reducing the pteropods to the rank of an order, which might be placed next to the *opistho-branches*.

ment on a plane surface would fail, for the affinities radiate in all directions, and the "net-work" to which Fabricius likened them, is as insufficient a comparison as the "chain" of older writers.*

CHAPTER VI.

NOMENCLATURE.

THE practice of using two names—generic and specific—for each animal, or plant, originated with Linnæus; therefore no scientific names date further back than his works. In the construction of these names, the Greek and Latin languages are preferred, by the common consent of all countries.

Synonyms. It often happens that a species is named, or a genus established, by more than one person, at different times, and in ignorance of each other's labours. Such duplicate names are called *synonyms;* they have multiplied amazingly of late, and are a stumbling-block and an opprobium in all branches of natural history.†

---

* The quinary arrangement of the molluscous classes reminds us of the eastern emblem of eternity—the serpent holding its tail in its mouth.

The following diagram is offered as an improved *circular system :—*

[FISHES.]
Di-branchiata.

Nucleo-		Tetra-
Opistho-		Proso-
Aporo-		Pulmo-
Pallio-		Lamelli-

Hetero-branchiata.
[ZOOPHYTES.]

† In Pfeiffer's *Monograph of the Helicidæ,* a family containing seventeen genera, no less than 330 *generic synonyms* are enumerated; to this list, Dr. Albers, of Berlin, has lately added *another hundred* of his own invention !

One very common estuary shell rejoices in the following variety of titles :—

> Scrobicularia piperata (*Gmelin sp*).
> Trigonella plana (*Da Costa*).
> Mactra Listeri (*Auct*).
> Mya Hispanica (*Chemnitz*).
> Venus borealis (*Pennant*).
> Lutraria compressa (*Lamarck*).
> Arenaria plana (*Megerle*).

As regards *specific* names, the earliest ought certainly to be adopted,—with, however, the following exceptions :—

1. MS. names; which are only admitted by courtesy.
2. Names given by writers antecedent to Linnæus.
3. Names unaccompanied by a description or figure.
4. Barbarisms; or names involving error or absurdity,*

It is also very desirable that names having a general (European) acceptation, should not be changed, on the discovery of earlier names in obscure publications.

With respect to *genera*,—those who believe in their real existence, as " ideas of the creating mind," will be disposed to set aside many random appellations, given to particular shells without any clear enunciation of their characters; and to adopt later names, if bestowed with an accurate perception of the grounds which entitle them to generic distinction.†

Authority for specific names. The multiplication of synonyms having made it desirable to place the *authority* after each

---

* This subject was investigated, and reported upon, by a committee of the British Association, in 1842; but the report was not sufficiently circulated.

† Several bad practices—against which there is, unhappily, no law—should be strongly discountenanced. First, the employment of names already in familiar use for other objects; such as *cidaris* (the title of a well-known genus of sea-urchins), for a group of spiral shells; and *arenaria* (a property of the botanists), for a bivalve. Secondly, the conversion of *specific* into *generic* titles, a process which has caused endless confusion; it has arisen out of the vain desire of giving new designations to old and familiar objects, and thus obtaining a questionable sort of fame.

name, another source of evil has arisen; for several naturalists (fancying that the *genus-maker*, and not the *species-maker*, should enjoy this privilege) have altered or divided almost every genus, and placed their signatures as the authorities for names given half a century or a century before, by LINNÆUS or BRUGUIERE.* British naturalists have disowned this practice, and agreed to distinguish, by the addition of "sp.," the authorities for those specific names whose generic appellations have been changed.

Types. The type of each genus *should be* that species in which the characters of its group are best exhibited, and most evenly balanced. (*Waterhouse.*) It has, however, been customary to take as the type, that species which the genus-maker placed first on his list; although by so doing there is risk of adopting an *aberrant* form, or one which very feebly represents the group, of which it is an obscure member.

* The authorities appended to specific names, are supposed to indicate an amount of work done in the determination and description of the species; when, therefore, the real author's name is suppressed, and a spurious one substituted, the case looks very like an attempt to obtain credit under false pretences.

ABBREVIATIONS.

Etym., etymology. *Syn.*, synonym. *Distr.*, distribution.

M.S., manuscript, i. e., *unpublished.*

Sp., species. Brit. M., (in the) British Museum.

Distr., Norway—New Zealand; including all intermediate seas.

Fossil, lias—chalk; implies that the genus existed in these, and all intervening strata. Chalk —; means that the genus commenced in the chalk, and has existed ever since.

Depth; — 50 fms.; genus found at all depths between low-water and 50 fathoms. A fathom is six feet.

$\frac{1}{4}$ one-fourth the real size; $\frac{4}{1}$ magnified four times.

Lat., breadth. *Long.*, length. *Alt.*, height or thickness.

Unc., (uncia) an inch. *Lin.*, (linea) a line, the $\frac{1}{12}$ of an inch.

Mill., millimetre, the twenty-fifth part of an inch.

MANUAL OF THE MOLLUSCA.

CLASS I. CEPHALOPODA.

THE cuttle-fishes, though excluded by dealers from the list of shell-fish, are the most remarkable, and, rightly considered, the most interesting of any; whilst their relatives, the *nautili* and *ammonites,* are unmatched for the symmetry and wondrous architecture of their pearly shells.

The principal locomotive organs of the *cephalopods,* are attached to the head, in the form of muscular arms or tentacles;* in addition to which, many have fins; and all can propel themselves by the forcible expulsion of water from their respiratory chamber.

Unlike most of the *mollusca,* they are symmetrical animals, having their right and left sides equally developed; and their shell is usually straight, or coiled in a vertical plane. The nautilus and argonaut alone (of the living tribes) have external shells; the rest are termed "naked cephalopods," because the shell is internal. They have powerful jaws, acting vertically, like the mandibles of birds; the tongue is large and fleshy, and part of its surface is sentient, whilst the rest is armed with recurved spines; their eyes are large, and placed on the sides of the head; their senses appear to be very acute. All are marine; and predatory, living on shell-fish, crabs, and fishes.

The nervous system is more concentrated than in the other *mollusca;* and the brain is protected by a cartilage. The respiratory organs consist of two or four plume-like gills, placed symmetrically on the sides of the body, in a large branchial cavity, opening forwards on the *under*† side of the head; in the middle of this opening is placed the *siphon* or *funnel.* The sexes are always distinct; but the males are much less numerous than the females, and in many species, at present unknown. They are divided into two orders, the names of which are derived from the number of the *branchiæ.*

ORDER I. DIBRANCHIATA, Owen.

Animal swimming; naked. *Head* distinct. *Eyes* sessile, prominent. *Mandibles* horny (Pl. I., fig. 2). *Arms* 8 or 10, provided with suckers. *Body* round or elongated, usually with a pair of *fins; branchiæ* two, fur-

* M. Schultze compares the arms of the cephalopods to the oral filaments of *myxine.*

† According to the established usage, we designate that the *under* or *ventral* side of the body, on which the funnel is placed. But if the cuttle fishes are compared with the nucleobranches, or the nautilus with the holostomatous gasteropods, their external analogies seem to favour an opposite conclusion.

hished with muscular ventricles; *ink-gland* always present; *parietes* of the *funnel* entire.

Shell internal (except in *argonauta*), horny or shelly, with or without air-chambers.

The typical forms of the cuttle-fishes were well described by Aristotle, and have been repeatedly examined by modern naturalists; yet, until Professor Owen demonstrated the existence of a second order of cephalopods, departing from all the abovementioned characters, it was not clearly understood how inseparably the organisation of the cuttle-fishes was connected with their condition as *swimming mollusca,* breathing by *two* gills.

The characters which co-exist with the two gills, are the internal rudimentary shell, and the substitution of other means of escape and defence, than those which an external shell would have afforded; viz.: powerful arms, furnished with suckers; the secretion of an inky fluid, with which to cloud the water and conceal retreat; more perfect organs of vision; and superadded branchial hearts, which render the circulation more vigorous.*

The *suckers* (*antlia* or *acetabula*), form a single or double series, on the inner surface of the arms. From the margin of each cup, the muscular fibres converge to the centre, where they leave a circular cavity, occupied by a soft *caruncle,* rising from it like the piston of a syringe, and capable of retraction when the sucker is applied to any surface. So perfect is this mechanism for effecting adhesion, that while the muscular fibres continue retracted, it is easier to tear away the limb than to detach it from its hold.† In the decapods, the base of the *piston* is surrounded by a horny dentated hoop; which in the uncinated calamaries, is folded, and produced into a long sharp claw.

The *ink-bag* (fig. 33), is tough and fibrous, with a thin silvery outer coat; it discharges its contents through a duct which opens near the base of the funnel. The ink was formerly used for writing (*Cicero*), and in the preparation of *sepia ;‡* and from its indestructible nature, is often found in a fossil state.

---

* In a few species, which have no fins, the arms are webbed. In the only kind which has an external shell, it is confined to the female sex, and is secreted by the membranes of the arms. It is now quite certain that such shells as those of the fossil *ammonites* and *orthocerata* would be incompatible with *dibranchiate* organization.

† "The complex, irritable mechanism, of all these suckers, is under the complete control of the animal. Mr. Broderip informs me that he has attempted, with a handnet, to catch an *octopus* that was floating by, with its long and flexible arms entwined round a fish, which it was tearing with its sharp hawk's bill; it allowed the net to approach within a short distance before it relinquished its prey, when, in an instant, it relaxed its thousand suckers, exploded its inky ammunition, and rapidly retreated under cover of the cloud which it had occasioned, by rapid and vigorous strokes of its circular web." (*Owen*.)

‡ Indian ink and sepia are now made of lamp-smoke, or of prepared charcoal.

The skin of the naked cephalopods is remarkable for its variously coloured vesicles, or pigment-cells. In *sepia* they are black and brown ; in the calamary, yellow, red, and brown ; and in the argonaut, and some octopods, there are blue cells besides. These cells alternately contract and expand, by which the colouring matter is condensed or dispersed, or perhaps driven into the deeper part of the skin. The colour accumulates, like a blush, when the skin is irri-tated, even several hours after separation from the body. During life, these changes are under the control of the animal, and give it the power of chang-ing its hue, like the chameleon. In fresh specimens, the *sclerotic plates* of the eyes have a pearly lustre ; they are sometimes preserved in a fossil state.

The *aquiferous pores* are situated on the back and sides of the head, on the arms (*brachial*), or at their bases (*buccal pores*).

The *mantle* is usually connected with the back of the head by a broad ("*nuchal*") muscular band ; but its margin is sometimes free all round, and it is supported only by cartilaginous ridges, fitting into corresponding grooves,* and allowing considerable freedom of motion.

The cuttle-fishes are nocturnal, or crepuscular animals, concealing them-selves during the day, or retiring to a lower region of the water. They in-habit every zone, and are met with equally near the shore, and in the open sea, hundreds of miles from land. They attain occasionally a much greater size than any other mollusca. MM. Quoy and Gaimard found a dead cuttle-fish in the Atlantic, under the equator, which must have weighed 2 cwt. when perfect ; it was floating on the surface, and was partly devoured by birds. Banks and Solander, also met with one under similar circumstances, in the Pacific, which was estimated to have measured six feet in length. (*Owen.*) The arms of the octopods are sometimes two feet long.† From their habits, it is difficult to capture some species alive, but they are frequently obtained, uninjured, from the stomachs of dolphins, and other fishes which prey upon them.

SECTION A. Octopoda.

Arms 8 ; suckers sessile. *Eyes* fixed, incapable of rotation. *Body* united to the head by a broad cervical band. *Branchial chamber* divided longitudinally by a muscular partition. *Oviduct* double ; no distinct nida-mental gland. *Shell* external and one-celled (*mono-thalamous*), or internal and rudimentary.

The Octopods differ from the typical cuttle-fishes in having only eight arms, without the addition of tentacles ; their bodies are round, and they sel-

---

* Termed the "apparatus of resistance," by D'Orbigny.

† *Denys Montfort*, having represented a "kraken octopod," in the act of scuttling a three-master, told M. Defrance, that if this were "swallowed," he would in his next edition represent the monster embracing the Straits of Gibraltar, or capsizing a whole squadron of ships. (*D'Orbigny.*)

dom have fins. They are the most eccentric or "aberrant" mollusks, superior in organization to all the rest, but manifesting some remarkable and unexpected analogies with the lowest classes of animals.

The males of some species of *octopus* and *eledone*, are similar to the females, but are comparatively scarce. Only the females of many others are known, and every specimen of the argonaut hitherto examined (amounting to many hundreds), has been of that sex. Dr. Albert Kölliker has suggested that the real males of the argonaut, and also of *octopus granulatus* and *tremoctopus violaceus* are the *hectocotyles*, previously mistaken for *parasitic worms*.

The *hectocotyle* of *octopus granulatus* was described by Cuvier,[*] who obtained several specimens from octopods captured in the Mediterranean. It is five inches in length, and resembles a detached arm of the octopus, its under surface being bordered with 40 or 50 pairs of alternate suckers.

The *hectocotyle* of *tremoctopus* was discovered by Dr. Kölliker, at Messina, in 1842, adhering to the interior of the gill-chamber and funnel of the poulpe; it is represented in Pl. I., fig. 3. The body is worm-like, with two rows of suckers on the ventral surface, and an oval appendage at the posterior end. The anterior part of the back is fringed with a double series of branchial filaments (250 on each side). Between the branchiæ are two rows of brown or violet *spots*, like the pigment cells of the *tremoctopus*. The suckers (40 on each side) closely resemble those of the *tremoctopus*, in miniature. Between the suckers are four or five series of *pores*, the openings of minute canals, passing into the abdominal cavity. The *mouth* is at the anterior extremity, and is minute and simple; the alimentary canal runs straight through the body, nearly filling it. The *heart* is in the middle of the back, between the branchiæ; it consists of an auricle and a ventricle, and gives origin to two large vessels. There is also an artery and vein on each side, giving branches to the branchial filaments. A *nerve* extends along the intestine, and one ganglion has been observed. The *oval sac* incloses a small but very long convoluted tube, ending in a muscular *vas deferens;* it contains innumerable *spermatozoa.*

The *hectocotyle* of the argonaut was discovered by *Chiaje*, who considered it a parasitic worm, and described it under the name of *trichocephalus acetabularis;* it was again described by *Costa*,[†] who regarded it as "a spermatophore of singular shape;" and lastly by Dr. Kölliker.[‡]

It is similar in form to the others, but is only seven lines in length, and has a filiform appendage in front, six lines long. It has two rows of alternate

[*] An. Sc. Nat. 1 Series, t. 18. p. 147. 1829.

[†] An. Sc. Nat. 2 Series, 7. p. 173.

[‡] Lin. Trans. Vol. 20, pt. 1, p. 9; and in his own zootomical *berichte*, where it is figured.

suckers, 45 on each side; but no *branchiæ*; the skin contains numerous changeable spots of red or violet, like that of the argonaut.*

According to the observations of Madame Power, "the newly hatched *argonaut* has no shell, and is quite unlike what it afterwards becomes; it is a sort of little worm, having two rows of suckers along its length, with a filiform appendage at one extremity, and a small swelling at the other. It might be supposed to represent an *extremely small brachial appendage*, from which the other parts were afterwards to be developed."† (*Kölliker.*)

FAMILY I. ARGONAUTIDÆ.

Dorsal arms (of the female) webbed at the extremity, secreting a symmetrical involuted shell. *Mantle* supported in front by a single ridge on the funnel.

Genus ARGONAUTA, Lin. Argonaut or paper sailor.

Etymology, *argonautai*, sailors of the ship Argo.

Synonyms, ocythoë (Rafinesque). Nautilus (Aristotle and Pliny).

Example, A. hians, Soland, pl. II., fig. 1. China.

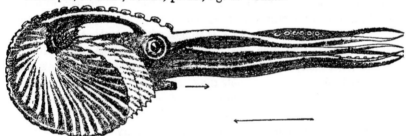

Fig. 32. *Argonauta argo L. swimming.*†

The *shell* of the argonaut is thin and translucent; it is not moulded on the body of the animal, nor is it attached by shell-muscles; and the unoccupied hollow of the spire serves as a receptacle for the minute clustered eggs. The argonaut sits in its boat with its siphon turned towards the keel,§ and its sail-shaped (dorsal) arms closely applied to the sides of the shell, as in fig. 32, where, however, they are represented as partially withdrawn, in order to show the margin of the aperture. It swims only by ejecting water from its fun-

* Similar instances of a permanently rudimentary condition of the male sex, occur amongst the lowest organized parasitic crustaceans; the males of *achtheres, lernæopoda, tracheliaster, &c.*, are frequently a thousand times smaller than the female, upon whom they live, and from whom they differ both in form and structure. Mr. Gosse has described a similar disparity of the sexes in *asplanchna*.

† An. Sc. Nat. 2 Series, vol. 16, p. 185.

‡ From a copy of Rang's figure, in Charlesworth's Magazine; one-fourth the natural size; the small arrow indicates the current from the *funnel*, the large arrow the direction in which the "sailor" is driven by the recoil.

§ Poli has represented it sitting the opposite way; the writer had once an argonaut shell with the nucleus *reversed*, implying that the animal had *turned quite round* in its shell, and remained in that position. The specimen is now in the York Museum.

el, and crawls in a reversed position, carrying its shell over its back like a snail. (*Madame Power and M. Rang.*)

It was the *nautilus* (*primus*) of Aristotle, who described it as floating on the surface of the sea, in fine weather, and holding out its sail-shaped arms to the breeze; a pretty fable, which poets have repeated ever since.

Distribution: 4 species of argonaut are known; they inhabit the open sea throughout the warmer parts of the world. Captain King took several from the stomach of a dolphin, caught upwards of 600 leagues from any land.

Fossil: A. hians is found in the sub-apennine tertiaries of Piedmont. This species is still living in the Chinese seas, but not in the Mediterranean.

FAMILY II. OCTOPODIDÆ.

Arms similar, elongated, united at the base by a web. *Shell* represented by two short styles, encysted in the substance of the mantle. (*Owen.*)

OCTOPUS, Cuvier. Poulpe.

Etym., *octo*, eight, *pous* (*poda*) feet.
Syn., *cistopus*. (*Gray.*)
Ex., O. tuberculatus Bl., pl. I., figs. 1 and 2 (mandibles).

Body oval, warty or cirrose, without fins; *arms* long, unequal; *suckers* in two rows; *mantle* supported in front by the branchial septum.

The octopods are the "polypi" of Homer and Aristotle; they are solitary animals, frequenting rocky shores, and are very active and voracious; the females oviposit on sea-weeds, or in the cavities of empty shells. In the markets of Smyrna and Naples, and the bazaars of India, they are regularly exposed for sale. "Although common (at St. Jago) in the pools of water left by the retiring tide, they are not very easily caught. By means of their long arms and suckers they can drag their bodies into very narrow crevices, and when thus fixed it requires great force to remove them. At other times they dart tail first, with the rapidity of an arrow, from one side of the pool to the other, at the same instant discolouring the water with a dark chesnut-brown ink. They also escape detection by varying their tints, according to the nature of the ground over which they pass. In the dark they are slightly phosphoescent." (*Darwin.*)*

Professor E. Forbes has observed that the octopus, when resting, coils its dorsal arms over its back, and seems to shadow forth the argonaut's shell.

Distr., universally found on the coasts of the temperate and tropical zones; 3 species are known; when adult they vary in length from 1 inch to 2 feet, according to the species.

PINNOCTOPUS, D'Orb. Finned octopus.

Body with lateral fins, united behind.

---

* Journal of a Voyage round the World. The most fascinating volume of travels published since Defoe's fiction.

The only known species, *P. cordiformis*, was discovered by MM. Quoy and Gaimard, on the coast of New Zealand; it exceeds 3 feet in length.

ELEDONE. (Aristotle.) Leach.

Type, E. octopodia, L.

Suckers forming a single series on each arm ; length 6 to 18 inches. *E. moschata* emits a musky smell.

Distr., 2 sp. Coasts of Norway, Britain, and the Mediterranean.

CIRROTEUTHIS, Eschricht. 1836.

Etym., *cirrus*, a filament, and *teuthis* a cuttle-fish.

Body with two transverse fins; *arms* united by a web, nearly to their tips; *suckers* in a single row, alternating with *cirri*. Length 10 inches. Colour violet. The only species (*C. Mülleri Esch.*) inhabits the coast of Greenland.

PHILONEXIS, D'Orb.

Etym., *philos*, an adept in *nexis*, swimming.

Type, P. atlanticus, D'Orb.

Arms free; suckers in two rows ; *mantle* supported by two ridges on the funnel. Total length, 1 to 3 inches.

Distr., 6 sp. Atlantic and Medit. Gregarious in the open sea; feeding on floating *mollusca*.

Sub-genus. *Tremoctopus* (Chiaje), pl. I., fig. 3.

Name from two large aquiferous pores (*tremata*) on the back of the head. *Arms* partly, or all webbed half-way up.

Distr., 2 sp. T. quoyanus and violaceus. Atlantic and Medit.

SECTION B. DECAPODA.

Arms 8.· *Tentacles* 2, elongated, cylindrical, with expanded ends. *Suckers* pedunculated, armed with a horny ring. *Mouth* surrounded by a buccal membrane, sometimes lobed and funished with suckers. *Eyes* moveable in their orbits. *Body* oblong or elongated, always provided with a pair of fins. *Funnel* usually furnished with an internal valve. *Oviduct* single. *Nidamental gland* largely developed. *Shell* internal; lodged loosely in the middle of the dorsal aspect of the mantle.

The arms of the decapods are comparatively shorter than those of the octopods; the dorsal pair is usually shortest, the ventral longest. The tentacles originate within the circle of the arms, between the third and fourth pairs; they are usually much longer than the arms, and in *cheiroteuthis* are six times as long as the animal itself. They are completely retractile into large subocular pouches in *sepia, sepiola,* and *rossia;* partly retractile in *loligo* and *sepioteuthis;* non-retractile in *cheiroteuthis.* They serve to seize prey which may be beyond the reach of the ordinary arms, or to moor the animal in safety during the agitation of a stormy sea.

The *shell* of the living decapods is either a horny "pen" (*gladius*) or a calcarious "bone" (*sepion*); not attached to the animal by muscles, but so loose as to fall out when the cyst which contains it is opened. In the genus *spirula*, it is a delicate spiral tube, divided into air-chambers by a series of partitions (*septa*). In the fossil genus *spirulirostra*, a similar shell forms the apex of a cuttle-bone; in the fossil *conoteuthis* a chambered shell is combined with a *pen;* and the *belemnite* unites all these modifications.

The decapods chiefly frequent the open sea, appearing periodically like fishes, in great shoals, on the coasts and banks. (*Owen, D'Orb.*)

FAMILY III. TEUTHIDÆ. CALAMARIES, OR SQUIDS.

Body, elongated; *fins* short, broad, and mostly terminal.

Shell, (gladius or *pen*) horny, consisting of three parts,—a shaft, and two lateral expansions or wings.

Sub-family A. *Myopsidæ*, D'Orb. *Eyes* covered by the skin.

LOLIGO. (*Pliny*) Lamarck. Calamary.

Syn., *teuthis* (Aristotle) Gray.

Type, L. vulgaris (sepia loligo L.) Fig. 1. Pl. I., fig. 6 (pen).

Pen, lanceolate, with the shaft produced in front; it is multiplied by age, several being found packed closely, one behind another, in old specimens. (Owen.)

Body tapering behind, much elongated in the males. *Fins* terminal, united, rhombic. *Mantle* supported by a cervical ridge, and by two grooves in the base of the funnel. *Suckers* in two rows, with horny, dentated hoops. *Tentacular club* with four rows of suckers. Length (excluding tentacles) from 3 inches to 2½ feet.

The calamaries are good swimmers; they also crawl, head-downwards, on their oral disk. The common species is used for bait, by fishermen, on the Cornish coast (Couch). Shells have been found in its stomach, and more rarely sea-weed (Dr. Johnston). Their egg-clusters have been estimated to contain nearly 40,000 eggs (Bohadsch).

Distr., 21 sp. in all seas. Norway—New Zealand.

Sub-genus. *Teudopsis*, Deslongchamps, 1835.

Etym., *teuthis*, a calamary and *opsis* like.

Type, T. Bunellii, Desl.

Pen, like *loligo*, but dilated and spatulate behind.

Fossil, 5 sp. Upper Lias, France, and Wurtemberg.

GONATUS, Gray.

Animal and *pen* like *loligo* in most respects. *Arms* with 4 series of cups, *tentacular club* with numerous small cups, and a single large sessile cup armed with a hook; funnel valveless.

Distr., a single species (*G. amœna*, Moller sp.) is found on the coast of Greenland.

SEPIOTEUTHIS, Blainville.

Type, S. sepioïdea, Bl. *Animal* like *loligo; fins* lateral, as long as the body. Length from 4 inches to 3 feet.

Distr., 13 sp., West Indies, Cape, Red Sea, Java, Australia.

BELOTEUTHIS, Münster.

Etym., *belos*, a dart and *teuthis.*

Type, B. subcostata, Münst. Pl. II., fig. 8., U. Lias, Wurtemberg.

Pen, horny, lanceolate; with a very broad shaft, pointed at each end, and small lateral wings.

Distr., 6 sp. described by Münster, considered varieties (differing in age and sex), by M. D'Orbigny.

GEOTEUTHIS, Münster.

Etym., *ge*, the earth (*i. e.* fossil) and *teuthis.*

Syn., belemnosepia (Agassiz.) belopeltis (Voltz) loligosepia Quenstedt.)*

Pen broad, pointed behind; shaft broad, truncated in front; lateral wings shorter than the shaft.

Fossil, 9 sp. U. Lias, Wurtemberg; Calvados; Lyme Regis. Several undescribed sp. in the Oxf. clay, Chippenham.

Besides the *pens* of this calamary the *ink-bag*, the muscular mantle, and the bases of the arms, are preserved in the Oxford clay. Some of the ink-bags found in the Lias are nearly a foot in length, and are invested with a brilliant nacreous layer; the ink forms excellent *sepia*. It is difficult to understand how these were preserved, as the recent calamaries " spill their ink" on the slightest alarm. (*Buckland*).

LEPTOTEUTHIS, Meyer.

Etym., *Leptos* thin, and *teuthis.*

Type, L. gigas Meyer, Oxford clay, Solenhofen.

Pen very broad and rounded in front, pointed behind; with obscure diverging ribs.

CRANCHIA, Leach, 1817.

Named in honour of Mr. J. Cranch, naturalist to the Congo expedition.

Type, C. scabra, Leach.

Body large, ventricose; fins small, terminal; mantle supported in front by a branchial septum. Length 2 inches. *Head* very small. *Eyes* fixed. *Buccal* membrane large, 8-lobed. *Arms* short, suckers in two rows. *Tentacular* clubs finned behind, cups in 4 rows. *Funnel* valved.

Pen long and narrow.

* These names must be set aside, being incorrect in themselves, and founded on a total misapprehension of the nature of the fossils.

Distr., 2 sp. W. Africa. In the open sea.

This genus makes the nearest approach to the octopods.

SEPIOLA. (Rondelet) Leach, 1817.

Ex., S. atlantica (D'Orb.) Pl. I., fig. 4.

Body short, purse-like ; mantle supported by a broad cervical band, and a ge fitting a groove in the funnel. *Fins* dorsal, rounded, contracted at the e. *Suckers* in 2 rows, or crowded, on the arms, in 4 rows on the tentacles. ngth 2 to 4 inches.

Pen, half as long as the back. *S. stenodactyla* (sepioloidea, D'Orb.) has pen.

Distr., 6 sp. Coasts of Norway, Britain, Medit., Mauritius, Japan, stralia.

Sub-genus. Rossia, Owen (*Fidenas?* Gray). *Mantle* supported by a vical ridge and groove. *Suckers* in 2 rows on the tentacles. Length 3 to 5 inches.

Distr., 6 sp. Regent Inlet, Britain, Medit., Manilla.

Sub-family B. *Oigopsidæ*, D'Orb.

Eyes naked. *Fins* always terminal, and united, forming a rhomb.

LOLIGOPSIS, Lam. 1811.

Etym., loligo, and *opsis*, like.

Type, L. pavo (Lesueur).

Body elongated, mantle supported in front by a branchial septum. *Arms* s't. *Cups* in 2 rows. *Tentacles* slender, often mutilated. *Funnel* valveless. *Pen* slender, with a minute conical appendix. Length from 6 to 12 inches.

Distr., pelagic. 8 sp. N. Sea, Atlantic, Medit., India, Japan, S. Sea.

CHEIROTEUTHIS, D'Orb.

Etym., *cheir*, the hand, and *teuthis*.

Type, C. veranii, Fér.

Mantle supported in front by ridges. *Funnel* valveless. *Ventral arms* v long. *Tentacles* extremely elongated, slender, with distant sessile cups o he peduncles, and 4 rows of pedunculated claws on their expanded ends.

Pen slender, slightly winged at each end. Length of the body 2 inches; to le tips of the arms 8 inches; to the ends of the tentacles 3 feet.

Distr., 2 sp. Atlantic, Medit. On gulf-weed, in the open sea.

HISTIOTEUTHIS, D'Orb.

Etym., *histion*, a veil; and *teuthis*.

Type, H. bonelliana, Fér. Length 16 inches.

Body short. *Fins* terminal, rounded. *Mantle* supported in front by ri s and grooves. *Buccal* membrane 6-lobed. *Arms* (except the ventral pa, webbed high up. *Tentacles* long, outside the web, with 6 rows of den- ta cups on their ends.

Pen short and broad.

Distr., 2 sp.   Mediterranean ; in the open sea.

ONYCHOTEUTHIS, Lichtenstein.   Uncinated calamary.

Etym., *onyx*, a claw, and *teuthis*.

Type, O. banksii, Leach. (= bartlingii ?) Pl. I., fig. 7 and fig. 8 (*pen*)

Syn, ancistroteuthis (Gray).   Onychia (Lesueur).

Pen narrow, with hollow, conical apex.

Arms with 2 rows of suckers. *Tentacles* long and powerful, armed with a double series of hooks ; and usually having a small group of suckers at the base of each club, which they are supposed to unite, and thus use their tentacles in conjunction.*   Length 4 inches to 2 feet.

The uncinated calamaries are solitary animals, frequenting the open sea, and especially the banks of gulf-weed (*sargasso*).   O. banksii ranges from Norway to the Cape and Indian ocean ; the rest are confined to warm seas. O. dussumieri has been taken swimming in the open sea, 200 leagues north of the Mauritius.

Distr., 6 sp.   Atlantic, Indian ocean, Pacific.

ENOPLOTEUTHIS, D'Orb.   Armed calamary.

Etym., *enoplos*, armed, and *teuthis*.

Type, E. smithii, Leach.

Syn., ancistrochirus and abralia (Gray), octopodoteuthis (Ruppell), verania (Krohn).

Pen lanceolate. *Arms* provided with a double series of horny hooks, concealed by retractile webs.   *Tentacles* long and feeble, with small hooks at the end.   Length (excluding the tentacles) from 2 inches to 1 foot ; but some species attain a larger size.   In the museum of the College of Surgeons there is an arm of the specimen of E. unguiculata, found by Banks and Solander in Cook's first voyage (mentioned at p. 64) supposed to have been 6 feet long when perfect.   The natives of the Polynesian Islands, who dive for shell-fish, have a well-founded dread of these formidable creatures. (*Owen.*)

Distr., 10 sp.   Medit., Pacific.

OMMASTREPHES, D'Orb.   Sagittated calamary.

Etym., *omma*, the eyes, and *strepho*, to turn.

Type, O. sagittatus, Lam.

Body cylindrical ; terminal fins large and rhombic. *Arms* with 2 rows of suckers, and sometimes an internal membranous fringe. *Tentacles* short and strong, with 4 rows of cups.

Pen, consisting of a shaft with three diverging ribs, and a hollow conical appendix.   Length from 1 inch to nearly 4 feet.

---

* The obstetric forceps of Professor Simpson were suggested by the suckers of the calamary.

The sagittated calamaries are gregarious, and frequent the open sea in all climates. They are extensively used in the cod-fishery off Newfoundland, and are the principal food of the dolphins and cachalots, as well as of the albatross and larger petrels. The sailors call them "sea-arrows" or "flying squids," from their habit of leaping out of the water, often to such a height as to fall on the decks of vessels. They leave their eggs in long clusters floating at the surface.

Distr., 14 recent sp.; similar *pens* (4 sp.) have been found fossil in the Oxford clay, Solenhofen; it may, however, be doubted whether they are generically identical.

FAMILY IV. BELEMNITIDÆ.

Shell consisting of a *pen*, terminating posteriorly in a chambered cone, sometimes invested with a fibrous *guard*. The air-cells of the *phragmo-cone* are connected by a *siphuncle*, close to the ventral side.

BELEMNITES, Lamarck. 1801.

Etym., belemnon, a dart.*

Ex., B. puzosianus, pl. II., fig. 5.

Phragmocone horny, slightly nacreous, with a minute globular nucleus at its apex; divided internally by numerous concave *septa*. *Pen* represented by two nacreous bands on the dorsal side of the phragmocone, and produced beyond its rim, in the form of sword-shaped processes (pl. II., fig. 5).† *Guard,* fibrous, often elongated and cylindrical; becoming very thin in front, where it invests the phragmocone.‡

Nearly 100 species of belemnites have been found in a fossil state, ranging from the lias to the gault, and distributed over all Europe. The *phragmocone* of the belemnite, which represents the terminal appendix of the calamaries, is

---

* The termination *ites* (from *lithos*, a stone) was formerly given to all *fossil* genera.

† The most perfect specimens known are in the cabinet of Dr. Mantell, and the British Museum; they were obtained by William Buy in the Oxford clay of Christian Malford, Wilts. The *last chamber* of a lias belemnite in the British Museum is 6 inches long, and 2½ inches across at the smaller end; a fracture near the siphuncle shows the *ink-bag*. The *phragmocone* of a specimen corresponding to this in size, measures 7½ inches in length.

‡ The specific gravity of the guard is identical with that of the shell of the recent pinna, and its structure is the same. Parkinson and others have supposed that it was originally a light and porous structure, like the cuttle bone; but the *mucro* of the sepiostaire, with which alone it is homologous, is quite as dense as the belemnite. We are indebted to Mr. Alex. Williams, M.R.C.S., for the following specific gravities of recent and fossil shells, compared with water as 1,000 :—

Belemnites puzosianus, Oxford clay	2,674
Belemnitella mucronata, chalk	2,677
Pinna, recent, from the Mediterranean	2,607
Trichites plottii, from the inferior oolite	2,670
Conus monile, recent	2,910
Conus ponderosus, Miocene, Touraine	2,713

E

divided into air-chambers, connected by a small tube (*siphuncle*), like the shell of the pearly nautilus. It is exceedingly delicate, and usually owes its preservation to the infiltration of calc. spar; specimens frequently occur in the lias, with the meniscus-shaped casts of the air-chambers loose, like a pile of watch-glasses. It is usually eccentric, its apex being nearest to the ventral side of the guard. The *guard* is very variable in its proportions, being sometimes only half an inch longer than the phragmocone, at others one or two feet in length. These variations probably depend to some extent on age and sex; M. D'Orbigny believes that the shells of the males are always (comparatively) long and slender; those of the females are at first short, but afterwards growing only at the points, they become as long in proportion as the others. The guard always exhibits (internally) concentric lines of growth; in *B. irregularis* its apex is hollow. The belemnites have been divided into groups by the presence and position of furrows in the surface of the guard.

SECTION I. Acœli (Bronn.) without dorsal or ventral grooves.

Sub-section 1. *Acuarii*, without lateral furrows, but often channelled at the extreme point.

Type., b. acuarius. 20 sp. Lias—Neocomian.

Sub-section 2. *Clavati*, with lateral furrows.

Type, b. clavatus. 3 sp. Lias.

SECTION II. Gastrocœli (D'Orb.) Ventral groove distinct.

Sub-section 1. *Canaliculati*, no lateral furrows.

Type, b. canaliculatus. 5 sp. Inf. oolite—Gt. oolite.

Sub-section 2. *Hastati*, lateral furrows distinct.

Type, b. hastatus. 19 sp. U. lias—Gault.

SECTION III. Notocœli (D'Orb.) with a dorsal groove, and furrowed on each side.

Type, b. dilatatus. 9 sp. Neocomian.

The belemnites appear to have been gregarious, from the exceeding abundance of their remains in many localities, as in some of the marlstone quarries of the central counties, and the lias cliffs of Dorsetshire. It is also probable that they lived in a moderate depth of water, and preferred a muddy bottom to rocks or coral-reefs, with which they would be apt to come in perilous collision. Belemnites injured in the life-time of the animal have been frequently noticed.

Belemnitella, D'Orb.

Syn., actinocamax, Miller (founded on a mistake.)

Type, B. mucronata, Sby. Pl. II., fig. 6.

Distr., Europe; N. America. 5 sp. U. greensand and chalk.

The *guard* of the belemnitella has a straight fissure on the ventral side of its alveolar border; its surface exhibits distinct vascular impressions. The

phragmocone is never preserved, but casts of the alveolus show that it was chambered, that it had a single dorsal ridge, a ventral process passing into the fissure of the guard, and an apical nucleus.

ACANTHOTEUTHIS (Wagner), Münster.

Etym., *acanthá*, a spine, and *teuthis*.

Syn., Kelæno (Munster.)   Belemnoteuthis?

Type, A. prisca, Rüppell.

Founded on the fossil hooks of a calamary, preserved in the Oxford clay of Solenhofen. These show that the animal had 10, nearly equal arms, all furnished with a double series of horny claws, throughout their length. A *pen* like that of the *ommastrephes* has been hypothetically ascribed to these arms, which may, however, have belonged to the *belemnite* or the *belemnoteuthis*.

BELEMNOTEUTHIS (Miller), Pearce, 1842.

Type, B. antiquus (Cunnington), fig. 33.

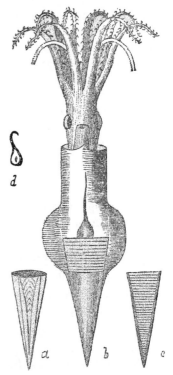

Shell consisting of a *phragmocone*, like that of the belemnite ; a horny dorsal *pen* with obscure lateral bands; and a thin fibrous *guard*, with two diverging ridges on the dorsal side.

Animal provided with *arms* and *tentacles* of nearly equal length, furnished with a double alternating series of horny hooks, from 20 to 40 pairs on each arm ; *mantle* free all round; *fins* large, medio-dorsal (much larger than in fig. 33).

Fossil in the Oxford clay of Chippenham. Similar horny claws have been found in the lias of Watchett; and a *guard* equally thin is figured in Buckland's Bridgewater Treatise, t. 44, fig. 14.

In the fossil calamary of Chippenham, the shell is preserved along with the muscular mantle, fins, ink-bag, funnel, eyes, and tentacles with their horny hooks ; all the specimens were discovered, and developed with unexampled skill, by William Buy, of Sutton, near Chippenham.

Fig. 33.   *Belemnoteuthis.**

* Fig. 33. *Belemnoteuthis antiquus*, ½, ventral side, from a specimen in the cabinet of William Cunnington, Esq., of Devizes. The last chamber of the phragmocone is preserved in this specimen. *a*, represents the dorsal side of an uncompressed phragmocone from the Kelloway rock, in the cabinet of J. G. Lowe, Esq.; *c*, is an ideal section of the same. Since this woodcut was executed, a more complete specimen has

CONOTEUTHIS, D'Orb.

Type, C. Dupinianus, D'Orb.   Pl. II., fig. 9.   *Neocomian*, France.

Phragmocone slightly curved.   *Pen* elongated, very slender.

This shell, which is like the pen of an ommastrephe, with a chambered cone, connects the ordinary calamaries with the belemnites.

FAMILY V.   SEPIADÆ.

Shell (cuttle-bone or *sepiostaire*) calcarious; consisting of a broad laminated plate, terminating behind in a hollow, imperfectly chambered apex (*mucro*).   *Animal* with elongated tentacles, expanded at their ends.

SEPIA (Pliny), Linnæus.

Type, S. officinalis, L.   Pl. I., fig. 5.

Syn., *belosepia*, Voltz.   (B. sepioïdea, pl. II., fig. 3, mucro only.)

Body oblong, with lateral fins as long as itself.   *Arms* with 4 rows of suckers.   *Mantle* supported by tubercles fitting into sockets on the neck and funnel.   Length 3 to 28 inches.

Shell as wide and long as the body; very thick in front, concave internally behind; terminating in a prominent *mucro*.   The thickened part is composed of numerous plates, separated by vertical fibres, which render it very light and porous.   T. Orbignyana, pl. II., fig. 2.

The cuttle-bone was formerly employed as an antacid by apothecaries; it is now only used as " pounce," or in casting counterfeits.   The bone of a Chinese species attains the length of 1½ feet. (*Adams*.)

The cuttle-fishes live near shore, and the *mucro* of their shell seems intended to protect them in the frequent collisions they are exposed to in swimming backwards. (*D'Orb.*)

Distr., 30 sp.   World-wide.

Fossil, 5 sp.   Oxf. clay, Solenhofen.   Several species have been founded on *mucrones* from the Eocene of London and Paris.   Pl. II., fig. 3.

SPIRULIROSTRA, D'Orb.

Type, S. Bellardii (D'Orb.)   Pl. II., fig. 4.   Miocene, Turin.

Shell, mucro only known; chambered internally; chambers connected by a ventral *siphuncle*; external spathose layer produced beyond the *phragmocone* into a long pointed beak.

BELOPTERA (Blainville) Deshayes.

Etym., *belos*, a dart, and *pteron*, a wing.

Type, B. belemnitoïdes, Bl.   Pl. II., fig. 7.

been obtained for the British Museum; the *tentacles* are not longer than the ordinary arms, owing, perhaps, to their partial retraction; this specimen will be figured in Dr. Mantell's "Petrifactions and their Teachings."   *d*, is a single hook, natural size; the specimens belonging to Mr. Cunnington and the late Mr. C. Pearce, show the large acetabular bases of the hooks.

Shell, mucro (only known) chambered and siphuncled; winged externally. *Fossil,* 2 sp. *Eocene.* Paris; Bracklesham

BELEMNOSIS, Edwards.

Type, B. anomalus, Sby. sp. Eocene. Highgate (unique.)
Shell, mucro, chambered and siphuncled; without lateral wings or elongated beak.

FAMILY VI. SPIRULIDÆ.

Shell entirely nacreous; discoidal; whirls separate, chambered (*polythalamous,*) with a ventral siphuncle.

SPIRULA, Lam., 1801.

Syn., lituus, Gray.
Ex., S. lævis (Gray.) Pl. I., fig. 9.
Body oblong, with minute terminal fins. *Mantle* supported by a cervical and 2 ventral ridges and grooves. *Arms* with 6 rows of very minute cups *Tentacles* elongated. *Funnel* valved.

Shell placed vertically in the posterior part of the body, with the involute spire towards the ventral side. The last chamber is not larger in proportion than the rest; its margin is organically connected; it contains the ink-bag.

The delicate shell of the spirula is scattered by thousands on the shores of New Zealand; it abounds on the Atlantic coasts, and a few specimens are yearly brought by the Gulf-stream, and strewed upon the shores of Devon and Cornwall. But the animal is only known by a few fragments, and one perfect specimen, obtained by Mr. Percy Earl on the coast of New Zealand.
Distr., 3 sp. All the warmer seas.

ORDER II. TETRABRANCHIATA.

Animal creeping; protected by an external shell.
Head retractile within the mantle. *Eyes* pedunculated. *Mandibles* calcarious. *Arms* very numerous. *Body* attached to the shell by adductor muscles, and by a continuous horny girdle. *Branchiæ* four. *Funnel* formed by the folding of a muscular lobe.

Shell external, camerated (poly-thalamous) and siphuncled; the inner layers and septa nacreous; outer layers porcellanous.*

It was long ago remarked by Dillwynn, that shells of the carnivorous gasteropods were almost, or altogether, wanting in the palæozoic and secondary strata; and that the office of these animals appeared to have been performed, in the ancient seas, by an order of cephalopods, now nearly extinct. Above 1,400 fossil species belonging to this order are now known by their shells; whilst their only living representative is the *nautilus pompilius,*

---

* The Chinese carve a variety of patterns in the outer opaque layer of the nautilus shell, relieved by the pearly ground beneath.

of which several specimens have been brought to Europe within the last few years.*

The shell of the tetrabranchiate cephalopods is an extremely elongated cone, and is either straight, or variously folded, or coiled.

It is *straight* in . . . . orthoceras . baculites.

    bent on itself in . . . . . ascoceras . ptychoceras.

    curved in . . . . cyrtoceras . toxoceras.

    spiral in . . . . trochoceras . turrilites.

    discoidal in . . . . . . gyroceras . crioceras.

    discoidal and produced in . lituites . . ancyloceras.

    involute in . . . . . . nautilus . . ammonites.

Internally, the shell is divided into cells or chambers, by a series of partitions (*septa*), connected by a tube or *siphuncle*. The last chamber is occupied by the animal, the rest are empty during life, but in fossil specimens they are often filled with spar. When the outer shell is removed (as often happens to fossils,) the edges of the *septa* are seen (as in Pl. III., figs. 1, 2.) Sometimes they form curved lines, as in *nautilus* and *orthoceras*, or they are *zig-zag*, as in *goniatites* (fig. 53,) or *foliaceous*, as in the ammonite, fig. 34.

Fig. 34. *Suture of an ammonite.†*

The outlines of the *septa* are termed *sutures;‡* when they are folded the elevations are called *saddles,* and the intervening depressions *lobes.* In *ceratites* (fig. 54) the *saddles* are round, the lobes *dentated;* in *ammonites* both lobes and saddles are extremely complicated. Broken fossils show that the *septa* are nearly flat in the middle, and folded round the edge (like a shirt-frill), where they abut against the outer shell-wall (fig. 37).

The *siphuncle* of the recent *nautilus* is a membranous tube, with a very thin nacreous investment ; in most of the fossils it consists of a succession of funnel-shaped, or bead-like tubes. In some of the oldest fossil genera, *actinoceras, gyroceras,* and *phragmoceras,* the siphuncle is large, and contains in

* The *frontispiece*, copied from Professor Owen's Memoir, represents the animal of the first nautilus, captured off the New Hebrides, and brought to England by Mr. Bennett; it is drawn as if lying in the section of a shell, without concealing any part of it. The woodcut, fig. 43, is taken from a more perfect specimen, lately acquired by the British Museum, in which the relation of the animal to its shell is accurately shown.

† *A. heterophyllus*, Sby., from the lias, Lyme Regis. British Museum. Only one side is represented ; the arrow indicates the dorsal saddle.

‡ From their resemblance to the sutures of the skull.

its centre a smaller tube, the space between the two being filled up with radiating plates, like the lamellæ of a coral. The position of the siphuncle is very variable; in the *ammonitidæ* it is *external*, or close to the outer margin of the shell (fig. 37). In the *nautilidæ* it is usually *central* (fig. 35), or *internal* (fig. 36).

 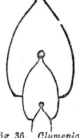 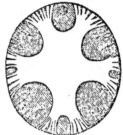

Fig. 35.   *Nautilus.*   Fig. 36.   *Clymenia.*   Fig. 37.   *Hamites.**

The *air-chambers* of the recent nautilus are lined by a very thin, living membrane; those of the fossil *orthocerata* retain indications of a thick vascular lining, connected with the animal by spaces between the beads of the siphuncle.†

The *body-chamber* is always very capacious; in the recent nautilus its cavity is twice as large as the whole series of air-cells; in the *goniatite* (fig. 39), it occupies a whole whirl, and has a considerable lateral extension; and in *ammonites communis* it occupies more than a whirl.

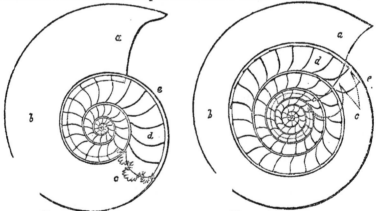

Fig. 38.   *Ammonites.*   Fig. 39.   *Goniatites.*‡

* Fig. 35. Nautilus pompilius, L.   Fig. 36. Clymenia striata, Münst., see pl. II., fig. 16.   Fig. 37. Hamites cylindraceus Defr., see fig. 58.

† The apocryphal genus *spongarium*, was founded on detached septa of an *orthoceras*, from the Upper Ludlow rock, in which the vascular markings distinctly radiate from the siphuncle. Mr. Jones, warden of Clun Hospital, has several of these in apposition.

‡ Fig. 38. Section of *ammonites obtusus*, Sby. lias, Lyme Regis; from a very young specimen.   Fig. 39. Section of *goniatites sphæricus*, Sby. carb. limestone, Bolland (in the cabinet of Mr. Tennant.)   The dotted lines indicate the *lateral extent* of the body-chamber.

The *margin of the aperture* is quite simple in the recent nautilus, and affords no clue to the many curious modifications observable in the fossil forms. In the *ammonites* we frequently find a dorsal process, or lateral projections, developed periodically, or only in the adult (fig. 55, and pl. III., fig. 5).

In *phragmoceras* and *gomphoceras* (figs. 40, 41) the aperture is so much contracted that it is obvious the animal could not have withdrawn its head into the shell like the nautilus.

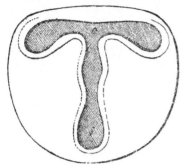 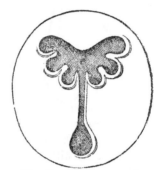

Fig. 40. *Gomphoceras.*   Fig. 41. *Phragmoceras.**

M. Barrande, from whose great work on the Silurian Formations of Bohemia these figures are taken, suggests that the lower part of the aperture (*s s*) which is almost isolated, may have served for the passage of the funnel, whilst the upper and larger space (*c c*) was occupied by the neck; the lobes probably indicate the position of the external arms.

The aperture of the pearly nautilus is closed by a disk or hood (fig. 43, *h*), formed by the union of the two dorsal arms, which correspond to the shell-secreting sails of the argonaut.

In the extinct *ammonites* we have evidence that the aperture was guarded still more effectively by a horny, or shelly *operculum*, secreted, in all probability, by these dorsal arms. In one group (*arietes*,) the operculum consists of a single piece, and is horny and flexible.† In the *round-backed* ammonites the operculum is shelly, and divided into two plates by a straight median suture (fig. 42). They were described in 1811, by Parkinson, who called them *tri-gonellites*, and pointed out the resemblance of their

Fig. 42.‡

---

* Fig. 40. *Gomphoceras Bohemicum* (Barrande), reduced view of the aperture; *s*, the siphonal opening. Fig. 41. *phragmoceras callistoma* (Barr.) both from the U. Silurian, Bohemia.

† This form was discovered by the late Miss Mary Anning, the indefatigable collector of the lias fossils of Lyme Regis, and described by Mr. Strickland, Geol. Journal, vol. I., p. 232. Also by M. Voltz, Mem. de l'Institute, 1837, p. 48.

‡ *Trigonellites lamellosus*, Park. Oxford clay, Solenhofen (and Chippenham,) associated with ammonites lingulatus, Quenstedt. (= A. Brightii, Pratt). From a specimen in the cabinet of Charles Stokes, Esq.

internal structure to the cancellated tissue of bones. Their external surface is smooth or sculptured ; the inner side is marked by lines of growth. Forty-five kinds are enumerated by Bronn ; they occur in all the strata in which ammonites are found, and a single specimen has been figured by M. D'Archiac, from the Devonian rocks of the Eifel, where it was associated with *goniatites*.*

˯ *Calcarious mandibles* or *rhyncholites* (F. Biguet) have been obtained from all the strata in which *nautili* occur ; and from their rarity, their large size and close resemblance to the mandibles of the recent nautilus, it is probable that they belonged only to that genus.† In the Muschelkalk of Bavaria one nautilus (*N. arietis*, Reinecke, = N. bidorsatus, Schlotheim,) is found, and two kinds of *rhyncholite ;* one sort, corresponding with the upper mandible of the recent nautilus, has been called "rhyncholites hirundo" (pl. II., fig. 11), the other, which appears to be only the lower mandible of the same species, has been described under the name of "conchorhynchus avirostris."‡

In studying the fossil *tetrabranchiata*, it is necessary to take into consideration the varying circumstances under which they háve been preserved. In some strata (as the lias of Watchett) the outer layer of the shell has disappeared, whilst the inner nacreous layer is preserved. More frequently only the outer layer remains ; and in the chalk formation the whole shell has perished. In the calcarious grit of Berkshire and Wiltshire the ammonites have lost their shells ; but perfect casts of the chambers, formed of calcarious spar, remain.§

Fossil *orthocerata* and *ammonites* are evidently in many instances *dead shells,* being overgrown with corals, serpulæ, or oysters ; every cabinet affords such examples. In others the animal has apparently occupied its shell, and prevented the ingress of mud, which has hardened all around it ; after this it has decomposed, and contributed to form those phosphates and sulphurets commonly present in the body-chamber of fossil shells, and by which the sediment around them is so often formed into a hard concretion.‖ In this state they are

* The *trigonellites* have been described by Meyer as bivalve shells, under the generic name of *aptychus ;* by Deslongchamps under the name of *Munsteria.* M. D'Orbigny regards them as cirripedes ! M. Deshayes believes them to be *gizzards* of the ammonites. M. Coquand compares them with *teudopsis ;* an analogy evidently suggested by some of the membranous and elongated forms, such as *T. sanguinolarius,* found with *am. depressus,* in the lias of Boll. Ruppell, Voltz, Quenstedt, and Zieten, regard the trigonellites as the *opercula of ammonites,* an opinion also entertained by many of the most experienced fossil collectors in England.

† M. D'Orbigny has manufactured *two genera of calamaries* out of these nautilus beaks ! (*rhynchoteuthis* and *palæoteuthis*). In the innumerable sections of *ammonites* which have been made, no traces of the mandibles have ever been discovered.

‡ *Lepas avirostris* (Schlotheim), described by Blainville as the beak of a brachiopod !

§ Called *spondylolites* by old writers.

‖ In the alum-shale of Whitby, innumerable *concretions* are found, which, when struck with the hammer, split open, and disclose an *ammonite.* See Dr. Mantell's "Thoughts on a Pebble," p. 21.

permeated by mineral water, which slowly deposits calcarious spar, in crystals, on their walls; or by acidulous water, which removes every trace of the shell, leaving a cavity, which at some future time may again become filled with spar, having the form of the shell, but not its structure. In some sections of *orthocerata*, it is evident that the mud has gained access to the air-cells, along the course of the blood-vessels; but the chambers are not entirely filled, because their lining membrane has contracted, leaving a space between itself and certain portions of the walls, which correspond in each chamber.

With respect to the purpose of the *air-chambers*, much ingenuity has been exercised in devising an explanation of their assumed *hydrostatic* function, whereby the nautilus can rise at will to the surface, or sink, on the approach of storms to the quiet recesses of the deep. Unfortunately for such poetical speculations, the nautilus appears on the surface, only *when driven up by storms*, and its sphere of action is on the *bed* of the sea, where it creeps like a snail, or perhaps lies in wait for unwary crabs and shell-fish, like some gigantic "sea-anemone," with outspread tentacles.

The tetrabranchs could undoubtedly swim, by their respiratory jets; but the discoidal nautili and ammonites are not well calculated, by their forms, for swimming; and the straight-shelled *orthocerata* and *baculites* must have held a nearly vertical position, head-downwards, on account of the buoyancy of their shells. The use of the air-chambers, is to render the whole animal (and shell) of nearly the same specific gravity with the water.* The object of the numerous partitions is not so much to sustain the pressure of the water, as to guard against the *collisions* to which the shell is exposed. They are most complicated in the *ammonites*, whose general form possesses least strength.† The purpose of the siphuncle (as suggested by Mr. Searles Wood) is to maintain the *vitality* of the shell, during the long life which these animals certainly enjoyed. Mr. Forbes has suggested that the inner courses of the *hamites*, broke off, as the outer ones were formed. But this was not the case with the *orthocerata*, whose long straight shells were particularly exposed to danger; in these the preservation of the shell was provided for by the increased size and strength of the siphuncle, and its increased vascularity. In *endoceras* we find the siphuncle thickened by *internal* deposits, until (in some of the very cylindrical species) it forms an almost solid axis.

The *nucleus* of the shell is rather large in the *nautili*, and causes an

---

* A *nautilus pompilius* (in the cabinet of Mr. Morris) weighs 1lb., and when the siphuncle is secured, it floats with a ½lb weight in its aperture. The animal would have displaced 2 pints (= 2½lbs) of water, and therefore, if it weighed 3lbs., the specific gravity of the animal and shell would scarcely exceed that of salt water.

† The siphuncle and lobed septa did not hold the animal in its shell, as Von Buch imagined: that was secured by the shell-muscles. The complicated sutures perhaps indicate lobed ovaries; they occur in genera, which must have produced very small eggs.

opening to remain through the shell, until the *umbilicus* is filled up with a callous deposit; several fossil species have always a hole through the centre.

In the *ammonites*, the *nucleus* is exceedingly small, and the whirls compact from the first.

It has been stated that the *septa* are formed periodically; but it must not be supposed that the shell-muscles ever become detached, or that the animal moves the distance of a chamber all at once. It is most likely that the *adductors* grow only in front, and that a constant waste takes place behind, so that they are always moving onward, except when a new septum is to be formed; the *septa* indicate periodic *rests*.

The consideration of this fact, that the nautilus must so frequently have an air-cavity between it and its shell, is alone sufficient to convince us, that the chambered cephalopods could not exist in very deep water. They were probably limited to a depth of 20 or 30 fathoms at the utmost.*

It is certain that the sexes were distinct in the *tetrabranchiata*, but since only the female of the living nautilus is known, we are left to conjecture how ar the differences observable in the shells, are dependant on sex. M. D'Orbigny, having noticed that there are two varieties of almost every kind of ammonite, —one compressed, the other inflated—naturally assumed that the first were the shells of male individuals (♂), the second of females (♀). Dr. Melville has made a similar suggestion with respect to the nautili; namely, that the umbilicated specimens are the males, the imperforated shells, females. This is rendered probable by the circumstance, that all the known specimens of *N. pompilius* were female, and that the supposed male (*N. macromphalus*) is very rare, as we have noticed amongst the male *dibranchiata*. Of the other recent species, both the presumed sexes (*N. umbilicatus* ♂ and *N. stenomphalus* ♀) are comparatively rare.

FAMILY I. NAUTILIDÆ.

Shell. Body-chamber capacious. *Aperture* simple. *Sutures* simple. *Siphuncle* central, or internal. (Figs. 35, 36.)

NAUTILUS, Breynius, 1732.

Shell involute or discoidal, few-whirled. *Siphuncle* central.

In the recent nautili, the shell is smooth, but in many fossil species it is corrugated, like the patent iron-roofing, so remarkable for its strength and lightness. (*Buckland.*) See pl. II., fig. 10.

---

* By *deep water*, naturalists and dredgers seldom mean more than 25 fathoms, a comparatively small depth, only found near coasts and islands. At 100 fathoms the pressure exceeds 265lbs. to the square inch. Empty bottles, securely corked, and sunk with weights beyond 100 fathoms, are always crushed, If filled with liquid, the cork is driven in, and the liquid replaced by salt water; and in drawing the bottle up again, the cork is returned to the neck of the bottle, generally in a reversed position. (*Sir F. Beaufort.*)

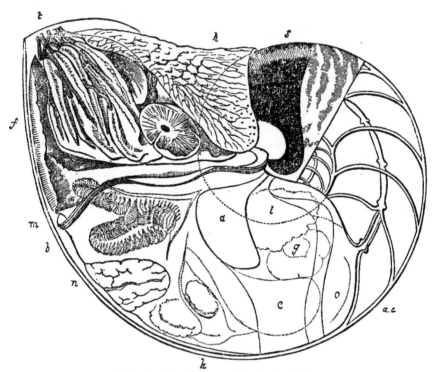

Fig. 43.   *Nautilus pompilius in its shell.**

The *umbilicus* is small or obsolete in the typical nautili, and the whirls enlarge rapidly.   In the palæozoic species, the whirls increase slowly, and are sometimes scarcely in contact.   The last *air-cell* is frequently shallower in proportion than the rest.

Animal.   In the recent nautilus, the *mandibles* are horny, but calcified to a considerable extent ; they are surrounded by a circular fleshy lip, external to which are four groups of *labial tentacles,* 12 or 13 in each group, they appear to answer to the *buccal membrane* of the calamary (fig. 1).   Beyond these, on each side of the head, is a double series of arms, or *brachial tentacles,* 36 in number ; the dorsal pair are expanded and united to form the hood, which closes the aperture of the shell, except for a small space on each side, which is filled by the second pair of arms.   The *tentacles* are lamellated

* This woodcut and 18 others illustrating the *tetrabranchiata,* are the property of Mr. Gray, to whom we are indebted for their use.   Fig. 43 represents the recent nautilus, as it appears on the removal of part of the outer shell-wall (from the specimen in the British Museum).   The *eye* is seen in the centre, covered by the hood (*h*): *t,* tentacles, nearly concealed in their sheaths ; *f,* funnel ; *m,* margin of the mantle, very much contracted ; *n,* nidamental gland ; *a, c,* air-cells and siphuncle ; *s,* portion of the shell ; *a,* shell-muscle,   The internal organs are indicated by dotted lines ; *b,* branchiæ ; *h,* heart and renal glands ; *c,* crop ; *g,* gizzard ; *l,* liver ; *o,* ovary.

on their inner surface, and are retractile within sheaths, or "digitations," which correspond to the eight ordinary arms of the cuttle-fishes ; their supe-riority in number being indicative of a lower grade of organization. Besides these there are four *ocular tentacles*, one behind and one in front of each eye ; they seem to be instruments of sensation, and resemble the tentacles of *doris* and *aplysia* (*Owen*). On the side of each eye is a hollow plicated process, which is not tentaculiferous. The *respiratory funnel* is formed by the folding of a very thick muscular lobe, which is prolonged laterally on each side of the head, with its free edge directed backwards, into the branchial cavity ; behind the *hood* it is directed forwards, forming a lobe which lies against the black-stained spire of the shell (fig. 43 *s*.)* Inside the funnel is a valve-like fold (fig. 44 *s*). The margin of the mantle is entire, and extends as far as the edge of the shell ; its substance is firm and muscular, as far back as the line of the shell-muscles and horny girdle, beyond which it is thin and transparent. The *shell-muscles* are united by a narrow tract, across the hollow occupied by the involute spire of the shell ; and are thus rendered horse-shoe shaped. The *siphuncle* is vascular ; it opens into the cavity containing the heart (*pe-ricardium*), and is most probably filled with fluid from that cavity. (*Owen*.)

Respecting the habits of the nautilus, very little is known, the specimen dissected by Professor Owen had it crop filled with fragments of a small crab, and its mandibles seem well adapted for breaking shells. The statement that it visits the surface of the sea of its own accord, is at present unconfirmed by observation, although the air cells would doubtless enable the animal to rise by a very small amount of muscular exertion.

Professor Owen gives the following passage, from the old Dutch naturalist, Rumphius, who wrote in 1705, an account of the rarities of Amboina. " When the nautilus floats on the water, he puts out his head and all his tenta-cles, and spreads them upon the water, with the poop of the shell above water ; but at the bottom he creeps in the reverse position, with his boat above him, and with his head and tentacles upon the ground, making a tolerably quick progress. He keeps himself chiefly upon the ground, creeping also sometimes into the nets of the fishermen ; but after a storm, as the weather becomes calm, they are seen in troops, floating on the water, *being driven up by the agitation of the waves*. This sailing, however, is not of long continuance ;

---

* The *funnel* is considered the homologue of the foot of the gasteropods, by Loven, a conclusion to which we cannot agree. The cephalopods ought to be compared with the *larval* gasteropods, in which the foot only serves to support an operculum ;—or with the floating tribes in which the foot is obsolete, or serves only to secrete a nida-mental raft (*ianthina*). However, on examining the nautilus preserved in the British Museum, and finding that the funnel was only part of a muscular collar, which ex-tends all round the neck of the animal, we could not avoid noticing its resemblance to the siphonal lappets of *paludina*, and to that series of lappets (including the *oper-culigerous lobe*) which surrounds the *trochus* (fig. 87).

for having taking in all their tentacles, they upset their boat, and so return to the bottom."

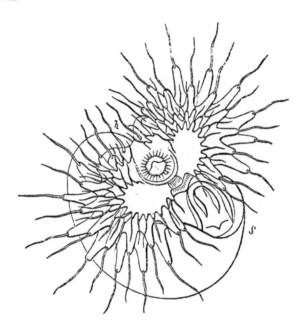

Fig. 44. *Nautilus expanded.* *

Distr., 2 or 4 sp.   Chinese seas, Indian ocean, Persian gulf.

Fossil, about 100 sp.   In all strata, S. and N. America (Chile). Europe, India (Pondicherry).

> *Sub-genus.   Aturia* (Bronn), = Megasiphonia D'Orb.

Type, N. zic-zac Sby.   Pl. II., fig. 12, London clay, Highgate.

Shell, sutures, with a deep lateral lobe; siphuncle nearly internal, large, continuous, resembling a succession of funnels.

Fossil, 4 sp.   Eocene, N. America, Europe, India.

Sub-genus?   Discites, McCoy.   *Whirls* all exposed; the last chamber sometimes produced.   L. silurian.—Carb : limestone.

Temnocheilus, McCoy.   Founded on the carinated sp. of the Carb. lime-stone.

Cryptoceras, D'Orb.   Founded on *N. dorsalis* Phil. and one other species, in which the siphuncle is nearly external.

* Ideal representation of the nautilus, when expanded, by Professor Lovén, who appears to have taken the details from M. Valenciennes memoir in the *Archives du Museum*, vol. 2, p. 257. *h*, hood. *s*, siphon. It is just possible, that when the nautilus issues from its shell, the gas contained in the last, incomplete, air-chamber, may expand; but this could not happen under any great pressure of water.

LITUITES, Breynius.

Etym., lituus, a trumpet.

Syn., Hortolus, Montf. (whirls separate.) Trocholites, Conrad.

Ex., L. convolvans, Schl. L. lituus, Hisinger.

Shell, discoidal; whirls close, or separate; last chamber produced in a straight line; siphuncle central.

Fossil, 15 sp. Silurian, N. America, Europe.

TROCHÓCERAS, Barrande, 1848.

Ex., T. trochoides, Bar.

Shell, nautiloid, spiral, depressed.

Fossil, 16 sp. U. Silurian, Bohemia.

Some of the species are nearly flat, and having the last chamber produced would formerly have been considered Lituites.

Fig. 45. *Clymenia striata, Munst.** Fig. 46. *C. linearis, Munst.*

CLYMENIA, Munster, 1832.

Etym., clymene, a sea-nymph.

Syn. Endosiphonites, Ansted. Sub-clymenia, D'Orb.

Ex., C. striata, pl. II., fig. 16 (Mus. Tennant).

Shell, discoidal; septa simple or slightly lobed; siphuncle internal.

Fossil, 43 sp. Devonian, N. America, Europe.

FAMILY II. ORTHOCERATIDÆ.

Shell, straight, curved, or discoidal; *body chamber* small; *aperture* contracted, sometimes extremely narrow (figs. 40, 41); siphuncle complicated.

It seems probable that the cephalopods of this family were not able to withdraw themselves completely into their shells, like the pearly nautilus; this was certainly the case with some of them, as M. Barrande has stated, for the siphonal aperture is almost isolated from the cephalic opening. The shell appears to have been often less calcified, but connected with more vascular parts than in the nautilus; and the siphuncle often attains an enormous development. In all this, there is nothing to suggest a doubt of their being *tetrabranchiate;* and the chevron-shaped coloured bands preserved on the *orthoceras anguliferus,*† sufficiently prove that the shell was essentially external.

---

* Fig. 45. Sutures of two species of Clymenia from Phillips' Pal. Fos., Devonshire.

† Figured by D'Archiac and Verneuil, Geol. Trans.

ORTHOCERAS, Breyn.

Etym., *orthos*, straight, and *ceras*, a horn.

Syn., cycloceras, McCoy.   Gonioceras, Hall.*

Ex. O. giganteum (diagram of a longitudinal section), pl. II, fig. 14.

Shell, straight; siphuncle central; aperture sometimes contracted.

Fossil, 125 typical sp. (D'Orb).†   L. Silurian—Trias; N. America, Australia, and Europe.

The *orthocerata* are the most abundant and wide spread shells of the old rocks, and attained a larger size than any other fossil shell.   A fragment of *O. giganteum*, in the collection of Mr. Tate of Alnwick, is a yard long, and 1 foot in diameter, its original length must have been 6 feet.   Other species, 2 feet in length, are only 1 inch in diameter, at the aperture.

Sub-genus I. *Cameroceras*, Conrad (= melia and thoracoceras, Fischer ?).

Siphuncle lateral, sometimes very large (*simple ?*).

Casts of these large siphuncles were called *hyolites* by Eichwald.

27 sp.   L. Silurian—Trias ? N. America and Europe.

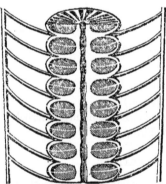
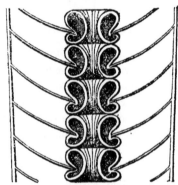

Fig. 47. *Actinoceras.*‡   Fig. 48. *Ormoceras.*

2. *Actinoceras* (Broun), Stokes. Siphuncle very large, inflated between the chambers, and connected with a slender central tube by radiating plates. 6 sp.   L. Silurian—Carb, N. America, Baltic, and Brit.

3. *Ormoceras*, Stokes. Siphuncular beads constricted in the middle (making the septa appear as if united to the centre of each).   3 sp.   L. Silurian, N. America.

4. *Huronia*, Stokes.   *Shell* extremely thin, membraneous or horny ? Siphuncle very large, central, the upper part of each joint inflated, connected

---

* *Theca* and *Tentaculites* are provisionally placed with the *Pteropoda*, they probably belong here.

† M Barrande has discovered 100 new species in the Upper Silurian rocks of Bohemia.

‡ Fig. 47. *Actinoceras Richardsoni*, Stokes. Lake Winipeg (diagram, reduced ¼). Fig. 48. *Ormoceras, Bayfieldi*, Stokes. Drummond Island, (from Mr. Stokes' paper, Geol. Trans.)

with a small central tube by radiating plates. 3 sp. L. Silurian. Drummond Island, Lake Huron.

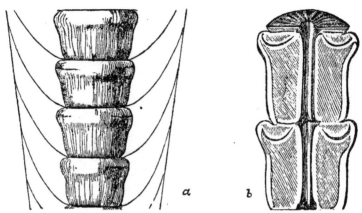

Fig. 49. *Huronia vertebralis.**

Numerous examples of this curious fossil were collected by Dr. Bigsby (in 1822), and by the officers of the regiments formerly stationed on Drummond Island. Specimens have also been brought home by the officers of many of the Artic expeditions. But with the exception of one formerly in the possession of Lieut. Gibson, 68., and another in the cabinet of Mr. Stokes, the siphuncle only is preserved, and *not a trace* remains of septa or shell wall. Some of those seen by Dr. Bigsby in the limestone cliffs, were 6 feet in length.

5. *Endoceras*, Hall (Cono-tubularia *Troost*). Shell extremely elongated, drical. Siphuncle very large, cylindrical, lateral; thickened internally by repeated layers of shell, or partitioned off by funnel-shaped diaphragms. 12 sp. Lower Silurian, New York.

6. Shell perforated by two distinct siphuncles? O. bisiphonatum Sby, Caradoc sandstone, Brit.

"Orthocerata with two siphuncles have been observed, but there has always appeared something doubtful about them. In the present instance, however, this structure cannot be questioned." (J. Sowerby.)

Small orthocerata of various species, are frequently found in the body chamber and open siphuncle of large specimens.† The *endoceras gemelliparum* and *proteiforme* of Hall, appear to be examples of this kind.

GOMPHOCERAS, J. Sby, 1839.

Etym., gomphos, a club, and *ceras*, a horn.

* Fig. 49. *Huronia vertebralis,* Stokes. *a*, from a specimen in the Brit. M., presented by Dr. Bigsby. The septa are added from Dr. Bigsby's drawing; they were only indicated in the specimen by "colourless lines on the brown limestone," *b*. represents a weathered section, presented to the Brit. Mus. by Captain Kellett and Lieutenant Wood of H.M.S. Pandora. The figures are reduced ½.

† Shells of *Bellerophon* and *Murchisonia* are found under the same circumstances.

Syn., Apioceras (Fischer).   Poterioceras (McCoy).
Type, G. pyriforme, Sby., fig. 51, and G. Bohemicum, Bar. fig. 40.

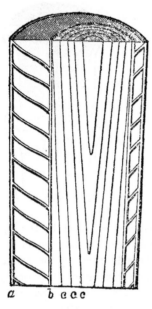
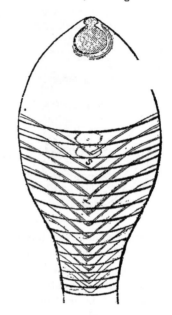

Fig. 50. *Endoceras.**   Fig. 51. *Gomphoceras.*†

Shell, fusiform or globular, with a tapering apex ; aperture contracted in
the middle ; siphuncle moniliform, sub-central.

Distr., 10 sp.   Silurian—Carb ; N. America, Europe.

ONCOCERAS, Hall.

Etym., *oncos*, a protuberance.
Type, O. constrictum, Hall.   Trenton limestone.
Shell, like a curved *gomphoceras* ; siphuncle external.
Distr., 3 sp.   Silurian, New York.

PHRAGMÓCERAS, Broderip.

Etym., *phragmos,* a partition, and *ceras*, a horn.
Type, P. ventricosum (Steininger sp.), pl. II., fig. 15.
Shell curved, laterally compressed ; *aperture* contracted in the middle
siphuncle, ventral, radiated. Ex., P. callistoma, Bar., fig. 41.
Distr., 8 sp.   U. Silurian—Devonian, Brit., Germany.

---

* Fig. 50.   Diagram of an *endoceras* (after Hall), *a*, shell-wall. *b.* Wall of siphuncle. *c c c.* Diaphragms (" embryo-tubes " of Hall).

† Fig. 51. *Gomphoceras pyriforme.* L. Ludlow rock, Mochtre hill, Herefordshire
(from Murch, Silur, syst., reduced ½). *s.* Beaded siphuncle.

CYRTÓCERAS, Goldf. 1833.

Etym., *curtos*, curved, *ceras*, horn.

Syn., Campulites, Desh. 1832 (including gyroceras). Aploceras, D'Orb. Campyloceras and trigonoceras, McCoy.

Ex., C. hybridum, volborthi and beaumonti (Barrande).

Shell, curved; *siphuncle* small, internal, or sub-central.

Distr., 36 sp.   L. Silurian, Carb—N. America, and Europe.

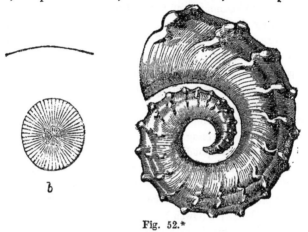

Fig. 52.*

GYRÓCERAS, Meyer, 1829.

Etym., *gyros*, a circle, and *ceras*.

Syn., Nautiloceras, D'Orb.

Ex., G. eifeliense, D'Arch., pl. II., fig. 13.   Devonian, Eifel.

Shell, nautiloid; whirls separate; siphuncle excentric, radiated.

Fossil, 17 sp.   U. Silurian—Trias? N. America, and Europe.

ASCOCERAS, Barrande, 1848.†

Etym., *ascos*, a leather bottle.

Shell, bent upon itself, like *ptychoceras*.

Distr., 7 sp.   U. Silurian, Bohemia.

FAMILY III.   AMMONITIDÆ.

Shell.   Body-chamber elongated; *aperture* guarded by processes, and closed by an operculum; *sutures* angulated, or lobed and foliated; *siphuncle* external (dorsal, as regards the shell).

The shell of the *ammonitidæ* has essentially the same structure with the nautilus.   It consists of an external porcellanous‡ layer, formed by the *collar*

---

* Fig. 52. *Gyroceras goldfussii* (= ornatum Goldf). *b*. Siphuncle of *G. depressum*, Goldf. sp. Devonian.   Eifel.   From M.M. D'Archiac and Verneuil.

† In Haidinger's Berichte.

‡ Its microscopic structure has not been satisfactorily examined; Prof. Forbes detected a punctate structure in one species.

of the mantle only; and of an internal nacreous lining, deposited by the whole
extent of its visceral surface.   There is an *ammonite* in the British Museum,
evidently broken and repaired during the life of the animal,* which shews
that the shell was deposited *from within.*   In some species of ammonites the
collar of the mantle forms prominent spines on the shell, which are too deep
for the visceral mantle to enter; they are therefore *partitioned off* (as in *A.
armatus*, Lias) from the body whirl and air cells, and not exhibited in *casts.*

The baculites, and ammonites of the section *cristati*, acquire when adult
a process projecting from the outer margin of their shell.   Certain other
ammonites (the *ornati*, *coronati*, &c.) form two *lateral* processes before they
cease to grow (pl. III., fig. 5).   As these processes are often developed in
very small specimens, it has been supposed that they are formed repeatedly
in the life of the animal (at each periodic rest), and are again removed when
growth recommences.   These small specimens, however, may be only dwarfs.
In one ammonite, from the inferior oolite of Normandy, the ends of these
lateral processes meet, "forming an arch over the aperture, and dividing
it into two outlets, one corresponding with that above the hood of the nautilus,
which gives passage to the dorsal fold of the mantle; the other with that
below the hood, whence issue the tentacles, mouth, and funnel; such a modification, we may presume, could not take place before the termination of the
growth of the individual."†   (Owen.)

M. D'Orbigny has figured several examples of deformed *ammonites*, in
which one side of the shell is scarcely developed, and the keel is consequently
lateral.   Such specimens probably indicate the partial atrophy of the branchiæ
on one side.   In the British Museum there are deformed specimens of *Am.
obtusus*, *amaltheus*, and *tuberculatus.*

Fig. 53.‡

* *A serpentinus* Schloth, U. Lias, Wellingboro. Rev. A. W. Griesbach.

† This unique and abnormal specimen is in the cabinet of S. P. Pratt, Esq.

‡ Fig. 53. *Goniatites sphericus*, Sby. Front and side views of a specimen from the
carb limestone of Derbyshire, in the cabinet of Mr. J. Tennant; the body-chamber
and shell-wall have been removed artificially.

GONIATITES, De Haan.

Etym., *gonia*, angles (should be written gonialites ?).
Syn., *aganides*, Montf.
Examples, G. Henslowi, pl. III., fig. 1., G. sphericus, fig. 53, and 39.
Shell, discoidal; sutures lobed; siphuncle dorsal.
Distr. 150 sp. Devonian—Trias, Europe.

BACTRITES, Sandberger (= stenoceras, D'Orb ?).

Shell, straight; sutures lobed. *Type*, B. subconicus, Sbger.
Distr., 2 sp. Devonian—Germany.

Fig. 54.*

CERATITES, De Haan.

Type, C. nodosus, pl. III., fig. 2.
Shell, discoidal; sutures lobed, the lobes crenulated. Fig. 54.
Distr., muschelkalk, 8 sp. Germany, France, Russia, Siberia.
Salt-marls (Keuper). 17 sp. S. Cassian, Tyrol.
M. D'Orbigny describes 5 shells from the gault and U. greensand as
eratites; but many ammonites have equally simple sutures, when young.

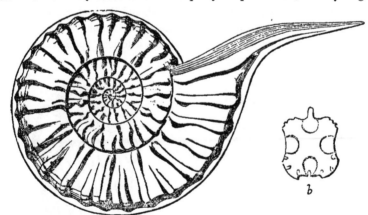

Fig. 55.†

AMMONITES, Bruguiere.

Etym., *ammon*, a name of Jupiter, worshipped in Libya under the form
f a ram. The ammonite is the *cornu ammonis* of old authors.

---

* Fig. 54. Suture of *ceratites nodosus* (Brug). The arrow in the dorsal lobe
oints towards the aperture.

† Fig. 55. *Ammonites rostratus*, Sby. From the U. green-sand of Devizes, in the
abinet of W. Cunnington, Esq. *b*, front view of one of its partitions.

Syn., orbulites Lam. planulites, Montf.

Shell, discoidal; inner whirls more or less concealed; septa undulated; sutures lobed and foliated; siphuncle dorsal.

Distr., 530 sp. Trias—chalk. Coast of Chili (D'Orb.) Santa Fe de Bogota (Hopkins), New Jersey, Europe, and S. India.

Capt. Alexander Gerard discovered ammonites similar to our L. oolitic species, in the high passes of the Himalaya, 16,200 feet above the sea.

Section A.   *Back, with an entire keel.*

1. *Arietes*,   L. oolites,   A. bifrons (pl. III., fig. 6), bisulcatus (pl. III., fig. 7).
2. *Falciferi*,   L. oolites,   A. serpentinus, radians, hecticus.
3. *Cristati*,   cretaceous,   A. cristatus, rostratus (fig. 55), varians.

B.   *Back crenated.*

4. *Amalthei*,   ool.   A. amaltheus, cordatus, excavatus.
5. *Rhothomagenses*, cret.   A. rhothomagensis (pl. III., fig. 4).

C.   *Back sharp.*

6. *Disci.*,   oolitic,   A. discus, clypeiformis.

D.   *Back channelled.*

7. *Dentati*,   { cret.   A. dentatus, lautus.
                    { ool.   A. Parkinsoni, anguliferus.

E.   *Back squared.*

8. *Armati*,   L. ool.   A. armatus, athletus, perarmatus.
9. *Capricorni*,   · L. ool.   A. capricornus, planicostatus.
10. *Ornati*,   ool.   A. Duncani, Jason (pl. III., fig. 5).

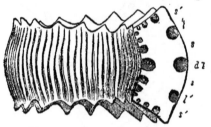

Fig. 56. *Ammonites coronatus.**

F.   *Back round, convex.*

11. *Heterophylli*, L. ool.   A. heterophyllus (fig. 34).
12. *Ligati*,   cret.   A. planulatus (pl. III., fig. 3).
13. *Annulati*,   ool.   A. annulatus, biplex, giganteus.
14. *Coronati*,   ool.   A. coronatus (fig. 56), sublævis.
15. *Fimbriati*,   ool.   A. fimbriatus, lineatus, hircinus.

* Fig. 56. Profile of ammonites coronatus, Brug. (reduced ½ from D'Orbigny) Kelloway rock, France. *d l.* dorsal lobe; *s s*, dorsal saddles; *l' l'* lateral lobes; *s' s'*. lateral saddles; accessory and ventral lobes. The number of accessory lobes increases with age.

16. *Cassiani*, 36 sp. of very variable form, and remarkable for the number and complexity of their lobes. Trias, Austrian Alps.

Fig. 57.*

Ex., A. Maximiliani (fig. 57), A. Metternichii.

CRIOCERAS, Leveille.

Etym., *krios*, a ram, and *ceras*, a horn.
Syn., tropæum, Sby.
Ex., C. cristatum, D'Orb. (pl. III., fig. 8).
Shell, discoidal; whirls separate.
Distr., 9 sp. Neocomian—Gault; Brit., France.

TOXOCERAS, D'Orb.

Etym., *toxon*, a bow, *ceras*, a horn.
Ex., T. annulare, D'Orb. (pl. III., fig. 12.)
Shell, bow-shaped; like an ammonite uncoiled.
Distr., 19 sp. Neocomian. Between this and *crioceras* and *ancyloceras* there are numerous intermediate forms.

ANCYLOCERAS, D'Orb.

Etym., *anculos*, incurved.
Ex., A. spinigerum (pl. III., fig. 10).
Shell, at first discoidal, with separate whirls; afterwards produced at a tangent and bent back again, like a hook or crosier.
Distr., 38 sp. Inf. oolite—chalk. S. America (Chile and Bogota), Europe.

SCAPHITES, Parkinson.

Etym., *scaphe*, a boat.
Ex., S. equalis (pl. III., fig. 9).
Shell, at first discoidal, with close whirls; last chamber detached and recurved.
Distr., 17 sp. Neocomian—chalk. Europe.

HELICOCERAS, D'Orb.

Etym., *helix* (*helicos*), a spiral, and *ceras*, horn.
Ex., H. rotundum, Sby, sp. pl. III., fig. 11 (diagram).

* Fig. 57. Am. Maximiliani Klipstein. (= A. bicarinatus Münst). Trias, Hallstadt (copied from Quenstedt). A, Profile shewing the numerous lobes and saddles. B, suture of one side; *v*, dorsal saddle.

Shell, spiral, sinistral; whirls separate.

Distr., 11 sp. Inf. oolite?—chalk. Europe.

TURRILITES, Lam.

Etym., *turris*, a tower, and *lithos*, a stone.

Shell, spiral, sinistral; aperture often irregular.

Distr., 27 sp. (Bronn). Gault—chalk. Europe.

The turrilite was perhaps *di-branchiate*, by the atrophy of the respiratory organs of one side. M. D'Orbigny includes in this genus particular specimens of certain *Lias ammonites* which are very slightly unsymmetrical; the same species occur with both sides alike. He also makes a genus (*heteroceras*) of two turrilites, in which the last chamber is somewhat produced and recurved. *T. reflexus* (Quenstedt, T. 20, fig. 16) has its apex inflected and concealed.

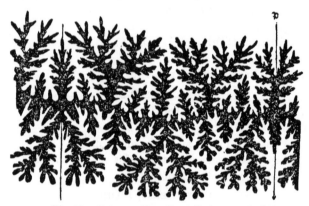

Fig. 58.　*Sutures of hamites cylindraceus, Defr.**

HAMITES, Parkinson.

Etym., *hamus*, a hook.

Ex., H. attenuatus, pl. III., fig. 15.

Shell, hook-shaped, or bent upon itself more than once, the courses separate.

Distr., 58 sp. Neocomian—chalk. S. America (Tierra del Fuego)—Europe.

The inner courses of this shell probably break away or are "decollated" in the progress of its growth (Forbes). M. D'Orbigny has proposed a new genus, *hamulina*, for the 20 neocomian species.

PTYCHOCERAS, D'Orb.

Etym., *ptyche*, a fold.

Ex., P. emericianum, D'Orb., pl. III., fig. 14.

---

* Fig. 58. Space between two consecutive sutures of the right side, from a specimen in the Brit. Mus. *a.* dorsal line. *b.* ventral. Baculite limestone, Fresville.

Shell, bent once upon itself; the two straight portions in contact.
Distr., 7 sp. Neocomian—chalk. Brit. France.

BACULITES, Lamarck.

Etym., baculus, a staff.
Ex., B. anceps. Pl. III., fig. 13.
Shell, straight, elongated; aperture guarded by a dorsal process.
Distr., 11 sp. Neocomian—chalk. Europe, S. America (Chile).

Baculina, D'Orb. B. Rouyana. Neoc., France. Sutures not foliated.

The chalk of Normandy has received the name of *baculite limestone*, from the abundance of this fossil.

CLASS II. GASTEROPODA.

The gasteropods, including land-snails, sea-snails, whelks, limpets, and the like, are the types of the *mollusca;* that is to say, they present all the leading features of molluscous organization in the most prominent degree, and make less approach to the appearance and condition of fishes than the cephalopods, and less to the crustaceans and zoophytes than the bivalves.

Their ordinary and characteristic mode of locomotion is exemplified by the common garden-snail, which creeps by the successive expansion and contraction of its broad muscular foot. These muscular movements may be seen following each other in rapid waves when a snail is climbing a pane of glass.

The *nucleobranches* are "aberrant" gasteropods, having the foot thin and vertical; they swim near the surface of the sea, in a reversed position, or adhere to floating sea-weed.

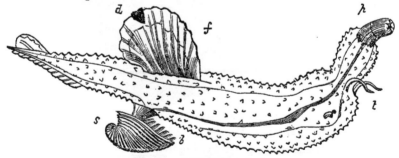

Fig. 59. *A nucleobranche.**

The gasteropods are nearly all unsymmetrical, the body being coiled up spirally, and the respiratory organs of the left side being usually atrophied. In *chiton* and *dentalium* the *branchiæ* and reproductive organs are repeated on each side.

---

* Fig. 59. *Carinaria cymbium*, L. sp. (after Blainville), Mediterranean; *p*, proboscis; *t*, tentacles; *b*, branchiæ; *s*, shell; *f*, foot; *d*, disk.

A few species of *cymba, litorina, paludina,* and *helix,* are viviparous; the rest are oviparous.

When first hatched the young are always provided with a shell, though in many families it becomes concealed by a fold of the mantle, or it is speedily and wholly lost.*

The gasteropods form two natural groups; one breathing air (*pulmonifera*), the other water (*branchifera*). The air-breathers undergo no apparent metamorphosis; when born, they differ from their parents in size only. The water-breathers have at first a small nautiloid shell, capable of concealing them entirely, and closed by an operculum. Instead of creeping, they swim with a pair of ciliated fins springing from the sides of the head; and by this means are often more widely dispersed than we should be led to expect from their adult habits; thus some sedentary species of *calyptræa* and *chiton* have a greater range than the "paper-sailor," or the ever-drifting oceanic-snail.

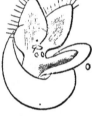

At this stage, which may fairly be compared with the larval condition of insects, there is scarcely any difference between the young of *eolis* and *aplysia,* or *buccinum* and *vermetus.* (M. Edw.)

Fig. 60.†

The development of the branchiferous gasteropods may be observed with much facility in the common river-snails (*paludina*); which are viviparous, and whose oviducts in early summer contain young in all stages of growth some being a quarter of an inch in diameter.

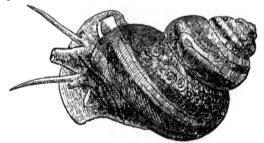

Fig. 61. *Paludina vivipara.*‡

Embryos scarcely visible to the naked eye have a well-formed shell, ornamented with epidermal fringes; a foot and operculum; and the head has long delicate tentacula, and very distinct black eyes.

---

* M. Lovén believes that the embryo shell of the nudibranches falls off at the time they acquire a locomotive foot.

† Fig. 60. Fry of Eolis (from Alder and Hancock); *o*, the operculum; the original s not larger than the letter o.

‡ Fig. 61. *Paludina vivipara* L. (original); the internal organs are represented as if seen through the shell. The ovary, distended with eggs and embryos, occupies the right side of the body whirl; the gill is seen on the left; and between them the termination of the alimentary canal. Surrey Docks, June, 1850.

The development of the pulmoniferous embryo is best seen in the transparent eggs of the fresh-water limneïds ; these are not hatched until the young
have passed the larval condition, and their ciliated head-lobes (or veil), are
superseded by the creeping disk, or foot.

The *shell* of the gasteropods is usually *spiral*, and univalve ; more rarely
tubular, or *conical*, and in one genus it is *multivalve*. The following are its
principal modifications :

A. Regularly spiral,

    a. elongated or turreted ; *terebra, turritella.*

    b. cylindrical ; *megaspira, pupa.*

    c. short ; *buccinum.*

    d. globular ; *natica, helix.*

    e. depressed ; *solarium.*

    f. discoidal ; *planorbis.*

    g. convolute ; aperture as long as the shell ; *cypræa, bulla.*

    h. fusiform ; tapering to each end, like *fusus.*

    i. trochi-form ; conical, with a flat base, like *trochus.*

    k. turbinated ; conical, with a round base, like *turbo.*

    l. few-whirled ; *helix hæmastoma.* Pl. XII., fig. 1.

    m. many-whirled ; *helix polygyrata.* Pl. XII., fig. 2.

    n. ear-shaped ; *haliotis.*

B. Irregularly spiral ; *siliquaria, vermetus.*

C. Tubular ; *dentalium.*

D. Shield-shaped ; *umbrella, parmophorus.*

E. Boat-shaped ; *navicella.*

F. Conical or limpet-shaped ; *patella.*

G. Multivalve and imbricated ; *chiton.*

The only symmetrical shells are those of *carinaria, atlanta, dentalium,*
and the limpets.*

Nearly all the spiral shells are *dextral*, or right-handed ; a few are constantly *sinistral*, like *clausilia ;* reversed varieties of many shells, both dextral and sinistral, have been met with.

The cavity of the shell is a single conical or spiral chamber ; no gasteropod has a multilocular shell like the nautilus, but spurious chambers are
formed by particular species, such as *triton corrugatus* (fig. 62), and *euomphalus
pentangulatus ;* or under special circumstances, as when the upper part of the
spire is destroyed.

Some spiral shells are complete tubes, with the whirls separate, or scarcely

* The curve of the spiral shells and their opercula, and also of the Nautilus, is *a
logarithmic spiral ;* so that to each particular species may be annexed a number, indicating the ratio of the geometrical progression of the dimensions of its whirls. Rev.
Moseley, " On geometrical forms of turbinated and discoid shells." *Phil. Trans.*
Lond. 1838. *Pt.* 2, *p.* 351.

in contact, as *scalaria, cyclostoma,* and *valvata ;* but more commonly the inner side of the spiral tube is formed by the pre-existing whirls (fig. 62).

The axis of the shell, around which the whirls are coiled, is sometimes open or hollow; in which case the shell is said to be perforated, or *umbilicated* (e. g. *solarium*). The perforation may be a mere chink, or fissure (*riam*), as in *lacuna ;* or it may be filled up by a shelly deposit, as in many *naticas*. In other shells, like the *triton,* the whirls are closely coiled, leaving only a pillar of shell, or *columella,* in the centre; such shells are said to be *imperforate.*

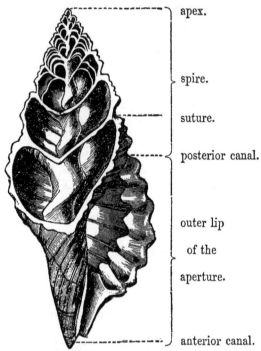

apex.

spire.

suture.

posterior canal.

outer lip

of the

aperture.

anterior canal.

Fig. 62. *Section of a spiral univalve.**

The *apex* of the shell presents important characters, as it was the *nucleus* or part formed in the egg; it is sinistral in the *pyramidellidæ,* oblique and spiral in the *nucleobranches* and *emarginulæ,* and mammillated in *turbinella pyrum* and *fusus antiquus.*

The apex is directed backwards in all except some of the *patellidæ,* in which it is turned forwards, over the animal's head. In the adult condition of some shells the apex is always truncated (or *decollated*), as in *cylindrella* and *bulimus decollatus ;* in others it is only truncated when the animals have lived

* Fig. 62. Longitudinal section of *triton corrugatus,* Lam., from a specimen in the cabinet of Mr. Gray. The upper part of the spire has been partitioned off many times successively.

in acidulous waters (e. g. *cerithidea* and *pirena*), and specimens may be obtained from more favorable situations with the points perfect.

The line or channel formed by the junction of the whirls is termed the *suture.*

The last turn of the shell, or *body-whirl,* is usually very capacious; in the females of some species the whirls enlarge more rapidly than in the males (e. g. *buccinum undatum*). The "base" of the shell is the opposite end to the apex, and is usually the front of the aperture.

The *aperture is entire* in most of the vegetable feeders (*holostomata*), but notched or produced into a *canal,* in the carnivorous families (*siphonostomata*); this canal, or siphon, is respiratory in its office, and does not necessarily indicate the nature of the food. Sometimes there is a posterior channel or canal, which is excurrent, or anal, in its function (e. g. *strombidæ* and *ovulum volva*); it is represented by the slit in *scissurella,* the tube of *typhis,* the perforation in *fissurella,* and the series of holes in *haliotis.*

The margin of the aperture is termed the *peristome;* sometimes it is continuous (*cyclostoma*), or becomes continuous in the adult (*carocolla*); very frequently it is "interrupted," the left side of the aperture being formed only by the body-whirl. The right side of the aperture is formed by the outer lip (*labrum*), the left side by the inner or columellar lip (*labium*), or partly by the body-whirl (termed the "wall of the aperture" by Pfeiffer).

The outer lip is usually thin and sharp in immature shells, and in some adults (e. g. *helicella* and *bulimulus*); but more frequently it is thickened; or reflected; or curled inwards (*inflected*), as in *cypræa;* or expanded as in pteroceras; or fringed with spines as in *murex.* When these fringes or expansions of the outer lip are formed periodically they are termed *varices.*

Lines of colour, or sculpture, running from the apex to the aperture are spiral or longitudinal, and others which coincide with the lines of growth are "transverse," as regards the whirls; but stripes of colour extending from the apex across the whirls are often described as "longitudinal" or "radiating," with respect to the entire shell.

Shells which are always concealed by the mantle are colourless, like *limax* and *parmophorus;* and those which are covered by the mantle-lobes when the animal expands, acquire a glazed or enamelled surface, like the cowries; when the shell is deeply immersed in the foot of the animal it becomes partly glazed, as in *cymba.* In all other shells there is an epidermis, although it is sometimes very thin and transparent.

In the interior of the shell the muscular impression is horse-shoe shaped, or divided into two scars; the horns of the crescent are turned towards the head of the animal.

The *operculum* with which many of the gasteropods close the aperture of their shell, presents modifications of structure which are so characteristic of the sub-genera, as to be worthy of particular notice. It consists of a horny layer, sometimes strengthened by the addition of calcarious matter on its exterior, and in its mode of growth it presents some resemblance to the shell itself. Its inner surface is marked by a muscular scar, whose lines bear no relation to the external lines of growth, and its form is unlike the muscular scar in the shell. It is developed in the embyro, within the egg, and the point from which it commences is termed the nucleus; many of the spiral and concentric forms fit the aperture of the shell with accuracy, the others only close the entrance partially, and in many genera, especially those with large apertures (e. g. *dolium, cassidaria, harpa, navicella*), it is quite rudimentary or obsolete.

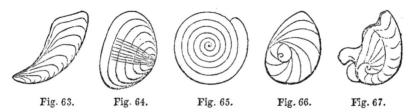

Fig. 63.   Fig. 64.   Fig. 65.   Fig. 66.   Fig. 67.

The operculum is described as—

Concentric, when it increases equally all round, and the nucleus is central or sub-central, as in *paludina* and *ampullaria* (pl. IX., fig. 26).

Imbricated or lamellar (fig. 64), when it grows only on one side, and the nucleus is marginal, as in *purpura, phorus*, and *paludomus*.

Claw-shaped, or unguiculate, (fig. 63, with the nucleus apical or in front), as in *turbinellus* and *fusus;* it is claw-shaped and serrated in *strombus* (fig. 69).

Spiral, when it grows only on one edge, and revolves as it grows; it is always *sinistral* in dextral shells.

Paucispiral, or few-whirled (fig. 66), as in *litorina*.

Sub-spiral, or scarcely spiral, in *melania*. Pl. VIII., fig. 25*.

Multispiral or many-whirled (fig. 65) as in *trochus*, where they sometimes amount to 20; the number of turns which the operculum makes is not determined by the number of whirls in the shell, but by the curvature of the opening, and the necessity that the operculum should revolve fast enough to fit it constantly (*Moseley*).

It is said to be *articulated* when it has a projection, as in nerita (fig. 67).

Too much importance, however, must not be attached to this very variable plate, as an aid to classification; it is present in some species of *voluta, oliva, conus, mitra*, and *cancellaria*, but absent in others; it is (indifferently) horny or shelly in the species of *ampullaria* and *natica;* in *paludina* it is concentric, in *paludomus lamellar*, in *valvata* spiral; in *solarium* and *cerithium*, it is *multispiral* or *paucispiral*.

Some of the gasteropoda can suspend themselves by glutinous threads, like *litiopa* and *rissoa parva*, which anchor themselves to sea-weeds (Gray), and *cerithidea* (fig. 68), which frequently leaves its proper element, and is found hanging in the air (Adams). A West India land-snail (*cyclostoma suspensum*) also suspends itself (Guilding). The origin of these threads has not been explained; but some of the *limaces* lower themselves to the ground by a thread which is not secreted by any particular gland, but derived from the exudation over the general surface of the body (Lister; D'Orbigny).

Fig. 68.

The division of this extensive class into orders and families, has engaged the attention of many naturalists, and a variety of methods have been proposed. Cuvier's classification was the first that possessed much merit, and several of his orders have since been united with advantage.

System of Cuvier.	System now adopted.
Class. GASTEROPODA.	
Order 1· Pectinibranchiata	
2. Scutibranchiata	Ord. *Prosobranchiata*, M. Edw.
3. Cyclobranchiata	
4. Tubulibranchiata	
5. Pulmonata	Ord. *Pulmonifera.*
6. Tectibranchiata	
7. Inferobranchiata	Ord. *Opisthobranchiata*, M. Edw.
8. Nudibranchiata	
Class. HETEROPODA.	Ord. *Nucleobranchiata.* Bl.

ORDER I. Prósobranchiáta.

Abdomen well developed, and protected by a shell, into which the whole animal can usually retire. *Mantle* forming a vaulted chamber over the back of the head, in which are placed the excretory orifices, and in which the branchiæ are almost always lodged. *Branchiæ* pectinated, or plume-like, situated (*proson*) in advance of the heart. *Sexes* distinct. (M. Edwards.)

SECTION A. Siphonostómata. Carnivorous Gasteropods.

Shell spiral, usually imperforate; aperture notched or produced into a canal in front. *Operculum* horny, lamellar.

Animal provided with a retractile proboscis; eye-pedicels connate with the tentacles; margin of the mantle prolonged into a siphon, by which water is conveyed into the branchial chamber; gills 1 or 2, comb-like, placed obliquely over the back. Species all marine.

FAMILY I. STRÓMBIDÆ. Wing-shells.

Shell with an expanded lip, deeply notched near the canal. *Operculum* claw-shaped, serrated on the outer edge.

Animal furnished with large eyes, placed on thick pedicels; tentacles slender, rising from the middle of the eye-pedicels. Foot narrow, ill adapted for creeping. Lingual teeth single; uncini, three on each side.

The strombs are carrion feeders, and, for molluscous animals, very active; they progress by a sort of leaping movement, turning their heavy shell from side to side. Their eyes are more perfect than those of the other gasteropods, or of many fishes.

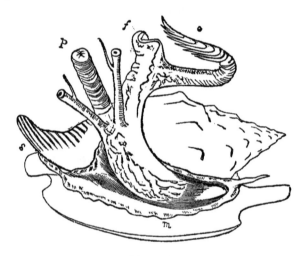

Fig. 69.*

STROMBUS, L. Stromb.

Etym., strombos, a top.

Type, S. pugilis. Pl. IV., fig. 1.

Shell rather ventricose, tubercular or spiny; spire short; aperture long, with a short canal above, and truncated below; outer lip expanded, lobed above, and sinuated near the notch of the anterior canal. Lingual teeth (*S. floridus*) 7 cusped; uncini, 1 tri-dentate, 2, 3 claw-shaped, simple.†

Distr., 60 species. West Indies, Mediterranean, Red Sea, India, Mau-

---

* Fig. 69. Strombus auris-Dianæ, L. (after Quoy and Gaimard), Amboina. *p,* proboscis, between the eye-pedicels; *f,* foot, folded up; *o,* operculum; *m,* border of the mantle; *s,* respiratory siphon.

† The lingual dentition of *strombus* resembles that of aporrhais, and is unlike that of the whelks; but it is more probable that aporrhais is the *representative* of strombus, than that it is very closely allied.

itius, China, New Zealand, Pacific, West America. On reefs, at low water, and ranging to 10 fathoms.

Fossil, 5 cretaceous species ; 3 sp. Miocene—. South Europe. There is a group of small shells in the eocene tertiary strata of England and France, nearly related to the living *S. fissurellus* L., some of which have been placed with *rostellaria*, because the notch in the outer lip is small, or obsolete. They probably constitute a sub-genus, to which Swainson's name *strombidia*, might be applied. *Example*, S. Bartonensis. Pl. IV., fig. 2.

The fountain-shell of the West Indies, *S. gigas*, L., is one of the largest living shells, weighing sometimes four or five pounds; its apex and spines are filled up with solid shell as it becomes old. Immense quantities are annually imported from the Bahamas for the manufacture of cameos, and for the porcelain works; 300,000 were brought to Liverpool alone in the last year, .850 (Mr. Archer).

PTERÓCERAS, Lam. Scorpion-shell.

Etym., *pteron*, a wing, and *ceras*, a horn.
Type, P. lambis. Pl. IV., fig. 3.
Shell like strombus when young ; outer lip, of the adult, produced into several long claws, one of them close to the spire, and forming a posterior anal.
Distr., 10 sp. India, China.
Fossil, nearly 100 sp. are enumerated by D'Orbigny, ranging from the as to the upper chalk ; many of them are more nearly related to aporrhaïs cerithiadæ).

ROSTELLARIA, Lam.

Etym., *rostellum*, a little beak.
Syn., fusus, Humphreys.
Example, R. curta. Pl. IV., fig. 4.
Shell with an elongated spire ; whirls numerous, flat ; canals long, the posterior one running up the spire ; outer lip more or less expanded, with only one sinus, and that close to the beak.
Distr., 5 sp. Red Sea, India, Borneo, China. *Range*, 30 fathoms.
Fossil, 70 sp. Neocomian — chalk (=aporrhaïs ?). 6 sp. Eocene—. Britain, France, &c.

The older tertiary species have the outer lip enormously expanded, and smooth-edged ; they constitute the section *hippochrenes* of Montfort (e. g. Rost. ampla, Solander. London clay).

Sub-genus? Spinigera, D'Orb. 1847. Shell like *rostellaria ;* whirls keeled ; keel developed into a slender spine on the outer lip, and two on each whirl, forming lateral fringes, as in *ranella.* Fossil, 5 sp. Inf. oolite—chalk. Britain, France.

F 3

SERAPHS, Montfort. (Terebellum, Lam.)

Etym., diminutive of *terebra*, an auger.

Type, S. terebellum (Linnæus sp.)=T. subulatum, Lam.　Pl. IV., fig. 5.

Shell smooth, sub-cylindrical; spire short or none; aperture long and narrow, truncated below; outer lip thin.

Distr., 1 sp.　China.　Philippines, 8 fms.　(Cuming.)

Fossil, 5 sp.　Eocene—.　London, Paris.

The animal of *terebellum* has an operculum like *strombus*; its eye-pedicels are simple, without tentacles (Adams).　In one fossil species, *T. fusiforme*, there is a short posterior canal, as in *rostellaria*.

FAMILY II.　MURICIDÆ.

Shell with a straight anterior canal; aperture entire behind.

Animal with a broad foot; eyes sessile on the tentacles, or at their base; branchial plumes 2.　*Lingual ribbon* long, linear; *rachis* armed with a single series of dentated teeth; *uncini*, single.　Predatory, on other *mollusca*.

MUREX (Pliny) L.

Types, M. palma-rosæ, Pl. IV., fig. 10.　M. tenuispina, Pl. IV., fig. 9.　M. haustellum, Pl. IV., fig. 8.　M. radix, pinnatus.

Shell ornamented with three or more continuous longitudinal varices; aperture rounded; beak often very long; canal partly closed; *operculum* concentric, nucleus sub-apical (Pl. IV., fig. 10); lingual dentition (M. erinaceus), teeth single, 3 crested; uncini single, curved.

Distr., 180 sp.　World-wide; most abundant on the W. coast of tropical America, in the Chinese Sea, West coast of Africa, West Indies; ranging from low water to 25 fathoms, rarely at 60 fathoms.

Fossil, 160 sp.　Eocene—.　Britain, France, &c.

A few of the species usually referred to this genus, belong to *pisania* and *trophon*.

The murices appear to form only one-third of a whirl annually, ending in a *varix*; some species form intermediate varices of less extent.　*M. erinaceus* a very abundant species on the coasts of the channel, is called "sting-winkle" by fishermen, who say it makes round holes in the other shell-fish with its beak.　See p. 27.　The ancients obtained their purple dye from species of *murex*; the small shells were bruised in mortars, the animals of the larger ones taken out.　(F. Col.)　Heaps of broken shells of the *M. trunculus* and caldron-shaped holes in the rocks may still be seen on the Tyrian shore. (Wilde.)　On the coast of the Morea, there is similar evidence of the employment of *M. brandaris* for the same purpose.　(M. Boblaye.)

TYPHIS, Montfort.

Etym., *typhos*, smoke.

Type, T. pungens. Pl. IV.. fig. 11.

Shell like murex; but having tubular spines between the varices, of which the last is open, and occupied by the excurrent canal.

Distr., 8 sp. Medit., W. Africa, Cape, India, W. America. —50 fms.

Fossil, 8 sp. Eocene—. London, Paris.

PISANIA, Bivon, 1832.

Etym., a native of (the coast near) *Pisa*, in Tuscany.

Syn., Pollia, Enzina, and Euthria (Gray).

Types, P. maculosa. Pl. IV., fig. 14 (Enzina) zonata. Pl. IV., fig. 15.

Shell with numerous indistinct varices, or smooth and spirally striated; canal short; inner lip wrinkled; outer lip crenulated.

Operculum ovate, acute; nucleus apical.

The *pisaniæ* have been usually confounded with *buccinum, murex,* and *ricinula*.

Distr., about 120 sp. W. Indies, Africa, India, Philippines, S. Seas, W America.

Fossil, ? sp. Eocene—. Brit., France, &c.

RANELLA, LAM. Frog-shell.

Syn., Apollon, Montfort and Gray.

Types, R. grauifera. Pl. IV., fig. 12. R. spinosa.

Shell with two rows of continuous varices, one on each side.

Operculum ovate, nucleus lateral.

Distr., 50 sp. Medit., Cape, India, China, Australia, Pacific, W. America.

Range, low-water to 20 fms.

Fossil, 23 sp. Eocene—.

TRITON. Lam.

Etym. *Triton*, a sea-deity. *Syn.*, persona (Montf. Gray).

Type, T. tritonis, L. sp. Pl. IV., fig. 13.

Shell with disconnected varices; canal prominent; lips denticulated.

Operculum ovate, sub-concentric.

Distr., 100 sp. W. Indies, Medit., Africa, India, China, Pacific, W. America. Ranging from low-water to 10 or 20 fathoms; one minute species has been dredged at 50 fathoms.

Fossil, 45 sp. Eocene—. Brit., France, &c. Chile.

The great triton (*T. tritonis*) is the conch blown by the Australian and Polynesian Islanders. A very similar sp. (*T. nodiferus*) is found in the Medit., and a third in the W. Indies.

FASCIOLARIA, Lam.

Etym., *fasciola*, a band.

Type, F. tulipa. Pl. V., fig. 1.

Shell fusiform, elongated; whirls round or angular; canal open; columellar lip tortuous, with several oblique folds. *Operc.* claw-shaped. F. gigantea of the S. Seas, attains a length of nearly two feet.

Distr., 16 sp. W. Indies, Medit., W. Africa, India, Australia, S. Pacific, W. America.

Fossil, 28 sp., U. chalk—. France.

TURBINELLA, Lam.

Etym., diminutive of *turbo*, a top.

Type, T. pyrum. Pl. V., fig. 2.

Shell thick; spire short; columella with several transverse folds. Operculum claw-shaped. Fig. 63. The shank-shell (*T. pyrum*) is carved by the Cingalese, and reversed varieties of it, from which the priests administer medicine, are held sacred.

Distr., 70 sp. W. Indies, S. America, Africa, Ceylon, Philippines, Pacific, W. America.

Fossil, 20 sp. Miocene—.

Sub-genera. Cynodonta (Schum.) T. cornigera. Pl. V., fig. 3.

Latirus (Montf.) T. gilbula. Pl. V., fig. 4.

Cuma (Humphr.) T. angulifera, inner lip with a single prominent fold operculum like *purpura*.

Lagena (Schum.) T. Smaragdula, L. sp. N. Australia.

CANCELLARIA, Lam.

Etym., *cancellatus*, cross-barred.

Type, C. reticulata. Pl. V., fig. 5.

Shell cancellated; aperture channelled in front: columella with several strong oblique folds; no operculum. The animals are vegetable feeders. (Desh.)*

Distr., 70 sp. W. Indies, Medit., W. Africa, India, China, California.

Fossil, 60 sp. Eocene—. Britain, France, &c.

TRICHOTROPIS, Broderip, 1829.

Etym., *Thrix*, (trichos) hair, and *tropis*, keel.

Type, T. borealis, Pl. VI., fig. 8. (= ? *Admete*, Phil., no operculum.)

Shell thin, umbilicated; spirally furrowed; the ridges with epidermal fringes; columella obliquely truncated; operc. lamellar, nucleus external.

Animal with a short broad head; tentacles distant, with eyes on the middle; proboscis long, retractile.

Lingual dentition similar to *strombus*; teeth single, hamate, denticulated; uncini 3: 1 denticulate 2 and 3 simple.

---

* *Cancellaria* and *trichotropis* form a small natural family connected with *cerithiadæ* and *strombidæ*.

Distr., 8 sp. Northern seas. U. States, Greenland, Melville Island, Behring's Straits, N. Brit. 15—80 fms.

Fossil, 1 sp. Miocene—. Brit.

PYRULA, Lam. Fig-shell.

Etym., diminutive of *pyrus*, a pear.

Syn., Ficula, Sw. Sycotypus, Br., Cassidula, Humph. Cochlidium, Gray.

Type, P. ficus. (Pl. V., fig. 6.)

Shell pear-shaped ; spire short ; outer lip thin ; columella smooth : canal long, open. No operculum in the typical species.

Distr., 39 sp. W. Indies, Ceylon, Australia, China, W. America.

Fossil, 30 sp. Neocomian—. Europe, India. Chile.

Pyrula ficus has a broad foot, truncated and horned in front ; the mantle forms lobes on the sides, which nearly meet over the back of the shell. Chinese seas, in 17—35 fms. water. (Adams.)

Sub-genera. Fulgur, Montf. P. perversa. (= *Pyrella*, Sw. P. spirillus.)

Rapana, Schum. P. bezoar, shell perforated. Operc. lamellar, nucleus external.

Myristica. Sw. P. melongena. Pl. V., fig. 7. Operc. pointed, curved.

FUSUS, Lam. Spindle-shell.

Syn., Colus, Humph. Leiotomus, Sw. Strepsidura, Sw.

Type, F. colus. Pl. V., fig. 8.

Shell fusiform ; spire many-whirled ; canal straight, long ; operculum ovate, curved, nucleus apical. Pl. V., fig. 9*.

Distr., 100 sp. World-wide. The typical sp. are sub-tropical. Australia, New Zealand, China, Senegal, U. States, W. America, Pacific.

Fossil, 320 sp. Bath oolite? Gault—Eocene—. Brit. &c.

Sub-genera, Trophon, Montf. F. magellanicus, Pl. IV., fig. 16. 14 sp. Antarctic and Northern seas. Brit. coast. 5—70 fathoms. *Fossil*, Chile, Brit.

Clavella, Sw. (cyrtulus, Hinds) body-whirl ventricose, suddenly contracted in front ; canal long and straight. Resembling a turbinella, without plaits. 2 sp. Marquesas, Panama. *Fossil*, Eocene. F. longævus (Solander), Barton, &c.

Chrysodomus, Sw. F. antiquus (var.) Pl. V., fig. 9. Canal short ; apex papillary ; lingual dentition like buccinum, 12 sp. Spitzbergen, Davis's Straits, Brit., Medit., Kamschatka, Oregon. Low water to 100 fms. *Fossil*, pliocene. Brit., Sicily.

Pusionella, Gray. F. pusio, L. sp. (=F. nifat, Lam.), columella keeled. Operc., nucleus internal, 7 sp. Africa, India. *Fossil*, tertiary. France.

Fusus colosseus and proboscidalis, Lam., are two of the largest living gasteropods. *Fusus (chrysodomus) antiquus*, called the red-whelk on the coasts of the channel, and " Buckie" in Scotland, is extensively dredged for the markets, being more esteemed than the *buccinum*. It is the "roaring buckie," in which the sound of the sea may always be heard. In the Zetland cottages it is suspended horizontally, and used for a lamp; the cavity containing the oil, and the canal the wick. (Fleming.) The reversed variety (F. contrarius, Sby) is found in the Medit., and on the coast of Spain; it abounds in the pliocene tertiary (crag) of Essex. The *fusus deformis*, a similar sp., found off Spitzbergen, is always reversed.

FAMILY III. BUCCINIDÆ.

Shell notched in front; or with the canal abruptly reflected, producing a kind of varix on the front of the shell.

Animal similar to *murex*; lingual ribbon long and linear, (fig. 16) rachidian teeth single, transverse, dentated in front; uncini single. Carnivorous.

BUCCINUM, L.   Whelk.

Etym., *buccina*, a trumpet, or triton's-shell.

Type, B. undatum. Pl. V., fig. 10.

Shell few whirled; whirls ventricose; aperture large; canal very short, reflected; operculum lamellar, nucleus external. (See *pisania*.)

Distr., 20 typical species. Northern and Antarctic seas. Low water to 100 fms. (Forbes). (B ? clathratum, 136 fms., off Cape.)

Fossil, 130 sp., including *pisania*, &c. Gault ?—Miocene—. Brit., France.

Fig. 70.   *Nidamental capsules of the Whelk.**

The whelk is dredged for the market, or used as bait by fishermen; it may be taken in baskets, baited with dead fish. Its nidamental capsules are aggregated in roundish masses, which, when thrown ashore, and drifted by the wind resemble corallines. Each capsule contains five or six young, which, when hatched, are like fig. 70, *b* : *a*, represents the inner side of a single capsule, shewing the round hole, from which the fry have escaped.

---

* Fig. 70. From a small specimen, on an oyster-shell, in the cabinet of Albany Hancock, Esq. The line at *b*, represents the length of the young shell.

Sub-genus. Cominella, Gray. *Ex. B. limbosum, purpura maculosa,* &c.
Operculum as in *fusus.* About 12 sp.

PSEUDOLIVA, Swainson.

Etym., named from its resemblance to *oliva,* in form.
Syn., sulco-buccinum, D'Orb. Gastridium (Gray), G. Sowerby.
Type, P. plumbea. Pl. V., fig. 12.
Shell globular, thick; with a deep spiral furrow near the front of the
body-whirl, forming, as in *monoceros,* a small tooth on the outer lip; spire
short, acute; suture channelled; inner lip callous aperture notched in front;
operculum? Animal unknown.
Distr., 6 sp.? W. America.
Fossil, 5 sp. Eocene. Brit., France, Chile.

? ANOLAX (Roissy), Conrad. Lea.

Etym., an aulax, without furrow.
Syn., buccinanops, D'Orb. Leiodomus, Sw. Bullia, Gray.
Types, A. gigantea, Lea. Buc. lævigatum. B. semiplicata, Pl. V., fig. 14.
Shell variable; like buccinum, pseudoliva, or terebra; sutures enamelled;
inner lip callous.
Animal without eyes; foot very broad; tentacles long and slender;
operculum pointed, nucleus apical.
Distr., 26 sp. Brazil, W. Africa, Ceylon, Pacific, W. America.
Fossil, 3 sp. Eocene—. N. America, France.

? HALIA, Risso.

Etym., halios, marine. *Syn.,* priamus, Beck.
Types, bulla helicoides (Brocchi). Miocene, Italy. Helix priamus (Meus-
chen). Coast of Guinea?
Shell like *achatina;* ventricose, smooth; apex regular, obtuse; operc. ?
The fossil species occurs with marine shells, and sometimes coated by a coral
(*lepralia*).

TEREBRA, Lamarck. Auger-shell.

Syn., acus, Humph. Subula, Bl. Dorsanum, Gray.
Type, T. maculata. Pl. V., fig. 13.
Shell long, pointed, many-whirled; aperture small; canal short; operc.
pointed, nucleus apical.
Animal blind, or with eyes near the summit of minute tentacles.
Distr., 109 sp., mostly tropical. Medit. (1 sp.) India, China, W. America.
Fossil, 24 sp. Eocene—. Brit., France, Chile.

EBURNA, Lamarck. Ivory-shell.

Etym., ebur, ivory. *Syn.,* latrunculus, Gray.

Type, E. spirata. Pl. V., fig. 11.

Shell umbilicated when young; inner lip callous, spreading and covering the umbilicus of the adult; *operculum* pointed, nucleus apical.

Distr., 9 sp. Red Sea, India, Cape, Japan, China, Australia. Solid, smooth shells, which have usually lost their epidermis, and are pure white, spotted with dark red; the animal is spotted like the shell. 14 fms. (Adams.)

NASSA, Lam. Dog-whelk.

Etym, *nassa*, a basket used for catching fish.
Syn., desmoulinsia and northia, Gray.
Type, N. arcularia. Pl. V., fig. 15.
Shell like buccinum; columellar lip callous, expanded, forming a tooth-like projection near the anterior canal. *Operc.* ovate, nucleus apical. Lingual teeth arched, pectinated; uncini, with a basal tooth.

The animal has a broad foot, with diverging horns in front, and two little tails behind. *N. obsoleta* (Say) lives within the influence of fresh water and becomes eroded. *N. reticulata*, *L.*, is common on the English shores, at low-water, and is called the dog-whelk by fishermen.

Distr., 68 sp. Low-water—50 fms. World-wide. Arctic, Tropical and Antarctic Seas.
Fossil, 19 sp. Eocene——. Brit., &c., N. America.
Sub-genus, cyllene, Gray. C. Oweni, Pl. V., fig. 17. Outer lip with a slight sinus near the canal; sutures channelled. W. Africa, Sooloo Islands, Borneo. *Fossil*, Miocene, Touraine.
Cyclonassa, Swainson. C. neritea, Pl. V., fig. 16.

PHOS, Montfort.

Etym., *phos*, light. *Syn.*, rhinodomus, Sw.
Type, P. senticosus, Pl. V., fig. 18.
Shell like nassa; cancellated; outer lip striated internally, with a slight sinus near the canal; columella obliquely grooved.

The animal has slender tentacles, with the eyes near their tips.

Distr., 30 sp. (Cuming.) Red Sea, Ceylon, Philippines, Australia, W. America.

? RINGICULA, Deshayes.

Etym., diminutive of *ringens*, from *ringo*, to grin.
Type, R. ringens, Pl. V., fig. 21.
Shell minute, ventricose, with a small spire; aperture notched, columella callous, deeply plaited; outer lip thickened and reflected.

Distr., 4 sp.? Medit., India, Philippines, Gallapagos.
Fossil, 9 sp., Miocene——. Brit., France. *Ringicula* is placed with *nassa*

by Mr. Gray, and Mr. S. Wood; it appears to us very nearly allied to *cinulia* (=*avellana,* D'Orb.) in *tornatellidæ.*

PURPURA (Adans), Lam. Purple.

Type, P. persica, Pl. VI., fig. 1.

Shell striated, imbricated or tuberculated; spire short; aperture large, slightly notched in front; inner lip much worn and flattened. Operc. lamellar, nucleus external. Pl. VI., fig. 2. Lingual dentition like murex erinaceus; teeth transverse, 3 crested; uncini small, simple.

Many of the *purpuræ* produce a fluid which gives a dull crimson dye; it may be obtained by pressing on the operculum. *P. lapillus* abounds on the British coast at low-water, amongst sea-weed; it is very destructive to mussel-beds (Fleming).

Distr., 140 sp. W. Indies, Brit., Africa, India, New Zealand, Pacific, Chile, California, Kamschatka. From low-water—25 fathoms.

Fossil, 30 sp. Miocene—. Brit., France, &c.

Sub-genus. Concholepas, Favan. C. lepas (Gmelin sp.) Pl. VI., fig. 3. Peru. The only sp. differs from purpura in the size of its aperture, and smallness of the spire.

? PURPURINA (Lycett, 1847). D'Orb.

Shell, ventricose, coronated; spire, short; aperture, large, scarcely notched in front.

Fossil, 9 sp., Bath-oolite. Brit. France. The type, *P. rugosa,* somewhat resembles *purpura chocolatum* (Duclos), but the genus probably belongs to an extinct group.

MONOCEROS, Lam.

Etym., monos, one; *ceras,* horn.

Syn., acanthina, Fischer. Chorus, Gray.

Type, M. imbricatum. Pl. VI., fig. 4 (Buc. monoceros, Chemn).

Shell, like purpura; with a spiral groove on the whirls, ending in a prominent spine on the outer lip. This genus is retained on account of its geographical curiosity; it consists of sp. of *purpura, lagena, turbinella, pseudoliva,* &c.

Distr., 18 sp. W. coast of America.

Fossil, tertiary. Chile.

M. gigantens (chorus) has the canal produced like *fusus. M. cingulatum* is a *turbinella,* and several sp. belong more properly to *lagena.*

PEDICULARIA, Swainson.

Type, P. sicula. Pl. VI., fig. 5 (*thyreus, Phil.*).

Shell very small, limpet-like; with a large aperture, channelled in front, and a minute, lateral spire. *Lingual dentition* peculiar; teeth single, hooked, denticulated; *uncini,* 3; 1, four-cusped, 2, 3, elongated, three-spined.

Distr., 1 sp. Sicily, adhering to corals. Closely allied to *purpura madreporarum*, Sby. Chinese Sea.

RICINULA, Lam.

Etym., dimunitive of *ricinus*, the (fruit of the) castor-oil plant.

Ex., R. arachnoïdes. Pl. VI., fig. 9 (=murex ricinus L.).

Shell, thick, tuberculated, or spiny; aperture contracted by callous projections on the lips. Operc. as in purpura.

Distr. 25 sp. India, China, Philippines, Australia, Pacific.

Fossil, 3 sp. Miocene—. France.

PLANAXIS, Lam.

Type, P. sulcata. Pl. VI., fig. 6. *Syn.*, quoyia and leucostoma.

Shell, turbinated; aperture notched in front; inner lip callous, channelled behind; operculum *subspiral* (quoyia) or semi-ovate. Pl. VI., fig. 7.

Distr., 11 sp. W. Indies, Red Sea, Bourbon, India, Pacific, and Peru.

Fossil, miocene?

Small coast shells, resembling periwinkles, with which Lamarck placed them.

MAGILUS, Montf., 1810.

Syn., campulote, Guettard, 1759. Leptoconchus, Rüppell.

Type, M. antiquus. Pl. V., figs. 19, 20.

Shell, when young, spiral, thin; aperture channelled in front; adult, prolonged into an irregular tube, solid behind; operculum lamellar.

Distr., 1 sp.? Red Sea. Mauritius.

The magilus lives fixed amongst corals, and grows upwards with the growth of the zoophytes in which it becomes immersed; it fills the cavity of its tube with solid shell, as it advances.

CASSIS, Lam. Helmet-shell.

Syn., bezoardica, Schum. Levenia, Gray. Cypræcassis, Stutch.

Type, C. flammea. Pl. VI., fig. 14.

Shell, ventricose, with irregular varices; spire, short; aperture long, outer lip reflected, denticulated; inner lip spread over the body-whirl; canal sharply recurved. Operculum small, elongated; nucleus in the middle of the straight inner edge.

Distr., 34 sp. Tropical seas; in shallow water. W. Indies, Medit., Africa, China, Japan, Australia, New Zealand, Pacific, Mexico.

Fossil, 36 sp. Eocene—. Chile, France.

The queen-conch (C. madagascariensis) and other large species, are used in the manufacture of shell cameos, p. 46. The periodic mouths (*varices*) which are very prominent, are not absorbed internally as the animal grows.

ONISCIA, Sowerby.

Etym., *oniscus*, a wood louse. *Syn.*, morum, Bolten.

Type, O. oniscus ; O. cancellata, pl. VI., fig. 15.

Shell, with a short spire, and a long narrow aperture, slightly truncated in front ; outer lip thickened, denticulated ; inner lip granulated.

Distr., 6 sp. W. Indies, China, Gallapagos. (20 fms.)

Fossil, 3 sp. Miocene.

CITHARA, Schumacher.

Etym., *cithara*, a guitar. *Syn.*, mangelia, Reeve (not Leach).

Type, cancellaria citharella, Lam. (cithara striata, Schum.)

Shell, fusiform, polished, ornamented with regular longitudinal ribs ; aperture linear, truncated in front, slightly notched behind ; outer lip margined, denticulated within ; inner lip finely striated. Operc.

Distr., above 50 sp. of this pretty little genus were discovered by Mr. Cuming, in the Philippine Islands.

CASSIDARIA, Lam.

Etym., *cassida*, a helmet.

Syn., morio, Montf. Sconsia, Gray.

Type, C. echinophora. Pl. VI., fig. 13.

Shell, ventricose ; canal produced, rather bent. No operculum.

Distr., 5 sp. Medit.

Fossil, 10 sp. Eocene—. Brit., France, &c.

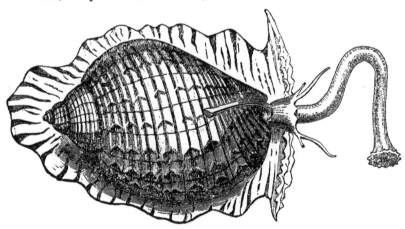

*Fig. 71.**

DOLIUM, Lam. The tun.

Type, D. galea. Pl. VI., fig. 12.

Shell, ventricose, spirally furrowed ; spire small ; aperture very large ; outer lip crenated. No operc.

Distr., 14 sp. Medit., Ceylon, China, Australia, Pacific.

* D. perdix, L. sp. ⅓ nat. size (after Quoy). Vanicoro, Pacific. The proboscis is exserted, and the siphon recurved over the front of the shell.

Fossil, 7 sp. (?Chalk. Brit.) Miocene—. S. Europe.

Sub-genus, malea, Valenc. (D. personatum) outer lip thickened and denti-culated ; inner lip with callous prominences.

HARPA, Lam. Harp-shell.

Type, H. ventricosa. Pl. VI., fig. 11. (=Buc. harpa, L.)

Shell, ventricose ; with numerous ribs, at regular intervals; spire small ; aperture large, notched in front. No operc.

The animal has a very large foot, with the front crescent-shaped, and divided by deep lateral fissures from the posterior part, which is said to sepa-rate spontaneously when the animal is irritated. Mostly obtained from deep-water, and soft bottoms.

Distr., 9 sp. Mauritius, Ceylon, Philippines, Pacific.

Fossil, 4 sp. Eocene—. France.

COLUMBELLA, Lam.

Etym., diminutive of *columba*, a dove.

Type, C. mercatoria. Pl. VI., fig. 10.

Shell, small ; with a long narrow aperture ; outer lip thickened (especi-ally in the middle), dentated ; inner lip crenulated. Operculum very small, lamellar.

Distr., 200 sp. Sub-tropical. W. Indies, Medit., India, Gallapagos, California. Small, prettily-marked shells ; living in shallow water, on sandy flats, or congregating about stones. (Adams.)

Fossil, 8 sp. Miocene—. (The Brit. sp. are *pisaniæ*).

Sub-genus. Columbellina, D'Orb. 4 sp. Cretaceous. France, India.

OLIVA, Lam. Olive, rice-shell.

Type, O. porphyria. Pl. VI., fig. 16. *Syn.*, strephona, Brown.

Shell, cylindrical, polished ; spire very short, suture channelled ; aper-ture long, narrow, notched in front ; columella callous, striated obliquely ; body whirl furrowed near the base. No operc. in the typical sp.

Animal, with a very large foot, in which the shell is half immersed ; mantle lobes large, meeting over the back of the shell, and giving off filaments which lie in the suture and furrow. The eyes are placed near the tips of the tentacles.

The olives are very active animals, and can turn over, when laid on their back ; near low water they may be seen gliding about or burying in the sands as the tide retires ; they may be taken with animal baits, attached to lines. They range downwards to 25 fms.

Distr., 117 sp. Sub-tropical, W. and E. America. W. Africa, India, China, Pacific.

Fossil, 20 sp. Eocene—. Brit., France, &c.

Sub-genera. Olivella, Sw. O. jaspidea, pl. VI., fig. 19.

Animal with small, acute frontal lobes. Operc. nucleus sub-apical.

Scaphula, Sw. O utriculus, pl. VI., fig. 18.

Frontal lobes large, rounded, operculate.

Agaronia, Gray. O. hiatula, pl. VI., fig. 17.

No eyes or tentacles. Frontal lobes moderate, acute.

ANCILLARIA, Lam.

Etym., ancilla, a maiden.

Types, A. subulata, pl. VI., fig. 20. A. glabrata, pl. VI. fig. 21.

Shell like oliva; spire produced, and entirely covered with shining enamel. Operc. minute, thin, pointed. Lingual teeth pectinated. Uncini simple, hooked.

Animal like oliva; said to use its mantle-lobes for swimming. (D'Orb.) In *A. glabrata,* a space resembling an umbilicus, is left between the callous inner lip and the body whirl.

Distr., 23 sp. Red Sea, India, Madagascar, Australia, Pacific.

Fossil, 21 sp. Eocene—. Brit., France, &c.

FAMILY IV. CONIDÆ, Cones.

Shell inversely conical; aperture long and narrow; outer lip notched at or near the suture; operculum minute, lamellar.

Animal, foot oblong, truncated in front; with a conspicuous (aquiferous?) pore in the middle. Head produced. Tentacles far apart. Eyes on the tentacles. Gills 2. Lingual teeth (*uncini?*) in pairs, elongate, subulate, or hastate.

Fig. 72.*

CONUS, L. Cone-shell.

Types, C. marmoreus, pl. VII., fig. 1. C. geographicus, antediluvianus, &c.

Shell conical, tapering regularly; spire short, many-whirled; columella smooth, truncated in front; outer lip notched at the suture; operculum pointed, nucleus apical.

Distr., 269 sp. All tropical seas. Medit., 2; Africa, 23; Red Sea, 5; Asia, 124; Australia, 16; Pacific, 25; Gallapagos, 3; W. America, 20; W. Indies and Brazil, 21.

Fossil, 80 sp. Chalk—. Brit., France, India, &c.

The cones range northward as far as the Mediterranean, and southward to the Cape; but are most abundant and varied in equatorial seas. They inhabit fissures and holes of rocks, and the warm and shallow pools inside coral-reefs, ranging from low water to 30 and 40 fathoms; they move slowly, and sometimes (C. aulicus) bite when handled; they are all predatory. (Adams.)

Sub-genus. Conorbis, Sw. C. dormitor, Pl. VII., fig. 2. Eocene—. Brit., France.

* Fig. 72. Lingual teeth of bela turricula (after Lovén).

PLEUROTOMA, Lam.

Etym., *pleura*, the side, and *toma*, a notch. *Syn.*, *turris*, Humph.

Types, P. Babylonica, Pl. VII., fig. 3. P. mitræformis, &c.

Shell fusiform, spire elevated ; canal long and straight ; outer lip with a deep slit near the suture. Operculum pointed, nucleus apical.

Distr., 430 sp. World-wide. Greenland, Brit., 17 ; Medit., 19 ; Africa, 15 ; Red Sea and India, 6 ; . China, 90 ; Australia, 15 ; Pacific, 0 ? W. America, 52 ; W. Indies and Brazil, 20. The typical sp. about 20 (China, 16 ; W. America, 4.) Low water to 100 fathoms.

Fossil. 300 sp. Chalk—. Brit., France, &c. Chile.

Sub-genera. *Drillia*, Gray. D. umbilicata, canal short.

Clavatula, Lam., canal short, operc. pointed, nucleus in the middle of the inner edge. C. mitra, Pl. VII., fig. 4.

Tomella, Sw., canal long ; inner lip callous near suture. T. lineata.

? *Clionella*, Gray. C. sinuata, Born sp. (= P. buccinoides) freshwaters, Africa.

Mangelia, Leach, (not Reeve). Apertural slit at the suture ; no operc., M. tæniata, Pl. VII., fig. 5. Greenland, Brit., Medit.

Bela, Leach. Operc. nucleus apical. B. turricula, Pl. VII., fig. 6.

Defrancia, Millet,* no operc. D. linearis, Pl. VII., fig. 7.

? *Lachesis*, Risso, L. minima, Pl. VII., fig. 8, apex mammillated ; operc. claw-shaped. Medit., S. Brit. In shallow water.

Daphnella, Hinds. D. marmorata. New Guinea. (Buc. junceum. L. clay).

FAMILY V. VOLUTIDÆ.

Shell turreted, or convolute ; aperture notched in front ; columella obliquely plaited. No operculum.

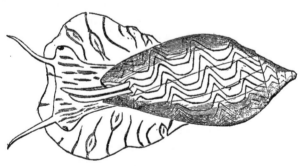

Fig. 73.†

Animal with a recurved siphon ; foot very large partly hiding the shell ;

* According to Mr. S. Hanley, *Defrancia* is synonymous with *Mangelia.*
† Fig. 73. V. undulata, Lam. ½ Australia (from Quoy and Gaimard),

mantle often lobed and reflected over the shell; eyes on the tentacles, or near their base. Lingual ribbon linear; *rachis* toothed; *pleuræ* unarmed.

VOLUTA, L. Volute.

Type, V. musica, Pl. VII., fig. 9.

Syn., cymbiola, harpula, Sw. Volutella, D'Orb. Scapha, &c., Gray.

Shell ventricose, thick; spire short, apex mammillated; aperture large, deeply notched in front; columella with several plaits. V. musica and a few others have a small operculum.

Animal, eyes on lobes at the base of the tentacles; siphon with a lobe on each side, at its base; lingual teeth 3 cusped.

V. vespertilio and *hebræa* fill the nuclei of their spires with solid shell. *V. brasiliana* forms nidamental capsules 3 inches long. (D'Orb.) In *V. angulata* the mantle is produced into a lobe on the left side, and overlaps the shell.

Distr., 70 sp. W. Indies, Cape Horn, W. Africa, Australia, Java, Chili.

Fossil, 80 sp. Chalk—. India, Brit., France, &c.

Sub-genera. Volutilithes, Sw. Spire pointed, many-whirled, columella plaits indistinct. V. spinosus, Pl. VII., fig. 10.

Living, 1 sp. (*V. abyssicola*), dredged at 132 fathoms; off the Cape. (Adams).

Fossil, Eocene. Brit., Paris.

Scaphella, Sw. Fusiform, smooth.

Ex., V. magellanica. *Fossil*, V. Lamberti, Crag, Suffolk.

Melo, Brod. Large, oval; spire short.

Type, M. diadema, Pl. VII., fig. 11. New Guinea, 8 sp.

CYMBA, Broderip. Boat-shell.

Syn., Yetus (Adans.) Gray.

Type, C. proboscidalis, Pl. VII., fig. 12, and fig. 74 (= V. cymbium, L.)

Shell like voluta; nucleus large and globular; whirls few, angular, forming a flat ledge round the nucleus.

The foot of the animal is very large, and deposits a thin enamel over the under side of the shell. It is ovo-viviparous, and the young animal is very large when born; the *nucleus* becomes partly concealed by the growth of the shell.

Distr., 10 sp. W. Africa, Lisbon.

MITRA, Lam. Mitre-shell.

Syn., turris, Montf. Zierliana, Gray. Tiara, Sw.

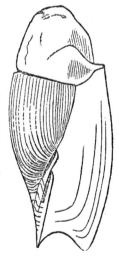

Fig. 74. *Cymba.*

Types, M. episcopalis, Pl. VII., fig. 13.   M. vulpecula, fig. 14.

Shell fusiform, thick; spire elevated, acute; aperture small, notched in front; columella obliquely plaited; operculum very small.

The animal has a very long proboscis; it emits a purple liquid, having a nauseous odour, when irrritated. The eyes are placed on the tentacles, or at their base.   Range, from low water to 15 fathoms, more rarely in 15—80 fathoms.

Distr., 350 sp. Philippines, India, Red Sea, Medit., W. Africa, Green-land (1 sp.), Pacific, W. America.   The extra-tropical species are minute. M. Greenlandica and M. cornea (Medit. sp.) are found together in the latest British Tertiaries (Forbes.)

Fossil, 90 sp.   Chalk—.   India, Brit., France, &c.

Sub-genera. *Imbricaria*, Schum. (conœlix, Sw.)

Shell, cone-shaped.   I. conica, Pl. VII., fig. 15.

Cylindra, Schum.   (Mitrella, Sw.)

Shell, olive-shaped.   C. crenulata, Pl. VII., fig. 16.

VOLVARIA, Lam.

Etym., *volva*, a wrapper.

Type, V. bulloïdes, Pl. VII., fig. 17.

Shell cylindrical, convolute; spire minute; aperture long and narrow; columella with 3 oblique plaits in front.

Fossil, 5 ? sp.   Eocene.   Brit., France.

MARGINÉLLA, Lam.

Etym., diminutive of *margo*, a rim.

Syn., porcellana (Adans.) Gray. Persicula, Schum.

Types, M. nubeculata, Pl. VII., fig. 18.   M. persicula, fig. 19.

Shell, smooth, bright; spire short or concealed; aperture truncated in front; columella plaited; outer lip (of adult) with a thickened margin. *Animal* similar to cypræa.

Distr., 90 sp. Tropical, W. Indies, Brazil, Medit. (1 small sp.) W. Africa, China, Australia.

Fossil, 30 sp.   Eocene—.   France, &c.

Sub-genus. *Hyalina*, Schum.   Outer lip scarcely thickened.

Type, voluta pallida, Mont., W. Indies.

FAMILY VI. CYPRÆIDÆ. Cowries.

Shell convolute, enamelled; spire concealed; aperture narrow, channelled at each end; outer lip (of adult) thickened, inflected.   No operculum.

Animal with a broad foot, truncated in front; mantle expanded on each side, forming lobes, which meet over the back of the shell; these lobes are usually ornamented with tentacular filaments; eyes on the middle of the ten-tacles or near their base; branchial plume single.   Lingual ribbon long,

)artly contained in the visceral cavity; *rachis* 1 toothed; *uncini* 3. The
:owries inhabit shallow water, near shore, feeding on zoophytes.

CYPRÆA, L. Cowry.

Etym., Cypris, a name of Venus.
Types, C. tigris, C. mauritiana, Pl. VII., fig. 20.

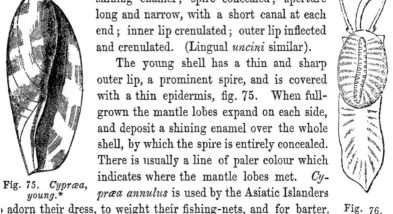

Shell ventricose, convolute, covered with
shining enamel; spire concealed; aperture
long and narrow, with a short canal at each
end; inner lip crenulated; outer lip inflected
and crenulated. (Lingual *uncini* similar).

The young shell has a thin and sharp
outer lip, a prominent spire, and is covered
with a thin epidermis, fig. 75. When full-
grown the mantle lobes expand on each side,
and deposit a shining enamel over the whole
shell, by which the spire is entirely concealed.
There is usually a line of paler colour which
indicates where the mantle lobes met. *Cypræa annulus* is used by the Asiatic Islanders

Fig. 75. *Cypræa, young.**

Fig. 76. *Trivia.†*

) adorn their dress, to weight their fishing-nets, and for barter.
pecimens of it were found by Dr. Layard in the ruins of Nimroud.
he money-cowrey (*C. moneta*) is also a native of the Pacific and Eastern
:as; many tons weight of this little shell are annually imported into this
)untry, and again exported for barter with the native tribes of Western
frica; in the year 1848 sixty tons of the money-cowry were imported into
iverpool; and in 1849 nearly three hundred tons were brought to the same
ace, according to the statement of Mr. Archer in the Industrial Exhibition.
[r. Adams observed the pteropodous fry of *C. annulus*, at Singapore, adhering
. masses to the mantle of the parent, or swimming in rapid gyrations, or
ith abrupt jerking movements by means of their cephalic fins.

Distr., 150 sp. In all warm seas (except E. coast S. America?) but
ost abundant in those of the old world. On reefs and under rocks at low
ater.

Fossil, 78 sp. Chalk—. India, Brit., France, &c.

Sub-genera. Cyprovula, Gray. C. capensis, Pl. VII., fig. 21. Aper-
iral plaits continued regularly over the margin of the canal.

Luponia, Gray. C. algoënsis, Pl. VII., fig. 22. Inner lip irregularly
.aited in front.

---

* Fig. 75. Cypræa testudinaria, L., young, China.
† Fig. 76. Trivia europæa, Mont. From the "British Mollusca," by Messis.
)rbes and Hanley.

Trivia, Gray.   C. europæa, Pl. VII., fig. 23 ;   fig. 76, and 15, B. Small shells with striæ extending over the back. (*Uncini ;* 1st denticulate 2, 3, simple.)

Distr., 30 sp.   Greenland, Brit., W. Indies, Cape, Australia, Pacific, W. America.

ERATO, Risso.

Etym., Erato, the muse of love-songs and mimicry. *Type*, E. lævis, Pl. VII., fig. 24.

Shell minute ; like *marginella ;* lips minutely crenulated. *Animal*, like *trivia.*

Distr., 8 sp.   Brit., Medit., W. Indies, China.

Fossil, 2 sp.   Miocene—. France, Brit. (Crag.)

OVULUM, Lam.

Etym., dimunitive of *ovum*, an egg. *Syn.*, amphiceras, Gronov.

Types, O. ovum, pl. VII., fig. 25.   O. gibbosa and verrucosa.

Shell, like *cypræa ;* inner lip smooth.

Distr., 36 sp.   Warm seas. W. Indies, Brit., Medit. China, W. America.

Fossil, 11 sp.   Eocene—.   France, &c.

Sub-genus, calpurna, Leach.' O. volva ("The weaver's shuttle"). Aperture produced into a long canal at each end. Foot narrow, adapted for walking on the round stems of the *gorgoniæ*, &c., on which it feeds. C. patula inhabits the S. coast of Britain, it is very thin, and has a sharp outer lip.

SECTION B. HOLOSTOMATA. Sea-Snails.

Shell, spiral or limpet shaped ; rarely tubular or multivalve : margin of the aperture entire. *Operculum*, horny or shelly, usually spiral.

Animal with a short non-retractile muzzle ; respiratory siphon wanting, or formed by a lobe developed from the neck (fig. 61), gills pectinated or plume-like, placed obliquely across the back, or attached to the right side of the neck ; neck and sides frequently ornamented with lappets and tentacular filaments. Marine or fresh-water. Mostly phytophagous.*

FAMILY I. NATICIDÆ.

Shell, globular, few-whirled ; spire small, obtuse ; aperture, semi-lunar ; lip, acute ; pillar often callous.

Animal, with a long retractile proboscis ; lingual ribbon linear ; *rachis*, 1 toothed ; *uncini*, 3 (similar to *trivia*, fig. 15, B.) ; foot very large ; mantle-lobes largely developed, hiding more or less of the shell. Species all marine.

---

* These " sections" are not very satisfactory, but they are better than any others yet proposed, and they are convenient, on account of the great extent of the order *proso-branchiata. Natica* and *scalaria* have a retractile proboscis. *Pirena* has notched aperture, and *aporrhais*, a canal.

NATICA (Adans.), Lamarck.

Syn., mammilla, Schm. Cepatia, Gray. Nacça, Risso.
Type, N. canrena, Pl. VIII., fig. 1.

Shell, thick, smooth; inner lip callous; umbilicus large, with a spiral callus; epidermis thin, polished; operculum sub-spiral.

Animal blind; tentacles connate with a head veil; front of the large foot provided with a fold (*mentum*), reflected upon and protecting the head; operc. lobe large, covering part of the shell; jaws horny; lingual ribbon short; branchial plume single.

The coloured markings of the naticæ are very indestructible; they are frequently preserved on fossils. The *naticæ* frequent sandy and gravelly bottoms, ranging from low water to 90 fathoms (Forbes). They are carnivorous, feeding on the smaller bivalves (Gould), and are themselves devoured by the cod and haddock. Their eggs are agglutinated into a broad and short spiral band, very slightly attached, and resting free on the sands.

Distr., 90 sp. Arctic seas, Brit., Medit., Caspian, India, Australia, China, Panama, W. Indies.

Fossil, 260 sp. Devonian—. S. America, N. America, Europe, India.

Sub-genera, naticopsis, M'Coy. N. Phillipsii. Shell imperforate; inner lip very thick, spreading. Operc. shelly (Brit. Mus.). Carb. limestone, 7 sp.

Operculum, horny.

Neverita, Risso. N. Alderi. Fig. 77.

Lunatia, Gray. N. Ampullaria. Perforation simple; epidermis dull, olivaceous. Northern seas.

Globulus, J. Sby. (Deshayesia,† Raulin; Ampullina, Desh. not Bl.) N. Sigaretina. Pl. VIII., fig. 2. Umbilicus narrow (rimate), lined by a thin callus. *Fossil*, eocene. Brit., Paris.

Polinices, Montf., (naticella Guild.) N. mammilla. Shell oblong; callus very large, filling the umbilicus.

Cernina, Gray. N. fluctuata. Pl. VIII., fig. 3. Globular, imperforate; inner lip callous, covering part of the body whirl.

Naticella, Müller. 19 sp. *Fossil*, Trias, S. Cassian.

---

* Fig. 77. Natica Alderi, Forbes. From an original drawing, communicated by Joshua Alder, Esq.

† *Deshayesia* was founded on a specimen with prominences on the pillar.

Sigaretus (Adans.), Lamarck.

Syn., cryptostoma, Bl. Stomatia, Browne.

Type, S. haliotoïdes. Pl. VIII., fig. 4.

Shell, striated; ear-shaped; spire minute; aperture very wide, oblique (not pearly). Operculum minute, horny, sub-spiral.

The flat species are entirely concealed by the mantle when living; the convex shells only partially, and they have a yellowish epidermis. The anterior foot lobe (*mentum*) is enormously developed.

Distr., 26 sp. W. Indies, India, China, Peru.

Fossil, 10 sp. Eocene—. Brit., France, S. America.

Sub-genus, naticina, Gray. N. papilla, pl. VIII., fig. 3. Shell ventricose, thin, perforated. W. Indies, Red Sea, China, N. Australia, Tasmania. *Eocene*, Paris.

Lamellaria, Montagu.

Etym., *lamella*, a thin plate.

Syn., marsenia, Leach. Coriocella, Bl.

Type, L. perspicua. Pl. VIII., fig. 6.

Shell ear-shaped; thin, pellucid, fragile; spire very small; aperture large, patulous; inner lip receding. No operc.

Animal much larger than the shell, which is entirely concealed by the reflected margins of the mantle; mantle non-retractile, notched in front; eyes at the outer bases of the tentacles. Lingual *uncini* 3, similar; or one very large.

Distr., 5 sp. Norway, Brit., Medit., New Zealand, Philippines.

Fossil, 2 sp. Miocene—. Brit. (Crag.)

Narica, Recluz.

Syn., vanicoro, Quoy. Merria, Gray. Leucotis, Sw.

Type, N. cancellata. Pl. VIII., fig. 8.

Shell thin, white, with a velvety epidermis; ribbed irregularly, and spirally striated; axis perforated. Operc. very small, thin.

Animal, eyes at the outer base of the tentacles; foot with wing-like lobes.

Distr., 6 sp. W. Indies, Nicobar, Vanikoro, Pacific.

Fossil, 4 sp. Gault— (D'Orb.) Brit., France.

Velutina, Fleming.

Etym., *velutinus*, velvety (from *vellus*, a fleece).

Type, V. lævigata. Pl. VIII., fig. 7.

Shell thin; with a velvety epidermis; spire small; suture deep; aperture very large, rounded; peristome continuous, thin. No operc.

Animal with a large oblong foot; margin of the mantle developed all round, and more or less reflected over the shell; gills 2; head broad; tentacles subulate, blunt, far apart; eyes on prominences at their outer bases. Carnivorous. Lingual dentition like trivia (fig. 15, B.).

Distr., 4 sp. Britain, Norway, N. America, Icy sea to Kamtschatka. iving on stones near low water, and ranging to 30 fms.

Fossil, 3 sp. Miocene—. Brit.

Sub-genus. Otina (Gray). V. otis. *Shell* minute, ear shaped. Animal ɩe velutina, but with a simple mantle, and very short tentacles. W. and W. Brit. coast; inhabiting chinks of rocks, between tide-marks (Forbes).

FAMILY II. PYRAMIDELLIDÆ.

Shell spiral, turreted; nucleus minute, sinistral; aperture small; columella ɩmetimes with one or more prominent plaits. *Operculum* horny, imbricated, ɩclus internal.

Animal with broad ear-shaped tentacles, often connate; eyes behind the ɩntacles, at their bases; proboscis retractile; foot truncated in front; tonɩe unarmed. Species all marine.

Several genera of fossil shells are provisionally placed in this order, from ɩeir resemblance to *eulima* and *chemnitzia*.* Tornatella, usually placed in or ɛar this family, is *opistho-branchiate.*

PYRAMIDÉLLA, Lam.

Etym., dimunitive of *pyramis*, a pyramid.

Syn., obeliscus, Humph. (P. dolabrata. Pl. VIII., fig. 11.)

Type, P. auris-cati. Pl. VIII., fig. 10.

Shell slender, pointed, with numerous plaited or level whirls; apex sinisɩal; columella with several plaits; lip sometimes furrowed internally. Operc. ɩdented on the inner side to adapt it to the columellar plaits.   The shell of ɩe typical pyramidellæ bears some resemblance to *cancellaria.*

Distr., 11 sp. W. Indies, Mauritius, Australia.

Fossil, 12 sp. Chalk?—. France, Brit.

ODOSTOMIA, Fleming, 1824.

Etym., *odous*, a tooth, and *stoma*, mouth.

Type, O. plicata, Pl. VIII., fig. 12.

Shell subulate or ovate, smooth; apex sinistral; aperture ovate; peristome ɩot continuous; columella with a single tooth-like fold; lip thin; operculum ɩorny, indented on the inner side.

Distr.,   sp. Brit., Medit., Red Sea, Australia.

Fossil, 15 sp.? Eocene—. Brit., France.

Very minute and smooth shells, having the habit of *rissoæ*, and like them ɩometimes found in brackish water. They range from low water to 40 fms. Ihe animal is undistinguishable from chemnitzia.

* " The *Pyramidellidæ* present subjects of much interest to the student of extinct ɩollusca; numerous forms, bearing all the aspect of being members of this family, ɩccur among the fossils of even the oldest stratified rocks. Many of them are gigantic ɩompared with existing species, and the group, as a whole, may be regarded, rather ɩs appertaining to past ages than the present epoch." (Forbes.)

CHEMNITZIA, D'Orbigny.

Etym., named in honour of Chemnitz, a distinguished conchologist of Nuremburg, who published seven volumes in continuation of Martini's " *Conchylien-Cabinet*," 1780-95.

Syn., turbonilla, Risso. Parthenia, Lowe. Pyramis and Jaminea, Br. Monoptigma, Gray. Amoura, Moller.

Type, C. elegantissima. Pl. VIII., fig. 13.

Shell slender, elongated, many-whirled; whirls plaited; apex sinistral; aperture simple; ovate; peristome incomplete; operculum horny, sub-spiral.

Animal, head very short, furnished with a long, retractile proboscis; tentacles triangular; eyes immersed at the inner angles of the tentacles; foot truncated in front, with a distinct *mentum*.

Distr., Brit. (4 sp.), Norway, Medit. Probably world-wide. Range from low water to 90 fms.

Fossil, 180 sp. Permian—. Brit., France, &c.

The " melaniæ" of the secondary rocks are provisionally referred to this genus. Those of the palæozoic strata to *loxonema*.

Sub-genus. Eulimella, Forbes. E. scillæ, Scacchi. 4 Brit. sp. Shell smooth and polished; columella simple; apex sinistral.

EULIMA, Risso, 1826.

Etym., *eulimia*, ravenous hunger. *Syn.*, pasithea, Lea.

Type, E. polita. Pl. VIII., fig. 14.

Shell small, white, and polished; slender, elongated, with numerous level whirls; obscurely marked on one side by a series of periodic mouths, which form prominent ribs internally; apex acute; aperture oval, pointed above; outer lip thickened internally; inner lip reflected over the pillar. Operculum horny, sub-spiral.

Animal, tentacles subulate, close, with the eyes immersed at their posterior bases; proboscis long, retractile; foot truncated in front, mentum bilobed; operc. lobe winged on each side; branchial plume single; mantle with a rudimentary siphonal fold.

The eulimæ creep with the foot much in advance of the head, which is usually concealed within the aperture, the tentacles only protruding. (Forbes.)

Distr., 15 sp. Brit., Medit., India, Australia, Pacific. In 5—90 fms. water.

Fossil, 40 sp. Carb. ?—. Brit., France, &c.

Sub-genus. Niso, Risso (=Bonellia, Desh.). N. terebellatus, Lam. sp. Axis perforated.

Fossil, 3 sp. Eocene —. Paris. *Distr.*, 5 sp. China, W. America (Cuming).

STYLINA, Fleming.

Ex., S. astericola. Pl. VIII., fig. 15. (Syn. stylifer, Brod.)

Shell, hyaline, globular or subulate, apex tapering, styliform, nucleus sinistral.

Animal with slender, cylindrical tentacles, and small sessile eyes at their outer bases; mantle thick, reflected over the last whirls of the shell; foot large, with a frontal lobe. Branchial plume single. Attached to the spines of sea-urchins, or immersed in living star fishes and corals.

Distr., 6 sp. W. Indies, Brit., Philippines, Gallapagos.

LOXONEMA, Phillips.

Etym., *loxos*, oblique, and *nema*, thread; in allusion to the striated surface of many species.

Shell elongated, many-whirled; aperture simple, attenuated above, effused below, with a sigmoidal edge to the outer lip.

Fossil, 75 sp. L. silurian—Trias. N. America, Europe.

MACROCHEILUS, Phillips.

Etym., *macros*, long, and *cheilos*, lip.

Shell, thick, ventricose, buccinoid; aperture simple, effuse below; outer lip thin, inner lip wanting, columella callous, slightly tortuous.

Type, M. arculatus, Schlotheim sp. Devonian. Eifel.

Fossil, 12 sp. Devonian—Carboniferous. Brit., Belgium.

FAMILY III. CERITHIADÆ. Cerites.

Shell spiral, elongated, many-whirled, frequently varicose; aperture channelled in front, with a less distinct posterior canal; lip generally expanded in the adult; operculum horny and spiral.

Animal with a short muzzle, not retractile; tentacles distant, slender; eyes on short pedicels, connate with the tentacles; mantle-margin with a rudimentary siphonal fold; tongue armed with a single series of median teeth, and three laterals or uncini; marine, estuary, or fresh-water.

CERITHIUM (Adans.). Bruguiere.

Etym., *ceration*, a small horn.

Type., C. nodulosum. Pl. VIII., fig. 16.

Shell turreted, many-whirled, with indistinct varices; aperture small, with a tortuous canal in front; outer lip expanded; inner lip thickened. Operculum horny, paucispiral. Pl. VIII., fig. 16*.

Distr., above 100 sp. World-wide, the typical species tropical. Norway, Brit., Medit., W. Indies, India, Australia, China, Pacific, Gallapagos.

Fossil. 460 sp. Trias—. Brit., France, U. States, &c.

Sub-genera. Rhinoclavis, Sw. C. vertagus. Canal long, bent abruptly operc., sub-spiral.

Bittium, Leach. C. reticulatum, Pl. VIII., fig. 17. Small northern species, ranging from low-water to 80 fathoms.

Triphoris, Deshayes. C. perversum, Pl. VIII., fig. 18. 30 sp. Norway—Australia. *Fossil.* Eocene—. Brit., France. Shell sinistral; anterior and posterior canals tubular. The third canal is only accidentally present, forming part of a varix.

Cerithiopsis, Forbes. C. tuberculare, Brit. Shell like *bittium;* proboscis retractile; operculum pointed, nucleus apical. Range 4—40 fms.

POTAMIDES, Brongniart. Fresh-water Cerites.

Etym., potamos, a river, and *eidos,* species.

Type., P. Lamarckii, Brong. (= Cerit. tuberculatum, Brard.)

Ex., P. mixtus. Pl. VIII., fig. 19.

Syn., tympanotomus, Klein, C. fuscatum, Africa. Pirenella, Risso, C. mammillatum, Pl. VIII., fig. 22.

Shell like cerithium, but without *varices,* in the very numerous typical fossil species; epidermis thick, olive-brown; operculum orbicular, many-whirled.

Distr., old world only? Africa, India. In the mud of the Indus they are mixed with sp. of ampullaria, venus, purpura, vulsella, &c. (Major W. E. Baker.)

Fossil (sp. included with cerithium) Eocene—. Europe.

Sub-genera. Cerithidea. Sw., C. decollata, Pl. VIII., fig. 24. Aperture rounded: lip expanded, flattened. Inhabit salt-marshes, mangrove swamps, and the mouths of rivers; they are so commonly out of the water as to have been taken for land-shells. Mr. Adams noticed them in the fresh-waters of the interior of Borneo, creeping on pontederia and sedges; they often suspend themselves by glutinous threads, fig. 78.

Fig. 78. *Cerithidea.**

Distr. India, Ceylon, Singapore, Borneo, Philippines, Port Essington.

Terebralia, Sw. Cerith, Telescopium, Pl. VIII., fig. 21.

Shell pyramidal; columella with a prominent fold, more or less continuous towards the apex; and a second, less distinct, on the basal front of the whirls (as in *nerinæa,* fig. 79). India, N. Australia.

T. telescopium is so abundant near Calcutta, as to be used for burning into lime; great heaps of it are first exposed to the sun, to kill the animals. They have been brought alive to England (Benson).

Pyrazus, Montf. Cerit, palustre, Pl. VIII., fig. 20.

Shell with numerous indistinct varices; canal straight, often tubular; outer lip expanded. India, N. Australia.

Cerith radulum and granulatum of the W. African rivers approach very nearly the fossil *potamides,* but they have numerous varices.

* C. obtusa, Lam. sp. copied from Adams.

Lampania, Gray (batillaria, Cantor).   Cerith, zonale. Pl. VIII., fig. 23.
Shell without varices, canal straight.   Chusan.

The fossil potamides decussatus, Brug., of the Paris basin, resembles this section, and retains its spiral red bands.

NERINÆA, Defrance.

Etym., nereis, a sea-nymph.
Ex., N. trachea. Fig. 79.

Shell elongated, many-whirled, nearly cylindrical; aperture channelled in front; interior with continuous ridges on the columella and whirls.

Fossil, 150 sp. Inf. oolite—U. chalk. Brit., France, Germany, Spain, and Portugal. They are most abundant, and attain the largest size to the south; and usually occur in calcarious strata, associated with shallow-water shells. (Sharpe.)

Sub-genera. 1. *Nerinæa.* Folds simple: 2—3 on the co lumella; 1—2 on the outer wall; columella solid, or perforated. Above 50 sp.

2. *Nerinella* (Sharpe), columella solid; folds simple; columellar, 0—1; outer wall 1.

3. *Trochalia* (Sharpe), columella perforated, with one fold; outer wall simple, or thickened, or with one fold; folds simple.

4. *Ptygmatis* (Sharpe), columella solid or perforated, usually with 3 folds; outer wall with 1—3 folds, some of them complicated in form.

Fig. 79.*

? FASTIGIELLA, Reeve.

Type., F. carinata, Reeve.
Shell like turritella; aperture with a short canal in front (Mus., Cuming, and Brit. M.).

APORRHAIS, Aldrovandus.

Etym., aporrhais (Aristotle) "spout-shell" from aporrheo, to flow away.
Syn., chenopus Philippi.
Type, A. pes-pelecani. Pl. IV., fig. 7, and fig. 80.

Shell with an elongated spire; whirls numerous, tuberculated; aperture narrow, with a short canal in front; outer lip of the adult expanded and lobed or digitated; operc. pointed, lamellar.

Animal with a short broad muzzle; tentacles cylindrical, bearing the eyes on prominences near their bases, outside; foot short, angular in front;

---

* Fig. 79. Nerinæa trachea, Desl., partly ground down to shew the form of the interior. Bath oolite, Ranville. Communicated by John Morris, Esq.

branchial plume single, long; lingual ribbon linear; teeth single, hooked, denticulated; uncini 3, the first transverse, 2 and 3 claw-shaped.

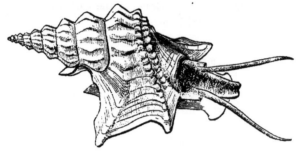

Fig. 80.*

Distr., 3 sp. Labrador, Norway, Brit., Medit. W. Africa. Range,— 100 fms.

Fossil; see *Pteroceras and Rostellaria;* above 200 species, ranging from the lias to the chalk, probably belong to this genus, or to genera not yet constituted.

STRUTHIOLARIA, Lam.

Etym., struthio, an ostrich (-foot), from the form of its aperture.

Type, S. straminea, Pl. IV., fig. 6.

Shell turreted; whirls angular; aperture truncated in front; columella very oblique; outer lip prominent in the middle, reflected and thickened in the adult; inner lip callous, expanded; operculum claw-shaped, curved inwards, with a projection from the outer, concave edge.

Animal with an elongated muzzle? tentacles cylindrical; eye-pedicels short, adnate with the tentacles, externally; foot broad and short. (Kiener.)

Distr., 5 sp. Australia and New Zealand; where alone it occurs sub-'fossil.

FAMILY IV. MELANIADÆ.

Shell spiral, turreted; with a thick, dark epidermis; aperture often channelled, or notched in front; outer lip acute; operculum horny, spiral. The spire is often extensively croded by the acidity of the water in which the animals live.

Animal with a broad non-retractile muzzle; tentacles distant, subulate; eyes on short stalks, united to the outer sides of the tentacles; foot broad and short, angulated in front; mantle-margin fringed; tongue long and linear, with a median and 3 lateral series of hooked multi-cuspid teeth. Often viviparous. Inhabiting fresh-water lakes and rivers throughout the warmer parts of the world. Only fossil in Britain.

---

* Fig. 80. Aporrhais pes-pelecani, L., from a drawing by Joshua Alder, Esq., n the "British Mollusca."

MELANIA, Lam.

Etym., *Melania*, blackness (from *melas*).
Type, M. amarula. Pl. VIII., fig. 25.
Syn. Thiara, Megerle. Pyrgula, Crist.

Shell turreted, apex acute (unless eroded); whirls ornamented with striæ or spines; aperture oval, pointed above: outer lip sharp, sinuous; operculum subspiral. Pl. VIII., fig. 25*.

Distr., 160 sp. S. Europe, India, Philippines, Pacific Islands. Distinct groups in the southern states of N. America.

Fossil, 25 sp. Eocene—. Europe (v. *chemnitzia*).

Sub-genera. *Melanàtria*, Bowdich. M. fluminea* Pl. VIII., fig. 26. Aperture somewhat produced in front; operculum with rather numerous whirls. This section includes some of the largest sp. of the genus, and is well typified by the fossil, M. Sowerbii (cerit. melanoides, Sby.) of the Woolwich sands. Old World, India, Philippines.

Vibex, Oken, V. fuscatus, Pl. VIII., fig. 29. V. auritus. W. Africa. Whirls spirally ridged, or muricated; aperture broadly channelled in front.

Ceriphasia, Sw., C. sulcata. N. America. Aperture like vibex; slightly notched near the suture.

Hemisinus, Sw., H. lineolatus. W. Indies. Aperture channelled in front.

Melafùsus, Sw. (Io, Lea. Glottella, Gray.) M. fluviatilis. Pl. VIII., fig. 27. U. States. Aperture produced into a spout in front.

Melàtoma, Anthony (not Sw.) M. altilis. Shell like anculotus; with a deep slit at the suture. U. States.

Anculotus, Say. A. præmorsus. Pl. VIII., fig. 28. Shell globular; spire very short; outer lip produced. U. States.

Amnicola, Anthony. A. isogona. Pl. IX., fig. 23. U. States.

? Pachystoma, Gray. M. marginata, Eocene. Paris. Peristome thickened externally, all round.

PALUDOMUS, Swainson.

Etym., *palus*, a marsh, and *domus*, home.
Syn., tanalia, Gray. Hemimitra, Sw.
Type, P. aculeatus, Gm. sp. Pl. IX., fig. 34.

Shell, turbinated, smooth or muricated; with wavy stains beneath the olive epidermis; spire small, usually eroded; operc. horny, lamellar, nucleus external. Animal like melania; mantle-margin fringed (Eydoux).

Distr., 10 sp. Ceylon (Himalaya?) in the mountain-streams, sometimes at an elevation of 6,000 feet. The Himalayan sp. (*melania conica*, Gray,

---

* This is a good section of *melania*, but Mr. Gray's type does not well represent it, being more like a pirena in the form of its aperture.

hemimitra retusa, Sw., and several others), referred to this genus, have a concentric operculum, like *paludina.*

MELANÓPSIS, Lam.

Types, M. buccinoides, M. costata. Pl. VIII., fig. 30.

Shell; body-whirl elongated; spire short and pointed; aperture distinctly notched in front; inner lip callous; operculum sub-spiral.

Distr., 20 sp. Spain, Asia Minor, New Zealand.

Fossil, 25 sp. Eocene—. Europe.

Sub-genus. Piréna, Lam. (faunus, Montf.) P. atra. Pl. VIII., fig. 31. Spire elongated, many whirled; outer lip of the adult produced.

Distr., 4 sp.? S. Africa, Madagascar, Ceylon, Philippines.

FAMILY V. TURRITELLIDÆ.

Shell tubular, or spiral; upper part partitioned off; aperture simple; operculum horny, many-whirled.

Animal with a short muzzle; eyes immersed, at the outer bases of the tentacles; mantle-margin fringed; foot very short; branchial plume single; tongue armed.

TURRITÉLLA, Lam.

Etym., diminutive of *turris,* a tower.

Syn., terebellum, torcula, zaria and eglisia, Gray.

Type, T. imbricata. Pl. IX., fig. 1.

Shell elongated, many-whirled, spirally striated; aperture rounded, margin thin; operculum horny, many-whirled; with a fimbriated margin.

Animal with long, subulate tentacles; eyes slightly prominent; foot truncated in front, rounded behind, grooved beneath; branchial plume very long; lingual ribbon minute; median teeth hooked, denticulated; uncini 3, serrulated. Carnivorous?

Distr., 50 sp. World-wide. Ranging from the Laminarian Zone to 100 fms. W. Indies, U. States, Brit. (1 sp.), Iceland, Medit., W. Africa, China, Australia, W. America.

Fossil, 170 sp., Neocomian—. Brit. &c., S. America, Australia.

Sub-genera. Proto, Defr., P. cathedralis, Pl. IX., fig. 3, aperture truncated below.

Mesalia, Gray, M. sulcata (var.) Pl. IX., fig. 2. Greenland—S. Africa.

Fossil, Eocene. Brit., France.

? ACLIS, Lovén.

Etym., A, without, *kleis,* a projection.

Syn., alvania, Leach (not Risso).

Type, A. perforatus, Mont. Pl. IX., fig. 4.

Shell minute, like turritella; spirally striated; aperture oval; outer lip prominent; axis slightly rimate; operculate.

Animal with a long retractile proboscis; tentacles close together, slender, inflated at the tips; eyes immersed at the bases of the tentacles; operc. lobe ample, unsymmetrical; foot truncated in front. Ranges to 80 fathoms water. 3 Brit. sp. Norway.

Fossil. ? sp., Miocene—. Brit. (Crag).

CÆCUM, Fleming.

Syn., corniculina, Münster. Brochus, Bronn. Odontidium, Phil.

Type, C. trachea, Pl. IX, fig. 5. Young sp., fig. 6.

Shell at first discoidal, becoming decollated when adult; tubular, cylindrical, arched; aperture round, entire; apex closed by a mammillated septum. Operc. horny, many-whirled. Lingual teeth, 0; uncini, 2, the inner broad and serrulated.

Distr., Brit., 2 sp., 10 fathoms. Medit.

Fossil, 4 sp. Eocene—. Brit., Castelarquato.

VERMETUS, Adanson. Worm-shell.

Syn., siphonium, Gray. Serpuloides, Sassi.

Types, V. lumbricalis, Pl. IX., fig. 7.

Shell tubular, attached; sometimes regularly spiral when young; always irregular in its adult growth; tube repeatedly partitioned off.; aperture round; operc. circular, concave externally.

Distr., Portugal, Medit., Africa, India.

Fossil, 12 sp. Neocomian—. Brit., France, &c.

? Sub-genus. *Spiroglyphus*, Daud. S. spirorbis Dillw. sp., irregularly tubular; attached to other shells, and half buried in a furrow which it makes as it grows. Perhaps an annelide?

SILIQUARIA, Brug.

Etym., *siliqua*, a pod.

Type, S. anguina, Pl. IX., fig. 8.

Shell tubular; spiral at first, irregular afterwards; tube with a continuous longitudinal slit.

Distr., 7 sp. Medit., N. Australia. Found in sponges.

Fossil, 10 sp. Eocene—. France, &c.

SCALARIA, Lam. Wentle-trap.

Etym., *scalaris*, like a ladder. *Type*, S. pretiosa, Pl. IX., fig. 9 (= T. scalaris, L.)

Shell, mostly pure white and lustrous; turreted; many-whirled; whirls round, sometimes separate, ornamented with numerous transverse ribs; aperture round; peristome continuous. Operc. horny, few-whirled.

Animal with a retractile proboscis-like mouth; tentacles close together, long and pointed, with the eyes near their outer bases; mantle-margin simple, with a rudimentary siphonal fold ; foot obtusely triangular, with a fold (*men-tum*) in front. Lingual dentition nearly as in *bulla ;* teeth 0 ; *uncini* nume-rous, simple; sexes distinct; predacious ? Range from low water to 80 fathoms. The animal exudes a purple fluid when molested.

Distr., nearly 100 sp. Mostly tropical. Greenland, Norway, Brit., Medit., W. Indies, China, Australia, Pacific, W. America.

Fossil, nearly 100 sp. Coral-rag—. Brit., N. America, Chile, India.

FAMILY VI. Litorinidæ.

Shell spiral, turbinated or depressed, never pearly ; aperture rounded ; peristome entire ; operculum horny, pauci-spiral.

Animal with a muzzle-shaped head, and eyes sessile at the outer bases of the tentacles; tongue long, armed with a median series of broad, hooked teeth, and 3 oblong, hooked uncini. Branchial plume single. Foot with a linear duplication in front, and a groove along the sole. Mantle with a rudimentary siphonal canal ; operc. lobe appendaged.

The species inhabit the sea, or brackish water, and are mostly litoral, feed-ing on algæ.

Litorina, Férussac. Periwinkle.

Etym., litus, the sea-shore.

Type, L. litorea, Pl. IX., fig. 10.

Shell turbinated, thick, pointed, few-whirled ; aperture rounded, outer lip acute, columella rather flattened, imperforate, operculum pauci-spiral, fig. 81. Lingual teeth hooked and tri-lobed; uncini hooked and dentated.

Distr., 40 sp. The periwinkles are found on the sea-shore, in all parts of the world. In the Baltic they live within the in-   Fig. 81.
fluence of fresh-water, and frequently become distorted; similar monstrosities are found in the Norwich crag.

The common sp. (*L. litorea*) is oviparous ; it inhabits the lowest zones of sea-weed between tide-marks. An allied sp. (*L. rudis*) frequents a higher region, where it is scarcely reached by the tide; it is viviparous, and the young have a hard shell before their birth, in consequence of which the species is not eaten. The tongue of the periwinkle is two inches long; its foot is divided by a longitudinal line, and in walking the sides advance alternately. The periwinkle and trochus are the food of the thrush, in the Hebrides, during winter.

Fossil, 10 sp? Miocene—. Brit., &c. It is probable that a large propor-tion of the oolite and cretaceous shells referred to *turbo,* belong to this genus, and especially to the section *tectaria.*

Sub-genera. Tectaria, Cuvier, 1817 (= Pagodella, Sw.) L. pagodus, Pl. IX., fig. 11. Shell muricated or granulated ; sometimes with an umbilical fissure. Operc. with a broad, membranous border. W. Indies, Zanzibar, Pacific.

Modulus, Gray (and nina, Gray) M. tectum, Pl. IX., fig. 13. Shell tro-chiform or naticoid; porcellanous; columella perforated; inner lip worn or toothed; operc. horny, many-whirled. *Distr.*, Philippines, W. America.

Fossarus (Adans.) Philippi. F. sulcatus, Pl. IX., fig. 12. *Syn.*, pha-sianema, Wood. Shell perforated; inner lip thin; operc. not spiral. *Distr.*, Medit. *Fossil*, 3 sp. Miocene—. Brit., Medit.

Risella, Gray. Lit., melanostoma, Pl. IX., fig. 14. Shell trochiform, with a flat or concave base; whirls keeled; aperture rhombic, dark or varie-gated, operc. pauci-spiral. Distr., N. Zealand.

SOLARIUM, Lam. Stair-case shell.

Etym., *solarium*, a dial.

Syn., architectoma, Bolten. Philippia, Gray. Helicocryptus, D'Orb?

Type., S. perspectivum, Pl. IX., fig. 15.

Shell orbicular, depressed; umbilicus wide and deep; aperture rhombic; peristome thin; operculum horny, sub-spiral.

The spiral edges of the whirls, seen in the umbilicus, have been fancifully compared to a winding stair-case.

Distr., 25 sp. Tropical seas. Medit., E. Africa, India, China, Japan, Australia, Pacific, W. America.

Fossil, 56 sp. Eocene—. Brit., &c. 26 other sp. (oolites—chalk,) are provisionally referred to this genus; the cretaceous sp. are *nacreous* (v. trochus).

Sub-genera. Torinia, Gray. T. cylindracea, operc. conical, multi-spiral, with projecting edges, fig. 82. Living, New Ire-land. *Fossil*, Eocene. Brit. Paris.

Omalaxis, Desh. (altered to *bifrontia*) S. bifrons, discoidal, the last whirl disengaged. 6 sp. Eocene, Paris, Brit.

? Orbis, Lea. Discoidal, whirls quadrate. *Fossil*, Eocene, America.

Fig. 82.*

? PHORUS, Montf. Carrier-shell.

Etym., *phoreus*, a carrier.

Syn., onustus, Humph., Xenophorus, Fischer.

Examples, P. conchyliophorus, Born. P. corrugatus, Pl. X., fig. 1.

Shell trochiform, concave beneath; whirls flat, with foliaceous or stellated margins, to which shells, stones, &c., are usually affixed; aperture very oblique, not pearly; outer lip thin, much produced above, receding far beneath. Operc. horny, imbricated, nucleus external (as in *purpura* and *paludomus*,) with the trans-verse scar seen through it, fig. 83. (Mus. Cuming.)

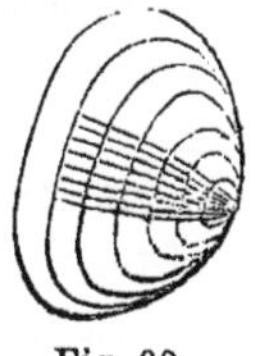

Fig. 83.

---

* Operculum of S. patulum, Lam. ¾, from Deshayes.

Fig. 83.

---

* Operculum of S. patulum, Lam. $\frac{2}{1}$, from Deshayes..

Animal with an elongated (non-retractile ?) proboscis ; tentacles long and slender, with sessile eyes at their outer bases ; sides plain ; foot narrow, elongated behind. (Adams.)　Related to *scalaria ?*

Most of the phori attach foreign substances to the margins of their shells, as they grow ; particular species affecting stones, whilst others prefer shells or corals.　They are called " mineralogists," and " conchologists," by collectors ; P. solaris and P. indicus are nearly or quite free from these disguises.　They are said to frequent rough bottoms, and to scramble over the ground, like the strombs, rather than glide evenly.

Distr., 9 sp.　W. Indies, India, Malacca, Philippines, China, W. America.

Fossil, 15 sp.　Chalk ?—Eocene—.　Brit., France.　Shells extremely like the recent *phorus*, are met with even in the carb. limestone.

LACUNA, Turton.

Etym., *lacuna*, a fissure.

Type, L. pallidula, Pl. IX., fig. 16.　*Syn.*, medoria, Gray.

Shell, turbinated, thin ; aperture semi-lunar ; columella flattened, with an umbilical fissure.　Operc. pauci-spiral.

Animal, operculigerous lobe furnished with lateral wings and tentacular filaments. Teeth, 5 cusped ; uncini 1, 2 dentated, 3 simple.　Spawn (*ootheca*) vermiform, thick, semicircular.　Range, low-water—50 fathoms.

Distr., Northern shores, Norway, Brit., Spain.　*Fossil*, 1 sp.　Glacial beds, Scotland.

? LITIOPA, Rang.

Etym., *litos*, simple, *ope*, aperture.

Type, L. bombix.　Pl. IX., fig. 24.

Shell minute, pointed ; aperture slightly notched in front ; outer lip simple, thin ; inner lip reflected.　Operc. spiral.

Distr., Atlantic, Medit., on floating sea-weed, to which they adhere by threads.　*Fossil*, 1 sp.　Miocene (Crag.).

RISSOA, Frémenville.

Etym., named after Risso,* a French zoologist.

Type, R. labiosa, Pl. IX., fig. 17.　*Syn.*, cingula, Flem.

Shell minute, white or horny ; conical, pointed, many-whirled ; smooth, ribbed, or cancellated ; aperture rounded ; peristome entire, continuous ; outer lip slightly expanded and thickened.　Operc. sub-spiral.

The animal has long, slender tentacles, with eyes on small prominences near their outer bases ; the foot is pointed behind ; the operculigerous lobe has a wing-like process and a filament (*cirrus*) on each side.　Lingual teeth single, sub-quadrate, hooked, dentated ; uncini 3 ; 1 dentated, 2, 3, claw-

---

* It is much to be regretted that some modern naturalists have tried to find out and bring into use the obscure genera of Risso, and the worthless fabrications of Mont-fort and Rafinesque, which had better have remained unknown.

shaped. They range from high-water to 100 fathoms, but abound most in shallow water, near shore, on beds of *fucus* and *zostera*.

Distr., about 70 sp. Universally distributed, but most abundant in the north temperate zone. N. America, W. Indies, Norway, Brit., Medit., Caspian, India, &c. *Rissoa parva* adheres to sea-weeds, by threads, like litiopa (Gray).

Fossil, 100 sp. Permian—. Brit., France, &c.

Sub-genera. Rissoina, D'Orb. Aperture channelled in front. Living and Fossil (10 sp. Bath oolite.— Brit.)=*Tuba*, Lea? America.

Hydrobia, Hartm. (=Paludinella, Lovén. Paludestrina, D'Orb.) *Shell* smooth; foot rounded behind; operc. lobe without filament. *Type*, litorina ulvæ, Pl. IX., fig. 18. *Fossil*, 10 sp. Wealden—. Brit., &c.

Syncera, Gray (Assiminea, Leach). S. hepatica. *Shell* like Hydrobia; tentacles connate with the eye pedicels, which equal them in length. Teeth 5—7 cusped; uncini 1, 2, dentated, 3 rounded. *Distr.*, brackish water. Brit., India.

Nematura, Benson. N. deltæ. Pl. IX., fig. 21. Aperture contracted; peristome entire. Operc. pauci-spiral. *Fossil*, eocene. Isle of Wight.

Jeffreysia, Alder (=Rissoëlla, Gray, MS.), J. diaphana. Shell minute, translucent. Operc. semilunar, imbricated, with a projection from the straight, inner side. (Pl. IX., fig. 19.) Head elongated, deeply cleft, and produced into two tentacular processes; mouth armed with denticulated jaws, and a spinous tongue; tentacles linear, eyes far behind, prominent, only visible through the shell; foot bi-lobed in front. 2 sp. Brit. On sea-weed, near low water (Alder).

SKENEA, Fleming.

Etym., named after Dr. Skene of Aberdeen; a cotemporary of Linnæus.

Syn., delphinoïdea, Brown.

Type, S. planorbis, Pl. IX., fig. 20.

Shell minute orbicular, depressed, few-whirled; peristome continuous, entire, round. Operc. pauci-spiral. Animal like rissoa, foot rounded behind. Found under stones at low-water, and amongst the roots of *corallina officinalis*.

Distr., ? sp. Northern seas. Norway, Brit.

? TRUNCATELLA, Risso. Looping-snail.

Type, T. truncatula. Pl. IX., fig. 25. (Mus., Hanley.)

Shell minute, cylindrical, truncated; whirls striated transversely; aperture oval, entire; peristome continuous. Operculum sub-spiral!

Animal with short, diverging triangular tentacles; eyes centrally behind; head bi-lobed; foot short, rounded at each end (Forbes).

The truncatellæ are found on stones and sea-weeds between tide-marks, and survive many weeks out of the water (Lowe). They walk by contracting the space between their lips and foot, like the geometric caterpillars (Gray). They are found semi-fossil, along with the human skeletons in the modern limestone of Guadaloupe.

Distr., 15 sp. W. Indies, Brit., Medit., Rio, Cape, Mauritius, Philippines, Australia, Pacific (Cuming).

? LITHOGLYPHUS, Megerle.

Type, L. fuscus. Pl. IX., fig. 22.

Shell naticoid, often eroded; whirls few, smooth; aperture large, entire; peristome continuous, outer lip sharp, inner lip callous; umbilicus rimate; epidermis olivaceous; operculum pauci-spiral.

Distr., sp. Europe, Oregon.

FAMILY VII. PALUDINIDÆ.

Shell conical or globular, with a thick, olive-green epidermis; aperture rounded; peristome continuous, entire; operculum horny or shelly, normally concentric.

Animal with a broad muzzle; tentacles long and slender; eyes on short pedicels, outside the tentacles. Inhabiting fresh-waters in all parts of the world.

PALUDINA, Lam. River-snail.

Etym., *palus* (paludis) a marsh. *Syn.*, viviparus, Gray.

Type, P. Listeri. Pl. IX., fig. 26. (P. vivipara, fig. 61.)

Shell turbinated, with round whirls; aperture slightly angular behind; peristome continuous, entire; operc. horny, concentric. *Animal* with a long muzzle, and very short eye-pedicels; neck with a small lappet on the left side, and a larger on the right, folded to form a respiratory siphon; gill comb-like, single; tongue short; teeth single, oval, slightly hooked and denticulated; uncini 3, oblong, denticulated. The paludinæ are viviparous; the shells of the young are ornamented with spiral rows of epidermal cirri.

Distr., 60 sp. Rivers and lakes throughout the N. hemisphere; Black sea, Caspian.

Fossil, 50 sp. Weald—. Brit., &c.

Sub-genus. Bithinia (Prideaux), Gray. B. tentaculata, Pl. IX., fig. 27. Shell small; operc. shelly. Animal oviparous; with only one neck-lappet, on the right side. The bithiniæ oviposit on stones and aquatic plants; the female lays from 30 to 70 eggs in a band of three rows, cleaning the surface as she proceeds; the young are hatched in three or four weeks, and attain their full growth in the second year (Bouchard).

AMPULLARIA, Lam. Apple-snail, or idol-shell.

Etym., *ampulla*, a globular flask.

Ex., A. globosa, Pl. IX., fig. 30. *Syn.*, pachylabra, Sw.

Shell globular, with a small spire, and a large ventricose body-whirl; peristome thickened and slightly reflected. Operc. shelly.

Animal with a long incurrent siphon, formed by the left neck-lappet; left gill developed, but much smaller than the right*; muzzle produced into

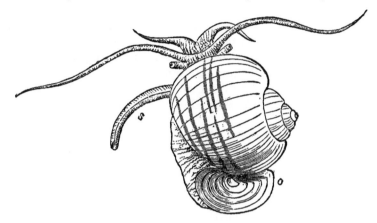

Fig. 84.†

two long tentacular processes; tentacles extremely elongated, slender. Inhabits lakes and rivers throughout the warmer parts of the world, retiring deep into the mud in the dry season, and capable of surviving a drought, or removal from the water for many years. In the lake Mareotis, and at the mouth of the Indus, ampullariæ are abundant, mixed with marine shells. Their eggs are large, inclosed in capsules, and aggregated in globular masses.

Distr., 50 sp. S. America, West Indies, Africa, India.

Sub-genera. *Pomus*, Humph. A. ampullacea. Operc. horny.

Marisa, Gray (ceratodes, Guilding). A. cornu-arietis. Pl. IX., fig. 31. Operc. horny. Shell discoidal.

Asolene, D'Orb. A. platæ. Animal without a respiratory siphon; operc. shelly. Distr., S. America.

Lanistes, Montf. A. bolteniana, L., Pl. IX., fig. 32. Shell reversed, umbilicated, peristome thin; operc. horny. Distr., W. Africa, Zanzibar, Nile.

Meladomus, Sw. Paludina olivacea, Sby. Shell reversed, imperforate; peristome thin; operc. horny.

? AMPHIBOLA, Schumacher.

Syn., ampullacera, Quoy. Thallicera, Sw.

---

* The ampullaria is said to have a pulmonic sac in addition to its gills (Gray, Owen), but we have not met with specimens sufficiently well preserved to exhibit it. would be very desirable to examine the amp. cornu-arietis, in which, probably, the gills are symmetrical, as in the cephalopods.

† Fig. 84. Ampullaria canaliculata, Lam. (from D'Orb.) South America. The branchial siphon (*s*) is seen projecting from the left side; *o*, operculum

Type, A. australis, Pl. IX., fig. 33.

Shell globular, with an uneven, battered, surface; columella fissured; outer lip channelled near the suture; operc. horny, sub-spiral. *Animal* without tentacles; eyes placed on round lobes; air-breathing; respiratory cavity closed, except a small valvular opening on the right side; a large gland occupies the position of the gill of paludina; sexes united (Quoy). Mr. Gray places this genus amongst the true *pulmonifera*.

Distr., 3 sp. Shores of New Zealand and the Pacific Islands. The living shells sometimes have *serpulæ* attached to them (Cuming). They are eaten by the New Zealanders.

VALVATA, Müller. Valve-shell.

Types, V. piscinalis, Pl. IX., fig. 28. V. cristata, Pl. IX., fig. 29.

Shell turbinated, or discoidal, umbilicated; whirls round or keeled; aperture not modified by the last whirl; peristome entire; operc. horny, multispiral.

Animal with a produced muzzle; tentacles long and slender, eyes at their outer bases; foot bi-lobed in front; branchial plume long, pectinated, partially exserted on the right side, when the animal is walking. Lingual teeth broad; uncini 3, lanceolate; all hooked and denticulated.

Distr., 6 sp. Brit., N. America.

Fossil, 19 sp. Wealden—. Brit., Belgium, &c.

FAMILY VIII. NERITIDÆ.

Shell thick, semi-globose; spire very small; cavity simple, from the absorption of the internal portions of the whirls; aperture semi-lunate; columellar side expanded and flattened; outer lip acute. Operculum shelly, subspiral, articulated.

At each end of the columella there is an oblong muscular impression, connected on the outer side by a ridge, on which the operculum rests; within this ridge the inner layers of the shell are absorbed.

Animal with a broad, short muzzle, and long slender tentacles; eyes on prominent pedicels, at the outer bases of the tentacles; foot oblong, triangular. Lingual dentition similar to the *turbinidæ*. Teeth 7; uncini very numerous.

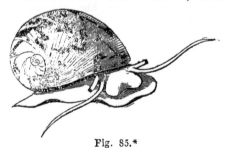

Fig. 85.*

* Fig. 85. Nerita polita, L. (from Quoy and Gaimard) New Ireland.

NERITA, L. Nerite.

Etym. Nerites, a sea-snail, from nereîs.

Type, N. ustulata, Pl. IX., fig. 35.

Shell thick, smooth or spirally grooved; epidermis horny; outer lip thickened and sometimes denticulated within; columella broad and flat, with its inner edge straight and toothed; operc. shelly, fig. 86.

Distr., 116 sp. Nearly all warm seas. W. Indies, Red Sea, Zanzibar, Philippines, Australia, Pacific, W. America, (Cuming).

Fig. 86.*

Fossil, 60 sp. Lias—. Brit. &c. The palæozoic nerites are referred by D'Orbigny to *turbo, natica,* &c. N. haliotis is a *pileopsis.*

Sub-genera. Neritoma, Morris, 1849. N. sinuosa, Sby. Portland stone, Swindon. (Mus., Lowe). Shell ventricose, thick; apex eroded; aperture with a notch in the middle of the outer lip. Casts of this shell are common, and exhibit the condition of the interior characteristic of all the *nerites ;* it was probably fresh-water.

Neritopsis, Grateloup. N. radula, Pl. VIII., fig. 9. Shell like *nerita ;* inner lip with a single notch in the centre.

Distr., 1 sp. Pacific. *Fossil,* 20 sp. Trias ? Brit., France, &c.

Velates, Montf. N. perversa, Gm. Pl. IX., fig. 36. Inner lip very thick and callous; outer lip prolonged behind, and partially enveloping the spire.

PILEOLUS, (Cookson) J. Sowerby.

Etym., pileolus, a little cap.

Type, P. plicatus, Pl. IX., fig. 37, 38.

Shell limpet-like above, with a sub-central apex; concave beneath, with a small semi-lunar aperture, and a columellar disk, surrounded by a broad continuous peristome.

Distr., marine; only known as fossils of the Bath oolite, Ancliffe, and Minchinhampton, 3 sp. *P. neritoides* is a neritina.

NERITINA, Lam. Fresh-water nerite.

Examples, N. zebra, Pl. IX., fig. 39. N. crepidularia, Pl. IX., fig. 40.

Shell rather thick at the aperture, but extensively absorbed inside; outer lip acute; inner straight denticulated; operc. shelly, with a flexible border; slightly toothed on its straight edge.

Animal like *nerita ;* lingual teeth ;—median, minute; laterals 3, 1 large, sub-triangular, 2, 3, minute; uncini about 60, first very large, hooked, denticulated; the rest equal, narrow, hooked, denticulated.

The neritinæ are small globular shells, ornamented with a great variety of black or purple bands and spots, covered with a polished horny epidermis

* Fig. 86. Operculum of N. peloronta. W. Indies.

They are mostly confined to the fresh waters of warm regions. One sp. (N. fluviatilis) is found in Brit. rivers, and in the brackish water of the Baltic. Another extends its range into the brackish waters of the N. American rivers. And the West Indian *N. viridis* and *meleagris,* are found in the sea.

N. crepidularia has a continuous peristome, and approaches *navicella* in form; it is found in the brackish waters of India. N. corona (Madagascar) is ornamented with a series of long tubular spines.

Distr., 76 sp. W. Indies, Norway, Brit., Black Sea, Caspian, India, Philippines, Pacific, W. America.

Fossil, 20 sp. Eocene—. Brit., France. &c.

NAVICELLA, Lam.

Etym., navicella, a small boat. *Type,* N. porcellana. Pl. IX., fig. 41.

Shell oblong, smooth, limpet-like; with a posterior, sub-marginal apex; aperture as large as the shell, with a small columellar shelf, and elongated lateral muscular scars ; operculum very small, shelly.

Distr., 18 sp. India, Mauritius, Moluccas, Australia, Pacific.

FAMILY IX., TURBINIDÆ.

Shell spiral, turbinated or pyramidal, nacreous inside; operculum calcarious and pauci-spiral, or horny and multi-spiral.

Animal with a short muzzle ; eyes pedunculated at the outer bases of the long and slender tentacles ; head and sides ornamented with fringed lobes and tentacular filaments (*cirri*) ; branchial plume single ; lingual ribbon long and linear, chiefly contained in the visceral cavity ; median teeth broad ; laterals 5, denticulated; uncini very numerous (sometimes nearly 100), slender, with hooked points (Fig. 15, A.).

Marine, feeding on sea-weeds (*algæ*).

The shells of nearly all the turbinidæ are brilliantly pearly, when the epidermis and outer layer of shell are removed ; many of them are used in this state for ornamental purposes.

TURBO, L. Top-shell.

Etym., turbo, a whipping-top.

Syn., batillus, marmorostoma, callopoma, &c. (Gray).

Type, T. marmoratus. Pl. X., fig. 2.

Shell turbinated, solid ; whirls convex, often grooved or tuberculated; aperture large, rounded, slightly produced in front; operculum shelly and solid, callous outside, and smooth, or variously grooved and mammillated, internally horny and pauci-spiral. In *T. sarmaticus* the exterior of the operculum is botryoidal, like some of the tufaceous deposits of petrifying wells.

Animal with pectinated head-lobes.

Distr., 60 sp. Tropical seas, W. Indies, Medit., Cape, India, China, Australia, New Zealand, Pacific, Peru.

Fossil, 360 sp. (including litorina) L. Silurian—. Universal.

PHASIANELLA, Lam. Pheasant-shell.

Syn., eutropia (Humphr.) Gray. Tricolea, Risso.

Type, P. australis. Pl. X., fig. 3.

Shell elongated, polished, richly coloured; whirls, convex; aperture oval, not pearly; inner lip callous, outer thin; operc. shelly, callous outside, sub-spiral inside.

Animal with long ciliated tentacles; head-lobes pectinated, wanting in the minute sp.; neck-lobes fringed; sides ornamented with 3 cirri; branchial plume long, partly free; foot rounded in front, pointed behind; its sides moved alternately in walking; lingual teeth even-edged.; laterals 5, hooked, denticulated; uncini about 70, gradually diminishing outwards, hooked and denticulated.

Distr., 25 sp. Australia, large sp. India, Philippines; small sp. Medit., Brit., W. Indies, very small sp.

Fossil, 70 sp. Devonian ?—. Europe.

The similarity of the existing Australian fauna, to that of the European oolites, strengthens the probability that some, at least, of these fossil shells are rightly referred to Phasianella.

IMPERATOR, Montf.

Type, I, imperialis, Pl. 10, fig. 4. *Syn.*, calcar.

Shell trochiform, thick, with a flat or concave base; whirls keeled or stellated; aperture angulated outside, brilliantly pearly; operc. shelly.

Distr., 20 sp. ? S. Africa, India, Australia, New Zealand.

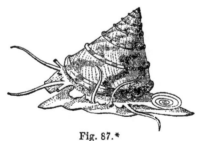

Fig. 87.*

TROCHUS, L.

Etym., *trochus*, a hoop.

Syn., cardinalia, tegula, and livona, Gray. Infundibulum, Montf. Chlorostoma, Sw. Trochiscus, Sby. Monilea, Sw.

Types, T. niloticus. Pl. X., fig. 5. T. zizyphinus. Fig. 87.

· * Fig. 87. *Trochus zizyphinus*, L., Pegwell Bay, Kent.

Shell pyramidal, with nearly a flat base ; whirls numerous, flat, variously striated ; aperture oblique, rhombic, pearly inside ; columella twisted, slightly truncated ; outer lip thin ; operculum horny, multi-spiral. Fig. 88 (T. pica).

Animal with 2 small or obsolete head-lobes between the tentacles ; neck lappets large : sides ornamented with lobes, and 3—5 cirri ; gill very long, linear ; lingual teeth 11, den-

Fig. 88.

ticulated ; uncini—90, diminishing outwards.

Distr., 150 sp. World-wide. Low-water to 15 fathoms ; the smaller species range nearly to 100 fathoms.

Fossil, 360 sp. Devonian—. Europe, N. America, Chile.

Sub-genera. Pyramis, Chemn., Tr. obeliscus, Pl. X., fig. 6, columella contorted, forming a slight canal.

Gibbula, Leach. Tr. magus, Brit. Shell depressed, widely umbilicated ; whirls tumid. Head-lobes largely developed ; lateral cirri 3.

Margarita, Leach. Tr. helicinus. Pl. X., fig. 7. Shell thin ; cirri 5 on each side. Distr., 17 sp. Greenland, Brit., Falkland Islands. Near low-water, under stones and sea-weed.

Elenchus, Humph. (= Canthiridus, Montf.) E. iris. Pl. X., fig. 8. Smooth, thin, imperforate, with a prominent base. Australia, N. Zealand. *F. iris* scarcely differs in form from Tr. zizyphinus ; *E. badius* is like a pearly phasianella ; and *E. varians* (bankivia, Menke) would be called a *chemnitzia*, if fossilized. Pl. X., fig. 9.

ROTELLA, Lamarck.

Etym., diminutive of *rota*, a wheel. (Syn., Helicina, Gray !)

Type, R. vestiaria. Pl. X., fig. 10.

Shell, lenticular, polished ; spire depressed ; base callous ; lingual teeth 13 ; uncini numerous, sub-equal.

Distr., 10 sp. India, Philippines, China, New Zealand.

MONODONTA, Lam.

Etym., *monos*, one, and *odous*, (odontos) a tooth.

Syn., labio, Oken. Clanculus, Montf. Otavia, Risso.

Types, M. labeo. Pl. X., fig. 11. M. pharaonis. Pl. X., fig. 12.

Shell, turbinated, few-whirled ; whirls spirally grooved and granulated ; lip thickened internally, and grooved ; columella toothed, more or less pro-minently and irregularly ; operc. horny, many-whirled.

Distr., 10 sp ? W. Africa, Red Sea, India, Australia.

Fossil, (included with trochus) Devonian—. Eifel.

DELPHINULA (Roissy), Lam.

Etym., diminutive of *delphinus*, a dolphin. (= Cyclostoma, Gray !)

Type, D. laciniata. Pl. X., fig. 13. (= T. delphinus, L.)

Shell orbicular, depressed ; whirls few, angulated, rugose, or spiny ; aperture round, pearly ; peristome continuous ; umbilicus open ; operculum horny, many-whirled. On reefs, at low-water.

Animal without head-lobes ; sides lobed and cirrated.

Distr., 20 sp. Red Sea, India, Philippines, China, Australia.

Fossil, 30 sp. ? Trias ?—Miocene—. Europe.

Sub-genera. Liotia, Gray. L. gervillii. Pl. X., fig. 14. Aperture pearly, with a regular, expanded border. Operc. multi-spiral, calcarious. *Distr.*, 6 sp. Cape, India, Philippines, Australia. *Fossil*, Eocene—. Brit., France.

Collonia, Gray, 1850. C. marginata. Pl. X., fig. 16. Peristome simple. Operc. calcarious, with a spiral rib on the outer side. *Distr.*, Africa. *Fossil*, Eocene—. Paris.

Cyclostrema, Marryat. C. cancellata, Pl. X., fig. 15. Shell nearly discoidal, cancellated, not pearly ; aperture round, simple ; umbilicus wide. Operc. spiral, calcarious. *Distr.*, 12 sp. Cape, India, Philippines, Australia, Peru. In 5—17 fathoms. *Serpularia*, Rœmer, has the whirls smooth and dis-united. Eocene, Paris.

ADEORBIS, Searles Wood.

Type, A. sub-carinatus. Pl. X., fig. 17.

Shell minute, not nacreous, depressed, few-whirled, deeply umbilicated ; peristome entire, nearly continuous, sinuated in its inner side, and slightly so externally. Operc. shelly, multi-spiral.

Distr., W. Indies—China. Low-water to 60 fathoms.

Fossil, 5 sp. Miocene—. Brit.

EUOMPHALUS, Sowerby.

Etym., *eu*, wide, and *omphalos*, umbilicus.

Syn., schizostoma, Bronn. Maclurea, Leseuer. Ophileta, Vanuxem. Platyschisma, McCoy.

Type, E. pentagonalis. Pl. X., fig. 18.

Shell depressed or discoidal ; whirls angular or coronated ; aperture polygonal ; umbilicus very large. Operc. shelly, round, multi-spiral (Salter).

Fossil, 80 sp., L. sil.—Trias. N. America, Europe, Australia.

Sub-genus. Phanerotinus, J. Sby. 1840, E. cristatus, Phil. Carb. limestone. Brit. Shell discoidal ; whirls separate ; outer margin sometimes foliaceous.

STOMATELLA, Lam.

Etym., diminutive of *stoma*, the aperture.

Type, S. imbricata. Pl. X., fig. 19.

Shell ear-shaped, regular ; spire small ; aperture oblong, very large and

H

oblique, nacreous; lip thin, even-edged; operc. circular, horny, multi-spiral. On reefs and under stones at low-water.

Distr., 20 sp. Cape, India, N. Australia, China, Japan, Philippines.

Sub-genus? *Gena*, Gray. Spire minute, marginal; no operculum. 16 sp. Red Sea, India, Seychelles, Swan River, Philippines (Adams).

BRODERIPIA, Gray.

Etym., named in honour of W. J. Broderip, Esq., the distinguished conchologist.

Type, B. rosea. Pl. X., fig. 20.

Shell minute, limpet-shaped, with a posterior sub-marginal apex; aperture oval, as large as the shell, brilliantly nacreous.

Distr., 3 sp. Philippines; Grimwood's Island, S. Seas (Cuming).

FAMILY X. HALIOTIDÆ.

Shell spiral, ear-shaped or trochiform; aperture large, nacreous; outer lip notched or perforated. No operculum.

Animal with a short muzzle and subulate tentacles; eyes on pedicels at the outer bases of the tentacles; branchial plumes 2; mantle-margin with a posterior (anal) fold or siphon, occupying the slit or perforation in the shell; operc. lobe rudimentary; lingual dentition similar to trochus.

In addition to the true haliotids, we have retained in this group such of the trochi-form shells as have a notched or perforated aperture.

HALIOTIS, L. Ear-shell.

Etym., *halios*, marine, and *ous* (otos) an ear.

Type, H. tuberculata, Pl. X., fig. 21.

Shell ear-shaped, with a small flat spire; aperture very wide, iridescent; exterior striated, dull; outer angle perforated by a series of holes, those of the spire progressively closed. Muscular impresssion horse-shoe shaped, the left branch greatly dilated in front. In *H. tricostalis* (padollus, Montf.) the shell is furrowed parallel with the line of perforations.

Animal with fimbriated head-lobes; side-lobes fimbriated and cirrated; foot very large, rounded. Lingual teeth;—median small; laterals single, beam-like; uncini about 70, with denticulated hooks, the first 4 very large.

The haliotis abounds on the shores of the Channel Islands, where it is called the ormer, and is cooked after being well beaten to make it tender. (Hanley); it is also eaten in Japan. It is said to adhere very firmly to the rocks, with its large foot, like the limpet. The shell is much used for inlaying, and other ornamental purposes.

Distr., 75 sp. Brit., Canaries, Cape, India, China, Australia, New Zealand, Pacific, California.

Fossil, 4 sp. Miocene—. Malta, &c.

Sub-genus? *Deridobranchus*, Ehrenberg, D. argus, Red Sea. **Shell** large and thick, like haliotis, but entirely covered by the thick, hard, plaited mantle of the animal.

STOMATIA (Helblin), Lamarck.

Etym., stoma, the aperture.

Type, S. phymotis, Pl. X., fig. 22.

Shell like haliotis, but without perforations, their place being occupied by a simple furrow; surface rugose, spirally ridged; spire small, prominent aperture large, oblong, outer margin irregular.

Distr., 12 sp. Java, Philippines, Torres Straits, Pacific. Under stones at low water (Cuming).

Fossil. M. D'Orbigny refers to this genus 18 sp., ranging from the L. Silurian to the chalk, N. America, Europe.

SCISSURELLA, D'Orb.

Etym., diminutive of *scissus,* slit.

Type, S. crispata, Pl. X., fig. 23. *Syn.,* anatomus, Montf.

Shell minute, thin, not pearly; body-whirl large; spire small; surface striated; aperture rounded, with a slit in the margin of the outer lip. Operculate.

Distr., 5 sp. Norway, Brit., Medit. In 7 fathoms water off the Orkneys, and in deep water east of the Zetland Isles.

Fossil, 4 sp. Miocene—. Brit., Sicily.

PLEUROTOMARIA, Defrance.

Etym., pleura, side, and *tome,* notch.

Type, P. anglica, Pl. X., fig. 24.

Shell, trochiform, solid, few-whirled, with the surface variously ornamented; aperture sub-quadrate, with a deep slit in its outer margin. The part of the slit which has been progressively filled up, forms a band round the whirls.

Fossil, 400 sp. Lower silurian—chalk. N. America, Europe, Australia. Specimens from clay strata retain their nacreous inner layers, those from the chalk and limestones have lost them, or they are replaced by crystalline spar. Pleurotomariæ with wavy bands of colour have been obtained in the carb. limestone of Lancashire. In this extensive group there are some species which rival the living turbines in magnitude and solidity, whilst others are as frail as ianthina.

Sub-genus. Scalites, Conrad (= raphistoma, Hall.) *E.g.,* S. angulatus and stamineus. L. silurian, New York. Shell thin; whirls angular, flat above (tabulated), 8 sp. L. silurian—carb. *Poly-tremaria,* D'Orb., is founded on P. catenata, Koninck, in which the margins of the slit are wavy, converting it into a series of perforations.

MURCHISONIA, D'Archiac.

Etym., named in honour of Sir Roderick I. Murchison.

Type, M. bilineata. Pl. X., fig. 25.

Shell elongated, many-whirled; whirls variously sculptured, and zoned like pleurotomaria; aperture slightly channelled in front; outer lip deeply notched.

The murchisoniæ are characteristic fossils of the palæozoic rocks; they have been compared to elongated pleurotomariæ, or to cerithia with notched apertures; the first suggestion is most probably correct.

Fossil, 50 sp.   L. silurian—Permian.   N. America, Europe.

TROCHOTOMA, Lycett.

Etym., trochus, and *tome,* a notch.

Syn., ditremaria, D'Orb.

Type, T. conuloides.   Pl. X., fig. 26.

Shell trochiform, slightly concave beneath; whirls flat, spirally striated, rounded at the outer angles; lip with a single perforation near the margin.

Fossil, 10 sp.   Lias—Coral Rag.   Brit., France, &c.

? CIRRUS, Sowerby.

Etym., cirrus, a curl.

Type, C. nodosus, Sby.   Min. Con. t. 141 and 219.

Shell sinistral, trochiform, base level; last whirl enlarging rather more rapidly, somewhat irregular.

Fossil, 2 sp.   Inf. oolite, Bath oolite.   Brit., France.

This genus was founded on a pleurotomaria, a euomphalus, and C. nodosus. (v. Min. Con.)   It is still doubtful what species may be referred to it.

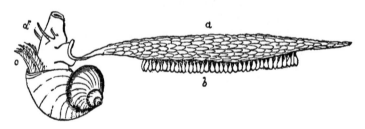

Fig. 89.*

IANTHINA, Lam.   Violet-snail.

Etym., ianthina, violet-coloured.

Type, helix ianthina L. (I. fragilis, Lam.)   Pl. X., fig. 27.

Shell thin, translucent, trochiform; nucleus minute, styliform, sinistral; whirls few, rather ventricose; aperture four-sided; columella tortuous; lip thin, notched at the outer angle. Base of the shell deep violet, spire nearly white.

Animal:—head large, muzzle-shaped, with a tentacle and eye-pedicel on

---

* Fig. 89. Ianthina fragilis, Lam. (from Quoy and Gaimard). Atlantic. *a* raft, *b* egg capsules, *c* gills, *d* tentacles and eye-stalks.

each side, but no eyes; foot small, secreting a float composed of numerous cartilaginous air-vesicles, to the under surface of which the ovarian capsules are attached. Lingual ribbon, rachis unarmed; uncini numerous, simple (like *scalaria*). Branchial plumes 2. Sexes separate.

Distr., 6 sp. Atlantic, Coral sea.

The ianthinæ, or oceanic-snails, are gregarious in the open sea, where they are found in myriads, and are said to feed on the small blue acalephæ (*velella*). They are frequently drifted to the southern and western British shores, especially when the wind continues long from the S.W.; in Swansea bay the animals have been found quite fresh. When handled they exude a violet fluid from beneath the margin of the mantle. In rough weather they are driven about and their floats broken, or detached, in which state they are often met with. The capsules beneath the further end of the raft have been observed to be empty, at a time when those in the middle contained young with fully formed shells, and those near the animal were filled with eggs. They have no power of sinking and rising in the water. The raft, which is much too large to be withdrawn into the shell, is an extreme modification of the operculum.

FAMILY XI. FISSURELLIDÆ.

Shell conical, limpet-shaped; apex recurved; nucleus spiral, often disappearing in the course of growth; anterior margin notched, or apex perforated; muscular impression horse-shoe shaped, open in front.

Animal with a well-developed head, a short muzzle, subulate tentacles, and eyes on rudimentary pedicels at their outer bases; sides ornamented with short cirri; branchial plumes 2, symmetrical; anal siphon occupying the anterior notch or perforated summit of the shell. Lingual dentition similar to trochus.*

FISSURELLA, Lam. Key-hole limpet.

Etym., diminutive of *fissura*, a slit.

Type, F. Listeri. Pl. XI., fig. 1.

Shell oval, conical, depressed with the apex in front of the centre and perforated; surface radiated or cancellated; muscular impression with the points incurved.

In very young shells the apex is entire and sub-spiral; but as the perforation increases in size it encroaches on the summit and gradually removes it. The key-hole limpets are locomotive; they chiefly inhabit the laminarian zone, but range downwards to 50 fms.

Distr., 120 sp. America, Brit., S. Africa, India, China, Australia. U. California—Cape Horn.

---

* Fissurella is the best gasteropod for comparison with the bivalves; its large gills, placed one on each side, and its symmetrical shell, pierced with a median orifice for the escape of the out-going branchial current, are unmistakeable indications of homologies with the lamelli-branchiata. See p. 48.

Fossil, 25 sp.   Carb. ? oolites—.   Brit., France.

Sub-genera. Pupillia, Gray.   F. apertura, Born. (= hiantula, Lam.)
Shell smooth, surrounded by a sharp white edge; perforation very large.
Distr., S. Africa.

Fissurellidæa, D'Orb.   F. hiatula, Lam. (=megatrema, D'Orb.)   Shell
cancellated; covered by the mantle of the animal.   3 sp.   Cape, Tasmania.

(*Macroschisma*, Sw.)   F. macroschisma.   Pl. XI., fig. 2.   Anal aperture
close to the *posterior* margin of the shell.   The animal is so much larger than
its shell, as to be compared to the *testacelle* by Mr. Cuming.   Distr., Philippines, Swan river.

Lucapina, Gray.   F. elegans, Gray (=aperta, Sby.).   Shell white, cancellated, margin crenulated; covered by the reflected mantle.   3 sp.   California.

PUNCTURELLA, Lowe.

Syn., cemoria, Leach.   Diadora, Gray.

Type, P. noachina.   Pl. XI., fig. 3.

Shell conical, elevated, with the apex recurved; perforation in front of the
apex, with a raised border internally; surface cancellated.

Distr., 2 sp.   Greenland, Boreal America, Norway, N. Brit., Tierra-del-
fuego.   In 20—100 fathoms water.

Fossil, in the glacial formations of N. Brit.

RIMULA, Defrance.

Etym., diminutive of *rima*, a fissure.   (Syn., Rimularia.)

Recent type, R. Blainvillii.   Pl. XI., fig. 4.

Shell thin and cancellated, with a perforation near the anterior margin.

Distr., several sp. found on sandy mud at low-water, or dredged in from
10—25 fms.   Philippines (Cuming).

Fossil, 3 sp.   Bath oolite—coral-rag.   Brit., France.

EMARGINULA, Lam.

Etym., dimunitive of *emarginata*, notched.

Type, E. reticulata.   Pl. XI., figs. 5, 6.

Shell oval, conical, elevated, with the apex recurved; surface cancellated;
anterior margin notched,   Muscular impression with recurved points.   The
nucleus (or shell of the fry) is spiral, and resembles *scissurella*.   The anterior
slit is very variable in extent.   The animal of Emarginula (and also of puncturella) has an isolated cirrus on the back of the foot, perhaps representing
the operculigerous lobe (Forbes).   Lingual dentition; median teeth sub-
quadrate; laterals 4, oblong, imbricated; uncini about 60, the first large and
thick, with a lobed hook, the rest linear, with serrulated hooks (Lovén).

Distr., 26 sp.   W. Indies, Brit., Norway, Philippines, Australia.   Range
from low-water to 90 fathoms.

Fossil, 40 sp.   Trias—.   Brit., France.

Sub-genus. Hemitoma, Sw. *Type*, E. octoradiata. (E. rugosa. Pl. XI., figs. 7, 8.) Shell depressed;. anterior margin slightly channelled.

PARMÓPHORUS, Blainville. Duck's-bill limpet.

Etym., *parme*, a shield, and *phoreus*, a bearer.
Type, P. australis. Pl. XI., fig. 9. *Syn.*, Scutus, Montf.
Shell lengthened-oblong, depressed; apex posterior; front margin arched. Muscular impression horse-shoe shaped, elongated. The shell is smooth and white, and permanently covered by the reflected borders of the mantle. The animal is black, and very large compared with the shell; its sides are fringed with short cirri, and its eyes sessile on the outer bases of thick tentacles; it is found in shallow-water, and walks freely (Cuming).
Distr., 10 sp. New Zealand, Australia, Philippines, Singapore, Red Sea, Cape.
Fossil, 3 sp. Eocene?—. Paris basin.

FAMILY XII. CALYPTRÆIDÆ. Bonnet-limpets.

Shell limpet-like, with the apex more or less spiral; interior simple, or divided by a shelly-process, variously shaped, to which the adductor muscles are attached.
Animal with a distinct head; muzzle lengthened; eyes on the external bases of the tentacles; branchial plume single. Lingual teeth single, uncini 3.

The bonnet-limpets are found adhering to stones and shells; most of them appear never to quit the spot on which they first settle, as the margins of their shells become adapted to the surface beneath, whilst some wear away the space beneath their foot, and others secrete a shelly base. Both their form and colour depend on the situation in which they grow; those found in the cavities of dead shells are nearly flat, or even concave above, and colourless. They are presumed to feed on the sea-weed growing round them, or on animacules; a *calyptræa*, which Professor Forbes kept in a glass, ate a small sea slug (*goniodoris*) which was confined with it. Both *calyptræa* and *pileopsis* sometimes cover and hatch their spawn in front of their foot (Alder and Clarke).

Mr. Gray arranges the bonnet-limpets next after the vermetidæ; their lingual dentition is like *velutina*.

CALYPTRÆA, Lam. Cup-and-saucer limpet.

Etym., *calyptra*, a (lady's) cap.
Syn., lithedaphus, Owen.
Types, C. equestris. Pl. XI., fig. 10. C. Dillwynnii, fig. 11.
Shell conical; limpet-shaped; apex posterior, with a minute, spiral nucleus; margin irregular; interior with a half-cup shaped process on the posterior side, attached to the apex, and open in front. Surface rugose or cancellated.

Animal with a broad muzzle; tentacles rather short; lanceolate; eyes on bulgings at the outer bases of the tentacles; mantle-margin simple, sides plain. Found under stones, between tide-marks, and in shallow water (Cuming).

Distr., 50 sp. W. Indies, Honduras, Brit., Medit., Africa, India, Philippines, China, Japan, New Zealand, Gallapagos, Chili.

Fossil, 30 sp. Carb? chalk—. Brit., France, &c.

Sub-genera. Crucibulum, Schum. (Dispotæa, Say., Calypeopsis, Less.) Ex. C. rudis, Pl. XI., fig. 12. Shell spinulose; internal cup entire; attached by one of its sides. Distr., W. America, Japan, W. Indies. Found on shells, with its base worn, or smoothed by a shelly deposit (Gray). Between this section and the next there are several intermediate forms.

Trochita, Schum. (Infundibulum, J. Sby., Galerus, Humph. Trochatella and Siphopatella, Lesson.) T. radians, Pl. XI., figs. 13, 14. (=Patella trochoides, Dillw.) T. sinensis, Pl. XI., fig. 15. Shell circular, more or less distinctly spiral; apex central; interior with a more or less complete subspiral partition. Distr., chiefly tropical, but ranges from Britain to New Zealand. *T. prisca* (McCoy) is found in the carb. limestone in Ireland; and several large species occur in the London clay and Paris basin. The recent C. sinensis — the "China-man's hat" of collectors—is found on the southern shores of England, and in the Mediterranean, in 5—10 fms. water (Forbes). Its lingual dentition is given by Lovén;— median teeth broad, hooked, denticulated; uncini 3, the first hooked and serrated, 2, 3 claw-shaped, simple.

CREPIDULA, Lam.

Etym., crepidula, a small sandal.

Type, C. fornicata, Pl. XI., fig. 16. *Syn.*, crypta, Humph.

Shell oval, limpet-like; with a posterior, oblique marginal apex; interior polished, with a shelly partition covering its posterior half.

The crepidulæ resemble the fresh-water navicellæ in form; but the internal ledge which mimics the columella of the nerite, is here the basis of the adductor muscles.

They are sedentary on stones and shells, in shallow water, and are sometimes found adhering to one another in groups of many successive generations. The specimens or species which live inside empty spiral shells are very thin, nearly flat, and colourless.

Distr., 40 sp. W. Indies, Honduras, Medit., W. Africa, Cape, India, Australia, W. America.

Fossil, 14 sp. Eocene—. France, N. America, Patagonia.

PILEOPSIS, Lam.   Bonnet-limpet.

Etym., pileos, a cap, and *opsis*, like.

Syn., capulus, Montf. Brocchia, Bronn.

Type, P. hungaricus, Pl. XI., fig. 17. P. militaris, Pl. XI., fig. 18.

Shell conical; apex posterior, spirally recurved; aperture rounded; muscular impression horse-shoe shaped.

Animal with a fringed mantle-margin; lingual teeth like *calyptræa*.

P. hungaricus (the Hungarian-bonnet) is found on oysters, in 5 to 15 fms. water; more rarely as deep as 80 fms., and then very small. P. militaris is extremely like a *velutina*.

Distr., 7 sp. W. Indies, Norway, Brit., Medit., India, Australia, California.

Fossil, 20 sp. Lias—. Europe.

Sub-genus. Amathina, Gray. A. tricarinata, Pl. XI., fig. 19. Shell depressed, oblong; apex posterior, not spiral, with three strong ribs diverging from it to the anterior margin.

Platyceras, Conrad (acroculia, Phil.). P. vetustus. Carb., limestone. Brit.

Fossil, 20 sp. Devonian—Trias. America, Europe.

HIPPONYX, Defrance.

Etym., hippos, a horse, and *onyx,* a hoof.

Type, H. cornucopia, Pl. XI., figs. 20, 21.

Shell thick, obliquely conical, apex posterior; base shelly, with a horse-shoe-shaped impression, corresponding to that of the adductor muscle.

Distr., 10 sp. W. Indies. Persian Gulf, Philippines, Australia, Pacific, W. America.

Fossil, 10 sp. U. chalk—. Brit., France, N. America.

Sub-genus. Amalthea, Schum. A. conica. Like hipponyx, but forming no shelly base; surface of attachment worn and marked with a crescent-shaped impression. Often occurs on living shells, such as the large turbines, and turbinellæ of the Eastern seas.

FAMILY XIII. PATELLIDÆ. Limpets.

Shell conical, with the apex turned forwards; muscular impression horse-shoe-shaped, open in front.

Animal with a distinct head, furnished with tentacles, bearing eyes at their outer bases; foot as large as the margin of the shell; mantle plain or fringed. Respiratory organ in the form of one or two branchial plumes, lodged in a cervical cavity; or of a series of lamellæ surrounding the animal, between its foot and mantle. Mouth armed with horny jaws, and a long ribbon-like tongue, furnished with numerous teeth, each consisting of a pellucid base and an opaque hooked apex.

The order *cyclo-branchiata* of Cuvier included the chitons and the limpets, and was characterised by the circular arrangement of the branchiæ. At a comparatively recent period it was ascertained that some of the patellæ (*acmæa*) had a free, cervical gill; whilst the chitons exhibited too many peculiarities to admit of being associated so closely with them. Professor

Forbes has very happily suggested that the cyclo-branchiate gill of patella is, in reality, a single, long branchial plume, originating on the left side of the neck, coiled backwards round the foot, and attached throughout its length. This view is confirmed by the circumstance that the gill of the sea-weed limpets (*nacellæ*) does not form a complete circle, but ends without passing in front of the animal's head.

PATELLA, L. Rock limpet.

Etym., *patella*, a dish. *Syn.*, helcion, Montf.

Ex., P. longicostata, Pl. XI., fig. 22.

Shell oval, with a sub-central apex; surface smooth, or ornamented with radiating striæ or ribs; margin even or spiny; interior smooth.

Animal with a continuous series of branchial lamellæ; mantle-margin fringed; eyes sessile, externally, on the swollen bases of the tentacles; mouth notched below. Lingual teeth 6, of which 4 are central, and 2 lateral; uncini 3.

The tongue of the common British limpet (P. vulgata) is rather longer than its shell; it has 160 rows of teeth, with 12 teeth in each row, or 1,920 in all (Forbes.) The limpets live on rocky coasts, between tide-marks, and are consequently left dry twice every day; they adhere very firmly, by atmospheric pressure (15lbs per square inch), and the difficulty of detaching them is increased by the form of the shell. On soft calcarious rocks, like the chalk of the coast of Thanet, they live in pits half an inch deep, probably formed by the carbonic acid disengaged in respiration; on hard limestones only the aged specimens are found to have worn the rock beneath, and the margin of their shell is often accommodated to the inequalities of the surrounding surface. These circumstances would seem to imply that the limpets are sedentary, and live on the sea-weed within reach of their tongues, or else that they return to the same spot to roost. On the coast of Northumberland we have seen them sheltering themselves in the crevices of rocks, whose broad surfaces, overgrown with nullipores, were covered with irregular tracks, apparently rasped by the limpets in their nocturnal excursions.*

The limpet is much used by fishermen for bait; on the coast of Berwickshire nearly 12,000,000 have been collected yearly, until their numbers are so decreased that collecting them has become tedious (Dr. Johnston). In the north of Ireland they are used for human food, especially in seasons of scarcity; many tons weight are collected annually near the town of Larne alone (Pattison).

On the western coast of S. America there is a limpet which attains the diameter of a foot, and is used by the natives as a basin (Cuming).

---

* If limpets are placed in stale water, or little pools exposed to the hot sun, they creep out more quickly than one would expect; the tracks they leave are very peculiar, and not likely to be mistaken when once seen.

Distr., 100 sp. Brit., Norway, &c. World-wide.

Fossil, above 100 sp. of patellidæ, including *acmæa*, L. silurian—. N. America, Europe.

Sub-genera. Nacella, Schum. (=patina, Leach.) Example, P. pellucida. Pl. XI., fig. 23. Shell thin; apex nearly marginal. Animal with the mouth entire below. Branchiæ not continued in front of the head. Found on the fronds and stalks of sea-weeds. Brit., Cape, Cape Horn.

Scutellina, Gray. S. crenulata. Shell with a broad margin, internally. 7 sp. Red Sea—Philippines—Pacific—Panama (Cuming).

Metoptoma, Phillips. M. pileus Ph. Shell limpet-like, side beneath the apex truncated. Resembling the posterior valve of a chiton. 7 sp. Carb. limestone. Brit.

ACMÆA, Eschscholtz.

Etym., acme, a point.

Syn., tectura, M. Edw. Lottia and scurria, Gray. Patelloida, Quoy.

Type, A. testudinalis. Pl. XI., fig. 24.

Shell like patella. *Animal* with a single pectinated gill; lodged in a cervical cavity, and exserted from the right side of the neck when the creature walks. Lingual teeth 3 on each side of the median line. Low-water to 30 fms. (Forbes.)

Distr., 20 sp. Norway, Brit., Australia, Pacific, W. America.

Sub-genera. Lepeta, Gray (= pro-pilidium, Forbes). Patella cæca, Müll. Shell minute, apex *posterior*. Animal blind. Brit. 30—90 fms.

Pilidium, Forbes. P. fulva, Müll. Brit. 20—80 fathoms water. Shell small, apex anterior. Animal blind; gills 2, not projecting; mantle even-edged. Both lepeta and pilidium have large single median teeth, with trilobed hooks; and 2 hooked uncini on each side.

GADINIA (Adanson), Gray.

Type, G. peruviana. Plate XI., fig. 26. *Syn.*, mouretia, Sby.

Shell conical; muscular impression horse-shoe shaped, the right side shortest, terminating at the siphonal groove.

Animal with a single cervical gill; tentacles expanded, funnel-shaped.

Distr., 8 sp. Medit., Red Sea, Africa, Peru.

Fossil, 1 sp. Sicily.

? SIPHONARIA, Blainville.

Type, S. sipho. Pl. XI., fig. 25.

Shell like patella; apex sub-central, posterior; muscular impression horse-shoe shaped, divided on the right side by a deep siphonal groove, which produces a slight projection on the margin.

Animal with a broad head, destitute of tentacles ; eyes sessile on prominent rounded lobes; gill ? single. The siphonariæ are found between tide-
marks, like limpets ; Mr. Gray places them with the pulmonifera, between
auriculidæ and cyclostomidæ.

Distr., 30 sp. Cape, India, Philippines, Australia, New Zealand, Pacific,
Gallapagos, Peru, Cape Horn (Cuming).

Fossil, 3 sp. Miocene —.

FAMILY XIV., DENTALIADÆ.   Tooth-shells.

DENTALIUM, L.

Type, D. elephantinum. Pl. XI., fig. 27.

Shell tubular, symmetrical, curved, open at each end, attenuated posteriorly ; surface smooth or longitudinally striated ; aperture circular, not
constricted.*

Animal attached to its shell near the posterior, anal orifice ; head rudimentary, eyes 0, tentacles 0 ; oral orifice fringed; foot pointed, conical,
with symmetrical side-lobes, and an attenuated base, in which is a hollow
communicating with the stomach. Branchiæ 2, symmetrical, posterior to the
heart ; blood red (Clarke) ; sexes united ?   Lingual ribbon wide, ovate;
rachis 1-toothed ; uncini single, flanked by single unarmed plates.

The tooth-shells are animal-feeders, devouring foraminifera' and minute
bivalves ; they are found on sand, or mud, in which they often bury themselves. The British sp. range from 10—100 fms. (Forbes.)

Distr., 30 sp. W. Indies, Norway, Brit., Medit., India.

Fossil, 70 sp. Devonian—. Europe, Chile.

FAMILY XV., CHITONIDÆ.

CHITON, L.

Etym., *chiton*, a coat of mail.

Ex., C. squamosus, spinosus, fascicularis, fasciatus. Pl. XI., figs. 28—31.

Shell composed of 8 transverse imbricating plates, lodged in a coriaceous
mantle, which forms an expanded margin round the body. The first seven
plates have posterior apices ; the eighth has its apex nearly in front. The
six middle plates are each divided by lines of sculpturing into a dorsal and
two lateral areas. All are inserted into the mantle of the animal by processes
(apophyses) from their front margins. The posterior plate is considered homologous with the limpet-shell, by Mr. Gray ; the other plates appear like
portions of its anterior slope, successively detached. The border of the mantle
is either bare, or covered with minute plates, hairs, or spines.

---

* D. gadus of Montagu is an annelide, belonging to the genus *ditrupa*.

Animal with a broad creeping disk like the limpet; proboscis armed with cartilaginous jaws, and a long linear tongue; lingual teeth 3; median small, laterals large, with dentated hooks; uncini 5, trapezoidal, one of them erect and hooked. No eyes, or tentacles. Branchiæ forming a series of lamellæ between the foot and the mantle, round the posterior part of the body. The heart is central, and elongated like the dorsal vessel of the annelides; the sexes are united; the re-productive organs are symmetrically repeated on each side, and have two orifices; the intestine is straight, and the anal orifice posterior and median.

Distr. More than 200 species are known; they occur in all climates throughout the world; most abundant on rocks at low-water, but frequently obtained by dredging in 10—25 fathoms water. Some of the small British species range as deep as 100 fms. (Forbes.) W. Indies, Europe, S. Africa, Australia, and New Zealand, California to Chiloë.

Fossil, 24 sp. Silurian—. Brit., Belgium, &c.

*Sub-genera.** *Chiton*, (Syn., lophurus, Poli. Radsia, callo-chiton, ischno-chiton, and lepto-chiton, Gray).

Ex., C. squamosus. Pl. XI., fig. 28. Border tessellated.

Distr. Brazil, W. Indies, Newfoundland, Greenland, Brit., Medit., Cape, Philippines, Australia, New Zealand, W. America.

Tonicia, Gray. C. elegans. Margin bare. *Distr.* Greenland, C. Horn, New Zealand, Valparaiso.

Acanthopleura, Guilding. C. spinosus. Pl. XI., fig. 29. Margin covered with spines, or elongated scales. *Syn.* Schizo-chiton, corephium, plaxiphora, onycho-chiton, enoplo-chiton, Gray. *Distr.* W. Indies, C. Horn, Falklands, Africa, Philippines, Australia, New Zealand, Valparaiso.

Mopalia, Gray. C. Hindsii. Border hairy. *Distr.*, W. America, Falkland Islands.

Katharina, Gray, C. tunicatus. Mantle covering all but the centre of the plates. *Distr.* New Zealand, W. America.

Cryptochiton, Gray,. " Saw-dust chiton." C. amiculatus. Valves covered with scaly epidermis. *Syn.*, cryptoconchus, Sw. Amicula, Gray. *Distr.*, California, New Zealand.

Acanthochites, Leach. C. fascicularis. Pl. XI., fig. 30. Border ornamented with tufts of slender spines, opposite the plates. *Distr.*, Brit., Medit. New Zealand.

Chitonellus, Lam. C. fasciatus, Quoy. Pl. XI., fig. 31. Border velvety; exposed portion of the plates small, distant; apophyses close to-

---

* The sub-genera of Mr. Gray are founded on the form of the *plates of insertion*; they are described in detail in the proceedings of the Zoological Society. Dr. Middendorf employs the number of the *branchial laminæ* for distinguishing the sections.

gether. *Distr.*, 10 sp. W. Indies, W. Africa, Philippines, Australia, Pacific, Panama. The chitonellæ are found in fissures of coral rock (Cuming).

Grypho-chiton, Gray. C. nervicanus.

Helminthochiton, Salter, 1847. H. Griffithii, Salter Geol. Journ. Plates sub-quadrate, not covered by the mantle ; apophyses widely separated. *Fossil.* Silurian. Ireland.

W. OSTELL. PRINTER, HART STREET, BLOOMSBURY.

A

MANUAL OF THE MOLLUSCA.

A

MANUAL OF THE MOLLUSCA;

OR,

RUDIMENTARY TREATISE

OF

RECENT AND FOSSIL SHELLS.

BY

S. P. WOODWARD, F.G.S.

ASSOCIATE OF THE LINNÆAN SOCIETY;
ASSISTANT IN THE DEPARTMENT OF MINERALOGY AND GEOLOGY
IN THE BRITISH MUSEUM; AND
MEMBER OF THE COTTESWOLDE NATURALISTS' CLUB.

ILLUSTRATED BY

A. N. WATERHOUSE AND JOSEPH WILSON LOWRY.

PART II.

———————

LONDON.

JOHN WEALE, 59, HIGH HOLBORN.

S m

MDCCCLIV.

LONDON:
PRINTED BY WILLIAM OSTELL,
HART STREET, BLOOMSBURY.

ERRATA AND ADDENDA.

Page   7 line 5 *for* " pterpoda" *read* " pteropoda."

—   " 13 *for* " brachiapoda" *read* " brachiopoda."

11   " 16 *for* " pector" *read* " pecten."

15   " 30 *for* " Mr. Robert" *read* " Mr. George Roberts;" the statement is undoubtedly correct.

22   " 16 *for* " slerotic" *read* " sclerotic."

25 Note. Striped muscular fibre has been observed in *Salpa*. (Huxley.)

28 line 8 erase the words " when withdrawn."

28 Fig. 16 *a*, anterior; *p*, posterior; *l*, lateral; *r*, rachidian.

30 line 27 erase " and by four in the brachiopoda."

39   " 22 the " tubular structure" of *pinna* is probably occasioned by the growth of a confervoid sponge between the laminæ. (Quekett.)

46   " 13 erase the word " cylindrella."

50   " 7 *for* " brachiopoda" *read* " opistho-branches."

52 erase lines 20—23, and see p. 245.

54 line 12 *see* Supplement.

65 M. Verany and H. Müller have shown that the *Hectocotyle* is developed in place of the right arm of the third pair of the male cephalopod, and *spontaneously* detached. See SUPPLE-MENT.

67 line 8 from bottom, *for* " dorsal" *read* " ventral."

68 *Tremoctopus* is a sub-genus of *Octopus*, not of *Philonexis*.

70 line 16 add " *Type*, Loligo Aalensis, Schubler."

71   " 14 *for* " Fidenas ? Gray" *read* " R. palpebrosa."

79 Note. *for* " the apocryphal genus *spongarium* was founded on" *read* " *most* of the so-called *spongaria* are."

89 Sub-genus 6, Diploceras (Salter). The shell is supposed to have resembled *Gonioceras*, and the external tube to be a simple cavity formed by the approximation of the lateral angles.

94 line 15 (and Pl. III. fig. 4) *for* " Rhothomagensis" *read* " Rotho-magensis, from *Rothomagum*, Rouen."

100   " 6 *for* " riam" *read* " rima."

105   " 8 *for* " Strombidia, Sw." *read* " Rimella, Ag."

106 erase line 3.

108 *Admete* (viridula) is a boreal form of Cancellaria, without plaits.

Page 108 Cuma (angulifera) and Rapana (p. 109) are Purpuræ.

115 Cithara, Schum. belongs to *Fam. Conidæ.*

127 line 15 *add Syn.* Polyphemopsis, Portlock.

128 " 2 *for* " Triphoris," *read* " Triforis."

— " 9 *for* " eidos, facies" *read* " ides, patronymic termination."

129 *Fastigiella; Fossil,* Eocene. Paris (*Cerithium rugosum,* Lam.)

131 *for* " Pachystoma, Gray" *read* " Chilostoma, Desh."

132 Remove *Aclis* to the Pyramidellidæ.

— line 3 from bottom, (and Pl. IX. fig. 4) *for* " A. perforata, Mont. MS." *read* " A. supranitida, Wood."

135 line 4 erase " Nina, Gray."

— " 6 *for* " many-whirled" *read* " few-whirled."

136 (and Pl. IX. fig. 24) *for* " Litiopa *bombix*" *read* " L. bombyx."

142 *Navicella* inhabits freshwaters, adhering to stones and plants.

145 line 30 *for* " Maclurea, Les." *read* " Straparollus, D'Orb."

154 line 6 from bottom, *for* " Pattison" *read* " R. Patterson."

155 *Metoptoma* is a sub-genus of *Pileopsis,* not Patella.

Exp. Plates. Pl. V. fig. 5, *for* " California" *read* " W. Indies."

" — fig. 7, *for* " China" *read* " W. Indies."

" VII. fig. 15, *for* " Philippines" *read* " Tahiti."

" XII. fig. 13, *for* " Australian Ids." *read* " Tahiti."

" — fig. 43, *for* " *Sby.* Philippines" *read* " Gray, ½ Jamaica.

Page 165 *Glandina;* the Lusitanian Bulimus Algirus belongs to this genus.

168 line 15 *insert* " devour" *before* " animal substances."

177 " 16 *for* " Megaloma" *read* " Lomastoma."

253 " 3 from bottom, erase " Ætheria has a large foot."

261 " 25 erase " *Aucella,* Keyserling;" it is a pearly shell, distinct from *Monotis* of Münster.

NOTICE.

In the long interval since the publication of the first part of this Manual, materials have so accumulated on the writer's hands, that it has been found impracticable to condense them within the space at first contemplated. The illustrations also have been more numerous than was originally expected, and occupy considerably more room. But although a SUPPLEMENT has become inevitable, the publisher has allowed an extra number of pages, in order to render the present part complete in itself. The writer hopes to make the Appendix more valuable by figures and descriptions of the animals of many hitherto undescribed Bivalve genera, the materials for which have already been placed at his disposal by Dr. J. E. Gray. The present part owes much to the assistance of Mr. Albany Hancock, of Newcastle; Mr. Thos. Davidson, F.G.S., and Mr. T. H. Huxley, F.R.S.

.

MANUAL OF THE MOLLUSCA.

PART II.

CLASS II. GASTEROPODA.—ORDER II. Pulmonifera.

THIS order embraces all the land-snails and other *mollusca* which breathe air. They are normal gasteropods, having a broad foot, and usually a large spiral shell; their breathing-organ is the simplest form of lung, and is like the branchial chamber of the sea-snails, but lined with a network of respiratory vessels. One large division of the land-snails is furnished with an operculated shell ; the rest are in-operculate, and sometimes shell-less.

The *pulmonifera* are closely related to the plant-eating sea-snails (*holostomata*), through *Cyclostoma*, and to the *nudibranches* by *Oncidium*. As a group, they are generally inferior to the sea-snails, on account of the comparative imperfection of their senses, and the union of the functions of both sexes in each individual.

SECTION -A. In-operculata.

The typical pulmonifera vary much in appearance and habits, but agree essentially in structure. Most of them have sufficiently large shells ; in the slugs, however, the shell is small and concealed, or rarely quite wanting. Snail-shells contain a larger proportion of animal matter than sea-shells, and their structure is less distinctly stratified (p. 40). In form, these shells represent many marine genera. The greater part are terrestrial, only some of the smaller families inhabit fresh-waters, or damp places near the sea. The respiratory orifice is small and valve-like,* to prevent too rapid desiccation in the land-snails, and to guard against the entry of water in the aquatic tribes. Land-snails are universally distributed; but the necessity for moist air, and the vegetable nature of their food, favour their multiplication in warm and humid regions; they are especially abundant in islands, whilst in hot and desert countries they appear only in the season of rain or dews. Their geological history is less complete than that of the purely marine orders; but

---

* Hence they are called *Adelo-pneumona* (concealed-lunged) by Gray.

I

their antiquity might be inferred from the distribution of peculiar genera in remote islands, associated with the living representatives of the ancient fauna of Europe. Fresh-water snails (*Limnæidæ*) occur in the English Weald, but fossil land-snails have not been found in strata older than the Tertiary in Europe, and then under forms generically, and even in one instance specifically, identical with living types of the new world (*Megaspira, Proserpina, Glandina*, and *Helix labyrinthica*). In the coal-strata of Nova Scotia, Sir Chas. Lyell has discovered a single specimen of a reversed and striated shell, apparently a *Clausilia*.

The *lingual dentition* of the pulmonifera confirms, in a remarkable manner, those views, respecting the affinities of the order, and its zoological value, which have been deduced from the more obvious characters afforded by the animal and shell. The operculated land-snails have seven-ranked teeth, like *Paludina* and *Litorina*. The in-operculated air-breathers have, without known exception, rows of very numerous, similar teeth, with broad bases, resembling tessellated pavement. Their crowns are recurved, and either aculeate or dentated. The lingual ribbon is very broad, often nearly as wide as it is long; and the number of teeth in a row (though usually a third less) is sometimes as great, or even greater, than the number of rows. The rows of teeth are straight or curved or angulated; when the rows are straight the teeth are similar in shape; curves indicate gradual changes, and angles accompany sudden alterations of form.

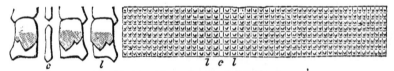

Fig. 90.   *Lingual teeth of Achatina.**

The absolute number of teeth is only a specific character, and is usually greatest in the larger species; but the *Helicellæ* have fewer teeth in proportion than the *Helices*, and Velletia has fewer than *Ancylus*. The anomalous genus *Amphibola* (p. 139) has an unusually broad tongue, armed with teeth similar to those of the snail.

Fig, 91.   *Lingual teeth of Amphibola.†*

About one-third the lingual membrane is spread over the tongue; the rest has its margins rolled together, and is lodged in a sac or dental canal, which

* Fragment of the lingual membrane of *Achatina fulica*, with central and lateral teeth more enlarged, from a specimen communicated by J. W. Laidlay, Esq.

† Part of the tongue of *Amphibola avellana*, from a preparation by J. W. Wilton, Esq., of Gloucester.

diverges downwards from the posterior part of the mouth, and terminates outside the buccal mass of muscles.[*]

The mode in which the tongue is used, may be seen by placing a *Limnæa* or *Planorbis* in a glass of water, inside which the green *conferva* has begun to grow; they will be observed incessantly cleaning off this film. The upper lip with its mandible is raised, the lower lip—which is horse-shoe shaped—expands, the tongue is protruded and applied to the surface for an instant, and then withdrawn; its teeth glitter like glass-paper, and in *Limnæa* it is so flexible, that frequently it will catch against projecting points, and be drawn out of shape slightly as it vibrates over the surface.

" The development of the (in-operculate) Pulmonifera has been worked out by Van Beneden and Windischmann,[†] by Oscar Schmidt,[‡] and by Gegenbaur;[§] the memoir, by the last named author, contains full information respecting *Limax* and *Clausilia,* and some important notices with regard to *Helix.*

" The yelk undergoes complete division. The first stage of development consists in the separation of the embryo into mantle and foot. The anterior. part of the body, in front of the mantle, dilates and forms a contractile sac—the homologue of the *velum* of marine gasteropods—which in *Doris, Polycera,* and *Æolis,* has been seen to exhibit similar contractions. (*Gegenbaur.*) To this contractile vesicle the name of *Yelk-sac* was given by Van Beneden and Windischmann, but it is a very different organ from the true Yelk-sac, which exists in the Cephalopoda alone among molluscs.

" A similar contractile dilatation exists at the end of the foot—and the contractions of this ' caudal' vesicle and of the ' vitellary' vesicle alternate, so as to produce a kind of circulation before the development of the heart.

" " The oral tentacles and parts about the mouth are the last to be completed.

" A peculiar gland exists during the embryonic period, attached to the parietes of the ' vitellary' vesicle, which Gegenbaur and Schmidt compare to a Wolffian body.

" Gegenbaur draws attention to the fact, that the first rudiment of the shell in *Limax, Clausilia* and probably *Helix,* is not secreted on the exterior of the mantle, as in other *gasteropoda;* but is deposited, in the form of calcarious granules, within its substance.

" Besides, therefore, the possession of Wolffian bodies, and of especial contractile organs, which subserve respiration and circulation during embryonic life—the terrestrial *gasteropoda* are further distinguished by the

[*] Thomson, An. Nat. Hist. Feb. 1851.
[†] Recherches sur l'embryogenie des Limaces. Müller's Archiv. 1841.
[‡] Ueber die Entwickelung von *Limax agrestis* Müller's Archiv, 1851.
[§] Beiträge zur Entwickelungs geschichte der Land-gasteropoden. Siebold and Kölliker's Zeitschrift, 1852.

peculiar mode of development of their shells—if the observations upon *Clausilia* and *Helix* may be extended to the rest. The first development of the shell within the substance of the mantle (a relation found hitherto only in the *Cephalopoda*) is up to the present time a solitary fact, without parallel among the other gasteropodous families." (*Huxley.*)

FAMILY I. HELICIDÆ.* Land-snails.

Shell external, usually well developed, and capable of containing the entire animal; aperture closed by an *epiphragm* during hybernation.†

Animal, with a short retractile head, with four cylindrical, retractile tentacles, the upper pair longest and bearing eye-specks at their summits. Body spiral, distinct from the foot; respiratory orifice on the right side, beneath the margin of the shell; reproductive orifice near the base of the right ocular tentacle; mouth armed with a horny, dentated, crescent-shaped upper mandible; lingual membrane oblong, central teeth in-conspicuous, laterals numerous, similar. (See Intr. p. 17.)

HELIX, L.‡

Type, H. pomatia, L. Roman snail. *Etym. Helix*, a coil.

Shell umbilicated, perforated or imperforate; discoidal, globosely-depressed or conoidal; aperture transverse, oblique, lunar or roundish; margins distinct, remote or united by callus.

Animal with a long foot, pointed behind; lingual teeth usually in straight rows, edge-teeth dentated.

Distr. including the sub-genera, above 1,200 sp. (several hundred sp. are undescribed). World-wide; ranging northward as far as the limit of trees, and southward to Tierra-del-fuego, but most abundant by far in warm and humid climates. M. D'Orbigny observed 6 sp. at elevations exceeding 11,000 feet, in S. America, and Layard found *H. gardeneri* at the height of 8,000 feet in Ceylon. The species of tropical and southern islands are mostly peculiar. Several of the smaller British species, and even the large garden-snail (*H. aspersa*), have been naturalised in the most remote colonies. The Neapolitans and Brazilians eat snails.

Fossil (extinct) sp. about 50. Eocene —. Europe.

Sections; Acavus, Montf. Shell imperforate. H. hæmastoma, Pl. XII. fig. 1.

Geotrochus (lonchostoma) Hasselt, Trochiform, flat beneath.

Polygyra, Say. Depressed, many-whirled. H. polygyrata, Pl. XII. fig. 2.

---

* The account of this family is chiefly taken from Dr. L. Pfeiffer's *Monographia Heliceorum.*

† The *epiphragm* is a layer of hardened mucus, sometimes strengthened with carbonate of lime; it is always minutely perforated opposite the respiratory orifice.

‡ The synonomy of the genus would fill several pages. See Intr. 1, p. 59.

Tridopsis, Raf. Aperture contracted by tooth-like projections. H. hirsuta, Pl. XII. fig. 5.

Carocolla, Lam. Peristome continuous. H. lapicida, Pl. XII. fig. 3.

Sub-genera. Anastoma, Fischer. (Tomigerus, Spix.) H. globulosa Pl. XII. fig. 4. Aperture of adult turned upwards, ringent; 4 sp. Brazil. *Hypostoma* (Boysii) Albers, is a minute Indian snail, in which the aperture is similarly distorted. *Lychnus* (Matheroni, Req.) has a similar shell, but no apertural teeth; 3 sp. occur in the Eocene Tertiary of the S. France.

Streptaxis, Gray. H. contusa, Pl. XII. fig. 6. Sub-globose, lower whirls receding from the axis of the upper; 24 sp. Brazil, W. Africa, Mascarene Ids. S. Asia.

Sagda, Beck. H. epistylium, Pl. XII. fig. 7. Imperforate, globosely conoid, close-whirled, aperture lamellate within, lip sharp; 3 sp. Jamaica.

Prosérpina (nitida) Guilding. Shell depressed, shining, callous beneath; aperture toothed inside; peristome sharp. *Distr.* 6 sp. Jamaica, Cuba, Mexico. *Fossil*, Eocene—. I. Wight (*F. Edwards*).

Helicella, Lam.* *Type*, H. cellaria, Pl. XII. fig. 8. Shell thin, depressed; peristome sharp, not reflected. Lingual edge-teeth aculeate. 90 sp.

Stenopus (cruentatus) Guild. *Syn.* Nanina (citrina) Gray; Ariophanta (lævipes, Pl. XII. fig. 9) Desm. Shell thin, polished; peristome thin, not reflected. Animal with the tail truncated and glandular, like *Arion;* mantle-margin produced, partly covering the shell. *Distr.* 70 sp. S. Asia and Ids. N. Zealand, Pacific Ids. W. Indies.

VITRINA, Draparnaud, Glass-snail.

Type, V. Draparnaldi, Pl. XII. fig. 28. *Syn.* Helicolimax, Fer.

Shell imperforate, very thin, depressed; spire short, last whirl large; aperture large, lunate or rounded, columellar margin slightly inflected, peristome often membranous.

Animal elongated, too large for complete retraction into the shell; tail very short; mantle reflected over the shell-margin, and furnished with a posterior lobe on the right side. Lingual teeth (of type) 100 rows of 75 each; marginal teeth with a single, long, recurved apex (*Thomson*). Occasionally animal-feeders, like the slugs.

V. Cuvieri and *Freycineti* (Helicarion Fer.) tail longer, more abruptly truncated, with a caudal gland like *arion*, mantle more developed.

Distr. 64 sp. Old World, 58; Greenland, 1; Brazil, 5.

Sub-genera. Daudebardia, Hartm. (Helicophanta, Fér.) V. brevipes, Pl. XII. fig. 29. Shell perforated, horizontally involute; aperture oblique, ample; 3 sp. Central Europe.

Simpulopsis (sulculosa) Beck; shell succinea-shaped. 5 sp. Brazil.

* For this group Mr. Gray formerly employed the name *Zonites*, given originally by Montfort to Helix Algira; in his later works he adopts *Helicella*.

SUCCINEA, Draparnaud. Amber-snail.

Type, S. putris, Pl. XII. fig. 23.

Syn. Cochlohydra, Fér. Helisiga (S. Helenæ) Less. Amphibulima (patula) Beck ; Pelta (Cumingii) Beck.

Shell imperforate, thin, ovate or oblong; spire small; aperture large, obliquely oval; columella and peristome simple, acute.

Animal large, tentacles short and thick, foot broad; lingual teeth like *helix;* S. putris has 50 rows, of 65 teeth each (*Thomson*). Inhabits damp places, but rarely enters the water.

Distr. 68 sp. Europe 5, Africa 3, India 1, Australia 1; Pacific Ids. 17, N. America 14, S. America 11, W. Indies 11. *Fossil.* Eocene, Brit.

Sub-genus. Omalonyx, D'Orb. O. unguis. Pl. XII. fig. 24. Shell oval, convex, translucent, spire nearly obsolete, margins sharp. Animal large, slug-like; shell placed on the middle of the back, with the mantle slightly reflected upon it all round. *Dist.* 2 sp. Bolivia; Juan Fernandez.

BULIMUS, Scopoli.

Etym. ? *Boulimos,* extreme hunger (in allusion to its voracity!)

Syn. Bulinus, Brod. (not Adans). *Type.* B. oblongus. Pl. XII. fig. 10.

Shell oblong or turreted; aperture with the longitudinal margins unequal, toothless or dentate; columella entire, revolute externally or nearly simple; peristome simple or expanded.

Animal like Helix. *B. ovatus* attains a length of 6 inches, and is sold in the market of Rio; it oviposits amongst dead leaves, the eggs have a brittle shell, and the young when hatched are an inch long. (See p. 54, fig. 31.)

Sections. Odontostomus (gargantuus) Beck, aperture toothed, 13 sp. Brazil.

Pachyotis, Beck (Caprella, Guild.) fig. 91.*

Partula, Fér. P. faba. Pl. XII. fig. 13, *Tahiti.* 26 sp. Asiatic, Australian and Pacific Ids. 24; S. America 2. The animal is ovo-viviparous.

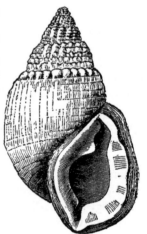

Gibbus (Lyonnetianus) Montf. Shell hump-backed; Mauritius, 2 sp.

Bulimulus, Leach. B. decollatus. Pl. XII. figs. 11, 12. Shell small, lip acute. Above 300 sp. England 3 sp.

Zua, Leach. Z. lubrica. Pl. XII. fig. 14. Shell polished, columella slightly truncated.

Azeca, Leach. A. tridens. Pl. XII. fig. 15. Shell polished, peristome thickened and toothed.

Fig. 91* *B. auris-vulpina.*

---

* Fig. 91. *Bulimus auris-vulpina,* Chemn. The great extinct land-snail of St.

Distr. 650 sp. Europe 30, Asia 130, Australia and Pacific Ids. 46, Africa 50, S. States 3, Tropical and S. American 330.

Fossil. 30 sp. Eocene —. Europe, S. Helena, Australia, W. Indies. B. Guadalupensis occurs in modern limestone, with human remains.

ACHÀTINA, Lamarck. Agate-shell.

Type, A. variegata, Pl. XII. fig. 22.

Syn. Cochlitoma, Fér. Columna, Perry. Subulina (octona) Beck. Liguus (virgineus) Montf. Cionella (acicula) Jeffr.

Shell imperforate, bulimiform; columella twisted, and truncated in front; aperture oval, angular above; peristome simple, acute.

Animal snail-like. The great African Achatinæ are the largest of all land-snails, attaining a length of 8 inches; their eggs exceed an inch in length, and have a calcarious shell.

Distr. 120 sp. Europe 9, Africa 38, Asia 8, tropical America 29.

Fossil. 14 sp. Eocene —. Europe; St. Helena.

Sub-genera. Glandina (voluta) Schum. (Oleacina, Bolten; Polyphemus, Montf.) *Shell* oblong, fusiform; aperture narrow, elliptical. *Animal* twice as long as the shell; eye tentacles deflected at the tips, beyond the eyes; vibracula much shorter, also deflected; lips elongated, tentacular. Frequents low and moist situations; in confinement one refused vegetable food, but at another snail. (*Say.*) 40 sp. W. Indies, Central America, Mexico, Florida. *Fossil.* Eocene —. *Glandina costellata.* I. Wight. (*F. Edwards.*)

Achatinella (vulpina) Sw. (Helicteres, Fér.) Columella twisted into a strong, tooth-like fold. Sandwich Ids. 25, Mariannes 2, Ceylon 1.

PUPA, Lamarck. Chrysalis-shell.

Type, P. uva. Pl. XII. fig. 16. *Syn.* Torquilla (juniperi) Studer.

Shell rimate or perforate, cylindrical or oblong; aperture rounded, often toothed;* margins distant, mostly united by a callous lamina.

Animal with a short foot, pointed behind; lower tentacles short.

Distr. 160 sp. Greenland 1, Europe 76, Africa 23, India 12, Pacific Ids. 2, N. America 30, S. America 5. *Fossil.* 40 sp. Eocene —. Europe.

Sub-genus. Vertigo, Müll. V. Venetzii. Pl. XII. fig. 17. Shell minute, sometimes sinistral. Animal with the oral tentacles rudimentary or obsolete. 12 sp. Old World.

CYLINDRÉLLA, L. Pfeiffer. Cylinder-snail.

Type, C. cylindrus. Pl. XII. fig. 20.†

Helena; from a specimen presented by Chas. Darwin, Esq. See "Journal of a Voyage round the World."

* Dr. Pfeiffer terms those teeth *parietal* which are situated on the body-whirl those on the outer lip *palatal,* and on the inner lip *columellar.*

† The figure is taken from a sp. in Mr. Cuming's cabinet, in which the empty apex, usually decollated, remains attached to the adult shell.

Syn. Brachypus, Guild.   Siphonostoma, Sw.

Shell cylindrical or pupiform, sometimes sinistral, many-whirled, apex of the adult truncated, aperture round, peristome continuous, expanded.

Animal similar to *clausilia;* foot short, oral tentacles minute.

Distr. 50 sp.   W. Indies 35, Mexico 5, Texas 2, S. America 1.

BALÈA, Prideaux.

Type, B. perversa.   Pl. XII. fig. 21.   *Syn.* Fusulus, Fitz.

Shell slender, usually sinistral, fusiform, multispiral, aperture ovate; peristome acute, margins unequal, wall of the aperture with one slight plait; columella simple.

Animal snail-like; teeth 20.20; rows 130 (*Thomson*).

Distr. 8 sp.   Norway, Hungary, New Granada, Tristan d'Acunha.   The British sp. is found, very rarely, in Porto Santo, only on the highest peak, at an elevation of 1,665 feet.   (*Wollaston.*)

Sub-genus. *Megaspira* (elatior) Lea.   Pl. XII. fig. 18.   Shell dextral, with the columella transversely plaited.   *Distr.* 1 sp.   Brazil.   *Fossil,* 1 sp. Eocene —.   Rheims.

TORNATELLINA, Beck.

Etym. Diminutive (or patronymic termination) of *tornatella.*

Type, T. bilamellata, Ant.   *Syn.* Strobìlus, Anton. Elasmatina, Petit.

Shell imperforate, ovate or elongated; aperture semi-lunar, margins unequal, disunited; columella twisted, truncated; inner lip 1-plaited.

Distr. 11 sp.   Cuba 1, S. America 2, Juan Fernandez 2, Pacific Ids. 5, N. Zealand 1.

PAXILLUS, A. Adams.

Type, P. adversus, Ad.   Borneo.

Shell small, pupiform, sinistral, rimate; spire pointed; aperture semi-ovate, ascending on the body-whirl; inner lip spreading, 1-plaited, outer lip expanded, notched in front.

CLAUSILIA, Draparnaud.

Etym. Dimin. of *clausum* a closed place.   *Syn.* Cochlodina, Fér.

Ex. C. plicatula, Drap. (=C. Rolphii, Leach).   Pl. XII. fig. 19.

Shell fusiform, sinistral; aperture elliptical or pyriform, contracted by lamellæ, and closed when adult by a moveable shelly plate (*clausium*) in the neck.

Animal with a short, obtuse foot; upper tentacles short, lower very small.   C. *bidens* has 120 rows of 50 teeth; C. nigricans 90 rows of 40 teeth each.

Distr. Above 200 sp.   Europe 146, Asia 48, Africa 4, S. America 3.

Fossil, 20 sp. Eocene —.   Brit. France.   Coal-strata, N. Scotia. (*Lyell.*) C. *maxima,* Grat. Miocene, Dax is two inches in length.

FAMILY II. Limacidæ. Slugs.

Shell small or rudimentary, usually internal, or partly concealed by the mantle, and placed over the respiratory cavity.

Animal elongated; body not distinct from the foot; head and tentacles retractile; tentacles 4, cylindrical, the upper pair supporting eyes; mantle small, shieldshaped; respiratory and excretory orifices on the right side.

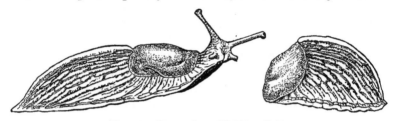

Fig. 92. *Limax Sowerbii Fér. Brit.*

LIMAX, L. Slug.

Type, L. maximus. Pl. XII. fig. 25. (L. cinereus, Müll.)

Shell internal, oblong, flat, or slightly concave beneath, nucleus posterior; margin membranous; epidermis distinct.

Animal, foot pointed and keeled behind; mantle shieldshaped, on the front of the back, granulated or marked with concentric striæ; respiratory orifice on the right side, near the posterior margin of the mantle; reproductive orifice near the base of the right ocular tentacle; lingual teeth tricuspid, those near the margin simple, aculeate.

The slugs are connected with the snails by *Vitrina;* their teeth are similar, but have more elongated cusps. The creeping-disk, or *sole* of the foot, extends the whole length of the animal; but they frequently lift up their heads, like the snails, and move their tentacles in search of objects above them. They often climb trees, and some can lower themselves to the ground by a mucous thread. When alarmed they withdraw their heads beneath the mantle, as in fig. 92. Slugs feed chiefly on decaying vegetable and animal substances; they oviposit at any time of the spring and summer when the weather is moist, and bury themselves in drought and frost. *Limax noctilucus,* Fér. (Phosphorax, Webb.) found in Teneriffe, has a luminous pore in the posterior border of the mantle.

Distr. 22 sp. Europe, Canaries, Sandwich Ids.

Fossil. Eocene —. Brit. The Ancylus ? latus, Edw. of the I. Wight appears to be a Limax.

Sub-genus. Geomalacus (maculosus) Allman. Ireland. *Shell* unguiform. *Animal* with a mucus gland at the extremity of the tail; respiratory orifice near the right anterior border of the mantle.

INCILARIA, Benson.

Type, I. bilineata, Cantor, Chusan. *Syn.* ? Meghimatium, Hasselt.

Animal elongated, tapering behind, entirely covered by a mantle; tentacles 4, the upper bearing eyes, the lower entire; respiratory orifice on the right side, near the front of the mantle. Lon. 1½ inches.

Philomycus (Raf.) Fér. = Tebennophorus, Binney, 1842, Bost. Soc. Journ. (Helix Carolinensis, Bosc) is also a slug with a long mantle.

ARION, Férussac. Land-sole.

Type, A. empiricorum, Fér. *Syn.* Limacella, Brard.

Shell oval, concave; or represented by numerous irregular calcarious granules.

Animal, slug-like; respiratory orifice on the right side, towards the front of the mantle; reproductive orifice immediately below it; tail rounded, slightly truncated, terminated by a mucus-gland. Lingual teeth, as in *limax*; A. empiricorum has 160 rows of 101 teeth each. The land-soles occasionally animal substances, such as dead worms, or injured individuals of their own species. They lay 70-100 eggs, between May and September, are 26-40 days hatching, and attain their full growth in a year; they begin to oviposit a month or two before that period. The eggs of *A. hortensis* are very phosphorescent for the first 15 days. (*Bouchard.*)

Distr. 6 sp. Europe. Norway, Brit. Spain, S. Africa.

Fossil. Newer Pliocene, Maidstone. (*Morris.*)

Plectrophorus (*corninus*, Bosc) Fér. 3 sp. Teneriffe; represented as having a small conical shell on the tail; probably an erroneous observation.

PARMACELLA, Cuvier.

Type, P. Olivieri, Cuv. *Etym. parma*, a small shield.

Syn. ? Peltella (Americana), Van Beneden.

Shell concealed, oblong, nearly flat, apex sub-spiral.

Animal vitrina-like, with an ample foot, pointed behind, and furnished with a mucus-pore; mantle small, shield-like, in the middle of the back, partly or entirely concealing the shell.

P. calyculata, Sby. (Cryptella, Webb,) Pl. 12, fig. 27, is patelliform, with an exposed papillary spire. *Distr.* 7 sp. S. Europe; Canary Ids. N. India.

Fig. 93. *Testacella haliotoides, Fer.* *

TESTACELLA, Cuvier.

Shell small, ear-shaped, situated on the posterior extremity of the body.

Animal, slug-like, elongated and tapering towards the head; back with

---

* Back view of a half-grown individual; side-view of shell on the tail, and front view of the head. From specimens communicated by Arthur Mackie, Esq., of Norwich.

2 principal lateral furrows, from which numerous vein-like grooves ramify; mantle not larger than the shell; respiratory orifice on the right side, beneath sub-spiral apex of the shell; reproductive orifice behind the right tentacle. The Testacella is subterranean in its habits, feeding on earth-worms, and visiting the surface only at night. Its lingual membrane is very large and wide, with about 50 rows of 20.20 teeth, which diminish rapidly in size towards the centre; each tooth is slender, barbed at the point, and slightly thickened at the base, and furnished with a projection on the middle of the posterior side.

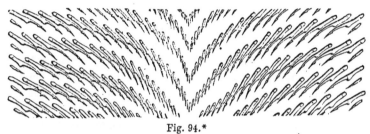

Fig. 94.*

Distr. 3 sp.   S. Europe; Canary Ids.   Brit. (introduced.)

FAMILY III.   ONCIDIADÆ.

Animal, slug-like, destitute of any shell, completely covered by a coriaceous mantle; tentacles cylindrical, retractile, with eyes at their extremities; foot much narrower than the mantle.

ONCIDIUM, Buchanan.

Type, O. Typhæ, Buch.   *Etym.* Diminutive of *Onkos*, a tubercle.

Animal oblong, convex, usually tuberculated; head with 2 retractile tentacles, bearing the eyes; mouth covered by a notched veil; no horny jaws; tongue broad, with above 70 rows of lingual teeth (in *O. celticum*), teeth 54.1.54;† the central teeth minute, triangular, with a single obtuse spine; laterals, slightly curved; heart opistho-branchiate; respiratory orifice posterior, distinct from the vent; sexes combined, ♂ organ under the right tentacle, ♀ at the posterior extremity of the body.

Distr. 16 sp.   Brit. Medit. Red Sea, Mauritius, Australia, Pacific.   The typical *Oncidia* live on aquatic plants, in the marshes of the warmer parts of the old world.   Those which frequent sea-shores have been separated under the name *Peronia*, Bl. (Onchis, Fér).   One species (*O. celticum*) is found

---

* Part of the lingual membrane of *T. haliotides*, from a preparation by Fisher Cocken, Esq., of Botesdale.   The dentition resembles that of *Ianthina*.

† This is a convenient mode of stating the number of lingual teeth in each row; it means that there is a single (symmetrical) tooth in the centre, and 54 lateral (un-*symmetrical) teeth on each side.   If the number of rows of teeth on the dental membrane is known, it may be added below, thus—*Peronia Mauritiana*, $\frac{80.1.80}{68.}$

on the coast of Cornwall, congregated in little groups, about a foot or two from the surface of the sea, where the waves break over them. They ascend and descend, so as to maintain their distance as the tides rise and fall; but will not bear long immersion in sea-water. (Couch.)

? *Buchanania* (*oncidioides*) Lesson. Named after Dr. F. Hamilton (Buchanan), the Zoologist of India. *Animal* oval, entirely covered by a simple mantle; respiratory orifice in the centre of the back; head with 4 tentacles, retractile beneath the mantle; foot oval, much smaller than the mantle; length 3½ inches. Coast of Chile. (Requires confirmation.)

VAGINULUS, Férussac.

Type, V. Taunaisii, Fér. *Syn.* Veronicella, Bl.

Animal elongated, slug-like, entirely covered by thick coriaceous mantle, smooth or granulated; head retractile under mantle; tentacles 4, upper pair slender, cylindrical, inflated at the tips and bearing eyes, lower pair short, bifid; foot linear, pointed behind; sexes united; ♂ orifice behind the right tentacle, ♀ midway on the right side, beneath the mantle: respiratory and excretory orifices at posterior extremity, between mantle and foot. Inhabits forests, in decayed wood and under leaves.

Distr. 6 sp. W. Indies, S. America, India, Philippines.

FAMILY IV. LIMNÆIDÆ.

Shell thin, horn-coloured; capable of containing the whole animal when retracted; aperture simple, lip sharp; apex sometimes eroded.

Animal with a short dilated muzzle; tentacles 2, eyes sessile at their inner bases; mouth armed with an upper mandible, tongue with teeth similar to *Helix.* The Limnæids inhabit fresh-waters, in all parts of the world; they feed chiefly on decaying leaves, and deposit their spawn in the form of oblong transparent masses, on aquatic plants and stones. They frequently glide beneath the surface of the water, shell downwards, and hybernate or restivate in the mud.

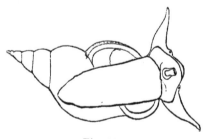

Fig. 95.

LIMNÆA,* Lamarck. Pond-snail.

Etym. Limnaios, marshy. *Type,* L. stagnalis, fig. 95. Pl. XII. fig. 30.

---

* Adjectives employed as names for shells should have the feminine termination.

Shell spiral, more or less elongated, thin, translucent; body-whirl large, aperture rounded in front; columella obliquely twisted.

Animal with a short, broad head; tentacles triangular, compressed; lingual teeth (*L. stagnalis*) 55.1.55, about 110 rows, central teeth minute, laterals bicuspid, the inner cusp largest. *L. peregra* feeds on the green freshwater algae; *L. stdgnalis* prefers animal substances.

Distr. 50 sp. . Europe, Madeira, India, China, N. America.

Fossil, 70 sp. Wealden —. Brit. France.

Sub-genus, Amphipeplea, Nilsson. A. glutinosa, Pl XII. fig. 31. *Shell* globular, hyaline. *Animal* with a lobed mantle, capable of expansion over the shell. Europe; Philippines.

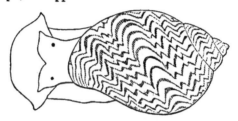

Fig. 96.

CHILINIA, Gray. Chilian-snail.

Ex. C. pulchra, D'Orb. fig. 96. *Syn.* Dombeya, D'Orb.

Shell oval, thin, ornamented with dark spots or wavy bands; columella thickened, with 1 or 2 strong prominent folds.

Distr., 14 sp. S. America; in clear running streams.

Fossil, 1 sp. Miocene, Rio Negro, Patagonia (D'Orb.)

PHYSA, Draparnaud.

Type, P. fontinalis, Pl. XII. fig. 32. *Etym. Physa*, a pouch.

Syn. Bulin, Adans. Rivicola, Fitz. Isidora, Ehr.

Shell ovate, sinistrally spiral, thin, polished; aperture rounded in front.

Animal with long slender tentacles; the eyes at their bases; mantle margin expanded and fringed with long filaments.

P. hypnorum (Aplexa, Fleming) has an elongated spire, and the mantle margin is plain. *Physopsis*, Krauss, S. Africa, has the base of the columella truncated. *Camptoceras* (terebra), Benson, India, has the whirls disunited, and the peristome continuous.

Distr. 20 sp. N. America, Europe, S. Africa, India, Philippines.

Fossil, 14 sp. Wealden —. Brit. France. The largest living sp. (P. Maugeræ, California) is 15 lines in length. A fossil sp. found at Grignon measures 26 lines, and another equally large occurs in India.

ANCYLUS, Geoffroy. River-limpet.

Etym. Ancylus (agkulos) a small round shield.

Type, A. fluviatilis, Müll. Pl. XII. fig. 33 (Patella lacustris, L.)

Shell conical, limpet-shaped, thin; apex posterior, sinistral; interior with a sub-spiral muscular scar.

Animal like Limnæa; tentacles triangular, with eyes at their bases; lingual teeth 37.1.37, in 120 rows, centrals small, laterals with long recurved hooks.

Distr. 14 sp. N. and S. America, Europe, Madeira. On stones and aquatic plants in running streams. *Fossil*, 8 sp. Eocene, Belgium.

Sub-genera, Velletia (oblonga, Lightf.) Gray. (Acroloxus, Beck) Shell and animal dextral; lingual teeth 40, in 75 rows. 3 sp. West Indies, Europe. *Fossil*, 2 sp. Eocene. Brit. France.

Latia (neritoides) Gray; shell limpet-like, interior with a transverse plate, turned up and notched on one side. N. Zealand.

PLANORBIS, Müller.

Syn. "Coret," Adans. *Type*, P. corneus, Pl. XII. fig. 34.

Shell discoidal, dextral, many-whirled; aperture crescentic, peristome thin, incomplete, upper margin projecting.

Animal with a short, round foot; head short, tentacles slender, the eyes at their inner bases; lingual teeth sub-quadrate, central and marginal bicuspid, laterals tricuspid; excretory orifices on left side of the neck.

Some species of *Planorbis* have the sutures and spire deeply sunk, and the umbilicus flattened; specimens occur with the spire elevated (fig. 97*). *P. contortus*, a minute species, has above 6,000 teeth, (*Cocken*). P. corneus secretes a purple fluid (Lister). *P. lacustris* (Segmentina, Fleming) has the whirls contracted, internally, by periodic septa, 3 in a whirl, with triradiate openings.

·Fig. 97.

Distr. 60 sp. N. America, Europe, India, China.

Fossil, 60 sp. Wealden —. Brit. France.

FAMILY V. AURICULIDÆ.

Shell spiral, covered with horny epidermis, spire short, body-whirl large; aperture elongated, denticulated; internal septum progressively absorbed.

Animal with a broad and short muzzle, tentacles 2, cylindrical, the eyes sessile behind them; mantle-margin thickened; orifices as in the snails; foot oblong; sexes united; mouth with a horny upper jaw; lingual teeth numerous, central series distinct, hooked, tricuspid. *A. livida* has about 31 laterals (Loven); another species examined by Mr. Wilton has 11 large laterals and about 100 smaller (*uncini*) on each side, gradually diminishing towards the edge, fig. 98, *c.* central teeth, *l.* laterals.

* *P. marginatus*, var. Rochdale, communicated by J. S. Gaskoin, Esq.

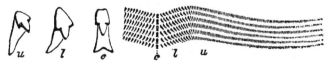

Fig. 98

The *Auriculæ* frequent salt-marshes, damp hollows, and places overflowed by the sea; they were long regarded as marine animals, and their shells confused with those of *Tornatella* and *Ringicula*.

AURICULA, Lamarck.

Type. A. Judæ. Pl. XII. fig. 35. *Etym. Auricula,* a little ear.

Syn. Cassidula, Fér (not Lam.) Marinula (pepita) King. Geovula, Sw.

Shell oblong, with thick, dark epidermis; spire obtuse; aperture long, narrow, rounded in front, with 2 or 3 strong folds on the inner lip; outer lip expanded and thickened.

Distr. 50 sp. Philippines, Celebes, Feejees, Australia, Peru.

Fossil, 20 sp. Neocomian —. France.

Fig. 99. *A. auris-felis.* (From Eyd. and Soul).

A. Judæ has truncated tentacles; the typical species are met with in the brackish-water swamps of tropical islands, on the roots of mangroves, and by small streams within the influence of the tide. One species has been observed by Mr. Adams in nearly 2 fathoms water.

Sub-genera, Polydonta, Fischer, *P. scarabæus,* Pl. XII. fig. 36. (Scarabus imbrium, Montf.) Shell oval, compressed; spire pointed many-whirled, with lateral varices; aperture toothed on both sides. *Distr.* 20 sp. India, Borneo, Celebes, Pacific Ids. Inhabits moist spots in woods near the sea, and is wholly terrestrial, feeding on decayed vegetables. (*Adams.*)

Pedipes (afra) Adans. *Shell* ovate, spirally striated, aperture denticulated on both sides; the animal loops in walking, like *truncatella.* *Distr.* W. Indies, Africa, Philippines, Pacific Ids. Under stones on the sea-shore.

Fossil, 5 sp. Eocene —. Brit. France.

CONOVULUS, Lamarck.

Type, C. coniformis, Brug. Pl. XII. fig. 37. (= Voluta coffea, L.?)

Syn. Melampus, Montf. Rhodostoma, Sw.

Shell obtusely cone-shaped, smooth; spire short, flat-whirled: aperture long, narrow; lip sharp, denticulated within; columella twisted in front; wall of the aperture with 1 or 2 spiral plaits.

Animal with short, tapering and rather compressed tentacles; foot divided transversely into two portions, advanced successively in walking.

Distr. W. Indies, Europe.　In salt-marshes and on the sea-shore.　The British species have thin ovate shells, with the spire moderately produced, and the aperture ·oval.　They form the sub-genus *Alexia.* (denticulata) Leach.　*Fossil.* Eocene.　Brit. France.

CARYCHIUM, Müller.

Type, C. minimum, Pl. XII. fig. 39.

Syn. Auricella, Hartm.

Shell minute, oblong, finely striated transversely; aperture oval, toothed, margins thickened, united by callus.

Animal with 2 blunt, cylindrical tentacles; eyes black, sessile, near together, behind the tentacles.

Distr. 3 sp.　Europe; N. America.　At the roots of grass in damp places, especially near the sea.

Fossil. Miocene —.　Europe.

The genus *Siphonaria,* described at p. 155, is supposed to be pulmoniferous, and to bear somewhat the same relation to *Auricula* that Ancylus does to *Limnaea.*　The lingual dentition is similar to Auricula; the centre teeth are distinct, the laterals numerous and hooked.

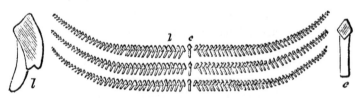

Fig. 100.*

SECTION B.　OPERCULATA.*

The Operculated land-snails are exceedingly like periwinkles (*litorinæ*), and chiefly differ from them in the situations they inhabit, and the medium respired.　They have a long truncated muzzle, 2 slender contractile tentacles, and the eyes are sessile on the sides of the head.‡　The mantle-margin is simple, and the pulmonary cavity is situated on the back of the neck, and quite open in front.　Lingual ribbon narrow; teeth 7-ranked.

---

　* *Siphonaria* sp. from the Cape; three rows of teeth, *c* central, *l* laterals, from a preparation by J. W. Wilton, Esq., of Gloucester.

　† *Phanero-pneumona* (open-lunged), Gray.　The account of this group is chiefly taken from the Catalogue prepared by my friend Dr. Baird.

　‡ The tentacles of the *helicidæ* are retractile, by inversion (p. 25) those of the *cyclostomidæ* are contractile only.

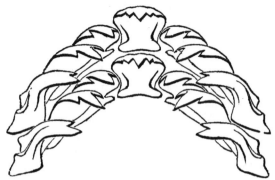

Fig. 101. *Lingual teeth of Cyclophorus.**

The sexes are distinct; the shell is spiral, and closed by an operculum, presenting many beautiful modifications of structure, characteristic of the smaller groups, which are often peculiar to limited regions, as in the *Helicidæ*. The oldest fossil species are found in the Eocene Tertiary.

FAMILY VI. CYCLOSTOMIDÆ.

Shell spiral, rarely much elongated, often depressed, spirally striated; aperture nearly circular; peristome simple. *Operculum* distinctly spiral.

Animal with the eyes on slight prominences at the outer bases of the tentacles; tentacles contractile only; foot rather elongated.

CYCLOSTOMA, Lamarck.

Etym. Cyclos circle, *stoma* mouth. *Type,* C. elegans, Pl. XII. fig. 40.

Syn. Leonia (mammillaris) and Lithidion, Gray.

Shell turbinated, thin, axis perforated; aperture oval; peristome continuous, simple, straight or expanded; epidermis very thin. Operculum shelly, pauci-spiral.

Animal with clavate tentacles; sole of the foot divided by a longitudinal groove, the sides moved alternately in walking; the end of the long muzzle is also frequently applied, as by the looping-snails (Truncatellæ), and used to assist in climbing.

Fig. 102. *Cyclostoma elegans, from Charlton, Kent.*

Distr. Above 80 sp. S. Europe; Africa, Madagascar. The only British

* *C. aquilum, Sby.* (original). From a specimen gathered by J. W. Laidlay, Esq. on the steps of the great idol-temple of Maulmein, Birmah.

sp. *C. elegans*, is found on calcarious soils; it ranges to the Canaries-and Algeria, and occurs fossil in the newer Tertiaries. Nearly half the species have the whirls spirally keeled, and have been distinguished under the name *Tropidophora*, by Troschel. They are found in Madagascar and the adjacent islands and coast of Africa. *Fossil*, 20 sp. Eocene, Europe.

Sub-genera. Otopoma (foliaceum), Gray. *Shell* sub-globose, umbilicated; peristome with an ear-like process covering part of the perforation. *Distr.* 15 sp. Arabia, Madagascar, China, New Ireland.

Choanopoma (lincina) Pfr. *Shell* often a little decollated; peristome usually double, the outer edge angularly expanded. *Lincina* (labeo) Br. has the last whirl produced. *Jamaicia* (anomala) C. B. Adams, has the operculum convex. *Distr.* 70 sp. W. Indies, and a few in Tropical America.

Cistula (fascia), Gray. = *Tudora* (megacheila), Gray. *Shell* ovate or elongated, apex usually decollated, peristome free; operculum with a thin shelly outer coat. *Chondropoma* (semilabre) Pfr. differs in the operculum being "sub-cartilaginous." *Distr.* About 70 sp. W. Indies; Tropical America, 8 sp.

Realia (hieroglyphica), Gray. = Hydrocæna (part) Parreyss, Omphalotropis, Pfr. Liarea (Egea), Gray. Bourciera (helicinæformis) Pfr. *Shell* turrited or turbinate, perforated; peristome simple, straight or expanded; operculum pauci-spiral, horny. *Distr.* 17 sp. Canaries, ? Mauritius, Pacific Ids. (Ecuador, *Bourciera.*)

Pomatias (maculatum), Studer. *Shell* slender, transversely striated; peristome reflected; operculum cartilaginous, concamerated within. *Distr.* 10 sp. S. Europe; Corfu.

? FERUSSINA, Grateloup.

Etym. named in honour of Baron Ferussac.

Type, F. anastomæformis, Gr. *Syn.* Strophostoma, Desh.

Shell rounded, depressed, umbilicated; whirls transversely striated above, spirally keeled below; aperture turned obliquely upwards, peristome simple, Operculum.?

Fossil, 1 sp. Miocene —. Dax; Turin.

CYCLOPHORUS, Montfort.

Etym. Cyclos, circle, *phoreus*, bearer.

Type, C. involutus, Pl. XII. fig. 41.

Shell depressed, openly umbilicated; aperture circular; peristome continuous, straight or expanded; epidermis thick; operculum horny, many-whirled.

Animal with long, slender pointed tentacles; foot broadly expanded, not grooved.

Distr. About 90 sp. India, Philippines, New Zealand, Pacific Ids. Tropical America. *C. gibbus,* Fér. (Alycaeus, Gray) has the last whirl distorted.

C. cornu-venatorium, Sby. (Aulopoma, Troschel) Ceylon, has the peristome free when adult; the operculum is larger than the aperture, and reflected over it.

Sub-genera. Pterocyclos (rupestris), Benson. Myxostoma and Steganostoma, Troschel. *Shell* depressed, nearly discoidal, widely umbilicated; peristome expanded, produced into a little wing at the suture; operc. sub-cartilaginous, spirally lamellated. *Distr.* 16 sp. India, Ceylon, Birmah, Borneo?

Cyclotus (fuscescens) Guilding (Aperostoma, Troschel). *Shell* depressed, widely umbilicated; operculum shelly, whirls numerous, with raised margins. *Distr.* 44 sp. W. Indies, Tropical America, India, Asiatic Ids. *Fossil.* Eocene, I. Wight (F. Edwards).

Leptopoma (perlucidum) Pfr. *Shell* turbinated, peristome simple, reflected; operc. membranous. *Distr.* 29 sp. Philippines, India, New Guinea, N. Zealand, Pacific Ids.

*Megaloma** (cylindraceum) Guild. (Farcimen, Troschel.) *Shell* oblong or pupa-shaped, scarcely perforated, aperture circular; operc. thin, horny, many-whirled, flat. *Distr.* 19 sp. West Indies, Tropical America, Canaries, India, Mauritius. *Fossil.* Eocene —. Paris and I. of Wight (*E. Forbes.*)

Craspedopoma (lucidum) Pfr. *Shell* turbinate, rimate, a little contracted near the aperture; operc. round, horny, many-whirled. *Distr.* 3 sp. Madeira, Palma. *Fossil.* Eocene —. I. Wight, Madeira.

Cataulus (tortuosus) Pfr. *Shell* pupa-shaped, with the base keeled, producing a channel in the front of the aperture; operc. circular, horny, the whirls easily separable. *Distr.* 6 sp. Ceylon.

Diplommatina (folliculus) Benson. *Shell* minute, (1 sp. sinistral) conical, with costulated whirls; peristome double; operc. horny, multispiral. *Distr.* 3 sp. India.

PUPINA, Vignard.

Type, P. bicanaliculata, Sby. Pl. XII. fig. 42. Australian Ids.

Shell sub-cylindrical, usually polished; aperture circular, peristome thickened, notched in front and at the suture; operc. membranous, narrowwhirled. P. *grandis*, Forbes, has a dull epidermis.

Distr. 8 sp. Philippines, New Guinea, New Ireland, Louisiades.

Sub-genus, Rhegostoma (nunezii) Hasselt. Aperture with a narrow channel in the middle of the columellar side. 6 sp. Philippines. Nicobar. In *R. lubricum* (Callia, Gray) the sinus is obsolete. *R. pupiniforme* (Pupinella, Gray) is perforated, and has a dull epidermis.

HELICINA, Lamarck.

Type, H. Neritella, Lam.

Syn. Oligyra, Say. Pachytoma, Sw. Ampullina, Bl. Pitonillus, Montf.

---

* Abridged from *Megaloma-stoma;* Swainson, who judiciously curtailed several preposterously long names, allowed this to remain.

Shell globose, depressed or keeled, callous beneath; aperture squarish or semi-lunar; columella flattened; peristome simple, expanded; operc. shelly or membraneous, squarish or semi-ovate, lamellar.

Animal like *Cyclophorus;* lingual teeth 3.1.3.   (Gray.)

Distr. 150 sp.   W. Indies, 50; Tropical America, 44; Pacific Ids., 26; Australian Ids. 3; Philippines, 7.

Sub-genera. Lucidella, (aureola) Gray.   Peristome more or less toothed internally; 8 sp. W. Indies, Tropical America.

Trochatella (pulchella), Sw.   *Shell* not callous beneath; peristome simple, expanded.   W. Indies 16 sp. Venezuela 1.

Alcadia, Gray. A. Brownei, Pl. XII. fig. 43. *Jamaica. Shell* helix-shaped, often velvety, callous beneath; columella flattened, straight; peristome slit in front; operc. shelly, semi-ovate, with a tooth-like process adapted to the slit in the peristome. *Distr.* 17 sp.   Cuba, Jamaica and Haiti.

STOASTOMA, C. B. Adams.

Etym. Stoa pillared, *stoma,* mouth.   *Type,* S. pisum, Ad.

Shell minute, globose-conic or depressed, spirally striated; aperture semi-oval; peristome continuous; inner margin straight, forming a small spiral keel round the umbilicus; operc. shelly, lamellar.

Distr. 19 sp.   Jamaica.   *S. succineum* (Electrina, Gray) has smooth whirls.   I. Opara, Polynesia.

FAMILY VII. ACICULIDÆ.

Shell elongated, cylindrical; operculum thin, sub-spiral.

Animal with the muzzle rather produced, slender and truncated; eyes sessile on the upper part of the head, behind the base of the slender tentacles; foot oblong, short, pointed behind.

ACICULA, Hartmann.

Type,. A. fusca, Pl. XII. fig. 44.   *Syn.* Acme and Acmaea, Hartm.*

Shell minute. slender, nearly imperforate; peristome slightly thickened, margins sub-parallel, joined by a thin callus; operc. hyaline.

Distr. 5 sp.   Brit. Germany, France; Vanicoro (on leaves). *A. fusca* is found in low, marshy situations, at the roots of grass; it occurs fossil in the Newer Pleiocene of Essex (J. Brown).

GEOMELANIA. Pfeiffer.

Type. G. Jamaicensis. Pfr.   *Etym. Ge,* the ground (i.e. terrestrial).

Shell imperforate, turreted; aperture entire, effused; peristome simple, expanded; margins joined, basal produced into a tongue-shaped process; operc. oval, pellucid, whirls few, rapidly enlarging.

Distr. 21 sp.   Jamaica.

---

* All given in the same year, 1821, the name *Acmaea* having been employed by Eschscholtz for a genus of limpets, *Acicula* has been retained by Pfeiffer and Gray for this land-shell.

ORDER III. Opistho-branchiata.

Shell rudimentary or wanting. *Branchiæ* arborescent or fasciculated, not contained in a special cavity, but more or less completely exposed on the back and sides, towards the rear (*opisthen*) of the body. Sexes united. (*M. Edwards*).

The molluscs of this order may be termed sea-slugs, since the shell, when it exists, is usually small and thin, and wholly or partially concealed by the animal. When alarmed or removed from their native element, they retract their gills and tentacles, and present such a questionable shape that the in-experienced naturalist will be likely enough to return them, with the refuse of the dredge, into the sea. Their internal structure presents many points of interest; in some the gizzard is armed with horny spines, or large shelly plates; in others the stomach is extremely complicated, its ramifications and those of the liver being prolonged into the branches of the respiratory organ. The tongue is always armed, but the number and arrangement of the lingual teeth is exceedingly variable, even in the same family; usually the dental membrane is broad and short, with many similar teeth in each row. The alimentary canal terminates more in the rear of the body than in the other univalve shell-fish.* The gills are behind the heart, and the auricle behind the ventricle; conditions which characterize the embryonic state of the mollusca generally.

Comparatively little is known of the geographical distribution of these animals; they have been found wherever the requisite search has been made, and are probably much more numerous than at present estimated. The shell-bearing genera flourished in the period when the secondary strata were deposited. The living species are chiefly animal-feeders, preying on other shell-fish and on zoophytes.

SECTION A. Tecti-branchiata.†

Animal usually provided with a shell, both in the larval and adult state; branchiæ covered by the shell or mantle; sexes united.

FAMILY I. Tornatellidæ.

Shell external, solid, spiral or convoluted, sub-cylindrical; aperture long and narrow; columella plaited; sometimes operculated.

Animal with a flattened, disk-like head, and broad obtuse tentacles; foot ample, furnished with lateral and operculigerous lobes.

---

* In the cuttle-fishes and pteropods it is bent upon itself *ventrally*, in the sea-snails *dorsally*, terminating in front, near its origin; the vascular system partakes of this flexure, and the gills are in advance of the heart. (*Huxley*.)

† *Mono-pleuro-branchiata. Bl. Pomato-branchia*, (from *poma*, a lid). Wiegm. The order *Tecti-branchiata* of Cuvier included only the family *Bullidæ*; it is here made to comprise the *Infero-branches* also; no object being gained by the multiplication of descriptive epithets.

The shells of this family are chiefly extinct, ranging from the period of the coal strata, and attaining their greatest development in the cretaceous age. *Tornatella* is essentially related to *Bulla*, but presents some resemblance to the *Pyramidellidæ* in its plaited and operculated aperture; in *Tornatina* the nucleus, or apex, is sinistral. The spiral striae which ornament many of the species, are punctate, as in the Bullidæ; and the outer lip often remarkably thickened, as in Auricula.

TORNATELLA, Lamarck.

Type. T. tornatilis, Pl. XIV. fig. 1. *Syn.* Actæon, Montf. (not Oken), Dactylus (solidulus) Schum. ? Monoptygma (elegans) Lea.

Shell solid, ovate, with a conical, many-whirled spire; spirally grooved or punctate-striate; aperture long, narrow, rounded in front; outer lip sharp; columella with a strong, tortuous fold; operculum horny, elliptical, lamellar.

Animal white; head truncated and slightly notched in front, furnished posteriorly with recumbent tentacular lobes, and small eyes behind them, near their inner bases; foot oblong, lateral lobes slightly reflected on the shell. Lingual teeth 12.12, similar, with long simple hooks.

Fig. 103.

Distr. 16 sp. U. States, Brit. Senegal, Red Sea, Philippines, Japan, Peru. T. tornatilis inhabits deep water, (—60 fms. *Forbes*).

Fossil, 70 sp. Trias — Lias —. N. America, Europe, S. India.

Sub-genera, Cylindrites (Llhwyd) Lycett. C. acutus, Sby. Pl. XIV. fig. 2. (A.) Shell smooth, slender, sub-cylindrical, spire small, aperture long and narrow, columella rounded, twisted, and directed slightly outwards. (B.) Shell oval, spire sunk, whirls with acute margins. Bath Oolite, Brit.

Acteonina, D'Orb. Tornatellæ "without columella plaits," 30 sp. Carb.—Portlandian, (including *Cylindrites*).

Acteonella, D'Orb. A. Renauxiana, Pl. XIV. fig. 3. Shell thick, cone-like or convoluted, spire short or concealed, aperture long and narrow, columella with 3 strong and regular spiral plaits in front. *Distr.* 11 sp. Chalk; Brit. France.

Acteon Cabanetiana, D'Orb. (*Itieria*, Matheron, 1842) Coral-rag, France, belongs to the genus *Nerinea* (D'Orb.) p. 129.

CINULIA. Gray.

Type, C. avellana, Pl. XIV. fig. 4. *Syn.* Avellana and Ringinella, D'Orb. *Shell* globular, thick, spirally groved and punctate, spire small; aperture narrow, rounded and sinuated in front; outer lip thickened and reflected; crenulated inside, columella with several tooth-like folds.

Fossil, 20 sp. Neocomian —Chalk. Brit. France.

RINGICULA, v. p. 112, Pl. V. fig. 21.

GLOBICONCHA, D'Orbigny.

Type, G. rotundata, D'Orb. *Fossil*, 6 sp. Chalk. France.

Shell ventricose, smooth, aperture crescent-shaped, simple, not toothed or thickened on the columellar side.

VARIGERA, D'Orbigny. 1850.*

Type, V. Guerangeri, D'Orb. *Fossil;* 8 sp. Neoc:—. Chalk. France.

Shell like *Globiconcha,* but with lateral varices.

TYLOSTOMA, Sharp. 1849.

Type, T. Torrubiæ, Sh. *Etym. Tulos,* a callosity, *stoma,* mouth.

Shell ventricose, smooth or punctate-striate, spire moderate, aperture ovate-lunate, pointed above, rounded in front; outer lip periodically (once or twice in a whirl) thickened inside and expanded, rising slightly; inner lip callous, spread over body-whirl.

Distr. 4 sp. L. Cretaceous rocks, Portugal.

? PTERODONTA, D'Orbigny.

Type, P. inflata, D'Orb. *Fossil*, 8 sp. Chalk. France.

Shell oblong, ventricose, spire elongated; aperture oval, lip slightly expanded, notched in front, and with a tooth-like ridge internally, remote from the margin.

? TORNATINA, A. Adams.

Type, T. voluta. Pl. XIV. fig. 5.

Shell cylindrical or fusiform, spire conspicuous, apex sinistral, suture channelled, columella callous, 1-plaited.

Animal with a broad, trigonal head, rounded in front; tentacular lobes triangular, with eyes at their outer bases; foot short, truncated in front.

Distr. 15 sp. W. Indies, U. States, Medit. Philippines, China, Australia. On sandy bottoms, ranging to 35 fms. (*Adams*).

Volvula, Adams (Bulla acuminata, Brug.) is a small convoluted shell, with the spire concealed, and the columella obsoletely folded; it is referred to *Cylichna* by Lovén, to *Ovulum* by Forbes. *Distr.* Brit. Medit. *Fossil.* Miocene —. Suffolk.

FAMILY II. BULLIDÆ.

Shell globular or cylindrical, convoluted, thin, often punctate-striated;

---

* The dates of M. D'Orbigny's genera, given in the *Prodrome de Paleontologie,* are dates of *invention;* the names were not published, in many instances, until years afterwards.

spire small or concealed; aperture long, rounded and sinuated in front; lip sharp. No operculum.

Animal more or less investing the shell; head a flattened disk,* with tentacular lobes, often united; eyes immersed in the centre of the disk, or wanting; foot oblong, furnished with a posterior lobe (*meta-podium*), and side-lobes (*epipodia*); gill single on the right side of the back, covered by the shell; mantle-margin simple or expanded, and enveloping the shell. Lingual dentition very various; central teeth often wanting, laterals single or numerous. Gizzard armed with calcarious plates. Sexes united.

The *Bullidæ* are animal-feeders; they are said to use their lateral lobes for swimming. About 150 recent species have been described by Mr. A. Adams in Sowerby's *Thesaurus Conchyliorum*. *Fossil* species date from the lower Oolites; one is found in the Aralo-Caspian formation.

BULLA, Lamarck. Bubble-shell.

Type, B. ampulla, Pl. XIV. fig. 6. *Syn*. Haminea (hydatis) Leach.

Shell oval, ventricose, convoluted, external or only partially invested by the animal; apex perforated; aperture longer than the shell, rounded at each end; lip sharp.

Animal with a large cephalic disk, truncated in front, bilobed behind, the lobes laminated beneath; eyes sub-central, immersed or wanting; lateral lobes very large, reflected on the sides of the shell, posterior lobe covering the spire; foot quadrate; gizzard furnished with 3 chiton-like plates; teeth. ?

Bulla naucum (*Atys*, Montf. *Alicula*, Ehr. *Roxania*, Leach). Pl. XIV. fig. 7; has the columella twisted, and the spire entirely concealed.

Distr. 50 sp. In all temperate and tropical seas, especially on sandy bottoms, ranging from low water to 25 or 30 fms.

Fossil, 70 sp. Ool. —. S. America, U.S. Europe.

Sub-genera ? Crypt-opthalmus (smaragdinus) Ehr. Red sea. *Shell* scarcely convolute, fragile, oval, convex, without spire or columella. *Animal* semi-cylindrical, head with short tentacular lobes, eyes small, concealed under the lateral margins of the head, mantle and lateral lobes enveloping the shell.

Phaneropthalmus, A. Adams. (Xanthonella, Gray) B. lutea, Quoy, New Guinea. *Shell* oval, convex, pointed behind, columella margin with a curved process. *Animal* long, cylindrical, head with short tentacular lobes, eyes in middle of disk, lateral lobes enveloping.

Linteria, A. Adams (Glauconella, Gray), Bulla viridis, Rang. Pl. XIV. fig. 7. *Shell* oval, widely open, showing the rudimentary internal spire.

---

* The cephalic expansion of the Bullidæ is formed by the fusion of the dorsal and oral tentacles. (*Cuvier.*) The tentacular lobes, or posterior part of the disk is supplied with nerves from the olfactory ganglia; the anterior portion of the disk receives branches from the labial nerve, which comes from the front margin of the cerebroid. (*Hancock.*)

Animal with a squarish, disk-like head, eyes sessile in the centre; mantle not investing; a posterior lobe; lateral lobes enveloping. (Pl.XIV. fig. 8, not 7).

ACERA, Müller.

Type, A. bullata, Pl. XIV. fig. 9. *Etym. Akeros*, hornless.

Shell thin, flexible, globosely-cylindrical, spire truncated, whirls channelled; aperture long, expanded and deeply sinuated in front, outer margin disunited at the suture; columella open, exposing the whirls.

Animal with a short and simple head-lobe, truncated in front and eyeless; lateral lobes nearly concealing the shell; lingual teeth hooked and serrulate, laterals about 40, narrow, claw-shaped; gizzard armed with horny teeth.

Distr. 7 sp. Greenland, Brit. Medit. Zanzibar, India, New Zealand.

A. bullata is found amongst weed, in 1—15 fms. water (*Forbes*).

CYLICHNA, Lovén.

Type, C. cylindracea, Pl. XIV. fig. 10. *Syn.* Bullina, Risso.

Shell strong, cylindrical, smooth or punctate-striate; spire minute or truncated; aperture narrow, rounded in front; columella callous, with one plait.

Animal short and broad, not investing the shell; head flattened, truncated in front, with sub-centrally immersed eyes, tentacular lobes more or less united; foot oblong, posterior and lateral lobes not much developed; gizzard armed; lingual teeth squarish, recurved and serrated, with 1 large and 5 or 6 small hooked laterals.

Distr. 20 sp. U. States, Greenland, Brit. Red Sea, Australia.

Fossil. Miocene —. Brit.

AMPHISPHYRA, Lovén.

Type. A. pellucida, Johnst. (*Amphi-sphyra*, double hammer.)

Syn. Utriculus (part) Brown. Rhizorus, Montf. Diaphana, Brown.

Shell small, thin, ovate, truncated, spire minute papillary, aperture long.

Animal entirely retractile into its shell; head wide, short, with lateral triangular tentacles; the eyes behind them minute, immersed; muzzle bi-lobed in front; foot oblong, truncated in front, notched behind; teeth 1.1.1, central quadrate, serrulate; laterals broad, hooked.

Distr. 5 sp. U. States, Norway, Brit. Borneo, Mexico.

'APLUSTRUM, Schumacher.

Type, Bulla aplustre, Pl. XIV. fig. 11. *Etym. Aplustre*, a ship's flag.

Syn. Bullina, Fér. Hydatina (physis) Schum. Bullinula (scabra) Beck.

Shell oval, ventricose, highly coloured; spire wide, depressed; aperture truncated in front; outer lip sharp.

Animal, with a very large foot, extending beyond the shell all round, and capable of enveloping it; a posterior lobe reflected on the spire; mantle not investing; tentacular lobes large, oval, ear-shaped; labial tentacles four; eyes

K

small, black, sessile at the inner bases of the tentacles; lingual teeth (*B. physis*) 13.0.13, serrated.

Distr. 10 sp.   U. States, W. Indies, Mauritius, Ceylon, China, Australia.

SCAPHANDER, Montfort.

Type, S. lignarius, Pl. XIV. fig. 12.   *Etym. Scaphe* boat, *aner*, man.

Shell oblong, convolute; spirally striated; aperture much expanded in front; spire concealed; epidermis thick; lingual teeth 1.0.1. crested.

Animal with a large oblong head, destitute of eyes; foot short and broad; lateral lobes reflected, but not enveloping the shell; gizzard of two large trigonal plates and a small narrow transverse plate (fig. 17).

Distr. 5 sp.   U. States, Norway, Brit. Medit. on sandy ground; 50 fms.
Fossil, 8 sp. Eocene —.   Brit. France.

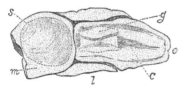

Fig. 104.   *Bullæa aperta.**

BULLÆA, Lamarck.†

Type, B. aperta, Pl. XIV. fig. 13.

Shell internal, white, translucent, oval, slightly convoluted, spire rudimentary.

Animal pale, slug-like; mantle investing the shell; head oblong; eyeless; foot broad; lateral lobes large, but not enveloping; tongue with 2 or 4 series of sickle-shaped *uncini*; gizzard with 3 longitudinal shelly plates. Egg capsules ovate, in single series on a long spiral thread; fry with a ciliated head-veil and an operculated, spiral shell, (*Lovén*).

Distr. 10 sp.   W. Indies, Greenland, Norway, Britain, Medit. Corea, Borneo.   *Fossil*, Eocene —.   France.

Sub-genus, Chelidonura, A. Adams, (Hirundella, Gray) B. hirundinaria, Quoy, Mauritius.   *Shell* concealed; outer lip produced posteriorly into a spur; columellar border inflected.   *Animal* with enveloping side lobes; mantle with two appendages behind, like the lateral processes of *Hyalaea*.

DORIDIUM, Meckel.

Etym. diminutive of *Doris*.   *Syn.* Acera, Cuv. Eidothea, Risso.

---

* From a specimen dredged at Folkstone; *o*, mouth, *c*, head, or cephalic disk, *l*, side-lobes of the foot, *m*, mantle,   The shell *s*, and gizzard *g*, are indistinctly seen through the translucent integuments.

† Gray adopts the pre-Linnean name *Philine* (Ascanius, 1762), and D'Orbigny the still older *Lobaria*, (Müller, 1741), names given to particular *species*, and not to *genera* as now understood.

Type, D. membranaceum, Meck. Medit.

Animal oblong, truncated behind, the angles produced and dilated or filiform; head ovate-oblong, retuse in front; side-lobes expanded, wing-like; mantle investing a rudimentary, membranous shell.

GASTROPTERON, Meckel.

Type, G. Meckelii, Bl. (Clio amate, Chiaje) Medit.

Animal shell-less, oval, with side-lobes developed into wing-like expansions meeting and uniting behind; cephalic disk triangular, obtuse in front, pointed behind, eyes centrally immersed; lingual teeth 5.1.5.; mantle? branchial plume exposed on the right side; reproductive orifice in front of the gill, excretory opening behind it. Lon. 1, lat. 2 inches.

Sormetus Adansonii, Bl. is described as semi-cylindrical, with sides grooved, head indistinct; shell unguiform, thin, and transparent.

Atlas (*Peronii*, Bl.) Lesueur. Head with 2 small tentacular lobes; body contracted in the middle; foot dilated circularly, and fringed at the margin.

FAMILY III. APLYSIADÆ.

Shell wanting, or rudimentary and covered by the mantle, oblong, trigonal, or slightly convoluted.

Animal slug-like, with distinct head, tentacles and eyes; foot long, drawn out into a tail behind; sides with extensive lobes, reflected over the back and shell; branchial plume concealed. Sexes united.

APLYSIA, Gmelin. Sea Hare.

Type. A. depilans, Pl. XIV. fig. 14. *Syn.* Siphonotus (geographicus) Ad.

Shell oblong, convex, flexible and translucent, with a posterior slightly incurved apex.

Animal oval, with a long neck and prominent back; head with 4 tentacles, dorsal pair ear-like with eyes at anterior lateral bases; mouth proboscidiform, with horny jaws, lingual teeth 13.1.13, hooked and serrated, about 30 rows; gizzard armed with horny spines; sides with ample lobes folding over the back, and capable of being used for swimming; gill in the middle of the back, covered by the shell, and by a lobe of the mantle which is folded posteriorly to form an excretory siphon.

Distr. 40 sp. W. Indies, Norway, Brit. Medit. Mauritius, China.

The Sea-hares are mixed feeders, living chiefly on sea-weed, but also devouring animal substances; they inhabit the laminarian zone, and oviposit amongst the weed in spring, at which time they are frequently gregarious (*Forbes*). They are perfectly harmless animals and may be handled with impunity. When molested they discharge a violet fluid from the edge of the internal surface of the mantle, which does not injure the skin, has but a faint smell, and changes to wine-red (*Goodsir*). In old times they were objects of superstitious dread, on account of their grotesque forms, and the imaginary properties of their fluid, which was held to be poisonous and to produce indelible stains.*

Fossil: one or two shells of the newest tertiary in Sicily have been doubt-fully referred to this genus.

Sub-genus, Aclesia (dolabrifera) Rang. *Shell* trapeziform. Side-lobes closely enveloping the body, leaving only a small dorsal respiratory opening, surface ornamented with filaments. W. Indies.

DOLABELLA, Lamarck.

Type. D. Rumphii, Pl. XIV. fig. 15. *Etym. Dolabella,* a small hatchet.

Shell hard, calcarious, trigonal, with a curved and callous apex.

Animal like Aplysia, with gill near posterior extremity of the body and lateral crests closely appressed, leaving only a narrow opening ; ornamented with branching filaments.

Distr. 12 sp. Medit. Mauritius, Ceylon, Society Ids. Sandwich Ids.

NOTARCHUS, Cuvier.

Type. N. Cuvieri, Bl. *Etym. Notos,* the back, *archos* vent.

Syn. Busiris (griseus) Risso, ? Bursatella (Leachii) Bl.

Animal shell-less, ornamented with filaments, sometimes dendritic, foot narrow, linear, lateral crests united, leaving only a narrow branchial slit ; gills not covered by an opercular mantle lobe.

Distr. 4 sp. Medit. Red Sea.

ICARUS, Forbes, 1843.

Type. I. Gravesii, F. *Syn. Lophocercus* (Sieboldtii) Krohn, 1847.

Shell like Bullæa ; convoluted, thin, ovate, covered with epidermis, outer lip separated at the suture, posterior angle inflected and rounded.

Animal slender, papillose ; tentacles 2, ear-shaped ; eyes sessile on sides of head; side-lobes reflected and partly covering the shell, united behind ; tail long and pointed.

LOBIGER, Krohn.

Type, L. Philippii, Pl. XIV. fig. 16. Sicily.

Shell oval, transparent, flexible, slightly convoluted ; covered with epidermis.

Animal slender, papillose, with 2 flattened, oval tentacles, and minute sessile eyes on the sides of the head ; shell exposed on the middle of the back, covering the plume-like gill ; sides with two pairs of rounded, dilated lobes, or natatory appendages, foot linear, tail long and slender.

---

* *Aplysia,* (from *a* and *pluo*) un-washable ; the *Aplysia* of the Greek Fishermen were sponges unfit for washing !

FAMILY IV. PLEUROBRANCHIDÆ.

Shell limpet-like or concealed, rarely wanting; mantle or shell covering the back of the animal; gill lateral, between the mantle-margin and foot; food vegetable, stomach extremely complicated.

PLEUROBRANCHUS, Cuvier.

Ex. P. membranaceus, Pl. XIV. fig. 17. *Etym. Pleura* side, *branchia* gill. *Syn.* Berthella (plumula) Bl. Oscanius (membr.) Gray.

Shell internal, large, oblong, flexible, slightly convex, lamellar, with a posterior, subspiral nucleus.

Animal oblong, convex; mantle covering the back aud sides, papillated, containing spicula; foot large, separated from the mantle by a groove; gill single, free at the end, placed on the right side between the mautle and foot; orifices near the base of the gill; head with 2 grooved tentacles, eyes at their outer bases; mouth armed with horny jaws and covered by a broad veil with tentacular lobes.

Distr. 20 sp. S. America, Norway, Brit. Medit. Red Sea.

Sub-genus ? Pleurobranchæa Meckel; P. Meckelii, Leve, Medit. *Syn.* Pleurobranchidium (maculatum), Quoy, S. Australia. Mantle-margin very narrow, not concealing the gill; dorsal tentacles ear-like, oral veil tentaculiform.

POSTEROBRANCHÆA, D'Orbigny.

Type, P. maculata, D'Orb. Coast of Chile.

Animal shell-less; oval, depressed, covered by a mantle broader than the foot; foot oblong, bi-lobed behind; branchial plume on the left side, projecting posteriorly; reproductive orifice in front of gill, excretory behind; proboscis covered by a broad bi-lobed veil; no dorsal tentacles.

RUNCINA, (Forbes) Hancock.

Type, R. Hancocki, Forbes. *Syn.* ? Pelta, Quatr. (not Beck.)

Animal minute, slug-like, with a distinct mantle; eyes sessile on the front part of the mantle; no tentacles; gills 3, slightly plumose, placed with the vent on the right side, at the hinder part of the back, beneath the mantle; gizzard armed; reproductive organs on the right side.

Distr. on *Confervæ* near high-water mark, Torbay.

UMBRELLA, Chemnitz. Chinese-umbrella shell.

Type, U. umbellata, Pl. XIV. fig. 18. *Syn.* Acardo, Lam. Gastroplax, Bl.

Shell limpet-like, orbicular, depressed, marked by concentric lines of growth; apex sub-central, oblique, scarcely raised; margins acute; inner surface with a central coloured and striated disk, surrounded by a continuous irregular muscular impression.

Animal with a very large tuberculated foot, deeply notched in front; mouth small, proboscidiform, retractile into the pedal notch, covered by a small lobed veil; dorsal tentacles ear-shaped, with large plicated cavities at their bases; eyes small, sessile between the tentacles; mantle not extending beyond the shell; gill forming a series of plumes beneath the shell in front and on the right side; reproductive organ in front of the dorsal tentacles; excretory orifice posterior, tubular.

Distr. 3 sp. Canaries, Medit. India, China, Sandwich Ids.

Fossil 2 sp. Eocene —. U. States, Sicily.

TYLODINA, Rafinesque.

Type, T. punctulata, Raf. (= citrina, Joannis) 3 sp. Medit. Norway.

Shell limpet-like, depressed, apex sub-central, with a minute spiral nucleus.

Animal oblong, foot truncated in front, rather pointed behind; dorsal tentacles ear-like, with eyes sessile at their inner bases; oral tentacles broad; branchial plume projecting posteriorly on the right side.

FAMILY V. PHYLLIDIADÆ.

Animal shell-less, covered by a mantle, branchial laminæ arranged in series on both sides of the body, between the foot and mantle. Sexes united.

PHYLLIDIA, Cuvier.

Type, P. pustulosa, Cuv. *Etym.* Diminutive of *Phyllon*, a leaf.

Animal oblong, covered with a coriaceous tuberculated mantle; dorsal tentacles clavate, retractile into cavities near the front of the mantle; mouth with two tentacles; foot broadly oval; gills forming a series of laminæ extending the entire length of both sides; excretory orifice in the middle line, near the posterior end of the back, or between the mantle and foot; reproductive organs on the right side; stomach simple, membranous.

Distr. 4 sp. Medit. Red Sea, India.

DIPHYLLIDIA, Cuvier.

Type, D. Brugmansii, Cuv. *Syn.* Pleurophyllidia, Chiaje. Linguella, Bl.

Animal oblong, fleshy; mantle ample; gills limited to the hinder two-thirds of the body; head with minute tentacles and a lobe-like veil; vent at the right side, behind the reproductive orifices; lingual teeth 30.1.30.

Distr. 4 sp. Norway, Brit. (*D. lineata*, Otto) Medit.

SECTION B. NUDIBRANCHIATA.

Animal destitute of a shell except in the embryo state; branchiæ always external, on the back or sides of the body; sexes united.

The Nudibranchiate sea-slugs are found on all coasts where the bottom is firm or rocky, from between tide-marks to a depth of 50 fathoms; a few species are pelagic, crawling on the stems and fronds of floating sea-weed. They have been found by Middendorff, in the Icy Sea, at Sitka, and in the sea of Ochotsk; in the tropical and southern seas they are abundant. No satisfactory account, however, has been published of any except the European, and especially the British species, which form the subject of an admirable monograph by Messrs. Alder and Hancock, in the transactions of the Ray Society. They require to be watched and drawn whilst living and active, since after immersion in spirits they lose both their form and colour. In some the back is covered with a *cloak* or mantle (?,) which contains calcarious spicula of various forms, sometimes so abundant as to form a hard shield-like crust.* The dorsal tentacles and gills pass through holes in the cloak somewhat like the " key-hole " in *Fissurella*. In others there is no trace of a mantle whatever. The eyes appear as minute black dots, immersed in the skin, behind the tentacles; they are well organized, and conspicuous in the young, but often invisible in the adult. The dorsal tentacles are laminated, like the antennæ of many insects (fig. 11, p. 23); they are never used as organs of touch, and are supplied with nerves from the olfactory ganglia. The nervous centres are often conspicuous by their bright orange colour; they are concentrated *above* the œsophagus; three pairs are larger than the rest, the *cerebroid* in front, the *branchial* behind, and the *pedal* ganglia at the sides. The cerebroid supplies nerves to the tentacles, mouth, and lips.

The *olfactory* ganglia are sessile on the front of the cerebroid (in *Doris*) or situated at the base of the tentacles (in *Æolis*). The *optic* ganglia are placed on the posterior border of the cerebroid; the auditory capsules are sessile on the cerebroid, immediately behind the eyes, they contain an agglomeration of minute *otolites* which are continually oscillating.† The *buccal* ganglia are below the œsophagus, united to the cerebroid by commissures, forming a ring; anterior to this a small ring is sometimes formed by the union of the 5th pair of nerves. The *pedal* ganglia (properly infra-œsophageal) are united laterally to the cerebroid and rarely meet below, but are united by commissures which form (together with those of the branchial centres) the 3rd ring, or *great nervous collar*. The *branchial* ganglia are united behind to the cerebroid, and sometimes blend with them; they supply the skin of the back, the rudimentary mantle, and the gills; beneath, and sessile on their front border is the single *visceral* ganglion. Besides this *excito-motory* system, (which includes the great centres, or brain, and the nerves of sensation and voluntary motion), the nudibranches possess a *sympathetic* system, consisting of innumerable minute ganglia, dotted over all the viscera, united by nerves forming plexuses, and connected in front with the buccal and branchial centres.‡

* According to Mr. Huxley, the "cloak" of the Dorids is not the equivalent of the *mantle*, but "has more relation to the *epipodium*."

† The auditory capsules of other Mollusca (excepting the Nucleobranches) are attached to the posterior side of the pedal (sub-œsophageal) ganglia.

‡ The *sympathetic system* supplies nerves to the heart and other organs which are independent of the will, and not ordinarily susceptible of pain; they are called "organic" nerves, as all the *vegetative* functions depend on them. Its existence in the

The digestive organs of the Nudibranches present two remarkable modifications : in *Doris* and *Tritonia* the liver is compact and the stomach a simple membranous sac; whilst in *Æolis* the liver is disintegrated, and its canals so large that the process of digestion must be chiefly carried on in them, and they are regarded as 'cœcal prolongations of the stomach; the cœca extend into a series of gill-like processes, arranged upon the back of the animal, which also contain part or the whole of the true liver; the gastric ramifications vary exceedingly in amount of complexity.

The vascular system and circulation of the nudibranchiate molluscs is incomplete. In *Doris* veins can be traced only in the liver and skin; the greater part of the blood from the arteries escapes into the visceral sinus and into a net-work of sinuses in the skin, from which it returns to the auricle by two lateral veins, without having circulated through the gills. The heart is contained in a *pericardium* to which is attached a small ventricle, or *portal* heart, for impelling blood to the liver ; the hepatic veins run side by side with the arteries and open into a circular vein, surrounding the vent, and supplying the gills. Only hepatic blood, therefore, circulates through the gills. In *Æolis* there are no special gills, but the gastro-hepatic papillæ are accompanied by veins which transmit blood to the auricle. The skin acts as an accessory breathing-organ ; it performs the function entirely in the *Elysiadæ*, and in the other families when by accident the branchiæ are destroyed. The water on the gills is renewed by ciliary action. The fry is provided with a transparent, nautiloid shell, closed by an operculum, and swims with a lobed head-veil fringed with cilia, like the young of most other gasteropods.—*Hancock* and *Embleton*, Phil. Trans. 1852. An. Nat. Hist. 1843.

FAMILY VI. DORIDÆ.* Sea-lemons.

Animal oblong; gills plume-like, placed in a circle on the middle of the back; tentacles two; eye-specks immersed, behind the tentacles, not always visible in the adult; lingual membrane with usually numerous lateral teeth, rachis often edentulous; stomach simple; liver compact; skin strengthened with spicula, more or less definitely arranged.

DORIS, L.

Etym. Doris, a sea-nymph. *Ex.* D. Johnstoni, Pl. XIII. fig. 1.

Animal oval, depressed ; mantle large, simple, covering the head and foot; dorsal tentacles 2, clavate or conical, lamellated, retractile within

Mollusca was first clearly demonstrated by M.M. Hancock and Embleton. The *excito-motory* system of the Mollusca corresponds with the *cerebro-spinal* system of the Vertebrata. .

* Contracted from *Dorididæ;* as the Greeks used Deucalides for *Deucaliontiades.* Ehrenberg divided the genus Doris into sections, by the number and form of the gills, characters of only specific importance.

cavities; gills surrounding the vent on the posterior part of the back, retractile into a cavity; head with an oral veil, sometimes produced into labial tentacles; mouth with a lower mandible, consisting of two horny plates, united near the front, and having 2 projecting points; lingual teeth numerous, central small, laterals similar, hooked and sometimes serrated (24-68 rows; 37-141 in a row; nidamental ribbon rather wide, forming a spiral coil of few volutions (p. 50, fig. 29.)

Sub-genus, Oncidoris (Bl. ?). D. bilamellata, Johnst. Back elevated, tuberculose; gills non-retractile; oral tentacles fused into a veil; buccal mass with a gizzard-like appendage; lingual teeth 2 in each row. (A. and H.)

D. scutigera (*Villiersia*) D'Orb. Rochelle; has the mantle more than usually strengthened with calcarious spicula.

The Dorids vary in length from 3 lines to more than 3 inches; they feed on zoophytes and sponges, and are most plentiful on rocky coasts, near low-water, but range as low as 25 fms. They occur in all seas, from Norway to the Pacific.

GONIODORIS, Forbes.

Etym. Gonia, an angle. *Type,* G. nodosa, Pl. XIII. fig. 2.

Animal oblong; tentacles clavate, laminated, non-retractile; mantle small, simple, exposing the head and foot. Spawn coiled irregularly.

Distr. Norway, Brit. (2 sp.) Medit. China. Between tide-marks.

TRIOPA, Johnston.

Type, T. clavigera, Pl. XIII. fig, 3. *Syn.* Psiloceros, Menke.

Animal oblong; tentacles clavate, retractile within sheaths; mantle margined with filaments; gills few, pinnate, around or in front of the dorsal vent. (A. and H.) Lingual teeth 8.1.8, or 8.0.8.

Distr. Norway, Brit. Low-water — 20 fms.

ÆGIRUS, Lovén.

Type, A. punctilucens, Pl. XIII. fig. 4. *Etym.* ? *Aix* (*aigos*) a goat.

Animal oblong or elongated, covered with very large tubercles; no distinct mantle; tentacles linear, retractile within prominent lobed sheaths; gills dendritic, placed around the dorsal vent. (A. and H.) Lingual teeth 17.0.17.

Distr. Norway, Brit. (2 sp.) France. Litoral zone.

THECACERA, Fleming.

Etym. Theke a sheath, *ceras* a horn. *Type,* T. pennigerum, Mont.

Animal oblong, smooth; tentacles clavate, laminated, retractile within sheaths; head with a simple frontal veil; gills pinnate, placed round the dorsal vent, and surrounded by a row of tubercles. (A. and H.)

Distr. Brit. 2 sp. Lon. ¼—½ inch. Found at low-water.

POLYCERA, Cuvier.

Etym. Polycera, many horns. *Type,* P. quadrilineata, Pl. XIII. fig. 5.

K 3

Animal oblong or elongated; tentacles laminated, non-retractile, sheath-less; head-veil bordered with tubercles or tentacular processes; gills with 2 or more lateral appendages. (A. and H.)

Distr. Norway, 5 sp. Brit. Red Sea. Within tide-marks and in deep water on corallines. The spawn is strap-shaped, and coiled on stones, in July and August. P. ocellata (*Plocamophorus,* Rüppell) has the cephalic tentacles branched.

IDÀLIA, Leuckart.

Etym. Idalia, Venus, from Mt. Idalium in Cyprus.

Syn. Euplocamus, Phil. Peplidium (Maderæ) Lowe.

' *Ex.* I. aspersa, Pl. XIII. fig. 6. Coralline zone.

Animal broadly oblong, nearly smooth, tentacles clavate or linear, with filaments at their base; head slightly lobed at the sides; mantle very small, margined with filaments; lingual teeth 2.0.2.

Distr. Norway, Brit. (4 sp.) Medit. Madeira.

ANCULA, Lovén.

Syn. Miranda, A. and H. *Type,* A. cristata, Alder.

Animal slender, elongated; mantle entirely adnate, ornamented with simple filaments; tentacles clavate, laminated; with filiform appendages at their base; labial veil produced on each side.

Distr. Norway, Brit. Lon. ½ inch.

CERATOSOMA (Gray), A. Adams.

Etym. Ceratois, horned, *soma,* body. *Type,* C. cornigerum, Ad.

Animal oblong, narrow, with two large and prominent horn-like processes on the posterior part of the back, behind the gills; gills 5, bipinnate; dorsal tentacles clavate, laminated, rising from rounded tubercles, non-retractile; head with short lateral processes: foot narrow.

Distr. Sooloo sea. (A. Adams.)

FAMILY VII. TRITONIADÆ.

Animal with laminated, plumose, or papillose gills, arranged along the sides of the back; tentacles retractile into sheaths; lingual membrane with 1 central and numerous lateral teeth; orifices on the right side.

TRITONIA, Cuvier.

Ex. T. plebeia, Pl. XIII. fig. 7.

Animal elongated; tentacles with branched filaments; veil tuberculated or digitated; gills in single series on a ridge down each side of the back; mouth armed with horny jaws; stomach simple; liver compact.

Distr. Norway, Brit. Under stones at low-water, — ∴5 fm. F. Hombergii, Cuv. found on the scallop-banks, attains a length exceeding 6 inches.

SCYLLÆA, L.

Type, S. pelagica, Pl. XIII. fig. 8.   *Etym. Scyllaea,* a sea-nymph.

Animal elongated, compressed; foot long, narrow and channelled, adapted for clasping sea-weed; back with 2 pairs of wing-like lateral lobes, bearing small tufted branchiæ on their inner surfaces; tentacles dorsal, slender, with lamellated tips, retractile into long sheaths; lingual teeth 24.1.24, denticulated; gizzard armed with horny, knife-like plates; orifices on the right side.

Distr. Atlantic, S. Brit. Medit.   On floating sea-weed.

Nerea (punctata) Lesson, New Guinea; 10 lines long, with ear-shaped tentacles, and 3 pairs of dorsal lobes.

TETHYS, L.

Etym. Tethys, the sea (personified.)   *Syn.* Fimbria, Bohadsch.

Type, T. fimbriata, L. Pl. XIII. fig. 9.

Animal elliptical, depressed; head covered by a broadly expanded, fringed disk, with 2 conical tentacles, retractile into foliaceous sheaths; gills slightly branched, a single row down each side of the back; reproductive orifices behind first gills, vent on right side, behind second gill; stomach simple.

Distr. 1 sp.   Medit.   Attains a foot in length, and feeds on other molluscs and crustaceans.   (*Cuvier.*)

? BORNELLA (Gray), A. Adams.

Type, A. Adamsii, Gray.   *Lon.* 4 inches.

Animal elongated; dorsal tentacles retractile into branched sheaths; head with stellate processes; back with two rows of cylindrical, branched, gastric processes, to which small dendritic gills are attached;* foot very narrow.

Distr. 2 sp.   Straits of Sunda, on floating weed; Borneo.

? DENDRONOTUS, A. and H.†

Etym. Dendron, a tree, *notos,* the back.

Type, D. arborescens, Pl. XIII. fig. 10.

Animal elongated; tentacles laminated; front of the head with branched appendages; gills arborescent, in single series down each side of the back; foot narrow; lingual teeth 10.1.10; stomach and liver ramified.

Distr. Icy sea; Norway, Brit.   On sea-weed and corallines; low-water —coralline zone.

? DOTO, Oken.

Etym. Doto, a sea-nymph.   *Ex.* D. coronata, Pl. XIII. fig. 11.

---

* This observation deserves further enquiry.

† This and the following genera are placed by Alder and Hancock in the family *Æolidæ;* they have a ramified stomach, but their external (*zoological*) characters agree better with *Tritonia* than *Æolis.*

Animal slender, elongated; tentacles linear, retractile into trumpet-shaped sheaths; veil small, simple; gills ovate, muricated, in single series down each side of the back; lingual membrane slender, with above 100 recurved, denticulated teeth, in single series; foot very narrow.

The stomach is ramified, and the liver is entirely contained in the dorsal processes, which fall off readily when the animal is handled, and are soon renewed.

Distr. Norway, Brit. On corallines in deep water — 50 fms.

? MELIBŒA, Rang.

Type, M. rosea, Rang; on floating weed, off the Cape.

Animal elongated, with a narrow, channelled foot and long slender tail; sides of the back with 6 pairs of tuberculated lobes, easily deciduous; tentacles cylindrical, retractile into long trumpet-shaped sheaths; head covered by a lobe-like veil; sexual orifices behind right tentacle, excretory behind first gill on the right side.

? LOMANOTUS, Verany.

Ex. L. marmoratus, Pl. XIII. fig. 12. *Syn.* Eumenis, A. and H.

Animal elongated, smooth; head covered with a veil; tentacles clavate, laminated, retractile into sheaths; gills filamentose, arranged along the sides of the back, on the wavy margins of the mantle; foot narrow, with tentacular processes in front; stomach ramified.

Distr. Brit. Medit. On corallines.

FAMILY VIII. ÆOLIDÆ.

Animal with papillose gills, arranged along the sides of the back; tentacles sheath-less, non-retractile; lingual teeth 0.1.0.; ramifications of the stomach and liver extending into the dorsal papillæ; excretory orifices on the right side; skin smooth, without spicula; no distinct mantle.

ÆOLIS, Cuvier.

Syn. Psiloceros, Menke. Eubranchus, Forbes. Amphorina, Quatref.

Type, Æ. papillosa, L. *Etym.* Æolis, daughter of Æolus.

Animal ovate; dorsal tentacles smooth, oval, slender; gills simple, cylindrical, numerous, depressed and imbricated; mouth with a horny upper jaw, consisting of two lateral plates, united above by a ligament; foot narrow; tongue with a single series of curved, pectinated teeth; spawn of numerous waved coils.

Sub-genera. Flabellina, Cuv. (Phyllodesmium, Ehr.) Body slender; dorsal tentacles laminated, buccal long; papillæ clustered; spawn multispiral. *Ex.* E. coronata, Pl. XIII. fig. 13. (also fig. 11, p. 23.)

Cavolina, Brug. (Montagua, Flem.) C. peregrina. Body lanceolate; tentacles smooth or wrinkled; papillæ in transverse, rather distant rows; spawn of 1 or 2 coils.

Tergipes, Cuv. T. lacinulata. Body linear; tentacles smooth; papillæ in a single row on each side; spawn kidney-shaped.

Distr. Norway, Brit. (33 sp.) U. States, Medit. S. Atlantic, Pacific. Found amongst rocks, at low-water; they are active animals, moving their tentacles continually, and extending and contracting their papillæ; they swim readily at the surface, inverted. They feed chiefly on sertularian zoophytes, and if kept fasting will devour each other; when irritated they discharge a milky fluid from their papillæ, which are very liable to fall off.

Glaucus, Forster.

Etym. *Glaucus*, a sea-deity. *Syn.* Laniogerus, Bl. Pleuropus, Raf.

Ex. G. Atlanticus, Pl. XIII. fig. 14.

Animal elongated, slender : foot linear, channelled; tentacles 4, conical; jaws horny; teeth in single series, arched and pectinated; gills slender, cylindrical, supported on 3 pairs of lateral lobes; stomach giving off large cœca to the tail and side lobes; liver contained in the branchial papillæ; sexual orifice beneath first dextral gill, vent behind second gill; spawn in a close spiral coil.

Distr. 6 sp. Atlantic, Pacific. Found on floating sea-weed; devours small sea-jellies, *Porpitæ* and *Velellæ*. (Bennet.)

Fiona, Alder and Hancock.

Type, F. nobilis, A. and H. *Syn.* Oithona, A. and H. (not Baird).

Animal elongated; oral and dorsal tentacles linear; mouth armed with horny jaws; gills papillary, clothing irregularly a sub-pallial expansion on the sides of the back, each with a membranous fringe running down its inner side.

Distr. Falmouth. Under stones at low-water. (Dr. Cocks.)

Embletonia, A. and H.

Etym. Dedicated to Dr. Embleton, of Newcastle.

Syn. Pterochilus, A. and H. ? Clœlia (formosa) Loven.

Type, E. pulchra, Pl. XIII. fig. 15.

Animal slender; tentacles 2, simple; head produced into a flat lobe on each side; papillæ simple, subcylindrical, in a single row down each side of the back.

Distr. Scotland (2 sp.) In the litoral and laminarian zones.

Calliopæa (bellula) D'Orb. Brest; has 2 rows of papillæ down each side of the back; cephalic lobes subulate; vent dextral. Lon. 3 lines.

Proctonotus, A. and H.

Type, P. mucroniferus, Pl. XIII. fig. 16. Dublin, shallow water.

Syn. Venilia, A. and H. Zephrina, Quatref.

Animal oblong, depressed, pointed behind; dorsal tentacles 2, linear, simple, with eyes at their base, behind; oral tentacles short; head covered by a small semilunar veil; mouth with horny jaws; gills papillose, on ridges down the sides of the back, and round the head in front; vent dorsal.

ANTIOPA, A. and H.

Type, A. splendida, A. and H.   *Syn.* Janus, Verany.

Animal ovate-oblong, pointed behind; dorsal tentacles lamellated, united at the base by an arched crest; head with a small veil and two labial tenta-cles; gills ovate, placed along the lateral ridges of the back and continuous above the head; vent central, posterior, sexual orifice at the right side; lingual teeth numerous. ?

Distr. Brit. Medit.

HERMÆA, Lovén.

Type, H. bifida, Pl. XIII. fig. 17.   Norway, Brit.

Animal elongated, tentacles folded longitudinally; gills numerous, papil-lose, arranged down the sides of the back; sexual orifice below right tenta-cles; vent dorsal, or sub-lateral, anterior.

ALDERIA, Allman.

Etym. Named after Joshua Alder, one of the authors of the Monograph on the British Nudibranchiate Mollusca.

Type, A. modesta, Pl. XIII. fig. 18.   Norway, S. Ireland and S. Wales.

Animal oblong, without tentacles; head lobed at the sides; gills papil-lose, arranged down the sides of the back; vent dorsal, posterior.

? *Stiliger* (ornatus) Ehrenberg; Red Sea.   Vent dorsal, anterior.

FAMILY IX.   PHYLLIRHOIDÆ.

Animal pelagic, foot-less (*apodal*), compressed, swimming freely with a fin-like tail; tentacles 2, dorsal; no branchiæ; lingual teeth in a single series; stomach furnished with elongated cœca; orifices on the right side; sexes united.

PHYLLIRHOE, Péron and Lesueur.

Etym. Phyllon, a leaf, *rhoë,* the wave.   *Syn.* Eurydice, Esch.

Type, P. bucephala, Péron.   *Distr.* 6 sp. Medit. Moluccas, Pacific.

Animal translucent, fusiform, with a lobed tail; muzzle round, truncated; jaws horny; lingual teeth 3.0.3.; tentacles long and slender, with short sheaths; intromittent organ long, bifid.

FAMILY X.   ELYSIADÆ.

Animal shell-less, limaciform, with no distinct mantle or breathing organ; respiration performed by the ciliated surface of the body; mouth armed with a single series of lingual teeth; stomach central, vent median, sub-central; hepatic organs branched, extending the length of the body and opening into the sides of the stomach; sexes united; male and ovarian orifices below the right eye; female orifice in the middle of the right side; heart with an auricle behind, and traces of an arterial and venous system, eyes sessile on the sides of the head, tentacles simple or obsolete.*

ELYSIA, Risso.

Type, E. viridis, Pl. XIII. fig. 19. *Syn.* Actæon, Oken.

Animal elliptical, depressed, with wing-like lateral expansions; tentacles simple, with sessile eyes behind them; foot narrow.

Distr. Brit. Medit. On *Zostera* and sea-weed, in the laminarian zone. *Placo-branchus* (ocellatus, Rang.) Hasselt, Java; described as 2 inches long, with four small tentacles; the lateral expansions much developed and meeting behind, the upper surface longitudinally plaited, and forming, when the side-lobes are rolled together, a sort of branchial chamber.

ACTEONIA, Quatrefages.

Ex. A. corrugata, Pl. XIII. fig. 20. British channel.

Animal minute, leach-like; head obtuse, with lateral crests proceeding from two short conical tentacles, behind which are the eyes.

CENIA, Alder and Hancock.

Type, C. Cocksii, Pl. XIII: fig. 21. *Etym. Cenia*, Falmouth. *Syn.* ? Fucola (rubra) (Quoy).

Animal limaciform, back elevated, head slightly angulated, bearing two linear dorsal tentacles, with eyes at their outer bases behind.

LIMAPONTIA, Johnston.

Type, L. nigra, Pl. XIII. fig. 22. *Syn.* Chalidis, Qu. Pontolimax, Cr.

Animal minute, leach-like; head truncated in front, with arched lateral ridges on which are the eyes; foot linear.

Distr. Norway, England and France, between half-tide and high-water, feeding on *Confervæ*, in the spring and summer; spawn in small pear-shaped masses, each with 50-150 eggs; fry with a transparent nautiloid shell, closed by an operculum.

ORDER IV. NUCLEOBRANCHIATA, Bl.†

The present order consists entirely of pelagic animals, which swim at the surface, instead of creeping on the bed of the sea. Their rank and affi-

---

* Order *Dermi-branchiata*, Quatref. (*Pelli-branchiata*, A. and H.) M. Quatrefages erroneously described the *Elysiadæ* as wanting both heart and blood vessels, like the Ascidian zoophytes; with them he associated the family *Æolidæ*, which he described as having a heart and arteries, but no veins, their office being performed by lacunæ of the areolar tissue. In both families the product of digestion (*chyle*) was supposed to be aërated in the gastric ramifications, by the direct influence of the surrounding water. To this group, which has been since abandoned, he applied the name *Phlebenterata*, (*phlebs*, a vein, *entera*, the intestines).

† So called because the respiratory and digestive organs form a sort of *nucleus* on the posterior part of the back. See fig. 105, *s. b.*, and Pl. XIV. fig. 24.

nities entitle them to the first place in the class; but their extremely aberrant form, and unusual mode of progression, have caused us to postpone
their description till after that of the ordinary and typical *gasteropoda*.

There are two families of nucleobranchiate mollusks; the *firolas* and
carinarias, with large bodies and small or no shells, and the *Atlantas*,
which can retire into their shells and close them with an operculum. Both
animal and shell are symmetrical, or nearly so; the nucleus of the shell is
minute and dextrally spiral.

The *nucleobranches* swim rapidly by the vigorous movements of their
fin-like tails, or by a fan-shaped ventral fin; and adhere to sea-weed by a
small sucker placed on the margin of the latter. Mr. Huxley has shown that
these organs represent the three essential parts of the foot in the most highly
developed sea-snails. The *sucker* represents the central part of the foot, or
creeping disk (*meso-podium*) of the snail and whelk; the ventral fin is
homologous with the anterior division of the foot, (*pro-podium*) which is very
distinct in *Natica* (p. 123), and in *Harpa* and *Oliva*; but is only marked by
a groove in *Paludina* and *Dolium* (fig. 71.) The terminal fin (or tail of
Carinaria) which carries the operculum of *Atlanta*, is the equivalent of the
operculigerous lobe (*meta-podium*) of the ordinary gasteropods, such as
Strombus (fig. 69).

The abdomen, or visceral mass, is small, whilst the anterior part of the
body (or *cephalo-thorax*, M. Edw.) is enormously developed. The proboscis
is large and cylindrical, and the tongue armed with recurved spines. The
alimentary canal of *Firola* is bent up at a right angle posteriorly on the
dorsal side; in *Atlanta* it is recurved, and ends in the branchial chamber. The heart is *proso-branchiate*, although in *Firola* the auricle is rather
above than in front of the ventricle, owing to the small amount of the dorsal
flexure.

The nucleobranches, and especially those without shells, "afford the most
complete ocular demonstration of the truth of MILNE EDWARDS' views with
regard to the nature of the circulation in the *mollusca*. · Their transparency
allows the blood-corpuscles to be seen floating in the general cavity of the
body—between the viscera and the outer integument—and drifting backwards to the heart; having reached the wall of the auricle they make their
way through its meshes as they best can, sometimes getting entangled
therein, if the force of the heart has become feeble. From the auricle they
may be followed to the ventricle, and thence to the aorta and pedal artery,
through whose open ends they pour into the tissues of the head and fin."
(*Huxley*.)

Such delicate and transparent creatures would hardly seem to need
any special breathing-organ, and in fact it is present or absent in species
of the same genus, and even in specimens of the same species. *Carinaria*
has fully-formed branchiæ; in *Atlanta* they are sometimes distinct, and

wanting in others; in *Firóloides* they are only indicated by a ciliated sub-spiral band. The larvae are furnished with a shell, and with ciliated *vela.* (Gegenbaur.)

The nucleobranches are *diœcious;* some individuals (of *Firola*) have a leaf-like appendage, others a long slender egg-tube depending from the oviduct, and regularly annulated.* The larvæ are furnished with a shell, and with ciliated *vela.* (Gegenbaur.)

The nervous system is remarkable for the wide separation of the centres. The buccal ganglia are situated considerably in front of the cephalic, and the *pedal ganglia* are far behind, so that the commissures which unite them are nearly parallel with the œsophagus. The *branchial ganglia* are at the posterior extremity of the body, as in the bivalves. The eyes are hour-glass shaped, and very perfectly organized; the auditory vesicles are placed behind, and connected with the cephalic ganglia, they each contain a round otolite, which sometimes seems to oscillate. (*Huxley.*)

FAMILY I. Firolidæ.

Animal elongated, cylindrical, translucent, furnished with a ventral fin, and a tail fin used in swimming; gills exposed on the posterior part of the back, or covered by a small hyaline shell. Mouth with a circular lip; lingual membrane with few rows of teeth : central teeth transversely elongated, with 3 recurved cusps; laterals 3 on each side, the first a transverse plate with a hooked apex, 2 and 3 sickle-shaped.†

Firola, Peron and Lesueur.

Type, F. Coronata, Forsk. Medit. *Syn.* Pterotrachæa, Forsk.

Animal fusiform, elongated, with a long, slender, proboscidiform head; fin narrowed at the base, furnished with a small sucker; tail elongated, keeled, sometimes pinnate; nucleus prominent; branchial processes numerous, conical, slender; tentacles 4, short and conical; eyes black and distinct, protected by a rudimentary eyelid; lingual ribbon oblong. The female *firolæ* have a long moniliform oviduct. *Anops Peronii,* D'Orb. described and figured as having no head (!) was probably a mutilated *Firola.* "Such specimens are very common, and seem just as lively as the rest." (Huxley.)

Distr. 8 sp. Atlantic, Medit. Pacific.

Sub-genus, Firoloides, Lesueur. (*Cerophora,* D'Orb.) F. Desmarestii, Les. Body cylindrical; head tapering, furnished with two slender tentacles; nucleus at the posterior extremity of the body, with or without small branchial filaments; egg-tube regularly annulated; tail fin small and slender, ventral fin without a sucker. *Distr.* 6 sp. Atlantic.

---

* We can only call to mind one other example of a segmented organ in the *mollusca;* viz. the penniform styles of *Teredo bipalmulata.*.

† The genus *Sagitta,* Q. and G. sometimes referred to this family, is an articulate animal. (Huxley.)

CARINÁRIA, Lamarck.

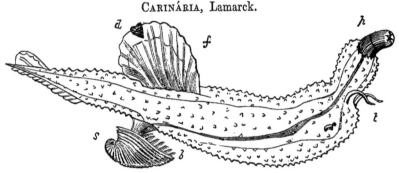

Fig. 105.*

Etym. Carina, a keel (or keeled vessel.)

Type. C. cymbium, L. fig. 105, Pl. XIV. fig. 19.

Shell hyaline, symmetrical, limpet-shaped, with a posterior sub-spiral apex and a fimbriated dorsal keel; nucleus minute, dextrally spiral.

Animal large, translucent, granulated; head thick, cylindrical; lingual ribbon triangular, teeth increasing rapidly in size, from the front backwards; tentacles long and slender, eyes near their base: ventral fin rounded, broadly attached, with a small marginal sucker; tail large, laterally compressed; nucleus pedunculated, covered by the shell, gills numerous, pinnate, projecting from beneath the shell.

Distr. 5 sp. Medit. and warmer parts of the Atlantic and Indian Oceans. They feed on small *Acalephæ*, and probably on the *pteropoda*; Mr. Wilton found in the stomach of a Carinaria two fragments of quartz rock, weighing together nearly 3 gr.

Fossil, 1 sp. Miocene. Turin.

CARDIÁPODA, D'Orbigny.

Ex. C. placenta, Pl. XIV. fig. 20.

Etym. Cardia, heart, *pous*, foot. *Syn.* Carinaroides, Eyd. and Souleyet. *Animal* like *Carinaria*. *Distr.* 5 sp. Atlantic.

Shell minute, cartilaginous; peristome expanded and bi-lobed in front, enveloping the spire behind.

FAMILY II. ATLANTIDÆ.

Animal furnished with a well-developed shell, into which it can retire; gills contained in a dorsal mantle-cavity; lingual teeth similar to *Carinaria*. *Shell* symmetrical, discoidal, sometimes closed by an operculum.

ATLANTA, Lesueur.

Type, A. Peronii, Pl. XIV. fig. 21-23. *Syn. Steira*, Esch.

Shell minute, glassy, compressed and prominently keeled; nucleus dex-

* Fig. 105. *p*. proboscis; *t*, tentacles; *b*, branchiæ; *s*, shell; *f*, foot; *d*, disk.

trally spiral; aperture narrow, deeply notched at the keel; operculum ovate, pointed, lamellar, with a minute, apical, dextrally spiral nucleus.

Animal 3-lobed; head large, sub-cylindrical; tentacles conical, with conspicuous eyes behind them; ventral fin flattened, fan-shaped, furnished with a small fringed sucker; tail pointed, operculigerous.

Distr. 15 sp. Warmer parts of the Atlantic, Canary Ids.

Sub-genus. Oxygyrus, Benson. *Syn.* Ladas, Cantraine; Helico-phlegma, D'Orb. *O.* Keraudrenii, Pl. XIII. figs. 24, 25. Shell milky, narrowly umbilicated on both sides; nucleus not visible; back rounded, keeled only near the aperture; body whirl, near the aperture, and keel cartilaginous; no apertural slit; operculum trigonal, lamellar. 2 sp. Atlantic. Medit.

The *Atlanta* was discovered by Lamanon, who supposed it to be the living analogue of the Ammonite. The operculum of *Oxygyrus* (Pl. XIII. fig. 25) is singularly like the *Trigonellites* (p. 80); that of Atlanta (fig. 22) is the only example of a *dextral* operculum to a dextral shell (p. 102).

PORCÉLLIA, Lévéille.

Ex. P. Puzosi, Pl. XIV. fig. 29.

Shell discoidal, many whirled; whirls keeled or coronated; nucleus spiral; aperture with a narrow dorsal slit.

Fossil, 10 sp. Devonian — Trias. Brit. Belgium

BELLÉROPHON, Monfort.

Ex. B. bi-carinatus, Lév. Pl. XIV. fig. 27. *Syn.* Euphemus, M'Coy.

Shell symmetrically convoluted, globular, or discoidal, strong, few-whirled; whirls often sculptured; dorsally keeled; aperture sinuated and deeply notched on the dorsal side.

Fossil, 70 sp. L. Silurian — Carb. N. America, Europe, Australia. The name *Bucania* was given by Hall to the species with exposed whirls; in B. expansus, Pl. XIV. fig. 28, the aperture of the adult shell is much expanded, and the dorsal slit filled up. (*Salter.*)

Bellerophina, D'Orb (not Forbes) is founded on the *Nautilus minutus.* Sby. Pl. XIV. fig. 26, a small globular shell, spirally striated, and devoid of *septa.* It is found in the *gault* of England and France.

CYRTOLITES, Conrad.

Type, C. ornatus, Pl. XIV. fig. 30.

Etym. Kurtos, curved, *lithos,* stone.

Shell thin, symmetrical, horn-shaped or discoidal, with whirls more or less separate, keeled and sculptured.

Fossil, 13 sp. L. Silurian — Carb, N. America, Europe.

? *Ecculiomphalus* (Bucklandi) Portlock, Pl. XIV. fig. 31. L. Silurian, Brit. U. States. Shell thin, curved, or discoidal with few widely separate whirls, slightly unsymmetrical, keeled.

Fig. 106.   *Maclurea Logani, (Salter) L. Silurian. Canada.*

? MACLUREA, Lesueur.

Named after Wm. Maclure, the first American geologist.

Shell discoidal, few whirled, longitudinally grooved at the back, and slightly rugose with lines of growth; dextral side convex, deeply and narrowly perforated; left side flat, exposing the inner whirls; operculum sinistrally sub-spiral, solid, with two internal projections (*t t*) one of them beneath the nucleus, very thick and rugose.

Fossil, 5 sp. L. Silurian. N. America; Scotland (Ayrshire, M'Coy).

This singular shell abounds in the "Chazy" limestone of the U. States and Canada; sections of it may be seen even in the pavement of New York; but specimens are very difficult to obtain. We are indebted to W. E. Logan, Esq., Geological Surveyor of Canada, for the opportunity of examining a large series of silicified specimens, and of figuring a perfect shell, with its operculum *in situ*. It has more the aspect of a bivalve, such as *Requienia Lonsdalii* (Pl. XVIII. fig. 12) than of a spiral univalve, but has no hinge. Many of the specimens are overgrown with a zoophyte, generally on the convex side only, rarely on both sides.

The Maclurea has been described as *sinistral ;* but its operculum is that of a dextral shell; so that the spire must be regarded as deeply sunk and the umbilicus expanded, as in certain species of *Planorbis:* unless it is a case conversely parallel to *Atlanta,* in which both shell and operculum have dextral nuclei. The affinities of *Maclurea* can only be determined by careful examination and comparison with allied, but less abnormal forms, associated with it in the oldest fossiliferous rocks; its relation to *Euomphalus* (p. 145) is not supported by the evidence of Mr. Logan's specimens.

CLASS III. PTEROPODA.

This little group consists of animals whose entire life is passed in the open sea, far away from any shelter, save what is afforded by the floating gulf-weed, and whose organization is specially adapted to that sphere of existence. In appearance and habits they strikingly resemble the fry of the ordinary sea-snails, swimming like them by the vigorous flapping of a pair of fins. To the naturalist ashore they are almost unknown ; but the voyager on the great ocean meets with them where there is little else to arrest his attention, and marvels at their delicate forms, and almost incredible numbers. They swarm in the tropics, and no less in arctic seas, where by their myriads the water is discoloured for leagues (*Scoresby*). They are seen swimming at the surface in the heat of the day, as well as in the cool of the evening. Some of the larger kinds have prehensile tentacles, and their mouths armed with lingual teeth, so that, fragile as they are, they probably feed upon still smaller and feebler creatures, (*e. g. entomostraca*). In high latitudes they are the principal food of the whale, and of many sea-birds. Their shells are rarely drifted on shore, but abound in the fine sediment brought up by the dredge from great depths. A few species occur in the tertiary strata of England and the continent; in the older rocks they are unknown, unless some comparatively gigantic forms (*conularia* and *theca*) have been rightly referred to this order.

In structure, the *Pteropoda* are most nearly related to the marine univalves, but much inferior to them. Their nervous *ganglia* are concentrated into a mass *below* the œsophagus; they have auditory vesicles, containing otolites; and are sensible of light and heat and probably of odours, although at most they possess very imperfect eyes and tentacles. The true foot is small or obsolete; in *cleodora* it is combined with the fins, but in Clio it is sufficiently distinct, and consists of two elements; in *Spirialis* the posterior portion of the foot supports an *operculum*. The fins are developed from the sides of the mouth or neck, and are the equivalents of the side-lappets (*epipodia*) of the sea-snails. The mouth of *Pneumodermon* is furnished with two tentacles supporting miniature suckers; these organs have been compared with the dorsal arms of the cuttle-fishes, but it is doubtful whether their nature is the same.* A more certain point of resemblance is the ventral flexure of the alimentary canal, which terminates on the under surface, near the right side of the neck. The pteropods have a muscular gizzard, armed with gastric teeth; a liver; a pyloric cœcum; and a contractile renal organ opening into the cavity of the mantle. The heart consists of an auricle and a ventricle, and is essentially *opistho-branchiate*, although sometimes affected by the general flexure of the body. The venous system is extremely incomplete. The respiratory organ, which is little more than a *ciliated surface*, is either situated at the extremity of the body and unprotected by a mantle, or included in a branchial chamber with an opening in front. The shell, when present, is symmetrical, glassy, and translucent, consisting of a dorsal and a ventral plate united, with an anterior opening for the head, lateral slits for long filiform processes of the mantle, and terminated behind in one or three points; in other cases it is conical, or spirally coiled and closed by a spiral operculum. The sexes are united, and the orifices situated on the right side of the neck. According to Vogt, the embryo Pteropod has deciduous *vela*,

* The figures of Eydoux and Souleyet represent them as being supplied with nerves from the *cephalic ganglia;* whereas the arms of the cuttle-fish, and all other parts or modifications of the foot, in the *mollusca*, derive their nerves from the *pedal ganglia* (Huxley).

like the sea-snails, before the proper locomotive organs are developed (*Huxley*).

From this it would appear that while the Pteropoda present some analogical resemblances to the *Cephalopoda*, and permanently *represent* the larval stage of the sea-snails, they are developed on a type sufficiently peculiar to entitle them to rank as a distinct group; not indeed of equal value with the *Gasteropoda*, but with one of its orders.

This group, the lowest of the univalve or encephalous orders, makes no approach towards the bivalves or *acephala*. Forskahl and Lamarck indeed compared *Hyalæa* with *Terebratula*; but they made the ventral plate of one answer to the dorsal valve of the other, and the anterior cephalic orifice of the pteropodous shell, correspond with the *posterior*, byssal foramen of the bivalve!

SECTION A. Thecosomata, Bl.*

Animal, furnished with an external shell; head indistinct: foot and tentacles rudimentary, combined with the fins; mouth situated in a cavity formed by the union of the locomotive organs; respiratory organ contained within a mantle-cavity.

FAMILY I. Hyaleidæ.

Shell straight or curved, globular or needle-shaped, symmetrical.

Animal with two large fins, attached by a columellar muscle passing from the apex of the shell to the base of the fins; body inclosed in a mantle; gill represented by a transversely plaited and ciliated surface, within the mantle cavity, on the *ventral* side; lingual teeth (of *Hyalea*) 1.1.1, each with a strong recurved hook.

Hyálea, Lamarck.

Etym. Hyalĕos, glassy. *Syn.* Cavolina, Gioeni not Brug.

Type, H. tridentata, fig. 107. Pl. XIV. fig. 32.

Shell globular, translucent; dorsal plate rather flat, produced into a hood; aperture contracted, with a slit on each side; posterior extremity tridentate. In H. trispinosa (*Diacria*, Gray) the lateral slits open into the cervical aperture.

Animal, with long appendages to the mantle, passing through the lateral slits of the shell; tentacles indistinct; fins united by a semicircular ventral lobe, the equivalent of the posterior element of the foot.

Distr. 19. sp. Atlantic, Medit. Indian Ocean.

Fossil, 5 sp. Miocene —. Sicily, Turin, Dax.

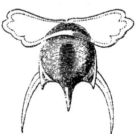

Fig. 107. H. tridentata.

* *Theke* a case, *soma* a body; several of the genera have no shells.

CLEODORA, Peron and Lesueur.

Syn. Clio, L. (part) not Müller. Balantium, Leach MS.

Type, C. pyramidata, Pl. XIV. fig. 33.

Shell pyramidal, 3 sided, striated transversely; ventral side flat, dorsal keeled; aperture simple, triangular, with the angles produced; apex acute.

Animal with rudimentary eyes; tentacles obsolete; mantle-margin with a siphonal (?) process; fins ample, united ventrally by a rounded lobe; lingual teeth 1.1.1. The transverse bars of the gill, the heart, and other organs are visible through the pellucid shell. In C. curvata and pellucida (*Pleuropus,* Esch.) the mantle is furnished with two long filaments on each side.

Distr. 12 sp. Atlantic, Medit. Indian Ocean, Pacific, C. Horn.

Fossil. Miocene —. Brit. (C. *infundibulum,* Crag.)

Sub-genus. Creseis, Rang. (Styliola, Lesueur). C. aciculata, Pl. XIV. fig. 34. Slender, conical, pointed, straight or curved. Fins rather narrow, truncate, with small tentacles projecting from their dorsal edges, and rudiments of the *mesopodium* on their surface; mantle-margin with a spiral process on the left side. M. Rang states that he has seen these pteropods clustering round floating seaweed. *Distr.* 5 sp. (like *Cleodora.*)

CUVIERIA, Rang.*

Dedicated to Baron Cuvier. *Type,* C. columnella, Rang, Pl. XIV. fig. 35.

Shell cylindrical, transparent; aperture simple, transversely ovate; apex acute in the young, afterwards partitioned off, and usually deciduous.

Animal with simple narrow fins, united ventrally by two small lobes; lingual teeth 1.1.1.

Distr. 4 sp. Atlantic, India, Australia.

Fossil 1 sp. (C. Astesana, Rang.) *Pliocene,* Turin.

Sub-genus, Vaginella, Daud. V. depressa, Pl. XIV. fig. 36. *Shell* oblong, with a pointed apex; aperture contracted, transverse. *Fossil,* 1 sp. *Miocene.* Bordeaux, Turin.

THECA, Morris. 1845.

Type, T. lanceolata. *Syn.* Creseis, Forbes.† Pugiunculus, Barr.

Shell straight, conical, tapering to a point, back flattened, aperture trigonal. Lon. 1-8 inches.

Fossil, 6 sp. *Silurian.* N. America, Brit., New South Wales.

PTEROTHECA, Salter.

Type, P. transversa, Portlock, 3 sp. L. Silurian; Ireland, Wales, Canada.

Shell bi-lobed, transversely oval, with a dorsal keel projecting slightly at each end; ventral plate small triangular.

---

* Under the name of "triptère," M.M. Quoy and Gaimard described the fragment of a pteropod, since ascertained to have been a *Cuvieria.*

† *Creseis Sedgwicki,* Forbes, is an orthoceras with very thin septa, belonging to the same group with (*Conularia*) *teres,* Sby. *Tentaculites,* Schl. is anellidous. (*Salter.*)

? CONULARIA, Miller.

Etym. Conulus, a little cone. *Type,* C. quadrisulcata, fig. 108.

Shell four-sided, straight, and tapering, the angles grooved, sides striated transversely, apex partitioned off.

Fossil, 15 sp.   Silurian — Carb.   N. America, Europe, Australia.

Sub-genus, Coleoprion (gracilis) Sandberger; Devonian, Germany. *Shell* round, tapering, sides obliquely striated, striæ alternating along the dorsal line.

EURYBIA, Rang. 1827.†

Etym. Eurybia, a sea-nymph.   Fig. 108.*

Ex. E. Gaudichaudi, Pl. XIV. fig. 37. (after Huxley.)

Animal globular ; fins narrow, truncated and notched at the ends, united ventrally by a small lobe (metapodium) ; mouth with two elongated tentacles, behind which are minute eye-peduncles and a two-lobed rudimentary foot *(mesopodium)* ; body inclosed in a cartilaginous integument, with a cleft in front, into which the locomotive organs can be retracted. Lingual teeth 1.0.1.

The animal has no proper gill, but Mr. Huxley has observed two ciliated circles surrounding the body, as in the larva of *Pneumodermon.*

Distr. 3 sp.   Atlantic, Pacific.

Sub-genus, Psyche, Rang.   P. globulosa, Pl. XIV. fig. 38.   *Animal* globular, with two simple oval fins.   *Distr.* 1 sp.   Off Newfoundland.

CYMBULIA, Peron and Lesueur.

Etym. Diminutive of *cymba,* a boat.

Type, C. proboscidea, Pl. XIV. fig. 39. (after Adams).

Shell cartilaginous, slipper-shaped, pointed in front, truncated posteriorly ; aperture elongated, ventral.

Animal with large rounded fins connected ventrally by an elongated lobe ; mouth furnished with minute tentacles; lingual teeth 1.1.1 ; stomach muscular, armed with two sharp plates.

Distr. 3 sp.   Atlantic, Medit. India Ocean.

TIEDEMANNIA, Chiaje.

Type, T. Neapolitana, Pl. XIV. fig. 40.   Named after Fr. Tiedemann.

Animal naked, transparent, fins united, forming a large rounded disk ; mouth central; tentacles elongated, connate; eye-tubercles minute.   Larva shell-bearing.   *Distr.* 2 sp.   Medit. Australia.

---

* Carboniferous limestone, Brit. Belgium.

† This name had been previously employed for four different genera of plants and animals.

FAMILY II. Limacinidæ.

Shell minute, spiral, sometimes operculate.

Animal with fins attached to the sides of the mouth, and united ventrally by an operculigerous lobe; mantle-cavity opening dorsally; excretory orifices on the right side.

The shells of the true *limacinidæ* are sinistral, by which they may be known from the fry of *Atlanta, Carinaria,* and most other Gasteropods.

Limacina, Cuvier.

Etym. Limacina, snail-like. *Syn.* Spiratella, Bl.

Ex. L. antarctica (drawn by Dr. Joseph Hooker), Pl. XIV. fig. 41.

Shell sub-globose, sinistrally spiral, umbilicated; whirls transversely striated; umbilicus margined; no operculum.

Animal with expanded fins, notched on their ventral margins; operc. lobe divided; lingual teeth 1.1.1.

Distr. 2 sp. Arctic and Antarctic Seas; gregarious.

Spirialis, Eydoux and Souleyet.

Ex. S. bulimoides, Pl. XIV. fig. 42. *Syn.* Heterofusus, Flem. Heliconoides, D'Orb. Peracle, Forbes. Scaea, Ph.

Shell minute, hyaline, sinistrally spiral, globose or turrited, smooth or reticulated; operculum thin, glassy, semilunar, slightly spiral, with a central muscular scar.

Animal with narrow, simple fins, united by a simple, transverse operculigerous lobe; mouth central, with prominent lips.

Distr. 12 sp. Greenland and Norway to C. Horn, Indian Ocean, Pacific.

? Cheletropis, Forbes.

Etym. Chele, a claw, *tropis,* a keel. *Syn.* Sinusigera, D'Orb.

Type, C. Huxleyi, Pl. XIV. fig. 43.

Shell dextrally spiral, imperforate, double-keeled; nucleus sinistral; aperture channelled in front; peristome thickened, reflected, with two claw-like lobes.

Animal pteropodous ? gregarious in the open sea.

Distr. 2 sp. S. America, S. E. Australia.

Another minute spiral shell, recently discovered, may be noticed here:

Macgillivrayia. Forbes.

Named after its discoverer, the Naturalist to H. M. S. Rattlesnake.

Type, M. pelagica, Pl. XIV. fig. 44.

Shell minute, dextrally spiral, globular, imperforate, thin, horny, translucent; spire obtuse; aperture oblong, entire; peristome thin, incomplete, operc. thin horny, concentric, nucleus sub-external.

L

Animal with 4 long tentacles, mantle with a siphonal process; foot ex-panded, truncated in front, furnished with a float after the manner of *Ianthina;* lingual dentition closely resembling *Jeffreysia.*

Distr. 2 sp. Taken in the towing-net off C. Byron, E. coast Australia, 15 miles from shore; floating, and apparently gregarious. (J. Macgillivray.) Mindoro. (Adams.)

SECTION B. GYMNOSOMATA, Bl.

Animal naked, without mantle or shell; head distinct; fins attached to the sides of the neck; gill indistinct.

FAMILY III. CLIIDÆ.

Body fusiform; head with tentacles often supporting suckers; foot small, but distinct, consisting of a central and posterior lobe; heart *opistho-bran-chiate;* excretory orifices distant, on the right side; lingual teeth (in *Clio*) 12.1.12, central wide, denticulated, uncini strongly hooked and recurved.

CLIO (L.)* Müller.

Etym. Clio, a sea-nymph. *Syn.* Clione, Pallas.

Type, C. borealis, Pl. XIV. fig. 45. (C. caudata, L. part.)

Head with 2 eye tubercles and 2 simple tentacula; mouth with lateral lobes, each supporting 3 conical retractile processes, furnished with numerous microscopic suckers; fins ovate; foot lobed. In swimming, the Clio brings he ends of its fins almost in contact, first above and then below. (*Scoresby.*)

Distr. 4 sp. Arctic and Antarctic Seas, Norway, India.

Sub-genus ? Cliodita (fusiformis), Quoy and Gaimard. Head supported on a narrow neck; tentacles indistinct. 3 sp. Cape, Amboina.

PNEUMODERMON, Cuvier.

Etym. Pneumon, lung (or gill), *derma,* skin.

Type, P. violaceum, Pl. XIV. fig. 46.

Body fusiform; head furnished with ocular tentacles; lingual teeth 4.0.4; mouth covered by a large hood supporting two small, simple, and two large acetabuliferous tentacles, suckers numerous, pedicillate, neck rather contracted; fins rounded; foot oval, with a pointed posterior lobe; excretory orifice situated near the posterior extremity of the body, which has small branchial processes and a minute, rudimentary shell.

---

* This name was employed by Linnæus for all the Pteropoda then known; his definition is most suited to the "northern clio," probably the only species with which he was personally acquainted. The first species enumerated in the Syst. Nat. is *C. caudata,* and reference is made to an indeterminable figure in Brown's Jamaica, and to Marten's account of the Spitzbergen mollusk (*C. borealis.*) In cases like this the rule is to adopt the practice of the next succeeding naturalist who defines the limits of the group more exactly.

In the fry of *Pneumodermon* the end of the body is encircled with ciliated bands. (Müller.)

Distr. 4 sp. Atlantic, India, Pacific Ocean.

Sub-genus ? *Spongiobranchæa*, D'Orbigny. S. Australis, Pl. XIV. fig. 47. Gill (?) forming a spongy ring at the end of the body; tentacles each with 6 rather large suckers. *Distr.* 2 sp. S. Atlantic (Fry of *Pneumodermon* ?). *Trichocyclus*, Eschscholtz, T. Dumerilii, Pl. XIV. fig. 48. *Animal* without acetabuliferous tentacles ? mouth proboscidiform; front of the head surrounded with a circle of cilia, and two others round the body.

? PELAGIA, Quoy and Gaimard.

Etym. Pelagus, the deep sea: (not = *Pelagia*, Peron and Les.)
Type, P. alba, Pl. XIV. fig. 49. Amboina.
Animal fusiform, truncated in front, rough; neck slightly contracted; fins small, fan-shaped.

CYMODOCÈA, D'Orbigny.

Etym. Kumodoke, a Nereid. *Type*, C. diaphana, Pl. XIV. fig. 50.
Animal fusiform, truncated in front, pointed behind; neck slightly contracted; fins 2 on each side, first pair large and rounded, lower pair ligulate; foot elongated; mouth proboscidiform. *Distr.* 1 sp. Atlantic.

CLASS IV. BRACHIOPODA, Cuvier, 1805,
(= Order *Pallio-branchiata*, Blainville, Prodr. 1814.)

The *Brachiopoda* are bivalve shell-fish which differ from the ordinary mussels, cockles, &c. in being always *equal-sided*, and never quite *equivalve*. Their forms are symmetrical, and so commonly resemble antique lamps, that they were called *lampades*, or "lamp-shells," by the old naturalists (Meuschen, 1787, Humphreys, 1797); the hole which in a lamp admits the wick, serves in the lampshell for the passage of the pedicle by which it is attached to submarine objects.*

The valves of the *Brachiopoda* are respectively dorsal and ventral; the ventral valve is usually largest, and has a prominent beak, by which it is attached, or through which the organ of adhesion passes. The dorsal, or smaller valve, is always free and imperforate. The valves are articulated by two curved teeth, developed from the margin of the ventral valve, and received by sockets in the other; this hinge is so complete that the valves cannot be separated without injury.† A few, abnormal genera, have no

---

* The principal modifications of external form presented by these shells, are given in plate 15; the internal structure of each genus is illustrated in the woodcuts, which are the same with those in Mr. Davidson's Introduction, and in the British Museum Catalogue. They are from original studies by the author, unless otherwise stated.

† The largest recent *Terebratula* cannot be opened more than ⅛ of an inch, except by applying force.

hinge; in *Crania* and *Discina* the lower valve is flat, the upper like a limpet; the valves of *Lingula* are nearly equal, and have been compared to a duck's bill. (Petiver).

Ventral valve.

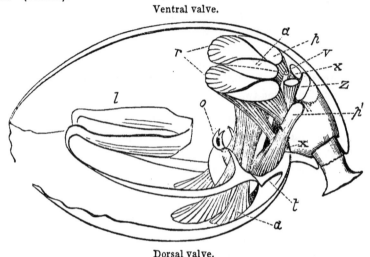

Dorsal valve.

Fig. 109.   *Muscular system of Terebratula.**

a. a. adductor-muscles; *r.* cardinal-muscles; *x.* accessory cardinals; *p.* ventral pedicle-muscles; *p.'* dorsal pedicle-muscles; *z.* capsular-muscles; *o.* mouth; *v.* vent; *l.* loop; *t.* dental socket.

The valves are both opened and closed by muscles; those which open the shell (*cardinales*) originate on each side the centre of the ventral valve, and converge towards the hinge-margin of the free valve, behind the dental sockets, where there is usually a prominent *cardinal process*.† The teeth form the *fulcrum* on which the dorsal valve turns.   The *adductor* muscles are four in number, and quite distinct in *Crania* and *Discina*; in *Lingula* the posterior pair are combined, and in *Terebratula* the four muscles are separate at their dorsal terminations, but united at their insertion in the centre of the larger valve.   The pedicle is fixed by a pair of muscles (each doubly-attached) to the dorsal hinge-plate, and by another pair to the ventral valve, outside the cardinal muscles.‡   In the hinge-less genera the contraction of the cardinal muscles must tend to slide the free valve forwards, and in *Crania* and *Discina* these muscles are attached to a prominent ventral

---

* *Waldheimia Australis, Quoy.* $\frac{2}{1}$.   From a drawing by Albany Hancock, Esq.

† The term "retractors" used at p. 8 is relinquished for the more appropriate term "cardinal muscles,'" given by Prof. King.   They are particularly interesting from their function, as antagonists of the adductor muscles, like the ligament of ordinary bivalves.

‡ The muscular system of *Terebratula* presents a considerable amount of resemblance to that of *Modiola* (fig. 177); the anterior and posterior pedal muscles may be compared to the dorsal and ventral pedicle muscles.

process, which renders them less oblique; the upper valve is restored to its place by two pairs of *retractor* sliding-muscles, which are perhaps the equivalents of the dorsal pedicle muscles of *Terebratula*.* The muscles are remarkably glistening and tendinous, except at their expanded ends, which are soft and fleshy; their impressions are often deep, and always characteristic; but difficult of interpretation from their complexity, their change of position, and the occasional suppression of some and combination of others.†

On separating the valves of a recent *Terebratula*, the digestive organs and muscles are seen to occupy only a very small space near the beak of the shell, partitioned off from the general cavity by a strong membrane, in the centre of which is placed the animal's mouth. The large cavity is occupied by the fringed arms, which have been already alluded to (page 8) as the characteristic organs of the class. Their nature will be better understood by comparing them with the lips and labial tentacles of the ordinary bivalves (pp. 24, 27, fig. 171, *p.p.*); they are in fact lateral prolongations of the lips supported on muscular stalks, and are so long as to require being folded or coiled up. In *Rhynchonella* and *Lingula* the arms are spiral and separate; in *Terebratula* and *Discina* they are only spiral at the tips, and are united together by a membrane, so as to form a lobed disk. It has been conjectured that the living animals have the power of protruding their arms in search of food; but this supposition is rendered less probable by the fact that in many genera they are supported by a brittle skeleton of shell. The internal skeleton consists of two spiral processes in the *Spiriferidæ* (fig. 132), whilst in *Terebratula* and *Thecidium* it takes the form of a *loop*, which supports the brachial membrane, but does not strictly follow the course of the arms. The mode in which the arms are folded is highly characteristic of the genera of *Brachiopoda;* the extent to which they are supported by a calcarious skeleton is of less importance, and liable to be modified by age. That margin of the oral arms which answers to the lower lip of an ordinary bivalve, is fringed with long filaments (*cirri*‡), as may be seen even in dry specimens of recent *Terebratulæ*. In some fossil examples the cirri themselves were supported by slender processes of shell;§ they cannot therefore be vibratile organs, but are probably themselves covered with microscopic cilia, like the oral tentacles of the ascidian polypes (*cilio-brachiata* of Farre). The anterior lip and inner margin of the oral arms is plain, and forms a

---

* In *Discina* one pair of the retractor muscles seems to be actually inserted in the pedicle. Mr. Hancock compares the pedicle muscles with the *retractors* of the Bryozoa; he objects to the hypothesis of the sliding movement of the valves.

† Prof. King has shown that the compound nature of a muscular impression is often indicated by the mode in which the vascular markings proceed from it (as in figs. 140, 145.)

‡ Called *cilia* at p. 8, but this term should be restricted to the microscopic organs which clothe the *cirri*.

§ *Spirifera rostrata* and *Terebratula pectunculoides*, in the British Museum.

narrow gutter along which the particles collected by the ciliary currents may be conveyed to the mouth. The object of the folding of the arms is obviously to give increased surface for the disposition of the *cirri.*

The mouth conducts by a narrow œsophagus to a simple stomach, which is surrounded by the large and granulated liver; the intestine of *Lingula* is reflected dorsally, slightly convoluted, and terminates between the mantle lobes on the *right* side (fig. 165). In *Orbicula* it is reflected ventrally, and passes straight to the right, ending as in *Lingula*. In *Terebratula, Rhynchonella,* and probably all the *normal* Brachiopoda, the intestine is simple and reflected ventrally, passing through a notch or foramen in the hinge-plate, and ending behind the ventral insertion of the adductor muscle (fig. 109, v.)*

The interior of the valves is lined by the two lobes of the mantle, which are often fringed with fine horny bristles (*setæ*); these are quite straight, brittle, and deeply implanted between the laminæ of the mantle; they serve to guard the opening of the valves. The mantle-lobes of the *Brachiopoda* are not only organs by which the shell is formed, they are also provided with large veins by which respiration is effected; in the *Terebratulidæ* there are two great venous trunks in the dorsal mantle-lobe, four in the ventral; in *Rhynchonella* and *Discina* the lobes are similar, and the *Orthidæ* have four large veins in the dorsal lobe and only two in the ventral. The first indication of a special breathing organ is presented by *Lingula,* in which the veins develope parallel rows of small vascular processes. (*Cuvier*.) The veins open into the visceral cavity,† which is itself a great vascular sinus. There are two organs which Prof. Owen regards as hearts, each consisting of an auricle and a ventricle, situated near the sides of the mouth in *Terebratula* ; but in *Lingula* (fig. 165, *h.*) they are more posterior, and quite at the sides. The ventricles propel the blood into the visceral and pallial arteries, and are therefore both branchial and systemic. The pallial arteries are very slender, and accompany the veins on their outer surfaces, forming linear impressions along the centre of the vascular markings in some fossil shells (fig. 141).

The *ova* of *Terebratula* are developed within the large veins, which they accompany as far as the secondary branches. In the *Rhynchonellidæ,* and probably in the extinct *Orthidæ,* the ovaria do not extend into the venous trunks, but occupy large sinuses on each side of the body; and in *Discina* and *Lingula* they (or the testes) fill the interstices of all the viscera, but do not appear to extend into the mantle. The ova are supposed to escape by two orifices, situated at the sides of the mouth in *Terebratula.* (Hancock.)

---

* The position at which the intestine terminates in the *Terebratulæ* and *Rhynchonellæ,* seems to necessitate the escape of the fæces by the umbonal opening ; in those extinct genera which have the foramen closed at an early age, there is still an opening between the valves (e. g. in *Uncites*) which has been mistaken for a byssal notch.

† The veins do not terminate in hearts as formerly supposed; the statement at p. 30, line 27, should be erased.   •

Recent *Discinæ* often have minute fry attached to their valves, and Mr. Suess, of Vienna, has noticed a specimen of the fossil *Stringocephalus*, which contained numerous embryo shells.

Nothing is yet known respecting the development of the *Brachiopoda*, but there can be no doubt that in their first stage they are free and able to swim about, until they meet with a suitable position. It is probable that in the second stage they all adhere by a byssus, which in most instances becomes consolidated, and forms a permanent organ of attachment. Some of the extinct genera (e. g. *Spirifera* and *Strophomena*) appear to have become free when adult, or to have fixed themselves by some other means. Four genera, belonging to very distinct families, cement themselves to foreign objects by the substance of the ventral valve.

The Lamp-shells are all natives of the sea. They are found hanging from the branches of corals, the under sides of shelving rocks, and the cavities of other shells. Specimens obtained from rocky situations are frequently distorted, and those from stony and gravelly beds, where there is motion in the waters, have the beak worn, the foramen large, and the ornamental sculpturing of the valves less sharply finished. On clay beds, as in the deep clay strata, they are seldom found; but where the bottom consists of calcarious mud they appear to be very abundant, mooring themselves to every hard substance on the sea-bed, and clustering one upon the other.

Some of the *Brachiopoda* appear to attain their full growth in a single season, and all, probably, live many years after becoming adult. The growth of the valves takes place chiefly at the margin; adult shells are more globular than the young, and aged specimens still more so. The shell is also thickened by the deposit of internal layers, which sometimes entirely fill the beak, and every portion of the cavity of the interior which is not occupied by the animal, suggesting the notion that the creature must have died from the plethoric exercise of the calcifying function, converting its shell into a mausoleum, like many of the ascidian zoophytes.

The intimate structure of the shell of the *Brachiopoda* has been investigated by Mr. Morris, Prof. King, and more recently by Dr. Carpenter; according to the last observer, it consists of flattened prisms of considerable length, arranged parallel to each other with great regularity, and obliquely to the surfaces of the shell, the interior of which is imbricated by their out-crop (fig. 110.) This structure only is found in the *Rhynchonellidæ*; but in most—perhaps all the other *Brachiopoda**—the shell is traversed by canals, from one surface

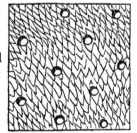

Fig. 110. *Terebratula.*

* The fossil shells of the older rocks are so generally pseudomorphous, or partake of the metamorphic character of the rock itself, that it is difficult to obtain specimens in a state fit for microscopic examination.

to the other, nearly vertically, and regularly, the distance and size of the perforations varying with the species. Their external orifices are trumpet-shaped, the inner often very small; sometimes they bifurcate towards the exterior, and in *Crania* they become arborescent. The canals are occupied by cœcal processes of the outer mantle-layer,* and are covered externally by a thickening of the epidermis. Mr. Huxley has suggested that these cœca are analogous to the vascular processes by which in many ascidians the *tunic* adheres to the *test;* the extent of which adhesion varies in closely allied genera. The large tubular spines of the *Productidæ* must have been also lined by prolongations of the mantle; but their development was more probably related to the maintenance of the shell in a fixed position, than to the internal economy of the animal. (*King.*) Dr. Carpenter states that the shell of the *Brachiopoda* generally contains less animal matter than other bivalves; but that *Discina* and *Lingula* consist almost entirely of a horny animal substance, which is laminar, and penetrated by oblique tubuli of extreme minuteness. He has also shown that there is not in these shells that distinction between the outer and inner layers, either in structure or mode of growth, which prevails among the ordinary bivalves; the inner layers only differ in the minute size of the perforations, and the whole thickness corresponds with the outer layer only in the *Lamellibranchiata*. The loop, or brachial processes, are always impunctate.

Of all shell-fish the *Brachiopoda* enjoy the greatest range both of climate, and depth, and time; they are found in tropical and polar seas; in pools left by the ebbing tide, and at the greatest depths hitherto explored by the dredge. At present only 70 recent species are known; but many more will probably be found in the deep-sea, which these shells mostly inhabit. The number of living species is already greater than has been discovered in any *secondary* stratum, but the vast abundance of fossil *specimens* has made them seem more important than the living types, which are still rare in the cabinets of collectors, though far from being so in the sea. Above 1,000 extinct species of Brachiopoda have been described, of which more than half are found in England. They are distributed throughout all the sedimentary rocks of marine origin from the Cambrian strata upwards, and appear to have attained their maximum, both of generic and specific development, in the Devonian age.* The oldest form of organic life at present known, both in the old and new world, is a *Lingula*. Some species (like *Atrypa reticularis*)

---

* Called the "lining membrane of the shell," by Dr. Carpenter. (Davidson Intr. Mon. Brach.) Mr. Quekett states that the perforations are closed externally by disks, surrounded by radiating lines, supposed to indicate the existence of vibratile cilia in the living specimens.

† The number of Devonian species amounts to 300; but these were not all living *at one time*, they are obtained from a whole series of deposits, representing a succession of periods.

extend through a whole "system" of rocks, and abound equally in both hemispheres; others (like *Spirifera striata*) range from the Cordillera to the Ural mountains. One recent Terebratula (*caput-serpentis*) made its appearance in the Miocene Tertiary; whilst others, scarcely distinguishable from it, are found in the Upper Oolite, and throughout the Chalk series and London Clay.*

FAMILY I. Terebratulidæ.

Shell minutely punctate; usually round or oval, smooth or striated; ventral valve with a prominent beak, and two curved hinge-teeth; dorsal valve with a depressed umbo, a prominent cardinal process between the dental sockets, and a slender shelly loop.

Animal attached by a pedicle, or by the ventral valve: oral arms united to each other by a membrane, variously folded; sometimes spiral at their extremities.

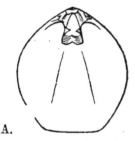 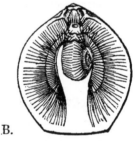

A.   B.

Fig. 111. *Terebratula vitrea, Born.*

Terebratula, (Llhwyd.) Brug. Lamp-shell.

Etym. Diminutive of *terebratus,* perforated.

Syn. Lampas, Humph. Gryphus, Muhlfeldt. Epithyris, Phil.

Types, T. maxillata, Pl. XV. fig. 1, (= Ter. minor-subrubra, Llhwyd. Anomia terebratula, L.) T. vitrea, fig. 3.

Shell smooth, convex; beak truncated and perforated; foramen circular; deltidium of two pieces, frequently blended; loop very short, simple, attached by its crura to the hinge-plate. (Fig. 111, A.)

Animal attached by a pedicle; brachial disk tri-lobed, centre lobe elongated and spirally convoluted. (Fig. 111, B.) The young of *T. diphya* (Pygope of Link) has bi-lobed valves, (Pl. XV. fig. 2.); when adult the lobes unite, leaving a round hole through the centre of the shell.

Distr. 1 sp. Medit. 90—250 fathoms on nullipore mud. (*Forbes.*)

Fossil, 100 sp. Devonian —. World-wide.

* The author has to ackowledge his obligation to Mr. Davidson for the use of the notes, drawings and specimens, assembled during the preparation of his great work on the British Fossil Brachiopoda, printed for the Palæontographical Society; to which work the student is referred for more copious descriptions and illustrations.

Sub-genera. Terebratulina (caput-serpentis) D'Orb. Pl. XV. fig. 3.
Fig. 112. *Shell* finely striated, auriculate, deltidium usually rudimental;

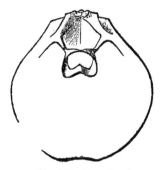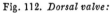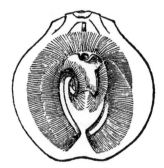

Fig. 112. *Dorsal valve:*   *Animal.* $\frac{2}{1}$

foramen incomplete; loop short, rendered annular in the adult by the union
of the oral processes. *Dist.* 7 sp. U. States, Norway, Cape, Japan. 10—
120 fms. *Fossil,* 20 sp. Oxfordian —. U. S. Europe.

Waldheimia (australis) King. Pl. XV. fig. 4 (p. 8, figs. 4, 5.) figs. 109,
113, 114.

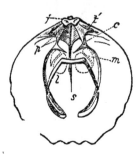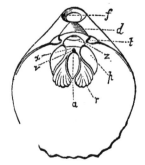

Fig. 113. *Dorsal valve.*   Fig. 114. *Ventral valve.*

Fig. 113. *j,* cardinal process; *t',* dental sockets; *p,* hinge-plate; *s,* septum; *c,*
crura of the loop; *l,* reflected portion of the loop; *m,* quadruple adductor-impression.
Fig. 114. *f,* foramen; *d,* deltidium; *t,* teeth; *a;* single adductor-impression; *r,*
cardinal muscles; *x,* accessory muscles; *p,* pedicle muscles; *v,* position of the vent;
z, attachment of pedicle-sheath.

Shell smooth or plaited, dorsal valve frequently impressed; foramen complete; loop elongated and reflected; septum (*s*) of smaller valve elongated.
Distr. 9 sp. Norway, Java, Australia, California, Cape Horn. Low-water—
100 fms. *Fossil,* 60 sp. Trias —. S. America, Europe. *Eudesia* (cardium) King, includes 1 recent, and 6 fossil species which are sharply plaited.
T. impressa (Pl. XV. fig. 5) is the type of a group which has the external
shape of *Terebratella.*

TEREBRATELLA, D'Orbigny.

Type, T. dorsata, Gmel. (= Magellanica, Chemn.) Pl. XV. fig. 7. Fig. 115.

Shell smooth or radiately plaited; dorsal valve longitudinally impressed; hinge-line straight, or not much curved; beak with a flattened area on each side of the deltidium; foramen large; deltidium incomplete; loop attached to the septum (*s*).

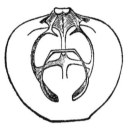

Fig. 115. *Terebratella.*

Animal like *Terebratula ;* the spiral lobe of the brachial disk becomes very diminutive in some species, and is obsolete in *Morrisia* and *T. Cumingii. Distr.* excluding subgenera, 16 sp. Cape Horn, Valparaiso (90 fms.), New Zealand, Japan, Ochotsk, Spitzbergen, Labrador. *Fossil*, 16 sp. Lias —. U.S. Europe. In *T. crenulata* and *Evansii*

Fig. 116. *Ter: Evansii.* Dav.

(fig. 116) the dorsal septum sometimes projects so far as to touch the opposite valve, but in other examples it remains undeveloped. (*Davidson.*)

Sub-genera. Trigonosemus (elegans) König. *Syn.* Delthyridæa (pectiniformis) M'Coy. Fissirostra, D'Orb. *Ex.* T. Palissii, Pl. XV. fig. 8. *Shell* finely plaited, beak prominent, curved, with a narrow apical foramen; cardinal area large, triangular; deltidium solid, flat; cardinal process very prominent. *Distr.* 5 sp. Chalk, Europe.

Lyra (Meadi), Cumberland, Min. Con. 1816. Pl. XV. fig. 6. *Syn.* Terebrirostra, D'Orb. Rhynchora, Dalman.* *Shell* ornamented with rounded ribs; beak very long, divided lengthwise internally, by the dental plates; loop doubly attached? *Distr.* 4 sp. cretaceous: Europe. Three species of similar form are found in the Trias of St. Cassian.

Magas. (pumila) Shy. Fig. 117. Shell smooth, conspicuously punctate, dorsal valve impressed, foramen angular, deltidium rudimentary; internal septum (*s*) prominent, touching the ventral valve; reflected portions of the loop disunited (*l*). 2 sp. U. Green-sand — Chalk. Europe. The recent *Ter. Cumingii*, of New Zealand,

Fig. 117. *M. pumila.* $\frac{2}{1}$

* The name *Rhynchora* was given by Dalman to the Ter. costata. Wahl. (= T. pectinata, L.) on the supposition that it was identical with Sowerby's *T. Lyra;* and as no specimen could be found with a long beak, an artificial one was manufactured for it, of which there is a cast in the Brit. M. The second species of "Rhynchora." *Ter. spatulata*, Wahl. has no beak whatever: in shape it is like an *Argiope*, but measures an inch each way. The ventral valve is a simple bent plate with the teeth at the angles; the dorsal valve is flat, with a very wide hinge-plate, and sockets at the angles, whilst a single septum projects from the centre, with portions of a loop attached.

resembles *Bouchardia* externally, but has the diverging processes of the loop as in Magas.

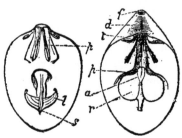

Fig. 118. *B. tulipa, Bl.**

Bouchardia (tulipa) Davidson, fig. 118. Beak prominent, with a minute apical foramen (*f*) deltidium blended with the shell (*d*) apophysis anchor-shaped, the septum (*s*) being furnished with two short lamellæ. Brazil, 13 fms.

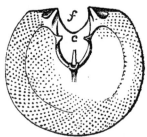

Fig. 119. *Animal.* $\frac{10}{1}$ *Dorsal valve.*†

Morrisia (anomioides, Scacchi) Davidson. Fig. 119. *Shell* minute, conspicuously punctate; foramen large, encroaching equally on both valves; hinge area small, straight; loop not reflected, attached to a small forked process in the centre of the valve. *Animal* with sigmoid arms, destitute of spiral terminations; cirri in pairs. *Distr.* 2 sp. Medit. 95 fms. (*Forbes.*) ? *Fossil.* 1 sp. Pliocene, Palermo.

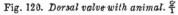
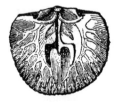

Fig. 120. *Dorsal valve with animal.* $\frac{2}{1}$ Fig. 121. *Dorsal valve.*

* The muscular impressions in *Bouchardia* have been compared with those of *Ter. Cumingii*, of which the animal is known. The large impressions (*r*) in the disk of the ventral valve appear to be formed by the cardinal muscles; *a*. by the adductor; *p*. by the pedicle muscles.

† Fig. 119. *c*. loop; *f*. pedicle notch; *o*. the ovaries. From the originals in Mr. Davidson's collection; magnified ten diameters.

Kraussia (rubra) Dav. Cape. Fig. 121. K. Lamarckiana, Dav. Australia. Fig. 120. *Shell* transversely oblong; hinge-line nearly straight; beak truncated, laterally keeled; area flat; foramen large, deltidium rudimentary; dorsal valve longitudinally impressed, furnished inside with a forked process rising nearly centrally from the septum; interior often strongly tuberculated. The apophysis is sometimes a little branched, indicating a tendency towards the form it attains in fig. 122. *Animal* with rather small oral arms, the spiral lobe very diminutive. *Distr.* 6 sp. S. Africa, Sydney, N. Zealand; low-water to 120 fms.

Fig. 122. *Animal.*   *Dorsal valve.*

? *Megerlia* (truncata) King, 1850. Pl. XV. fig. 9. Fig. 122. *Loop* trebly attached; to the hinge-plate by its crura, and to the septum by processes from the diverging and reflected portions of the loop. *Distr.* 2 sp. Medit. Philippines. These species belong to the same natural group with *Kraussia*.

? *Kingena* (lima) Dav. Cretaceous, Europe, Guadaloupe. Valves spinulose; loop trebly attached.

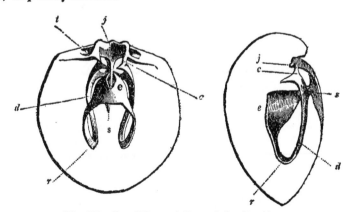

Fig. 123. *Ter.* (*Kingena*) *lima*; (*after* Davidson.)

t. dental sockets; *j.* cardinal process, *c.* crura; *d.* diverging processes of loop; *r*, reflected portion; *e.* third attachment of loop; *s.* dorsal septum.

? *Ismenia* (pectunculus) King.   Coral rag, Europe.   Valves ornamented with corresponding ribs; loop trebly attached.

? *Waltonia* (Valenciennei) Dav. New Zealand.   Perhaps the fry of *Ter. rubicunda*, with the reflected part of the loop wanting.

Fig. 124. *Argiope decollata.* $\frac{4}{1}$ Fig. 125. *A. Neapolitana*, Sc.* $\frac{8}{1}$

ARGIOPE, Eudes Deslongchamps.

Etym. Argiope, a nymph.   *Syn.* Megathyris, D'Orb.

Type, A. decollata, Pl. XV. fig. 10.   Fig. 124—126.

Shell minute, transversely oblong or semi-ovate, smooth or with corresponding ribs; hinge line wide and straight, with a narrow area to each valve; foramen large, deltidium rudimentary; interior of dorsal valve with one or more prominent, sub-marginal septa; loop two or four-lobed, adhering to the septa, and more or less confluent with the valve.

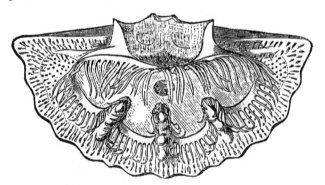

Fig. 126. *A. decollata,* $\frac{40}{1}$; dorsal valve with the animal, from a specimen dredged by Prof. Forbes in the Ægean.   The oral aperture is seen in the centre of the disk.

Animal with oral arms folded into two or four lobes, united by membrane, forming a brachial disk fringed with long cirri: mantle extending to the margins of the valves, closely adherent.

Distr, 4 sp.   N. Brit. Madeira, Canaries, Medit.   40—105 fathoms.

Fossil. 5 sp.   U. Greensand —.   Europe.

* Interiors of dorsal valves magnified, from the originals in Coll. Davidson.

Fig. 127. *T. radians.*   Fig. 128. *T. Mediterraneum.* * $\frac{4}{1}$

THECIDIUM, Defrance.

Etym. Thekidion, a small pouch.   *Type,* T. radians, Pl. XV. fig. 11.

Shell small, thick, punctate, attached by the beak; hinge-area (*h*) flat; deltidium (*d*) triangular, indistinct: *dorsal valve* (fig. 127) rounded, depressed; interior with a broad granulated margin; cardinal process prominent, between the dental sockets; oral processes united, forming a bridge over the small and deep visceral cavity; disk grooved for the reception of the loop, the grooves separated by branches from a central septum; loop often unsymmetrical, lobed, and united more or less intimately with the sides of the

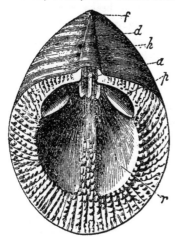

grooves: *ventral valve* (fig. 129) deeply excavated, hinge-teeth prominent; cavities for the adductor (*a*) and pedicle muscles (*p*) small; disk occupied by two large smooth impressions of the cardinal muscles, bordered by a vascular line. *Animal* (fig. 128) with elongated oral arms, folded on themselves and fringed with long cirri; mantle extending to the margin of the valves and closely adherent; epidermis distinct.

T. radians is the only un-attached species, it is supposed to be fixed by a pedicle when young (D'Orb.)

T. hieroglyphicum, Pl. XV. fig. 12, has a very complicated interior; whilst in several others there are but two brachial lobes.

Fig. 129. *T. radians,* $\frac{4}{1}$.

The Liassic species form the subject of a monograph by M. Eugene Deslongchamps; they are often minute, and attached in numbers to sea-urchins, corals, and terebratulæ.

Distr. 1 sp. Medit.   *Fossil,* 27 sp.   Trias —.   Europe.

* Dorsal valve with the animal, magnified.   Coll. Davidson.

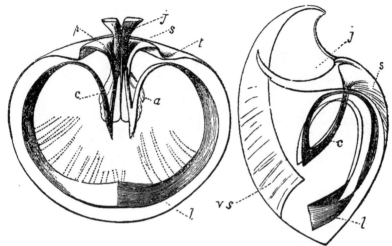

Fig. 130. *Dorsal valve.*   *Profile.**

a, adductor; *c*, crura; *l*, loop; *j*, cardinal process; *p*, hinge-plate; *s*, dorsal septum;
v. s. ventral septum; *t*, dental sockets.

? Stringocephalus, Defrance.

Etym. Strinx (*stringos*) an owl, *cephale* the head.†
S. Burtini, Pl. XV. fig. 13.   Fig. 130, 131.   Devonian, Europe.

Shell punctate; sub-orbicular, with a prominent beak: *ventral* valve
with a longitudinal septum (*v. s.*) in the middle; hinge-area distinct; foramen
large and angular in the young shell, gradually surrounded by the deltidium
and rendered small and oval in the adult; deltidium
composed of three elements; teeth prominent; *dorsal*
valve depressed, cardinal process (*j*) very prominent, sometimes touching the opposite valve, its extremity forked
to receive the ventral septum (*v.s.*); hinge-plate (*p*) supporting a shelly loop, after the manner of *Argiope*.

Fig. 131.‡

FAMILY II.   Spiriferidæ.

Shell furnished internally with two calcarious spiral processes (*apophyses*)
directed outwards, towards the sides of the shell, and destined for the support
of the oral arms ; which must have been fixed immoveably; the spiral lamellæ

---

* The loop (which was discovered by Prof. King) has a distinct suture in the
middle; the dotted lines proceeding from its inner edge are added from a drawing by
Mr. Suess, and represent what he regards as shelly processes for supporting a membranous disk.   They may be portions of spirals, whose outer whirls are confluent.

† Internal casts of *Producta gigantea* are called "owl-heads" by quarrymen in the
North of England.   (*Sowerby*).

‡ Fig. 131. Young shell, magnified 4 diameters; *h*, hinge area; *b*, deltidium:
p, pseudo-deltidium.

are sometimes spinulose, indicating the existence of rigid cirri, especially on the front of the whirls; valves articulated by teeth and sockets.

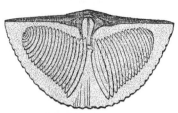 

Fig. 132. *Dorsal;*   *Ventral valve.*½

SPIRIFERA, Sowerby.

Type, S. striata, Sby. fig. 132.   *Syn.* Trigonotreta, König. Choristites. Fischer. Delthyris, Dalman. Martinia &c. M'Coy.

Shell impunctate,* transversely oval or elongated, tri-lobed, beaked, bi-convex, with a dorsal ridge and ventral furrow; hinge-line wide and straight; area moderate, striated across; foramen angular, open in the young, afterwards progressively closed; *ventral* valve with prominent hinge-teeth, and a central muscular scar, consisting of the single adductor flanked by two cardinal impressions: *dorsal* valve with a small cardinal process, a divided hinge-plate, and two conical spires directed outwards and nearly filling the cavity of the shell; crura united by an oral loop. The shell and spires are sometimes silicified, in limestone, and may be developed by means of acid. In *S. mosquensis* the dental plates are prolonged nearly to the front of the ventral valve.

Distr. 200 sp. L. Sil. — Trias. Arctic America — Chile, Falkland Ids. Europe; China; Thibet; Australia; Tasmania. In China these and other fossils are used as medicine.

Sub-genera. Spiriferina, D'Orb. S. Walcotti, Pl. 15, f. 14. *Shell* punctate, external surface spinulose; foramen covered by a pseudo-deltidium; interior of ventral valve with a prominent septum, rising from the adductor scar. *Distr.* 6 sp. Trias — L. Oolites. Brit. France, Germany, S. America.

Cyrtia, Dalman. C. exporrecta, Pl. XV. fig. 15. *Shell* impunctate, pyramidal, beak prominent, area equiangular, deltidium with a small tubular foramen. *Fossil,* 7 sp. Silur. — Trias. Europe. In *C. Buchii, heteroclyta, calceola*, &c. the shell is punctate.

ATHYRIS, M'Coy.

Etym. *A*, without, *thuris*, a door.† (i. e. deltidium). *Syn.* Spirigera, D'Orb. Cleiothyris, King (not Phil.)

---

* Prof. King attributes this to metamorphism; *S. Demarlii.* Bouch. from the Devonian limestone, is punctate. (*Carpenter*).

† Sometimes employed, *incorrectly*, in the sense of a door-*way* or foramen.

Types, A. concentrica, Buch.   A. Roissyi, fig. 133, 134.   A. lamellosa,
Pl. XV. fig. 16.

Shell impunctate, transversely oval, or sub-orbicular, bi-convex, smooth,
or ornamented with squamose lines of growth, sometimes developed into
wing-like expansions, (fig. 134*); hinge-line curved, area obsolete, foramen

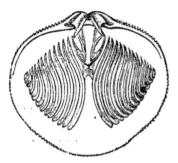
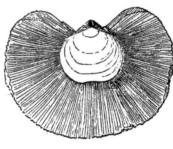

Fig. 133. *Interior of dorsal valve.*   Fig. 134. *Specimen with fringe.*

round, truncating the beak, deltidium obsolete; hinge-plate of dorsal valve
with four muscular cavities, perforated by a small round foramen, and supporting
a small complicated loop (?) between the spires; spires directed outwards,
crura united by a prominent oral loop.

The foramen in the hinge-plate occupies the situation of the notch
through which the intestine passes in the recent *Rhynchonellæ;* in *A. concentrica* a slender curved tube is sometimes attached to the foramen, beneath
the hinge-plate. *A. tumida* has the hinge-plate merely grooved, and the
byssal foramen is angular.

Fossil, about 20 sp. Silurian — Lias.   N. and S. America; Europe.

Sub-genus? Merista, Suess. Ter. scalprum, Rœmer,
(A. cassidea, Quenst.   Sp. plebeia. Ph.) Silurian —
Devonian; Europe. *Shell* impunctate, dental plates
(*v*) and dorsal septum (*d*) supported by arched plates
(" shoe-lifter" processes, of King) which readily detach, leaving cavities (as in fig. 135); spiral arms have
been observed in all the species.

Fig. 135. *Merista.*

RETZIA, King.

Dedicated to the distinguished Swedish naturalist, Retzius.

Type, Ter. Adrieni, Vern. *Ex.* R. serpentina, Carb. L. Belgium. Fig. 136

Shell punctate, terebratula-shaped; beak truncated by a round foramen
rendered complete by a distinct deltidium: hinge-area small, triangular,
sharply defined; interior with diverging shelly spires.

Fossil, about 20 species. Silurian — Trias. S. America. U. S. Europe·

---

* The spurious genus *Actinoconchus* (M'Coy) was founded on this character;
similar expansions are formed by species of Atrypa, Camarophoria, and Producta.

Prof. King first pointed out the existence of calcarious spires in several *Terebratulæ* of the older rocks, and others have been discovered by MM. Quenstedt, De Koninck, and Barrande. In form they resemble Terebratulina, Eudesia, and Lyra.

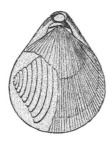

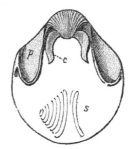

Fig 136. *Retzia serpentina, D. K.*   Fig. 137, *Uncites gryphus.*

UNCITES, Defrance.

Type, U. gryphus, Pl. XV. fig. 17.   Fig 137.   *Fossil*, Devonian. Europe.

Shell impunctate; oval, bi-convex, with a long incurved beak; foramen apical, closed at an early age; deltidium, large, concave; spiral processes directed outwards; no hinge-area.

The large, concave deltidium of Uncites so much resembles the channel formed by the dental plates of *Pentamerus*, that Dalman mistook the shell for a member of that genus. The discovery of internal spires, by Prof. Beyrich, shows that it only differs from *Retzia* in being impunctate and destitute of hinge-area. Some of the specimens have corresponding depressions in the sides of the valves (fig. 137, *p*) forming pouches which do not communicate with the interior.

FAMILY III. RHYNCHONELLIDÆ.

Shell impunctate, oblong, or trigonal, beaked; hinge-line curved; no area; valves articulated, convex, often sharply plaited; foramen beneath the beak, usually completed by a deltidium, sometimes concealed; hinge-teeth supported

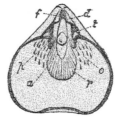

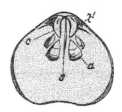

Fig. 138. *R. nigricans.*   Fig. 139. *Ventral:*   Dorsal.

Fig. 138. Dorsal valve with the animal; *a*, adductor muscles: *i*, intestine.

Fig. 139. R. psittacea, interiors. *s*, septum; *f*, foramen; *d*, deltidium; *t*, teeth; *t'*, sockets; *c*, oral lamellæ; *a*, adductor impressions; *r*, cardinal; *p*, pedicle muscles; *o*, ovarian spaces.

by dental plates; hinge-plate deeply divided, supporting oral lamellæ, rarely provided with spiral processes; muscular impressions grouped as in *Terebratula;* vascular impressions consisting of two principal trunks in each valve, narrow, dichotomising, angular, the principal posterior branches inclosing ovarian spaces.

Animal (of *Rhynchonella*) with elongated spiral arms, directed inwards, towards the concavity of the dorsal valve; alimentary canal terminating behind the insertion of the adductor in the ventral valve; mantle not adhering, its margin fringed with a few short setæ.

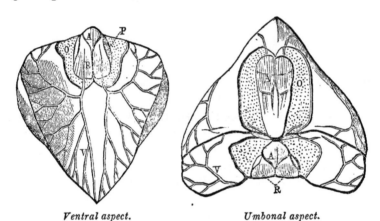

Ventral aspect.   *Umbonal aspect.*

Fig. 140. *Rh. acuminata, internal casts.*

Fig. 140. Umbonal aspect, with the dorsal valve above (Coll. Prof. King). Ventral aspect (Coll. Prof. Morris). A, adductor; R, cardinal; P, pedicle; V, vascular; O, ovarian impressions.

RHYNCHONELLA.   Fischer.

Syn. Hypothyris, Phil. Hemithyris (psittacea) D'Orb. Acanthothyris (spinosa) D'Orb. Cyclothyris (latissima) M'Coy. Trigonella (part) Fischer (not L. nor Da Costa).

Types, R. acuta, Pl. XV. fig. 18 : furcillata, fig. 19 : spinosa, fig. 20 : acu· minata, fig. 140 : nigricans, fig. 138 ; psittacea, fig. 139 (p. 8, fig. 3).

Shell trigonal, acutely beaked, usually plaited; dorsal valve elevated in front, depressed at the sides ; ventral valve flattened, or hollowed along the centre, hinge plates supporting two slender curved lamellæ; dental plates diverging.

The foramen is at first only an angular notch in the hinge-line of the ventral valve, but the growth of the deltidium usually renders it complete in the adult shell; in the cretaceous species it is tubular. In *R. acuminata* and many other palæozoic examples, the beak is so closely incurved as to allow no space for a pedicle. Both the recent *Rhynchonellæ* are black; *R. octoplicata* of the Chalk sometimes retains six dark spots.

Distr. 2 sp. *R. psittacea,* Labrador (low water ?) Hudson's Bay, 100 fms.: Melville Id. Sitka; Icy Sea. *R. nigricans,* New Zealand, 19 fms.

Fossil, 250 sp. L. Silurian —. N. and S. America, Europe, Thibet, China.

Sub-genera. ? *Porambonites,* Pander. P. æquirostris, Schl. *Shell* impunctate; surface minutely pitted; each valve with a minute hinge-area and indications of two septa; foramen angular, usually concealed. *Distr.* 4 sp. L. Silurian. Russia, Portugal.

Camarophoria, King. T. Schlotheimi, Buch. Figs. 141, 142. Ventral valve with converging dental plates (*d*) supported on a low septal ridge (*s*); dorsal valve with a prominent septum (*s*) supporting a spoon-shaped central process (*v*); oral lamellæ long and slender (*o*). Foramen angular, cardinal process distinct (*j*). *Fossil,* 9 sp. ? Carb. — Permian (Magnesian limestone). Germany; England.

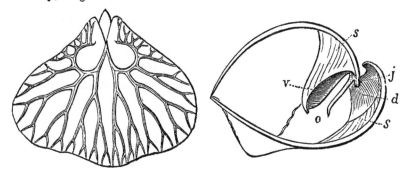

Fig. 141. Internal cast.*   Fig. 142. Section.

PENTAMERUS, Sowerby.

Etym. Pentameres, 5-partite.

Syn. Gypidia (conchydium) Dalman.

Type, P. Knightii, Pl. XV. fig. 22. Fig. 143.

Shell impunctate, ovate, ventricose, with a large incurved beak; valves usually plaited; foramen angular; no area or deltidium; dental plates (*d*) converging, trough-like, supported on a prominent septum (*s*); dorsal valve with two contiguous longitudinal septa (*s s*) opposed to the plates of the other valve.

Oral lamellæ have been detected by Mr. Salter in *P. liratus;* in *P.* ? *brevirostris* (Devonian, Newton) the dorsal valve has a long trough-like process supported by a single low septum.

Fossil, 20 sp. Arctic America, U. S. Europe.

* Ventral side of cast, showing the V shaped cavity of the dental plates, and the impressions of branchial veins, accompanied by arteries; (after King.)

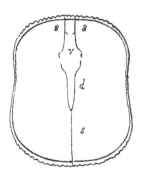

Fig. 143. *Longitudinal;*   *Transverse section.*

The relations of the animal to the shell, in such a species as *P. Knightii* can only be inferred by comparison with other species in which the internal plates are less developed, and with other genera, such as *Cyrtia* and *Camarophoria.* In fig. 143, the small central chamber (*v*) must have been occupied by the digestive organs, the large lateral spaces (*d s*) by the spiral arms : it is doubtful whether any muscles were attached to these plates ; in *Porambonites* the adductor impression is situated beyond the point to which the dental plates converge, and in *Camarophoria* the muscular impressions occupy the same position as in *Rhynchonella.*

ATRYPA, Dalman.

Syn. Cleiothyris, Phillips. Spirigerina, D'Orb.* Hipparionyx, Vanuxem. *Type,* A. reticularis, Pl. XV. fig. 17.   Figs. 144, 145.

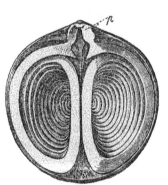
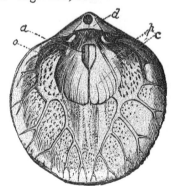

Fig. 144, *Dorsal valve.*   Fig. 145, *Ventral valve; interiors.*

p, hinge-plate; *a*, impressions of adductor muscle; *c*, cardinal muscle   *p*, pedicle muscle ; *o*, ovarian sinus; *d*, deltidium.

Shell impunctate : oval, usually plaited and ornamented with squamose lines of growth ; dorsal valve gibbose ; ventral depressed in front ; beak

* The term *Atrypa* (*a,* without, *trupa,* foramen) is objectionable, like all Dalman's names; but M. D'Orbigny has made no improvement by proposing *Spirigerina*, in addition to Spirifera, Spirigera, and Spiriferina!

small, often closely incurved: foramen round, sometimes completed by a deltidium, often concealed: dorsal valve with a divided hinge-plate, supporting two broad spirally coiled lamellæ; spires vertical, closely appressed, and directed towards the centre of the valve; teeth and impressions like *Rhynchonella*.

The shells of this genus differ from *Rhynchonella* chiefly in the calcification of the oral supports, a character of uncertain value.

Fossil, 15 sp. L. Silurian — Trias. America (Wellington Channel! Falkland Ids.), Europe, Thibet.

FAMILY IV. ORTHIDÆ.*

Shell transversely oblong, depressed, rarely foraminated; hinge-line wide and straight; beaks inconspicuous; valves plano-convex, or concavo-convex, each with a hinge-area (*h*) notched in the centre; *ventral* valve with prominent teeth (*t*); muscular impressions occupying a saucer-shaped cavity with a raised margin; adductor (*a*) central; cardinal and pedicle impressions (*r*) conjoined, lateral, fan-like: *dorsal* valve with a tooth-like cardinal-process between two curved brachial processes (*c*); adductor impression (*a*) quadruple: vascular impressions consisting of six principal trunks in the dorsal valve, two in the ventral, the external branches turned outwards and backwards inclosing wide ovarian spaces (*o*). Indications have been observed, in several genera, of horizontally-coiled spiral arms; the space between the valves is often very small. The shell-structure is punctate, except in a few instances, where the original texture is probably obliterated.

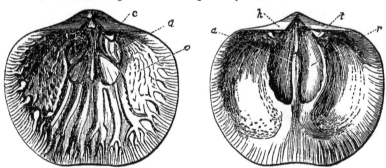

Dorsal valve.†   *Ventral valve.*
Fig. 147. *Orthis, striatula. Devonian, Eifel.*

ORTHIS, Dalman.

Etym. Orthos, straight. *Type*, O. rustica, Pl. XV. fig. 23.

Syn. Dicœlosia (biloba) King. Platystrophia (biforata) King. Gonambonites (inflexa) Pander. Orthambonites (calligramma) Pander.

---

* The names of the Families are formed from those of the *typical* genera, by substituting *idæ* for the last syllable of the genitive case.

† From a specimen presented by M. De Koninck to the British Museum; internal casts of this fossil were called *hysterolites* by old authors.

Shell transversely oblong, radiately striated or plaited, bi-convex, hinge-line narrower than the shell, cardinal process simple, brachial processes tooth-like, prominent and curved.

Fossil, 100 sp. L. Silurian — Carb. Arctic America, U.S. S. America, Falkland Ids. Europe, Thibet.

? *Sub-genera, Orthisina,* D'Orb. O. anomala, Schl. Fig. 148. *Syn.* Pronites (ascendens) and Hemipronites, Pander. *Shell* impunctate ? widest at the hinge-line; cardinal notch closed, byssal notch (*fissure*) covered by a convex pseudo-deltidium, sometimes perforated by a small round foramen. *Fossil.* L. Silurian, Europe.

O. *pelargonatus* (Streptorhynchus, King) from the Magnesian limestone, *O. senilis,* Carb limestone, and some Devonian species, have the beak twisted, as it if had been attached; there is no foramen.

Fig. 148, *Orthisina.*

STROPHOMENA, Blainville.*

Etym. Strophos bent, *mene* crescent.

Ex. S. rhomboidalis, Pl. XV. fig. 24. (= Leptæna depressa, Sby.)

Syn. Leptæna (depressa) Dalman. Leptagonia, M'Coy. Enteletes, Fischer.

Shell semi-circular, widest at the hinge-line, concavo-convex, depressed, radiately striated; area double; ventral valve with an angular notch, progressively covered by a convex pseudo-deltidium; umbo depressed, rarely (?) perforated, in young shells, by a minute foramen (fig. 149, *e*); muscular depressions 4, central pair narrow, formed by the adductor: external pair (*m*) fan-like, left by the cardinal and pedicle muscles; *dorsal* valve with a bi-lobed cardinal process, between the dental sockets, and four depressions for the adductor muscle.

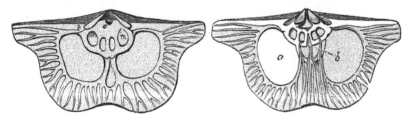

Fig. 149. *Ventral valve.*   *Dorsal valve.*

Interior of S. analoga, Carb. limestone (after King).

e, foramen; *t,* teeth; *o,* ovarian spaces; *b,* brachial pits?

* The name *Strophomena* (rugosa) was originally given by Rafinesque to some unknown or imaginary fossil; it has, however, been adopted both in America and Europe for the group typified by *S. alternata* and *planumbona.*

There are no apparent brachial processes in the dorsal valve of *Stro-phomena*, and it is possible that the spiral arms may have been supported at some point near the centre of the shell (*b*) as in *Producta ; S. rhomboidalis* occasionally exhibits traces of spiral arms, in the ventral valve. *S. latissima* Bouch. has plain areas, like *Calceola.*

The valves of the Strophomenas are nearly flat until they approach their full growth, they then bend abruptly to one side ; the dorsal valve becomes concave in S. alternata and rhomboidalis, whilst in S. planumbona and euglypha it be· comes convex; these distinctions are not even sub-generic.

Fossil, 100 sp. L. Silurian — Carb. N. America, Europe, Thibet.

S. demissa, Conr. (Stropheodonta,

Fig. 150. *Leptæna.* $\frac{2}{1}$

A, hinge areas; v, ventral, B, interior of dorsal valve.

Hall). *S. Dutertrii,* and several other species have a denticulated hinge-line.

Sub-genera ? Leptæna (part) Dalman. L. transversalis, fig. 150. (Plec-tambonites, Pander.) Valves regularly curved ; dorsal concave, thickened, muscular impressions elongated. *Fossil,* L. Silurian —Lias. N. America, Europe. The lias Leptænas resemble *Theoidia* internally ; they are free shells, with sometimes a minute foramen at the apex of the triangular delti-dium ; *L. liassina,* Pl. XV. fig. 25.

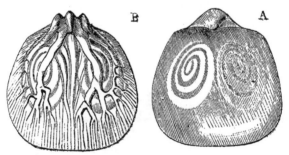

Fig, 145. *Producta ? Leonhardi,* $\frac{?}{1}$.[*]

Koninckia, Suess. Producta Leonhardi, Wissm. (*P. alpina,* Schl.) fig. 145. Trias, St. Cassian. *Shell* orbicular, concavo-convex, smooth ; valves articu· lated ? closely appressed ; ventral valve convex, dorsal concave ; beak in-curved, no hinge-area nor foramen ? interior of each valve furrowed by two spiral lines of four volutions, directed inwards, and crossing the vascular impressions ; umbo with 3 diverging ridges. The small spiral cavities, once occupied by the arms, and now filled with spar, may be seen in specimens with both valves, by holding them to the light. Mr. Suess of Vienna states

[*] A, Translucent specimen; B, interior of dorsal valve.

M

that he has found traces of very slender spiral lamellæ occupying the furrows. This curious little shell most resembles the Triassic *Leptæna dubia* (Producta) Münster (= *Crania Murchisoni*, Klipst. !)

DAVIDSONIA, Bouchard.

Dedicated to the author of the Monograph of British Fossil Brachiopoda. *Type*, D. Verneuili, Bouch. Fig. 151. Devonian, Eifel.

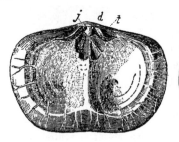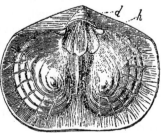

Fig. 151. *Dorsal valve.*   *Ventral valve,* $\frac{2}{1}$.

Shell solid, attached by outer surface of the ventral valve to rocks, shells, and corals; valves plain, articulated; *ventral* valve with a wide area (*h*); foramen angular, covered by a convex deltidium (*d*): disk occupied by two conical elevations, obscurely grooved by a spiral furrow of 5-6 volutions; *dorsal valve* with two shallow lateral cavities; vascular impressions consisting of two principal sub-marginal trunks, in each valve, with diverging branches; cardinal and adductor impressions distinct. The furrowed cones undoubtedly indicate the existence of spiral arms, similar to those of *Atrypa* (fig. 144), but destitute of calcified supports. The mantle-lobes seem to have continued depositing shell until the internal cavity was reduced to the smallest possible limit.

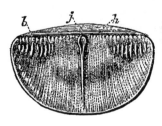

Fig. 152. *Dorsal valve.*   *Ventral valve.*

? CALCEOLA, Lamarck.

Etym. Calceola, a slipper. *Type*, C. sandalina, Pl. XV. fig. 26. Fig. 152.

Shell thick, triangular; valves plain, not articulated: *ventral* valve pyramidal; area large, flat, triangular, with an obscure central line; hinge-line straight, crenulated, *dorsal* valve flat, semi-circular, with a narrow area (*h*). a small cardinal process (*j*), and two lateral groups of small apophysary (?) ridges (*b*); internal surface punctate-striate. *Fossil*, Devonian, Eifel, Brit.

The supposed Carboniferous species (*Hypodema*, D.K.) is, perhaps, related to *Pileopsis*. *Calceola* is shaped like *Cyrtia*, and its hinge-area resembles that of some Strophomenas.

FAMILY V. PRODUCTIDÆ.

Shell concavo-convex, with a straight hinge-line; valves rarely articulated by teeth; closely appressed, furnished with tubular spines; ventral valve convex; dorsal concave; internal surface dotted with conspicuous, funnel-shaped punctures; *dorsal* valve with a prominent cardinal process; brachial processes (?) sub-central; vascular markings lateral, broad and simple; adductor impressions dendritic, separated by a narrow central ridge; *ventral* valve with a slightly notched hinge-line; adductor scar central, near the umbo; cardinal impressions lateral, striated.

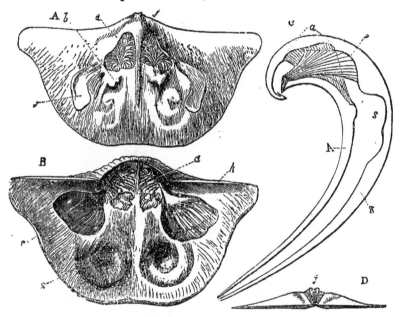

Fig. 153. *Producta gigantea,* ¼ *Carb. limestone.*
A, interior of dorsal valve; B, interior of ventral valve, with the umbo removed; C, ideal section of both valves; D, hinge-line of A; *j*, cardinal process; *a*, adductor; *r*, cardinal muscles; *b*, oral processes ?; *s*, hollows occupied by the spiral arms; *v*, vascular impressions; *h*, hinge-area.

PRODUCTA, Sowerby.

Type, P. gigantea, Sby, = *Anomia producta*, Martin.

Ex. P. horrida, Pl. XV. fig. 27. P. proboscidea, Pl. XV. fig. 28.

Shell free, auriculate, beak large and rounded; spines scattered; hinge area in each valve linear, indistinct; no hinge-teeth; cardinal process lobed, striated; vascular impressions simple, curved; ventral valve deep, with two rounded or sub-spiral cavities in front. These shells may have been attached

M 2

by a pedicle when young, the impressions of the pedicle-muscle blending with those of the hinge-muscles (*c*) in the ventral valve. A few species appear to have been permanently fixed. *P. striata* is irregular in its growth, elongated and tapering towards the beak, and occurs in numbers packed closely together. *P. proboscidea* seems to have lived habitually in cavities, or half-buried in mud, as suggested by M. D'Orbigny; its ventral valve is prolonged several inches beyond the other, and has its edges rolled together and united, forming a large permanently open tube for the brachial currents. The large spines are most usually situated on the ears of the ventral valve, and may have served to moor the shell; being tubular they were permanently susceptible of growth and repair. Although edentulous, the dorsal valve must have turned on its long hinge-line with as much precision as in those genera which are regularly articulated by teeth.

Fossil, 60 sp. Devonian — Permian. N. and S. America, Europe, Spitzbergen, Thibet, Australia.

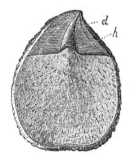
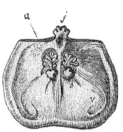

Fig. 154. *Exterior.*   *Interior.*

Sub-genus, Aulosteges, Helmersen. A. Wangenheimii, Vern. fig. 154. Permian, Russia. *Shell* like Producta; ventral valve with a large flat triangular hinge-area (*h*), with a narrow convex pseudo-deltidium (*d*) in the centre: beak a little distorted, as if attached when young; dorsal valve slightly convex near the umbo; interior as in *Producta (longi-spina.)*

STROPHALOSIA, King.

Ex. S. Morrisii, King. fig. 155.

Syn. Orthothrix, Geinitz.

Shell attached by the umbo of the ventral valve; sub-quadrate; covered with long slender spines; valves articulated, dorsal moderately concave, ventral convex, each with a small area; fissure covered; vascular impressions conjoined, reniform.

Fossil, 8 sp. Devonian — Trias. Europe; Himalaya (Gerard).

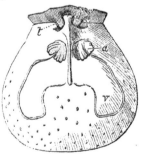

Fig. 155. *S. Morrisii.*

CHONETES, Fischer.

Ex. C. striatella, Pl. XV. fig. 29. *Etym.* *Chone*, a cup.

Shell transversely oblong, with ·a wide and straight hinge-line; area double; valves radiately striated, articulated; hinge-margin of ventral valve with a series of tubular spines; fissure covered; interior punctate-striate; vascular impressions (*v*) very small. (*Davidson*).

Fossil, 24 sp. Silurian — Carboniferous. Europe, N. America, Falkland Ids.

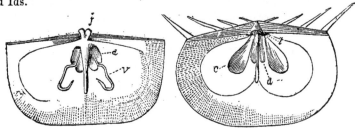

Fig. 156. *Dorsal valve.*   *Ventral valve.**

FAMILY VI. CRANIADÆ.

Shell orbicular, calcarious, hinge-less; attached by the umbo, or whole breadth of the ventral valve, rarely free; dorsal valve limpet-like; interior of each valve with a broad granulated border; disk with four large muscular impressions, and digitated vascular impressions; structure punctate.

Animal with free spiral arms, directed towards the concavity of the dorsal valve, and supported by a nose-like prominence in the middle of the lower valve; mantle extending to the edges of the valves, and closely adhering, its margins plain. (Fig. 159.)

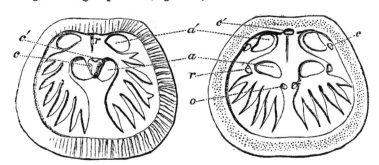

Fig. 157. *Ventral valve.*   Fig. 158. *Dorsal valve.*

Crania anomala, Muller. $\frac{2}{1}$ *Zetland.*

a, anterior adductors; *a'*, posterior adductors; *c*, protractor sliding muscles; *c'*, cardinal muscle; *r*, *o*, retractor sliding muscles.

* Interiors of two sp. of *Chonetes* from Nehou and the Eifel, after Davidson; *a*, adductor: *c*, cardinals.

CRANIA, Retzius.

Etym. *Kraneia,* capitate.   *Type,* Anomia craniolaris, L.

Ex. C. Ignabergensis, Pl. XV. fig. 30.   C. anomala, figs. 157—159.

Syn. Criopus, Poli.   Orbicula (anomala) Cuvier, = O. Norvegica, Lam.

Shell smooth or radiately striated; umbo of dorsal valve sub-central : of ventral valve sub-central, marginal, or prominent and cap-like, with an obscure triangular area traversed by a central line.

The large muscular impressions of the attached valve are sometimes convex, in other species deeply excavated; those of the upper valve are usually convex, but in *C. Parisiensis* the anterior (central) pair are developed as prominent diverging apophyses.   In *C. tripartita,* Münster, the nasal process divides the fixed valve into three cells.*

C. Ignabergensis is equivalve, and either quite free or very slightly attached.   *C. anomala* is gregarious on rocks and stones in deep water, both in the North Sea and Mediterranean (40—90 fathoms, *living;* 150 fms. *dead;* Forbes) : the animal is orange-coloured, and its labial arms are thick, fringed with cirri, and disposed in a few horizontal gyrations (fig. 159.)

Distr. 5 sp. Spitzbergen, Brit. Medit. India, New S. Wales. — 150 fms. *Fossil,* 28 sp.   L. Silurian —.   Europe.

C. antiquissima, Eichw. (Pseudo-crania M'Coy) is free, and has the internal border of the valves smooth; the branchial impressions blend in front. *Spondylobolus craniolaris,* M'Coy, is a small and obscure fossil, from the L. Silurian shale of Builth.   The upper valve appears to have been like *Crania,* the lower to have had a small grooved beak, with blunt, tooth-like processes at the hinge-line.

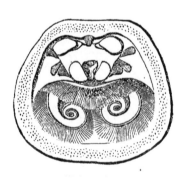
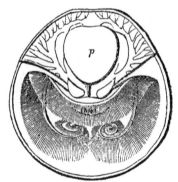

Fig. 159. *Crania.*†   Fig. 160. *Discina.*‡

* M. Quenstedt has placed the Oolitic Cranias in *Siphonaria!*

† Dorsal valve with the animal, seen by removing the mantle.

‡ The animal as seen on the removal of part of the lower mantle-lobe, the extremities of the labial arms are displaced forwards, in order to show their spiral terminations : *p,* is the expanded surface of the pedicle; the mouth is concealed by the overhanging cirri.   The mantle-fringe is not represented.

FAMILY VII. DISCINIDÆ.

Shell attached by a pedicle, passing through a foramen in the ventral valve; valves not articulated; minutely punctate.

Animal with a highly vascular mantle, fringed with long horny setæ : oral arms curved backwards, returning upon themselves, and ending in small spires directed downwards, towards the ventral valve.

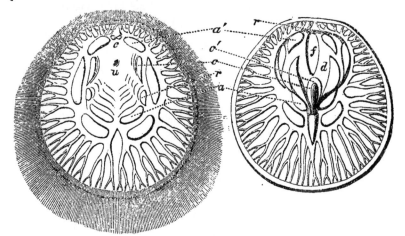

Fig. 161. *Dorsal.*   Fig. 162. *Ventral lobe.*

Discina lamellosa, Brod. $\frac{2}{1}$.

u, umbo ; *f*, foramen; *d*, disk; *a*, anterior adductors; *a'*, posterior adductors ; *c, c'*, protractor sliding muscles; *r*, retractor muscles. The mantle-fringe is not represented in fig. 162.

DISCINA, Lamarck.

Syn. Orbicula, Sby (not Cuvier*).   Orbiculoidea (elliptica) D'Orb.

Type, D. lamellosa, Pl. XV. fig. 31.   (= D. ostreoides, Lam.)

Shell orbicular, horny; upper valve limpet-like, smooth or concentrically lamellose, apex behind the centre; lower valve flat or conical, with a sunk and perforated disk on the posterior side; interior polished; lower valve with a central prominence in front of the foramen.

Animal transparent; mantle lobes distinct all round; labial folds united, not extensile; alimentary canal simple, bent upon itself ventrally, and terminating between the mantle-lobes on the right side.   There are four distinct adductor muscles, as in *Crania ;* and the same number of sliding muscles, viz. two pairs for the protraction and two for the retraction of the dorsal valve, but some of these are probably inserted in the pedicle.   The oral cirri are extremely tender and flexible, contrasting with the stiff and brittle setæ of the mantle, which are themselves more like the bristles of certain anne-

---

* The *Orbicula* of Cuvier was the *Patella anomala*, Müll (= Crania) as pointed out by Dr. Fleming, in the " History of British Animals," 1828.

lides (e. g. the sea-mouse, *Aphrodite*). The relation of the animal to the perforate and imperforate valves is shown to be the same as in *Terebratula*, by the labial fringe; but the only process which can *possibly* have afforded support to the oral arms, is developed from the centre of the ventral valve, as in *Crania*. Baron Ryckholt has represented a Devonian fossil from Belgium, with a fringed border; but if this shell is the *Crania obsoleta* of Goldfuss, the fringe must belong to the shell, and not to the mantle.

Distr. 7 sp. W. Africa, Malacca, Peru, Panama.

Fossil, 29 sp. Silurian —. Europe, U. States, Falkland Ids. The (27) Palæozoic and secondary species constitute the genus *Orbiculoidea*, D'Orb. (*Schizotreta*, Kutorga.) In some species the valves are equally convex, and the foramen occupies the end of a narrow groove.

Sub-genus, Trematis, Sharpe. (= Orbicella, D'Orb.) *T. terminalis*, Emmons. Valves convex, superficially punctate; dorsal valve with a thickened hinge-margin (and three diverging plates, indicated on casts; *Sharpe.*) *Fossil*, 14 sp. L. and U. Silurian. N. America, Europe.

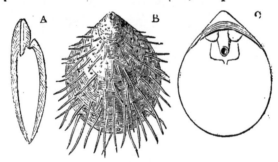

Fig. 163.   Fig. 164. *Exterior.*   Fig. 163, a, *Interior.*

SIPHONOTRETA, Verneuil.

Etym. *Siphon* a tube, *tretos* perforated.

Types, S. unguiculata, Eichw. fig. 163, 163, a. S. verrucosa, fig. 164.

Shell oval, bi-convex, slightly beaked, conspicuously punctate, or spiny; beak perforated by a tubular foramen; hinge-margins thickened; ventral valve with four close adductor scars surrounding the foramen. The spines are tubular, and open into the interior of the shell by prominent orifices. (*Carpenter.*) S. *anglica*, Morris, has moniliform spines.

Fossil, 6 sp. L. and U. Silurian. Brit. Bohemia, Russia.

? *Acrotreta* (sub-conica) Kutorga, L. Silurian, Russia. Shaped like *Cyrtia*, with an apical foramen; no hinge.

FAMILY VIII. LINGULIDÆ.

Shell oblong or orbicular, sub-equivalve, attached by a pedicle passing out between the valves; texture horny, minutely tubular.

Animal with a highly vascular mantle, fringed with horny setæ; oral arms thick, fleshy, spiral, the spires directed inwards, towards each other; valves opened and closed by sliding muscles.

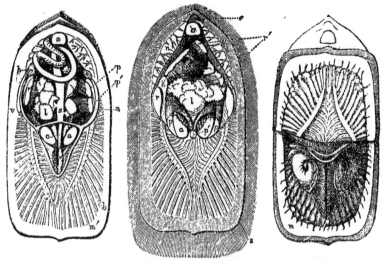

Fig. 165. *Dorsal.**   166. *Ventral.*   167. *Ventral.*

Lingula anatina, Lam (original). *Syn.* Patella unguis, L. (part.)

a a, anterior adductors; *a'*, posterior adductor; *p p*, external protractors; *p'p'*, central protiactors; *r r*, anterior retractors; *r'r'r'*, posterior retractors; *c*, capsule of pedicle; *n n*, visceral sheath; *o*, œsophagus; *s*, stomach; *l*, liver; *i*, intestine; *v*, vent; *h h*, auricles; *h'*, left ventricle; *b*, branchial vessels; *m'*, mantle margin; *m*, inner lamina of mantle-margin retracted, showing bases of setæ; *s*, setæ.

LINGULA, Bruguière.

Etym. Lingula, a little tongue.   *Type*, L. anatina, Pl. XV: fig. 32.

Shell oblong, compressed, slightly gaping at each end, truncated in front, rather pointed at the umbones; dorsal valve rather shorter, with a thickened hinge-margin, and a raised central ridge inside.

Animal with the mantle-lobes firmly adhering to the shell, and united to the epidermis, their margins distinct, and fringed all round; branchial veins giving off numerous free, elongated, narrow loops from their inner surfaces; visceral cavity occupying the posterior half of the shell, and surrounded by a strong muscular sheath; pedicle elongated, thick; adductor muscles 3, the posterior pair combined; two pairs of retractors, the posterior pair unsym-

* In fig 165 a small portion of the liver and visceral sheath have been removed, to show the course of the stomach and intestine. In some specimens the whole of the viscera, except a portion of the liver, are concealed by the ovaries. In fig. 167, the front half of the ventral mantle-lobe is raised, to show the spiral arms; the black spot in the centre is the mouth, with its upper and lower lips, one fringed, the other plain. The mantle-fringe has been omitted in figs. 165-7.

metrical, one of them dividing; protractor sliding muscles, two pairs; stomach long and straight, sustained by inflections of the visceral sheath; intestine convoluted dorsally, terminating between the mantle-lobes on the right side; oral arms disposed in about six close whirls, their cavities opening into the prolongation of the visceral sheath in front of the adductors.

Observations on the living Lingula are much wanted; the oral arms probably extended as far as the margins of the shell; and the pedicle, which is often nine inches long in preserved specimens, is doubtless much longer, and contractile when alive. The shell is horny and flexible, and always of a greenish colour.

Distr. 7 sp. India, Philippines, Moluccas, Australia, Feejees, Sandwich Ids. W. America.

Fossil, 34 sp. L. Silurian —. N. America, Europe, Thibet.

Lingulæ existed in the British Seas as late as the period of the Coralline Crag. The recent species have been found at small depths, and even at low-water half buried in sand. *L. Davisii*, L. Silurian, Tremadoc, has a pedicle-groove like *Obolus*, fig. 168. (Salter).

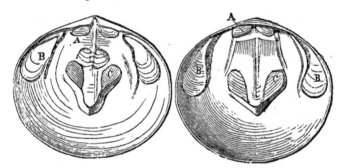

Fig. 168. *Ventral valve.*   Fig. 169. *Dorsal valve.*
Obolus Davidsoni (Salter). Wenlock limestone, Dudley.
A, posterior adductors; B, sliding muscles; C, Anterior adductors.
The *pedicle-scar* in the centre of fig. 168 has no letter.

OBOLUS, Eichwald.

Syn. Ungula, Pander; Aulonotreta, Kutorga.

Etym. Obolus, a small Greek coin. *Type*, O. Apollinis, Eichw.

Shell orbicular, calcario-corneous, depressed, sub-equivalve, smoot ; hinge-margin thickened inside, and slightly grooved in the ventral valve; posterior adductor impressions separate; anterior pair sub-central; impressions of sliding-muscles lateral. Fig. 168, 169 (after Davidson.)

Fossil, 4 sp. L. and U. Silurian. Sweden, Russia, England, U. States.

CLASS V. CONCHIFERA, LAMARCK.
(*Lamelli-branchiata*, Blainville.)

The bivalve shell-fish, or *Conchifera*, are familiar to every one, under the form of oysters, scallops, mussels, and cockles.* They come next to the univalves (*gasteropoda*) in variety and importance, and though less numerous specifically, are far more abundant individually.† The bivalves are all aquatic, and excepting a few widely-dispersed and prolific genera, are all inhabitants of the sea; they are found on every coast, and in every climate, ranging from low-water mark to a depth of more than 200 fathoms.

In their native element the Oyster and Scallop lie on one side, and the lower valve is deeper and more capacious than the upper; in these the foot is wanting, or else small, and not used for locomotion. Most other bivalves live in an erect position, resting on the edges of their shells, which are of equal size. Those which move about much, like the river-mussel, maintain themselves nearly horizontally,‡ and their keel-shaped foot is adapted for ploughing through sand or mud. The position of those bivalves which live half-buried in river-beds or at the bottom of the sea, is often indicated by the darker colour of the part exposed; or by deposits of tufa, or the growth of sea-weed on the projecting ends of the valves.

In *Nucula* and some others the foot is deeply cleft, and capable of expanding into a disk, like that on which the snails glide: whilst in the mussel, pearl-oyster, and others which habitually spin a *byssus*, the foot is finger-like and grooved.

The burrowing species have a strong and stout foot with which they bore vertically into the sea-bed, often to a depth far exceeding the length of their valves; these never voluntarily quit their abodes, and often become buried and fossilized in them. They most usually burrow in soft ground, but also in coarse gravel, and firm sands and clays; one small *modiola* makes its hole in the cellulose tunic of Ascidians, and another in floating blubber.

The boring shell-fish have been distinguished from the mere burrowers, perhaps without sufficient reason, for they are found in substances of every degree of hardness, from soft mud to compact limestone, and the method employed is probably the same. §

The means by which bivalves perforate stone and timber has been the subject of much inquiry, both on account of its physiological interest, and the desire to obtain some remedy for the injuries done to ships and piers and breakwaters. The ship-worm (*teredo*) and some allied genera; perforate timber only; whilst the *pholas* bores into a variety of materials, such as

* They are the *Dithyra* of Aristotle and Swainson, and constitute the second or sub-typical group in the quinary system.

† It has been stated that the predatory *mollusca* are more numerous than the vegetable-feeders; but it is not so with the individuals constituting the species.

‡ This is the position in which they are always figured in English books, being best suited for the comparison of one shell with another.

§ See the admirable memoir by Mr. Albany Hancock, in the An. Nat. Hist. for October, 1848.

chalk, shale, clay, soft sandstone and sandy marl, and decomposing *gneiss ;**
it has also been found boring in the peat of submarine forests, in wax, and in
amber.† It is obvious that these substances can only be perforated alike
by *mechanical* means ; either by the foot or by the valves, or both together,
as in the burrowing shellfish. The *pholas* shell is rough, like a file, and
sufficiently hard to abrade limestone; and the animal is able to turn from
side to side, or even quite round in its cell, the interior of which is often
annulated with furrows made by the spines on the front of the valves. The
foot of the *pholas* is very large, filling the great anterior opening of the
valves ; that of the ship-worm is smaller, but surrounded with a thick collar,
formed by the edges of the mantle, and both are armed with a strong *epithelium.* The foot appears to be a more efficient instrument than the shell in
one respect, inasmuch as its surface may be renewed as fast as it is worn
away.‡ (*Hancock.*)

The mechanical explanation becomes more difficult in the case of another
set of shells, *lithodomus, gastrochæna, saxicava,* and *ungulina,* which bore
only into calcarious rocks, and attack the hardest marble, and still harder
shells (fig. 25, p. 42). In these the valves can render no assistance, as they
are smooth, and covered with *epidermis ;* neither does the foot help, being
small and finger-like, and not applied to the end of the burrow. Their power
of movement also is extremely limited, their cells not being cylindrical, whilst
one of them, *saxicava,* is fixed in its crypt by a *byssus.* These shell fish have
been supposed to dissolve the rock by chemical means (*Deshayes*), or else to
wear it away with the thickened anterior margins of the mantle. (*Hancock.*) §

The holes of the *lithodomi* often serve to shelter other animals after the

* There is a specimen from the coast of France, in the Brit. Museum.

† Highgate resin, in the cabinet of Mr. Bowerbank.

‡ The final polish to some steel goods is said to be given by the *hands* of work-
women. In Carlisle Castle they point to the rude impression of a hand on the
dungeon wall, as the work of FERGUS M'IVOR, in the two years of his solitary imprisonment.

§ All attempts to detect the presence of an acid secretion have hitherto failed, as
might be expected ; for the hypothesis of an acid solvent supposes only a very feeble
but continuous action, such as in nature always works out the greatest results in the
end. See Liebig's Organic Chemistry, and Dumas and Boussingault on the "Balance
of Organic Nature." Intimately connected with this question are several other
phenomena; the removal of portions of the interior of univalves, by the animal
itself, as in the genera *Conus, Auricula,* and *Nerita* (fig. 24, p. 40); the perforation of
shells by the tongues of the carnivorous gasteropods ;and the formation of holes in
wood and limestone by limpets. Some facts in surgery also illustrate this subject,
(1) dead bone is removed when granulations grow into contact with it : (2) if a hole is
bored in a bone, and an ivory peg driven into it, and covered up, so much of the peg as
is imbedded in the bone will be removed. (*Paget.*) The "absorption" of the fangs
of milk-teeth, previous to shedding, is well-known. In these cases the removal of the
bone earth is effected without the development of an acid, or other disturbance of the
neutral condition of the circulating fluid.

death of the rightful owners; species of *Modiola, Arca, Venerupis,* and *Coralliophaga,* both recent and fossil, have been found in such situations, and mistaken for the real miners.*

The boring shellfish have been called "stone-eaters" (*lithophagi*) and "wood-eaters" (*xylophagi*), and some of them at least are obliged to swallow the material produced by their operations, although they may derive no sustenance from it. The ship-worm is often filled with pulpy, impalpable sawdust, of the colour of the timber in which it worked. (*Hancock.*) No shellfish deepens or enlarges its burrow after attaining the full-growth usual to its species (p. 43).

The bivalves live by filtering water through their gills.† Whatever particles the current brings, whether organic or inorganic, animal or vegetable, are collected on the surface of the breathing-organ and conveyed to the mouth. In this manner they help to remove the impurities of turbid water.‡ The mechanism by which this is effected may be most conveniently examined in a bivalve with a closed mantle, like the great *Mya* (fig. 170), which lives in the mud of tidal rivers, with only the ends of its long combined siphons exposed at the surface.§ The siphons can be extended twice the length of the shell, or drawn completely within it; they are separated, internally, by a thick muscular wall. The branchial siphon (*s*) has its orifice surrounded by a double fringe; the exhalent siphon (*s'*) has but a single row of tentacles; these organs are very sensitive, and if rudely touched the orifices close and the siphon itself is rapidly withdrawn. When unmolested, a current flows steadily into the orifice of the branchial siphon, whilst another current rises up from the exhalent tube. There is no other opening in the mantle except a small slit in front (*p*) through which the foot is protruded. The body of the animal occupies the centre of the shell (*b*), and in front of it is the mouth (*o*) furnished with an upper and a lower lip, which are prolonged on each side into a pair of large membranous palpi (*t*). The gills (*g*) are placed two on each side of the body, and are attached along their upper, or dorsal margins; behind the body they are united to each other and to the siphonal partition. Each gill is composed of two laminæ, divided internally into a series

* Fossil univalves (*trochi*) occupying the burrows of a *pholas,* were discovered by Mr. Bensted in the Kentish-rag of Maidstone. See Mantell's Medals of Creation. M. Buvignier has found several species of *Arca* fossilized in the burrows of *lithodomi.*

† It seems scarcely necessary to remark that the bivalves do not feed upon prey caught between their valves. Microscopists are well aware that sediment taken from the alimentary canal of bivalve shellfish contains the skeletons of animalcules and minute vegetable organisms, whose geometrical forms are remarkably varied and beautiful; they have also been obtained (in greater abundance than ordinary) from mud filling the interior of fossil oyster-shells.

‡ When placed in water coloured with indigo, they will in a short time render it clear, by collecting the minute particles and condensing them into a solid form.

§ Alder and Hancock on the branchial currents of *Pholas* and *Mya.* An. Nat. Hist. Nov. 1851.

of parallel tubes, indicated outside by transverse lines; these tubes open into longitudinal channels at the base of the gills, which unite behind the posterior adductor muscle at the commencement of the exhalent siphon (*c*). Examined by the microscope, the gill laminæ appear to be a network of blood-vessels whose pores opening into the gill-tubes, are fringed with vibratile *cilia*. These microscopic organs perform most important offices; they create the currents of water, arrest the floating particles, and mould them, mixed with the viscid secretion of the surface, into threads, in the furrows of the gill, and propel them along the grooved edge of its free margin, in the direction of the mouth; they are then received between the palpi in the form of ravelled threads. (*Alder* and *Hancock*.)

In *Mya*, therefore (and in other burrowers), the cavity of the shell forms a closed branchial chamber, and the water which enters it by the respiratory siphon can only escape by passing through the gills into the dorsal channels, and so into the exhalent siphon. In the river-mussel the gills are not united to the body, but a slit is left by which water might pass into the dorsal channel, were it not for the close apposition of the parts under ordinary circumstances (fig. 171, *b*). The gills of the oyster are united

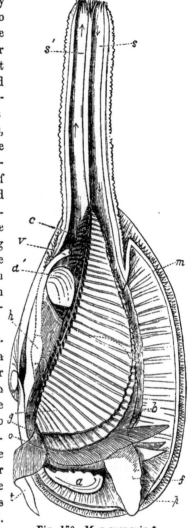

Fig. 170. *Mya arenaria.**

throughout, by their bases, to each other and to the mantle, completely separating the branchial cavity from the *cloaca*. In *Pecten* the gills and mantle are free, but the "dorsal channels" still exist, and carry out the filtered water. ·

---

* *Mya arenaria*, L. (original, from specimens obtained at Southend, and communicated by Miss Hume). The left valve and mantle lobe and half the siphons are removed. *a, a'*, adductor muscles; *b*, body; *c*, cloaca; *f*, foot; *g*, branchiæ; *h*, heart; *m*, cut edge of the mantle; *o*, mouth; *s, s'*, siphons; *t*, labial tentacles; *v*, vent. The arrows indicate the direction of the currents; the four rows of dots at the base of the gills are the orifices of the branchial tubes, opening into the dorsal channels.

In some genera the gills subserve a third purpose; the oviducts open into the dorsal channels. and the eggs are received into the gill-tubes and retained there until they are hatched. In the river-mussel the outer gills only receive the eggs, with which they are completely distended in the winter months (Fig. 171, *o, o*). In *Cyclas* the inner gills form the *marsupium*, and only from 10 to 20 of the fry are found in them at one time; these remain until they are nearly a quarter the length of the parent.*

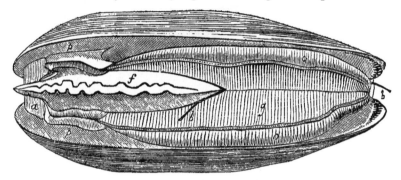

Fig. 171. River-mussel. (*Anodon cygneus* ♀)†

The *valves of the Conchifera* are bound together by an elastic *ligament*, and articulated by a hinge furnished with interlocking teeth. The shell is closed by powerful adductor muscles, but opens spontaneously by the action of the ligament, when the animal relaxes, and after it is dead.

Each valve is a hollow cone, with the apex turned more or less to one side; the apex is the point from which the growth of the valve commences, and is termed the beak, or *umbo* (p. 37.) The beaks (*umbones*) are near the hinge, because that side grows least rapidly, sometimes they are quite marginal; but they always tend to become wider apart with age. The beaks are either straight, as in *Pecten;* curved as in *Venus;* or spiral, as in *Isocardia* and *Diceras.* In the latter case each valve is like a spiral univalve, especially those with a large aperture and small spire, such as *Concholepas;* it is the left valve which resembles the ordinary univalve, the right valve being a *left-handed spiral* like the reversed gasteropods. When one valve is spiral and the other flat, as in *Chama ammonia* (fig. 185), the resemblance to an operculated spiral univalve becomes very striking (see p. 47).

* Some other particular respecting the organization and development of bivalve shell-fish are given in the introductory chapter. For an account of their vascular system see Milne-Edwards, An. Sc. Nat. 1847, Tom. VIII. p. 77.

† The valves are forcibly opened and the foot (*f*) contracted; *a*, anterior adductor-. muscle, much stretched; *p, p*, palpi; *g*, inner gills; *o, o*, outer gills distended with spawn; *b, b*, a bristle passed through one of the dorsal channels.

The relation of the shell to the animal may be readily determined, in most instances, by the direction of the *umbones*, and the position of the ligament. The umbones are turned towards the front, and the ligament is posterior; both are situated on the back, or dorsal side of the shell. The *length* of a bivalve is measured from the anterior to the posterior side, its *breadth* from the dorsal margin to the base, and its *thickness* from the centres of the closed valves.*

Dorsal margin.

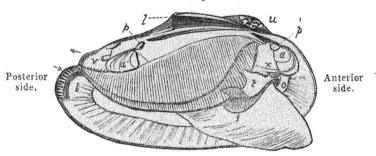

Posterior side.

Anterior side.

Ventral margin, or base.

Fig. 172. *Unio pictorum*, L. (original) with the right valve and mantle-lobe removed; *a, a*, adductor muscles; *p. p*, pedal muscles; *x*, accessory pedal muscle; *u*, umbo; *l*, ligament; *b*, branchial orifice; *v*, anal opening; *f*, foot; *o*, mouth; *t*, palpi.

The *Conchifera* are mostly *equivalve*, the right and left valves being of the same size and shape, except in the *Ostreidæ* and a few others. In *Ostrea*, *Pandora* and *Lyonsia* the right valve is smallest; in *Chamostrea* and *Corbula*, the left; whilst the *Chamaceæ* follow no rule in this respect.

The bivalves are all more or less *inequilateral*, the anterior being usually much shorter than the posterior side. *Pectunculus* is nearly equilateral, and in *Glycimeris* and *Solemya*, the anterior is much longer than the posterior side. The front of the smaller Pectens is shewn by the byssal notch; but in the large scallops, oysters and *Spondyli*, the only indication of the position of the animal is afforded by the large internal muscular impression, which is on the *posterior* side. The ligament is sometimes between the umbones, but s never anterior to them. The *siphonal impression*, inside the shell, is always posterior.

Bivalves are said to be *close*, when the valves fit accurately, and *gaping*

---

* Linnæus and the naturalists of his school, described the front of the shell as the back, the left valve as the right, and *vice versa*. In those works which have been compiled from "original descriptions" (instead of specimens) sometimes one end, sometimes the other, is called *anterior;* and the *length* of the shell is sometimes estimated in the direction of the length of the animal, but just as frequently in a line at right angles to it.

when they cannot be completely shut. In *Gastrochæna* (Pl. XXIII. fig. 15,) the opening is anterior, and serves for the passage of the foot ; in *Mya* it is posterior and siphonal; in *Solen* and *Glycimeris* both ends are open. In *Bysso-arca* (Pl. XVII. fig, 13,) there is a ventral opening formed by corresponding notches in the margin of the valves, which serves for the passage of the byssus; in *Pecten*, *Avicula*, and *Anomia*, (fig. 176 *s*) the byssal notch (or *sinus*) is confined to the right valve.

The *surface* of bivalve shells is often ornamented with ribs which *radiate* from the umbones to the margin, or with *concentric* ridges, which coincide with the lines of growth. Sometimes the sculpturing is oblique, or wavy; in *Tellina fabula* it is confined to the right valve. In many species of *Pholas*, *Teredo* and *Cardium* the surface is divided into two areas by a transverse furrow, or by a change in the direction of the ribs. The *lunule* (see fig. 14, p. 26,) is an oval ·space in *front* of the beaks; it, is deeply impressed in *Cardium retusum*, L. *Astarte excavata* and the genus *Opis*. When a similar impression exists behind the beaks it is termed the *escutcheon*.*

The *ligament* of the *Conchifera* forms a substitute for the muscles by which the valves of the *Brachiopoda* are opened. It consists of two parts, the ligament properly so called, and the *cartilage*; they exist either combined or distinct, and sometimes one is developed and not the other. The external ligament is a horny substance, similar to the *epidermis* which clothes the valves; it is usually attached to ridges on the posterior hinge-margins, behind the umbones, and is consequently stretched by the closing of the valves. The ligament is large in the river-mussels, and small in the Mactras and Myas, which have a large internal cartilage; in *Arca* and *Pectunculus* the ligament is spread over a flat, lozenge-shaped area, situated between the umbones, and furrowed with cartilage grooves. In *Chama* and *Isocardia* the ligament splits in front, and forms a spiral round each umbo. The *Pholades* have no ligament, but the anterior adductor is shifted to such a position on the hinge-margin that it acts as a hinge·muscle. (Pl. XXIII, fig. 13.)

The internal ligament, or *cartilage*, is lodged in furrows formed by the ligamental plates, or in pits along the hinge-line; in *Mya* and *Nucula* it is contained in a spoon-shaped process of one or both valves. . It is composed of elastic fibres placed perpendicularly to the surfaces between which it is contained, and is slightly iridescent when broken; it is compressed by the closing of the valves, and tends forcibly to open them as soon as the pressure of the muscles is removed. The name *Amphidesma* (double ligament) was given to certain bivalves, on the supposition that the separation of the carti-

---

* Only those technical terms which are used in a *peculiar sense* are here referred to ; for the rest, any Dictionary may be consulted, especially *Roberts's Etymological Dictionary of Geology*, by Longman and Co.

lage from the ligament was peculiar to them. The cartilage-pit of many of the *Anatinidæ* is furnished internally with a moveable ossicle.

The ligament is frequently preserved in fossil shells, such as the great Cyprinas and Carditas of the London Clay, the Unios of the Wealden, and even in some lower Silurian bivalves.

All bivalves are clothed with an *epidermis* (*v.* p. 40) which is organically connected with the margin of the mantle. It is developed to a remarkable extent in *Solemya* and *Glycimeris* (Pl. XXII. fig. 13, 17), and in *Mya* it is continued over the siphons and closed mantle-lobes, making the shell appear *internal*.

The *interior* of bivalves is inscribed with characters borrowed directly from the shell-fish, and affording a surer clue to its affinities than those which the exterior presents. The structure of the *hinge* characterizes both families and genera, whilst the condition of the respiratory and locomotive organs may be to some extent inferred from the muscular markings.

The margin of the shell on which the ligament and teeth are situated, is termed the *hinge-line*. It is very long and straight in *Avicula* and *Arca*, very short in *Vulsella*, and curved in most genera. The locomotive bivalves have *generally* the strongest hinges, but the most perfect examples are presented by *Arca* and *Spondylus*. The central teeth, those immediately beneath the *umbo*, are called hinge (or *cardinal*) teeth; those on each side are *lateral* teeth. Sometimes lateral teeth are developed, and not cardinal teeth (*Alasmodon; Kellia*): more frequently the hinge-teeth alone are present. In young shells the teeth are sharp and well-defined; in aged specimens they are often thickened, or even obliterated by irregular growth (*Hippopodium*) or the encroachment of the hinge-line (*Pectunculus*). Many of the fixed and boring shells are *edentulous*.*

The *muscular impressions* are those of the adductors, the foot and byssus, the siphons, and the mantle (see p. 26.)

The *adductor impressions* are usually simple, although the muscles themselves may be composed of two elements,† as in *Cytherea chione* (fig. 14, p. 26) and the common oyster. The impression of the posterior adductor in *Spondylus* is double (Pl. XVI. fig. 15). In *Pecten varius* (fig. 173, *a, a,*) large independent impressions are formed by the two portions of the adductor, and in the *left* valve there is a third impression (*p*) produced by the foot, which in the byssiferous pectens is a simple conical muscle with a broad base.

---

* The dentition of bivalve shells may be stated thus:—cardinal teeth, 2.3 or $\frac{2}{3}$ —meaning 2 in the *right* valve, 3 in the *left;* lateral teeth 1—1, 2—2, or 1 anterior and 1 posterior in the *right* valve, 2 anterior and 2 posterior lateral teeth in the left valve.

† Compare the shell of *modiola*, Pl. XVII. fig. 5, with the woodcut, fig. 177.

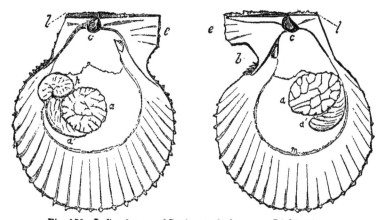

Fig. 173. *Left valve.*   (*Pecten varius*):   *Right valve.*

a, a, adductor; *p,* pedal impression; *m,* pallial line: *l,* ligamental margin; *c, c,* cartilage; *e, e,* anterior ears; *b.* byssal sinus.

In the *left* valve of *Anomia* there are four distinct muscular impressions
(fig. 175). Of these, the small posterior spot alone is produced by the *adductor,* and corresponds with the solitary impression in the right valve.

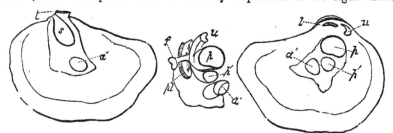

Fig. 176. *Right valve.*   Fig. 174.   Fig. 175. *Left valve.* *

The adductor itself (fig. 174 *a'*) is double. The large central impression (*p*)
is produced by the muscle of the *plug* (the equivalent of the *byssal* muscle in
Pinna and *Modiola*). The small impression within the umbo (*u*) and the
third impression in the disk (*p'*) (wanting in *Placunomia*) are caused by the
retractors of the foot.

The term *monomyary,* employed by Lamarck to distinguish the bivalves
with one adductor, applies only to the *Ostreidæ,* part of the *Aviculidæ,* and
to the genera *Tridacna* and *Mülleria.*

The *dimyary* bivalves have a second adductor, near the anterior margin,

---

* Fig. 176. Right valve of *Anomia ephippium,* L. *l,* ligamental process; *s,* sinus.
Fig. 175, Left valve; *l,* ligament pit. Fig. 174. Muscular system, from a drawing
communicated by A. Hancock, Esq. *f,* the foot; *pl,* the plug. The muscle *p* is
generally described as a portion of the *adductor;* but it is certain, from a comparison
of this shell with *Carolia* and *Placuna,* that *a'* represents the entire adductor, and *p*
the byssal muscle.

which is small in *Mytilus* (fig. 30), but large in *Pinna.* The *retractor* muscles of the foot (already alluded to at p. 26) have their fixed points near those of the adductors; the anterior pair are attached within the umbones (fig. 177, *u, u,*) or nearer the adductor, as in *Astarte,* and *Unio* (fig. 172). The posterior pair (*p' p'*) are often close to the adductor, and leave no separate impression. The *Unionidæ* have two additional retractors of the foot, attached laterally behind the anterior adductors; in *Leda, Solenella,* and a few others, this lateral attachment forms a line extending from the anterior adductor backwards into the umbonal region of the shell. (See Pl. XVII. fig. 21, 22.)

In those shellfish like *Pinna* and the mussel, which are permanently moored by a strong *byssus,* the foot (*f*) serves only to mould and fix the threads of which it is formed. The fibres of the foot-muscles pass chiefly to the byssus (*b*), and besides these two additional muscles (*p, p*) are developed. In *Pinna, Modiola* and *Dreissena* the byssal muscles are equal to the great adductors in size.

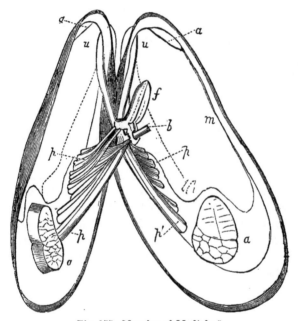

Fig. 177. Muscles of Modiola.*

In a few rare instances the muscles are fixed to prominent *apophyses.* The *falciform processes* of *Pholas* and *Teredo* (Pl. XXIII. fig. 19, 26) are developed for the attachment of the foot-muscle; the posterior muscular

* Fig, 177. Muscular system of *Modiola modiolus*, L. from a drawing communicated by A. Hancock, Esq. *aa*, anterior, *a'a'* posterior adductors; *uu* and *p'p'*, pedal muscles; *pp*, byssal muscles; *f*, foot; *b*, byssus; *m*, pallial line.

ridge of *Diceras* and *Cardilia* resembles a lateral tooth, and in the extinct genus *Radiolites* both adductors were attached to large tooth-like processes of the opercular valve; but, as a rule, the muscles deposit less shell than the mantle, and their impressions deepen with age.

The *pallial line* (fig. 177, *m*) is produced by the muscular fibres of the mantle-margin; it is broken up into irregular spots in the monomyary bivalves, and in *Saxicava* and *Panopæa Norvegica*.

The *siphonal impression*, or *pallial sinus* (fig. 14, p. 26,) only exists in those shells which have retractile siphons; its depth is an index to their length. The large combined siphons of *Mya* (fig. 170) are much longer than the shell; and those of some *Tellinidæ* three or four times its length, yet they are completely retractile. The small siphons of *Cyclas* and *Dreissena* cause no inflection of the pallial line. The *form* of the sinus is characteristic of genera and species.

In the *umbonal area* (within the pallial line) there are sometimes furrows produced by the viscera, which may be distinguished from the muscular markings by absence of polish and outline. (See *Lucina*, Pl. XIX. fig. 6.)

Fossil bivalves are of constant occurrence in all sedimentary rocks; they are somewhat rare in the oldest formations, but increase steadily in number and variety through the *secondary* and *tertiary* strata, and attain a maximum of development in existing seas.

Some families, like the *Cyprinidæ* and *Lucinidæ* are more abundant fossil than recent; whilst many genera, and one whole family (the *Hippuritidæ*), have become extinct. The determination of the affinities of fossil bivalves is often exceedingly difficult, owing to the conditions under which they occur. Sometimes they are found in pairs, filled up with hard stone; and frequently as casts, or moulds of the interior, giving no trace of the hinge, and very obscure indications of the muscular markings. Casts of single valves are more instructive, as they afford impressions of the hinge.[*]

Another difficulty arises from the frequent destruction of the nacreous or lamellar portion of the fossil bivalves, whilst the cellular layers remain. The *Aviculidæ* of the chalk have entirely lost their pearly interiors; the *Spondyli, Chamas*, and *Radiolites* are in the same condition, their inner layers are gone and no vacancy left, the whole interior being filled with chalk. As it is the inner layer alone which forms the hinge, and alone receives the impressions of the soft parts, the true characters of the shells could not be determined from such specimens. Our knowledge of the extinct *Radiolite* is derived from natural moulds of the interior, formed before the dissolution of

---

[*] These impressions may be conveniently moulded with *gutta-percha*. M. Agassiz published a set of plaster-casts of the interiors of the genera of recent shells, which may be seen in the Brit. Museum. [*Memoire sur les moules des Mollusques, vivans et fossiles, par L. Agassiz, Mem. Soc. Sc. Nat. Neuchatel, t.* 2.].

the inner layer of shell, or from specimens in which this layer is replaced by spar.

The necessities of geologists have compelled them to pay very minute attention to the markings in the interior of shells, to their microscopic texture, and every other available source of comparison and distinction. It must not, however, be expected that the entire structure and affinities of molluscous animals can be predicated from the examination of an internal mould or a morsel of shell, any more than that the form and habits of an extinct quadruped can be inferred from a solitary tooth or the fragment of a bone.*

The *systematic arrangement* of the bivalves now employed is essentially that of Lamarck, modified, however, by many recent observations. The families follow each other according to *relationship*, and not according to absolute rank; the *Veneridæ* are the highest organized, and from this culminating point the stream of affinities takes two courses, one towards the Myas, the other in the direction of the oysters; groups analogically related to the *Tunicaries* and *Brachiopoda*.

SECTION A.   Asiphonida.

a. Pallial line simple: Integro-pallialia.

Fam. 1. Ostreidæ.
2. Aviculidæ.
3. Mytilidæ.

4. Arcadæ.
5. Trigoniadæ.
6. Unionidæ.

SECTION B.   Siphonida.

7. Chamidæ.
8. Hippuritidæ.
9. Tridacnidæ.
10. Cardiadæ.

11. Lucinidæ.
12. Cycladidæ.
13. Cyprinidæ.

b. Pallial line sinuated: Sinu-pallialia.

14. Veneridæ.
15. Mactridæ.
16. Tellinidæ.
17. Solenidæ.

18. Myacidæ.
19. Anatinidæ.
20. Gastrochænidæ.
21. Pholadidæ.

The characters which have been most relied on for distinguishing these groups and the genera of bivalves are the following, stated nearly in the order of their value:—

1. Extent to which the mantle-lobes are united.
2. Number and position of muscular impressions.
3. Presence or absence of a *pallial sinus*.
4. Form of the foot.
5. Structure of the *branchiæ*.

* *Etudes Critiques sur les Mollusques Fossiles, par L. Agassiz, Neuchatel,* 1840.

6. Microscopic structure of the shell. (*v.* p, 38.)
7. Position of the *ligament*, internal or external.
8. Dentition of the hinge.
9. Equality or inequality of the valves.
10. Regularity or irregularity of form.
11. Habit;—free, burrowing or fixed.
12. Medium of respiration, fresh or salt-water.

A few exceptions may be found, in which one or other of these characters does not possess its usual value.* Such instances serve to warn us against too implicit reliance on *single characters.* Groups, to be *natural*, must be based on the consideration of all these particulars—on "the totality of the animal organization." (Owen).

SECTION A. Asiphonida.

Animal unprovided with respiratory siphons; mantle-lobes free, or united at only one point which divides the branchial from the exhalent chamber (*cloaca*); pallial impression simple.

Shell usually pearly or sub-nacreous inside; cellular externally; pallial line simple or obsolete.

FAMILY I. Ostreidæ.

Shell inequivalve, slightly inequilatural, free or adherent, resting on one valve; beaks central, straight; ligament internal; epidermis thin; adductor impression single, behind the centre; pallial line obscure; hinge usually edentulous.

Animal marine; mantle quite open; very slightly adherent to the edge

* 1. *Cardita* and *Crassatella* (Fam. 13) have the mantle more open, whilst in *Iridina* (6), and especially in *Dreissena* (3) it is more closed than in the most nearly allied genera.

2. *Mulleria* (6) and *Tridacna* (9) are monomyary.

3. *Leda* (4) and *Adacna* (10) have a pallial sinus; *Anapa* (16) has none.

4. The form of the foot is usually characteristic of the families; but sometimes it is *adaptively* modified.

5. *Diplodonta* (11) has four gills.

6. Pearly structure is variable even in species of the same genus.

7. *Crassatella* (13) and *Semele* (16) have an internal ligament; in *Solenella* and *Isoarca* (4) it is external.

8. *Anodon* (16), *Adacna, Serripes* (10), and *Cryptodon* (11) are edentulous.

9. *Corbula* (18) and *Pandora* (19) are more inequivalve than their allies; *Chama arcinella* (7) is equivalve.

10. *Hinnites* (1), *Ætheria* (6), *Myochama* and *Chamostrea* (19) are irregular.

11. *Pecten* is free, byssiferous, or fixed: *Arca* free or byssiferous. This character varies with *age* and *locality* in the same species. It does not always depend on the form of the foot, as *Ætheria*, though fixed, has a large foot, and *Lithodomus* and *Ungulina*—boring shells—have the foot like *Mytilus* and *Lucina.*

12. *Novaculina* is a river *Solen*, and *Scaphula* a fresh-water *Arca.*

of the shell; foot small and byssiferous, or obsolete; gills crescent-shaped, 2 on each side; adductor muscle composed of two elements, but representing only the *posterior* shell-muscle of other bivalves.

OSTREA, L. Oyster.

Syn. Amphidonta and Pycnodonta, Fischer. Peloris, Poli.

Type, O. edulis, L. *Ex.* O. diluviana, Pl. XVI. fig. 1.

Shell irregular, attached by the left valve; upper valve flat or concave, often plain; lower convex, often plaited or foliaceous, and with a prominent beak; ligamental cavity triangular or elongated; hinge toothless; structure sub-nacreous, laminated, with prismatic cellular substance between the margins of the laminæ.

Animal with the mantle-margin double, finely fringed; gills nearly equal, united posteriorly to each other and the mantle-lobes, forming a complete branchial chamber; lips plain; palpi triangular, attached; sexes distinct.*

Distr. 60 sp. Tropical and temperate seas. Norway, Black Sea, &c.

Fossil, 200 sp. Carb. —. U. States, Europe, India.

The interior of recent oyster-shells has a slightly nacreous lustre; in fossil specimens an irregular cellular structure is often very apparent on decomposed or fractured surfaces. Fossil oysters which have grown upon *Ammonites, Trigoniæ,* &c. frequently take the form of those shells.

In the "cock's-comb" oysters both valves are plaited; *O. diluviana* sends out long root-like processes from its lower valve. The "Tree oyster" (*Dendrostrea,* Sw.) grows on the root of the mangrove. Oyster shells become very thick with age, especially in rough water; the fossil oyster of the Tagus (*O. longirostris*) attains a length of two feet. The greatest enemy of oyster-banks is a sponge (*Cliona*), which eats into the valves, both of dead and living shells; at first only small round holes, at irregular intervals, and often disposed in regular patterns, are visible; but ultimately the shell is completely mined and falls to pieces. Natural oyster-banks usually occur in water several fathoms deep; the oysters spawn in May and June, and the fry ("spats") are extensively collected and removed to artificial grounds, or tanks, where the water is very shallow; they are then called "natives," and do not attain their full growth in less than 5 or 7 years, whilst the "sea-oysters" are full-grown in 4 years. Native oysters do not breed freely, and many sometimes die in the spawning season; they are also liable to be killed by frost. The season is from August 4 to May 12. From 20 to 30,000 bushels of "natives" and 100,000 bushels of sea-oysters are annually sent to the London market. Many other species of oysters are eaten in India, China, Australia, &c. "Green oysters" are those which have fed on *con-*

---

* The course of the alimentary canal in the common oyster is incorrectly represented by Poli, and copied in the Crochard ed. of Cuvier.

fervæ in the tanks. *Sub-genera. Gryphæa,*
Lamarck. G. incurva, Sby (section) fig.
178. Free, or very slightly attached; left
valve with a prominent, incurved umbo;
right valve small, concave. *Fossil,* 30 sp.
Lias — Chalk. Europe, India.

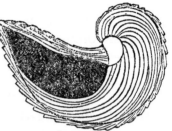

    Exogyra, Sby. E. conica, Pl. XVI.
fig. 2. *Shell* chama-shaped, attached by
the left valve; umbones sub-spiral, turned
                            Fig. 178. Gryphæa.
to the posterior side (i. e. reversed); right valve opercular. *Fossil,* 40 sp.
L. Oolite — Chalk. U. States; Europe.

ANOMIA, L.

Etym. Anomios, unequal. *Ex.* A. Achæus, Pl. XVI. fig. 3.

Syn. Fenestrella, Bolten; Cepa, Humph. Aenigma, Koch.

Shell sub-orbicular, very variable, translucent, and slightly pearly
within, attached by a plug passing through a hole or notch in the right
valve: upper valve convex, smooth, lamellar or striated; interior with a
sub-marginal cartilage-pit, and four muscular impressions, 3 sub-central, and
one in front of the cartilage (see fig. 175, p. 249): lower valve concave,
with a deep, rounded notch in front of the cartilage process; disk with a
single (adductor) impression.

Animal with the mantle open, its margins with a short double fringe;
lips membranous; palpi elongated, fixed, striated on both sides; gills 2 on
each side, united posteriorly, the outer laminæ incomplete and free; foot small,
cylindrical, subsidiary to a lamellar and more or less calcified byssal plug,
attached to the upper valve by three muscles; adductor muscle behind the
byssal muscles, small, composed of two elements; sexes distinct; ovary extending into the substance of the lower mantle-lobe

In *A. pernoides,* from California, there is an anterior (*pedal*) muscular
impression in both valves.

"There is no relationship of *affinity* between *Anomia* and *Terebratula,*
but only a resemblance through formal *analogy;* the parts which seem identical are not homologous." (Forbes).

The Anomiæ are found attached to oysters and other shells, and frequently
acquire the form of the surfaces with which their growing margins are in
contact. They are not edible.

Distr. 20 sp. N. America, Brit. Black Sea, India, Australia, W. America,
Icy sea. Low-water — 100 fms.

Fossil, 30 sp. Oolite —, Chile, U. States, Europe.

Sub-genera. Placunomia (Cumingii) Broderip. *Syn.* Pododesmus, Phil.
P. macroschisma, Pl. XVI. fig. 4. Upper valve with only two muscular
impressions; the pedal scar radiately striated; the byssal plug is often fixed

N

in the lower valve, and its muscle becomes (functionally) an adductor. *Distr.* 12 sp. W. Indies, Brit. (*P. patelliformis*), New Zealand, California, Behring's sea, Ochotsk. — 50 fms.

Limanomia (Grayana) Bouchard. Shell eared like *Lima. Fossil,* 4 sp. Devonian ; Boulonnais, China ?

PLACUNA, Solander. Window-shell.

Etym. Plakous a thin cake. *Ex.* P. sella, Pl. XVI. fig. 5.

Shell sub-orbicular, compressed, translucent, free, resting on the right valve ; hinge area narrow and obscure ; cartilage supported by two diverging ridges in the right valve and corresponding grooves in the left ; muscular impressions double, the larger element round and central, the smaller distinct and crescent shaped, in front of it.

The Placunæ are very closely allied to *Anomia ;* and many intermediate forms may be traced. The shell of each consists entirely of sub-nacreous, plicated laminæ, peculiarly separable, and occasionally penetrated by minute tubuli. (*Carpenter.*) *P. sella,* called, from its shape, the "saddle-oyster," is remarkably striated. In *P. placenta,* Pl. XVI. fig. 6, the anterior carti-lage ridge is only half so long as the other, which appears to be connected with the economy of the shell when young; in specimens 1 inch across, there is a pedal impression below the cartilage grooves of the upper valve, and a shallow sinus in the margin of the lower valve, indicating a slight byssal attachment at that age.

Distr. 4 sp. Scinde, N. Australia, China.

Sub-genera. Carolia, Cantraine 1835, (after Prince Charles Bonaparte.) *Syn.* Hemiplacuna, G. Sby. *Type,* C. placunoides, Pl. XVI. fig. 7. *Shell* like *Placuna ;* hinge, when young, like Anomia, with a byssal plug passing through a small deep sinus in front of the cartilage process, which is closed in the adult. *Distr.* 3 sp. (Brit. Mus.) Tertiary, Egypt, America ?

Placunopsis, Morris and Lycett. P. Jurensis, Rœmer. Sub-orbicular, upper valve convex, radiately striated, or taking the form of the surface to which it adheres; lower valve flat; ligamental groove sub-marginal, trans-verse ; muscular impression large, sub-central. *Fossil,* 4 sp. Lower Oolites, Europe.

PECTEN, O. F. Müller. Scallop.

Etym. Pecten, a comb. *Type,* P. maximus (Janira, Schum.)

Syn. Argus, Poli. Discites, Schl. Amusium, Muhlfeldt.

Shell sub-orbicular, regular, resting on the right valve, usually orna-mented with radiating ribs; beaks approximate, eared; anterior ears most prominent; posterior side a little oblique; right valve most convex, with a notch below the front ear; hinge-margins straight, united by a narrow ligament; cartilage internal, in a central pit; adductor impression double, obscure; pedal impression only in the left valve, or obsolete (fig. 173).

Animal with the mantle quite open, its margins double, the inner pendent like a curtain (*m*) finely fringed; at its base a row of conspicuous round black eyes (*ocelli*) surrounded by tentacular filaments; gills (*br*) exceedingly delicate, crescent-shaped, quite disconnected posteriorly having separate excurrent canals; lips foliaceous; palpi truncated, plain outside, striated within; foot finger-like, grooved, byssiferous in the young.

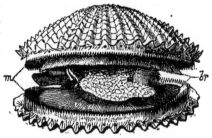

Fig. 179. Pecten varius.*

The Scallop (*P. maximus*) and "quin" (*P. opercularis*) are esteemed delicacies; the latter covers extensive banks, especially on the N. and W. of Ireland, in 15 to 25 fm. water. The scallop ranges from 3—40 fms.: its body is bright orange, or scarlet, the mantle fawn-colour, marbled with brown; the shell is used for "scalloping" oysters, formerly it was employed as a drinking cup, and celebrated as such in Ossian's "hall of shells." An allied species has received the name of "St. James's shell" (*P. Jacobœus*); it was worn by pilgrims to the Holy-land, and became the badge of several orders of knighthood.†

Most of the Pectens spin a byssus when young, and some, like *P. varius*, do so habitually; *P. niveus* moors itself to the fronds of the tangle (*Laminaria.*)

The Rev. D. Landsborough observed the fry of *P. opercularis*, when less than the size of a sixpence, swimming in a pool of sea-water left by the ebbing of the tide. "Their motion was rapid and zig-zag; they seemed, by the sudden opening and closing of their valves, to have the power of darting like an arrow through the water. One jerk carried them some yards, and then by another sudden jerk they were off in a moment on a different tack."

The shell of Pecten and the succeeding genera consists almost exclusively of membranous laminæ, coarsely or finely corrugated. It is composed of two very distinct layers, differing in colour (and also in texture and destructibility), but having essentially the same structure. Traces of cellularity are sometimes discoverable on the external surface; *P. nobilis* has a distinct prismatic-cellular layer externally. (*Carpenter.*)

---

* The Pectens do not open so wide as here represented; their "curtains" remain in contact at one point on the posterior side, separating the branchial from the exhalent currents.

† When the monks of the ninth century converted the fisherman of Gennesarat into a Spanish warrior, they assigned him the scallop-shell for his "cognizance."— *Moule's Heraldry of Fish.*

Sub-genera, Neithea, Drouet. P. quinque-costatus and other fossil sp. with concavo-convex valves and distinct hinge-teeth; the inner layers of these shells are wanting in all specimens from the English chalk.

Pallium, Schum. P. plica, Pl. XVl. fig. 8. Hinge obscurely toothed.

Hinnites (Cortesii) Defr. P. pusio, Pl. XVI. fig. 10. Shell regular and byssiferous when young; afterwards cementing its lower valve and becoming more or less irregular. *Distr.* 2 sp. *Fossil,* Trias ? Miocene —, Europe.

Hemipecten, A. Adams. H. Forbesianus, Pl. XVI. fig. 9. Shell hyaline, posterior ears obsolete, anterior prominent; right valve flat, byssal sinus deep; structure permeated by microscopic tubuli, as in *Lima.*

Distr. 120 sp. World-wide; Nova-Zembla — C. Horn; — 200 fms. *Fossil,* 450 sp. (including Aviculo-pecten). Carb. —. World-wide.

LIMA, Bruguiere.

Etym. Lima, a file. *Ex.* L. squamosa, Pl. XVI. fig. 11. (Ostrea lima, L.)

Syn. Plagiostoma (Llhwyd) Sby. P. cardiiforme, Pl. XVI. fig. 12.

Shell equivalve, compressed, obliquely oval; anterior side straight, gaping, posterior rounded, usually close; umbones apart, eared; valves smooth, punctate-striate, or radiately ribbed and imbricated; hinge area triangular, cartilage pit central; adductor impression lateral, large, double; pedal scars 2, small.

Animal, mantle-magins separate, inner pendent, fringed with long tentacular filaments, ocelli inconspicuous; foot finger-like, grooved; lips with tentacular filaments, palpi small, striated inside; gills equal on each side, distinct.

The shell is always white; its outer layer consists of coarsely-plicated membranous lamellæ; the inner layer is perforated by minute tubuli, forming a complete network. (*Carpenter.*)

The Limas are either free or spin a byssus; some make an artificial burrow when adult, by spinning together sand or coral-fragments and shells, but the habit is not constant. (*Forbes.*) The burrows of *L. hians* are several times longer than the shell, and closed at each end. (*Charlesworth.*) " This species is pale or deep crimson, with an orange mantle; when taken out of its nest it is one of the most beautiful marine animals to look upon, it swims with great vigour, like the scallop, by opening and closing its valves, so that it is impelled onwards or upwards in a succession of jumps. The filaments of the fringe are easily broken off, and seem to live many hours after they are detached, twisting themselves like worms." (*Landsborough.*) *L. spinosa* has conspicuous ocelli, and short filaments.

Sub-genera, Limatula, S. Wood. L. sub-auriculata, Pl. XVI. fig. 13. Valves equilateral; 8 sp. Greenland — Brit. *Fossil,* Miocene —. Europe.

Limæa, Bronn. L. strigilata, Pl. XVI. fig. 14.* Hinge minutely

---

* After Bronn; the figure in Brocchi does not show the teeth.

toothed. *Fossil*, 4 sp. Lias — Pliocene. The recent *Limæa ? Sarsii* (Lovén) Norway (= L. crassa of the Ægean ?) has the mantle-border plain. Some of the larger recent sp. have obscure lateral teeth.

Distr. 20 sp. Norway, Brit. W. Indies, Canaries, India, Australia; 1—150 fms. The largest living sp. (*L. excavata*, Chemn.) is found on the coast of Norway.

Fossil, 200 sp. Carb. ? Trias —. U. States, Europe, India. The so-called *Plagiostoma spinosum* is a Spondylus.

SPONDYLUS, (Pliny) L. Thorny-oyster.

Type, S. gædaropus, L. *Ex.* S. princeps, Pl. XVI. fig. 15.

Syn. Dianchora, Sby. Podopsis, Lam. Pachytes, Defr.

Shell irregular, attached by the right valve, radiately ribbed, spiny or foliaceous; umbones remote, eared; lower valve with a triangular hinge-area, cartilage in a central groove, nearly or quite covered; hinge of 2 curved interlocking teeth in each valve; adductor impression double.

Animal, with the mantle open and gills separate, as in *Pecten*; lips foliaceous, palpi short; foot small, cylindrical, truncated.

In aged specimens the circular portion of the muscular scar exhibits dendritic vascular markings. The lower valve is always most spiny and least coloured; in some sp. (like *S. imperialis*) the shell is scarcely, if at all, attached by its beak or spines. The inner shell-layer is very distinct from the outer, and always wanting in fossil specimens from calcarious rocks, then called *Dianchoræ*. Specimens from the Miocene of St. Domingo, which have lost this layer, contain a loose mould of the original interior. Water-cavities are common in the inner layer, the border of the mantle having deposited shell more rapidly than the umbonal portion. (*Owen*, Mag. Nat. Hist. 1838, p. 409.)

Distr. 30 sp. W. Indies, Canaries, Medit. India, Torres Straits, Pacific, W. America:—105 fms.

Fossil, 45 sp. Inf. Oolite ? Neocomian —. Europe, U. States, India.

Sub-genus, Pedum, Brug. P. spondyloides, Pl. XVI. fig. 16. *Shell* thin, smooth, compressed, attached by a byssus passing through a deep notch in the right valve. Inhabits coral-reefs, where it is found half-imbeded; Red Sea, Indian Ocean, Mauritius, Chinese Seas.

PLICATULA, Lamarck.

Etym. Plicatus, plaited.

Type, P. cristata, Pl. XVI. fig. 17.

Shell irregular, attached by the umbo of the right valve; valves smooth or plaited; hinge-area obscure; cartilage quite internal; hinge-teeth, 2 in each valve; adductor scar simple.

Distr. 6 sp. W. Indies, India, Philippines, Australia, W. America.

Fossil, 40 sp. Trias —. U. S. Europe, Algeria, India.

P. Mantelli (Lea) Alabama, has the valves eared.

FAMILY II. AVICULIDÆ. Wing-shells.

Shell inequivalve, very oblique, resting on the smaller (right) valve, and attached by a byssus; epidermis indistinct: outer layer prismatic-cellular, (fig. 180) interior nacreous; posterior muscular impression large, sub-central, anterior small, within the umbo; pallial line, irregularly dotted; hinge-line straight, elongated; umbones anterior, eared, the posterior ear wing-like; cartilage contained in one or several grooves; hinge edentulous, or obscurely toothed.

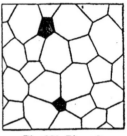

Fig. 180, Pinna.*

Animal with the mantle-lobes free, their margins fringed; foot small, spinning a byssus; gills 2 on each side, crescent-shaped, entirely free (*Desh.*) or united to each other posteriorly, and to the mantle (as in the Oyster, and not as in *Pecten*).

The wing-shells, or pearl-oysters, are natives of tropical and temperate seas; there are no living species in northern latitudes, where fossil forms are very numerous.

AVICULA (Klein) Bruguiere.

Etym. Avicula, a little bird. *Type,* A. hirundo, Pl. XVI. fig. 18.

Shell obliquely oval, very inequivalve; right valve with a byssal sinus beneath the anterior ear; cartilage pit single, oblique; hinge with 1 or 2 small cardinal teeth, and an elongated posterior tooth, often obsolete; posterior muscular impression (adductor and pedal) large, sub-central; anterior (pedal scar) small, umbonal.

Animal (of meleagrina) with mantle-lobes united at one point by the gills, their margins fringed and furnished with a pendent curtain; curtains fringed in the branchial region, plain behind; foot finger-like, grooved; byssus often solid, cylindrical, with an expanded termination; pedal muscles 4, posterior large in front of the adductor; adductor composed of 2 elements; retractors of the mantle forming a series of dots, and a large spot near the adductor; lips simple: palpi truncated; gills equal, crescentic, united behind the foot. (Brit. M.)

Distr. 25 sp. Mexico, S. Brit. Medit. India, Pacific:—20 fms.

Fossil, 300 sp. L. Silurian —. World-wide.

Sub-genera, Meleagrina, Lam. M. margaritifera, Pl. XVI. fig. 19. The "pearl-oysters" are less oblique than the other *aviculæ,* and their valves are flatter and nearly equal; the posterior pedal impression is blended with that of the great adductor. They are found at Madagascar, Ceylon, Swan

---

* The cellular structure may be seen with a hand-lens, in the thin margin of the shell, by holding it up to the light; or on the edges of broken fragments.

R. Panama, &c. Manilla is the chief port to which they are taken. There are three principal kinds, which are worth from £2 to £4 per cwt.: 1. the silver-lipped, from the Society Ids. of which about 20 tons are annually imported to Liverpool; 2. the black-lipped, from Manilla, of which 30 tons were imported in 1851; 3. a smaller sort from Panama, 200 tons of which are annually imported; in 1851 a single vessel brought 340 tons. (T. C. Archer.) These shells afford the "mother-o'-pearl" used for ornamental purposes; and the "oriental" pearls of commerce (p. 38). Mr. Hope's pearl, said to be the largest known, measures 2 inches long, 4 round, and weighs 1800 grains.* Pearl-oysters are found in about 12 fathom water; the fisheries of the Persian Gulf and Ceylon have been celebrated from the time of Pliny.

Malleus, Lam. M. vulgaris, Pl. XVI. fig. 20. The "hammer-oyster" is remarkable for its form, which becomes extremely elongated with age; both ears are long, and the umbones central. When young it is like an ordinary Avicula, with a deep byssal notch in the right valve. 6 sp. China, Australia.

Vulsella, Lam. V. lingulata, Pl. XVI. fig. 21. *Syn.* Reniella, Sw. Shell oblong, striated, sub-equivalve; umbones straight, earless. Often found imbedded in living sponges. *Distr.* 3 sp. Red Sea, India, Australia, Tasmania. *Fossil*, 4 sp. U. Chalk —. Brit. France.

Pteroperna, Lycett, 1852. P. costatula, Desl. *Shell* with a long posterior wing; hinge-line bordered by a groove; anterior teeth numerous, minute; posterior 1 or 2, long, nearly parallel with the hinge-margin. *Fossil*, 3 sp. Bath oolite; Brit. France.

? *Aucella* (Pallasii) Keyserling, 1846. (*Monotis*, Münster, not Bronn.) Very inequivalve; left umbo prominent, earless; right valve small and flat, with a deep sinus beneath the small anterior ear. *Fossil*, Permian — Gault. Europe. "In *A. cygnipes* we find no trace of prismatic cellular structure or nacre, but the coarsely corrugated and somewhat tubular structure of the Pectens." (*Carpenter*.)

Ambonychia (bellistriata) Hall, 1847. Nearly equivalve, gibbose, oblique, obtusely winged. *A. vetusta* (Inoceramus, Sby.) is concentrically furrowed; the right valve has a small anterior ear (usually concealed) separated by a deep and narrow sinus. *Fossil*, 12 sp. L. Silurian — Carb. U. S. Europe.

? *Cardiola* (interrupta) Broderip, 1844. Equivalve, gibbose, obliquely oval, radiately ribbed; beaks prominent; hinge-area short and flat. *Fossil*, 17 sp. U. Silurian — Dev. U. S. Europe.

? *Eurydesma* (cordata) Morris; Devonian? N. S. Wales. *Shell* equivalve,

---

* Sections of oriental pearls exhibit very fine concentric laminæ surrounding a grain of sand, or some such extraneous matter; the nacreous lustre has been attributed to the diffraction of light from the out-cropping edges of the laminæ, but Dr. Carpenter has shown that it may result from the minute plication of a *single* lamina. (See fig. 23, p. 38.)

sub-orbicular, ventricose, very thick near the beaks; ligamental area long, wide, sub-internal; byssal groove close to the umbo; right valve with a large, blunt hinge-tooth; adductor impression single, placed anteriorly; pallial line dotted.

Pterinea (lævis) Goldf. 1832. *Shell* thick, rather inequivalve, very oblique and broadly winged; beaks anterior; sinus shallow; hinge-area long, straight, narrow, striated lengthwise; anterior teeth few, radiating; posterior teeth laminar, elongated; anterior (pedal) scar deep, posterior (adductor) impression large, very eccentric. *Fossil*, 25 sp. L. Sil. — Carb. U. S. Europe, Australia. Pteronites (angustatus) M'Coy, 1844, is thinner and has the teeth, &c. less developed.

Monotis, Bronn, 1830. M. salinaria, Schl. *Trias*, Hallein. Obliquely oval, compressed, radiated; anterior side short, rounded; posterior slightly eared. *Syn.* ? *Halobia* (salinarum) Br. 1830. *Trias*, Hallstadt. Semi-oval, radiated, compressed, with a shallow sinus in front, hinge-line long and straight.

POSIDONOMYA, Bronn.

Syn. Posidonia, Br. 1828. (not König). *Poseidôn*, Neptune.
Type, P. Becheri, Pl. XVI. fig. 22.
Shell thin, equivalve, compressed, earless, concentrically furrowed; hinge-line short and straight, edentulous.
Fossil, 50 sp. L. Silurian — Trias. U. S. Europe.

? AVICULO-PECTEN, M'Coy, 1852

Type, Pecten granosus, Sby. Min. Con. t. 574.
Shell inequivalve, sub-orbicular, eared; hinge-areas flat, with several long, narrow cartilage furrows, slightly oblique on each side of the umbones; right valve with a deep and narrow byssal sinus beneath the anterior ear; adductor impression large, simple, sub-central; pedal scar small and deep, beneath the umbo.
Fossil (see Pecten). L. Silurian — Carb. Spitzbergen — Australia.

GERVILLIA, Defrance.

Etym. Dedicated to M. Gerville, a French naturalist.
Ex. G. anceps, Pl. XVII. fig. 1.
Shell like Avicula; elongated: anterior ear small, posterior wing-like: area long and flat, cartilage pits several, wide apart; hinge-teeth obscure, diverging posteriorly.
Fossil, 30 sp. Carb. — Chalk. Europe.
Sub-genus ? *Bakewellia*, King. B. ceratophaga, Schl. *Fossil*, 5 sp. Permian, Brit. Germany, Russia. *Shell* small, inequivalve, cartilage pits 2—5; hinge with anterior and posterior teeth; anterior muscular impression and pallial line distinct.

PERNA, Bruguiere.

Etym. Perna, a shell-fish (resembling a *gammon*) Pliny.

Syn. Melina, Retz. Isognomon, Klein. Pedalion, Solander.

Type, P. ephippium, L. Pl. XVII. fig. 2.

Shell nearly equivalve, compressed, sub-quadrate; area wide, cartilage pits numerous, elongated, close-set; right valve with a byssal sinus; muscular impression double.

The Pernas vary in form like the *Aviculæ*; some are very oblique, some very inequivalve, and many fossil sp. have the posterior side produced and wing-like. In some Tertiary Pernas the pearly layer is an inch thick.

Distr. 16 sp. Tropical seas; W. Indies — India — W. America.

Fossil, 30 sp. Trias —. U. States, Chile, Europe.

Sub-genera, Crenatula, Lamk. C. viridis, Pl. XVI. fig. 24. *Shell* thin, oblong, compressed; byssal sinus obsolete; cartilage pits shallow, crescent-shaped. *Distr.* 5 sp. N. Africa, Red Sea — China; in sponges.

Hypotrema, D'Orb. 1853. H. rupellensis (= ? Pulvinites Adansonii, Defr. 1826); Coral-rag, Rochelle. *Shell* oblong, inequivalve; right valve flat or concave, with a round byssal foramen near the hinge; left valve convex, with a muscular impression near the umbo; hinge-margin broad, curved, with about 12 close-set transverse cartilage grooves.

INOCERAMUS, Sowerby (1814).

Etym. Is (inos) fibre, *Keramos* shell.

Ex. I. sulcatus, Pl. XVII. fig. 3. *Syn.* Catillus, Brongn.

Shell inequivalve, ventricose, radiately or concentrically furrowed, umbones prominent; hinge-line straight, elongated; cartilage pits transverse, numerous, close-set.

This genus differs from *Perna* chiefly in form. *I. involutus* has the left valve spiral, the right opercular. *I. Cuvieri* attains the length of a yard. Large flat fragments are common both in the chalk and flints, and are often perforated by the *Cliona.* Hemispherical pearls have been found developed from their inner surface, and spherical pearls of the same prismatic-cellular structure occur detached, in the chalk. (*Wetherell.*) The *Inocerami* of the gault are nacreous.

Fossil, 40 sp. Lias — Chalk. S. America, U.S. Europe, Algeria, Thibet.

PINNA, L.

Etym. Pinna, a fin or wing. *Type,* P. squamosa, Pl. XVI. fig. 23.

Shell equivalve, wedge-shaped; umbones quite anterior; posterior side truncated and gaping; ligamental groove linear, elongated; hinge edentulous; anterior adductor scar apical, posterior sub-central, large, ill-defined; pedal scar in front of posterior adductor.

Animal with the mantle margin doubly fringed; foot elongated, grooved, spinning a powerful byssus, attached by large triple muscles to the centre of each valve; adductors both large; palpi elongated; gills long.

N 3

Distr. 30 sp. U. States, S. Brit. Medit. Australia, Pacific, Panama.

Fossil, 50 sp. Devonian —. U. S. Europe, S. India.

The shell of the *Pinna* attains a length of two feet; when young it is thin, brittle, and translucent, consisting almost entirely of prismatic cell-layers; the pearly lining is thin, divided, and extends less than halfway from the beak. Some fossil Pinnas crumble under the touch into their component fibres. The living sp. range from extreme low-water to 60 fms; they are moored vertically, and often nearly buried in sand, with knife-like edges erect. The byssus has sometimes been mixed with silk, spun, and knitted into gloves, &c. (Brit. Mus.) A little crab which nestles in the mantle and gills of the Pinna, was anciently believed to have formed an alliance with the blind shellfish, and received the name of Pinna-guardian (*Pinnoteres*) from Aristotle; similar species infest the Mussels and *Anomiæ* of the British coast.

Sub-genus, Trichites, (Plott) Lycett. T. Plottii, Llhwyd. ("Pinni-gene," Saussure.) *Shell* thick, inequivalve, somewhat irregular, margins undulated. *Fossil,* 5 sp. Oolitic strata of England and France. Fragments an inch or more in thickness are common in the Cotteswolde-hills; full-grown individuals are supposed to have measured a yard across.

FAMILY III. Mytilidæ. Mussels.

Shell equivalve, oval or elongated, closed, umbones anterior, epidermis thick and dark, often filamentose; ligament internal, sub-marginal, very long; hinge edentulous; outer shell layer obscurely prismatic-cellular;* inner more or less nacreous; pallial line simple; anterior muscular impression small and narrow, posterior large, obscure.

Animal marine or fluviatile, attached by a byssus; mantle-lobes united between the siphonal openings; gills two on each side, elongated, and united behind to each other and to the mantle, dorsal margins of the outer and inner-most laminæ free; foot cylindrical, grooved.

The shells of this family exhibit a propensity for concealment, frequently spinning a nest of sand and shell-fragments, burrowing in soft substances, or secreting themselves in the burrows of other shells.

Mytilus, L. Sea-mussel.

Ex. M. smaragdinus, Pl. XVII. fig. 4.

Shell wedge-shaped, rounded behind; umbones terminal, pointed; hinge-teeth minute or obsolete; pedal muscular impressions two in each valve, small, simple, close to the adductors.

Animal with the mantle-margins plain in the anal region, and projecting slightly; branchial margins fringed; byssus strong and coarse; gills nearly equal; palpi long and pointed, free.

---

* A thin layer of minute cells may frequently be detected immediately under the epidermis. (*Carpenter.*)

The common edible mussel frequents mud-banks which are uncovered at low-water; the fry abound in water a few fathoms deep; they are full-grown in a single year. From some unknown cause they are, at times, extremely deleterious. The consumption of mussels in Edinburgh and Leith is estimated at 400 bushels (=400,000 mussels) annually; enormous quantities are also used for bait, especially in the deep sea fishery, for which purpose 30 or 40 millions are collected yearly in the Frith of Forth alone. (*Dr. Knapp.*) Mussels produce small and inferior pearls. At Port Stanley, Falkland Ids. Mr. Macgillivray noticed beds of mussels which were chiefly dead, being frozen at low-water. *M. bilocularis* (Septifer, Recluz) has an umbonal shelf for the support of the anterior adductor, like *Dreissena ;* it· is found at Mauritius and Australia. *M. exustus* (Brachydontes, Sw.) has the hinge-margin denticulated continuously.

Distr. 50 sp. World-wide. Ochotsk, Behring's Sea, Russian Ice-meer; Black Sea, C. Horn, Cape, New Zealand.

Fossil, 80 sp. Permian —. U. S. Europe, S. India.

? MYALINA, Koninck, 1842.

Types, M. Goldfussiana, Kon. Carb. M. acuminata, Sby. Permian.

Shell equivalve, mytili-form; beaks nearly terminal, septiferous internally; hinge-margin thickened, flat, with several longitudinal cartilage-grooves; muscular impressions 2; pallial line simple.

Fossil 6 sp. Carb. — Permian. Europe. The ligamental area resembles that of the recent Arca obliquata, Chemn. India.

MODIOLA, Lam. Horse-mussel.

Etym. Modiolus, a small measure, or drinking-vessel.

Ex. M. tulipa, Pl. XVII. fig. 5. M. modiolus, p. 250, fig. 177.

Shell oblong, inflated in front: umbones anterior, obtuse: hinge toothless; pedal impressions 3 in each valve, the central elongated; epidermis often produced into long beard-like fringes.

Animal with the mantle-margin simple, protruding in the branchial region; byssus ample, fine; palpi triangular, pointed.

The *Modiolæ* are distinguished from the Mussels by their habit of burrowing, or spinning a nest. Low-water—100 fms.

Distr. 50 sp. chiefly tropical; *M. modiolus,* Arctic seas — Brit.

Fossil, 130 sp. Silurian ? Lias —. U. S. Europe, Thibet, S. India.

Sub-genera. Lithodomus, Cuv. M. lithophaga, Pl. XVII. fig. 7. *Shell* cylindrical, inflated in front, wedge-shaped behind; epidermis thick and dark; interior nacreous.* *Distr.* 12 sp. W. Indies — New Zealand, *Fossil,*

---

* The outer shell-layer has a tubular structure; the tubes are excessively minute, seldom branching, oblique and parallel. (*Carpenter.*)

16 sp. Bath oolite —. Europe, U. S. The "date-shell" bores into corals, shells, and the hardest limestone rocks (fig. 25, p. 42); its burrows are shaped like the shell, and do not admit of free rotatory motion. The animal, which is eaten in the Medit. is like a common mussel; in *L. patagonicus* the siphons are produced. Like other burrowing shellfish, they are luminous. Perforations of *Lithodomi* in limestone cliffs, and in the columns of the Temple of Serâpis at Puteoli, have afforded conclusive evidence of changes in the level of sea-coasts in modern times. (*Lyell's Principles of Geology.*)

Crenella, Brown. C. discors, Pl. XVII. fig. 8. (Lanistes, Sw. Modiolaria, Beck.) *Shell* short and tumid, partly smooth, and partly ornamented with radiating striæ; hinge-margin crenulated behind the ligament; interior brilliantly nacreous. *Animal* with the anal tube and branchial margins prominent. *Distr.* Temperate and arctic seas; Nova Zembla, Ochotsk, Brit. New Zealand. Low-water — 40 fms. Spinning a nest, or hiding amongst the roots of sea-weed and corallines. *M. marmorata,* Forbes, burrows in the test of *Ascidia. Fossil,* U. Green-sand —. Europe.

Modiolarca (trapezina) Gray; Falkland Ids. — Kerguelen, attached to floating sea-weed; mantle-lobes united, pedal opening small, foot with an expanded sole, front adductor round. *M.? pelagica,* Pl. XVII. fig. 6. is found burrowing in floating blubber, off the Cape. (*Forbes.*)

? *Mytilimeria* (Nuttallii) Conrad. *Shell* irregularly oval, thin, edentulous, gaping posteriorly; umbones sub-spiral; ligament short, semi-internal. *Distr.* California; animal gregarious, forming a nest.

Modiolopsis (mytiloides) Hall, 1847 (= Cypricardites, part, Conrad. Lyonsia, part, D'Orb.) *Shell* like modiola, thin and smooth, front end somewhat lobed; anterior adductor scar large and oval. *Fossil,* Silurian, U. S. Europe.

? *Orthonotus* (pholadis) Conrad. L. Silurian, New York. *Shell* elongated, margins parallel, umbones anterior, back plaited.*

DREISSENA, Van Beneden.

Etym. Dedicated to Dreyssen, a Belgian physician.
Syn. Mytilomya, Cantr. Congeria, Partsch. Tichogonia, Rossm.
Type, D. polymorpha, Pl. XVII. fig. 9. (Mytilus Volgæ, Chemn.)
Shell like *Mytilus,* without its pearly lining; inner layer composed of large prismatic cells; umbones terminal; valves obtusely keeled; right valve with a slight byssal sinus; anterior adductor supported on a shelf within the beak; pedal impression single, posterior.

---

* Hall and Salter employ the name *Orthonotus* for such shells as *Solen constrictus,* Sandb. Devonian, Germany; *Sanguinolites anguliferus,* M'Coy, U. Silurian, Kendal; and *Solenopsis minor,* M'Coy, Carb. limestone, Ireland. M. D'Orbigny has mistaken the plaits for teeth, and placed the genus with *Nucula.* The recent *M. plicata,* Lam. from Nicobar Ids. has the same long straight back and plaited dorsal region.

Animal with the mantle closed; byssal orifice small; anal siphon very small, conical, plain, branchial prominent, fringed inside; palpi small, triangular; foot-muscles short and thick, close in front of the posterior adductor.

D. polymorpha is a native of the Aralo-Caspian rivers; in 1824 it was observed by Mr. J. Sowerby in the Surrey docks, to which it appears to have been brought with foreign timber, in the

Fig. 181. Dreissena.

holds of vessels. It has since spread into the canals and docks of many parts of the country, and has been noticed in the iron water-pipes of London, incrusted with a ferruginous deposit. (*Cunnington.*)

Fossil. 10 sp. Eocene —. Brit. Germany.

FAMILY IV. ARCADÆ.

Shell regular, equivalve, with strong epidermis; hinge with a long row of similar, comb-like teeth; pallial line distinct; muscular impressions sub-equal. Structure corrugated, with vertical tubuli in rays between the ribs or striæ. (*Carpenter.*)

Animal with the mantle open; foot large, bent, and deeply grooved; gills very oblique, united posteriorly to a membranous septum.

ARCA, L.

Etym. Arca, a chest. *Type*, A. Noæ, Pl. XVII. fig. 12.

Ex. A. granosa, Pl. XVII. fig. 10. A. pexata, fig. 11. A. zebra, fig. 13.

Shell equivalve or nearly so, thick, sub-quadrate, ventricose, strongly ribbed or cancellated; margins smooth or dentated, close or sinuated ventrally; hinge straight, teeth very numerous, transverse; umbones anterior, separated by a flat, lozenge-shaped ligamental area, with numerous cartilage-grooves; pallial line simple; posterior adductor-impression double; pedal scars 2, the posterior elongated.

Animal with a long pointed foot, heeled and deeply grooved; mantle furnished with ocelli; palpi 0; gills long, narrow, less striated externally, continuous with the lips: hearts two, each with an auricle.

The name *Bysso-arca* was chosen unfortunately, by Swainson, for the *typical* species of the genus, in which the byssal orifice is sometimes very large (Pl. XVII. fig. 13). The byssus is a horny cone, composed of numerous thin plates, occasionally becoming solid and calcarious; it can be cast off and re-formed with great rapidity. (*Forbes.*) The Arcas with close valves have the left valve a little larger than the right, and more ornate.

The Bysso-arks secrete themselves under stones at low-water, in crevices of rocks, and the empty burrows of boring mollusks; they are often much worn and distorted.

Distr. 130 sp. World-wide, most abundant in warm sea; low water —

230 fms. (*A. imbricata*, Poli). Prince-Regent Inlet (*A. glacialis*) *A. sca-phula, Benson*, is found in the Ganges and its branches, from Calcutta to Humeerpoor on the Jumna, 1000 miles from the sea.

Fossil, 200 sp. L. Silurian —. U. S. Europe; S. India.

CUCULLÆA, Lamarck.

Etym. Cucullus, a cowl. *Type*, C. concamerata, Pl. XVII. fig. 14.

Shell sub-quadrate, ventricose; valves close, striated; hinge-teeth few and oblique, parallel with the hinge-line at each end; posterior muscular impression bounded by an elevated ridge.

Distr. 1 sp. Mauritius, Nicobar, China.

Fossil, 100 sp. L. Silurian —. N. America, Patagonia, Europe.

Sub-genus, Macrodon, Lycett. M. Hirsonensis, Pl. XVII. fig. 15. *Shell* with a few oblique anterior teeth and one or more long laminar posterior teeth. The Ark-shells of the Palæozoic and secondary strata have their anterior teeth more or less oblique, like *Arca*, the posterior teeth parallel with the hinge-line like *Cucullæa ;* their valves are close or gaping below; their umbones frequently sub-spiral; and the hinge-area is often very narrow; and in some species only the posterior moiety is visible.

PECTUNCULUS, Lam.

Type, P. pectiniformis, Pl. XVII. fig. 16. (Arca pectunculus, L.)

Shell orbicular, nearly equilateral, smooth or radiately striated; umbones central, divided by a striated ligamental area; hinge with a semicircular row of transverse teeth; adductors sub-equal; pallial line simple; margins crenated inside.

Animal with a large crescent-shaped foot, margins of the sole undulated; mantle open, margins simple, with minute ocelli; gills equal, lips continuous with the gills.

Distr. 50 sp. W. Indies, Brit. India, N. Zealand, W. America: ranging from 8 to 60, rarely 120 fathoms.

Fossil, 70 sp. Neocomian —. U. S. Europe: S. India.

The teeth of *Pectunculus* and *Arca* increase in number with age, by additions to each end of the hinge-line, but sometimes the central teeth are obliterated by encroachments of the ligament.

LIMOPSIS, Sassi, 1827.

Type, L. aurita, Pl. XVII. fig. 17. *Syn.* Trigonocœlia, Nyst.

Shell orbicular, convex, slightly oblique; ligamental area with a triangular cartilage-pit in the centre; hinge with 2 equal, curved series of transverse teeth.

Distr. 1 sp. Red Sea (Nyst.)

Fossil, 17 sp. Bath-oolite —. U. States; Europe.

NUCULA, Lam.

Etym. Diminutive of *nux*, a nut. *Ex.* N. Cobboldiæ, Pl. XVII. fig. 18.

Shell trigonal, with the umbones turned towards the short *posterior* side; smooth or sculptured, epidermis olive, interior pearly, margins crenulated; hinge with prominent internal cartilage-pit, and a series of sharp teeth on each side; pallial line simple.

Animal with the mantle open, its margins plain; foot large, deeply fissured in front, forming when expanded a disk with serrated margins; mouth and lips minute, palpi very large, rounded, strongly-plaited inside and furnished with a long convoluted appendage; gills small, plume-like, united behind the foot to the branchial septum.

The Nucula uses its foot for burrowing, and Prof. Forbes has seen it creep up the side of a glass of sea-water. The labial appendages protrude from the shell at the same time with the foot. *N. mirabilis*, Adams, from Japan, is sculptured like the extinct *N. Cobboldiæ*.

Distr. 70 sp. U. S. Norway, Cape, Japan, Sitka, Chile. On coarse bottoms, from 5—100 fms.

Fossil, 100 sp. L. Silurian ? —. Trias —. America, Europe, India.

Sub-genera. Nuculina, D'Orb.* 1847. N. miliaris, Pl. XVII. fig. 19. *Shell* minute; teeth few, in one series, with a posterior lateral tooth. *Eocene*, France. Nucinella (ovalis) Searles-Wood, 1850 (= Pleurodon, Wood, 1840) a minute shell from the Coralline crag of Suffolk, is described as having an external ligament.

? *Stalagmium* (margaritaceum) Conrad, 1833 = Myoparo costatus, Lea. *Eocene*, Alabama. ? *S. Nystii*, Galeotti (Nucunella, D'Orb. *Eocene*, Belgium. *Shell* like *Limopsis*; ligamental area narrow, wholly posterior.

ISOARCA, Münster, 1842.

Type, I. subspirata, M. Oxford Clay; France, Germany.

Shell ventricose; beaks large, anterior, often sub-spiral; ligament entirely external; hinge-line curved, with two series of transverse teeth, smallest in the centre; pallial line simple.

I. Logani (Ctenodonta) Salter, L. Silurian, Canada. is 3 inches long and has the ligament preserved.

Fossil, 14 sp. L. Silurian — Chalk. N. America; Europe.

Sub-genus. Cucullella, M'Coy. C. antiqua, Sby. U. Silurian, Herefordshire. *Shell* elliptical, with a strong rib behind the anterior adductor impression.

LEDA, Schumacher.

Etym. Leda, in Greek myth. mother of Castor and Pollux.

Syn. Lembulus (Leach) Risso. *Ex.* L. caudata, Pl. XVII. fig. 20.

Shell resembling *Nucula;* oblong, rounded in front, produced and pointed

---

* *N. donaciformis*, Parreyss, from the White Nile, is a crustacean! (*Estheria*).

behind; margins even; pallial line with a small sinus; umbonal area with a linear impression joining the anterior adductor.

Animal furnished with two partially-united, slender, unequal, siphonal tubes (*Forbes*); gills narrow, plume-like, deeply laminated, attached throughout; mantle-margin with small ventral lobes forming by their apposition a third siphon.

Distr. 30 sp.   Northern and Arctic Seas, 10—180 fms.   Siberia, Melville Id. Mass. Brit. Medit. Cape, Japan, Australia.

Fossil, 110 sp.   U. S. Europe; S. India.

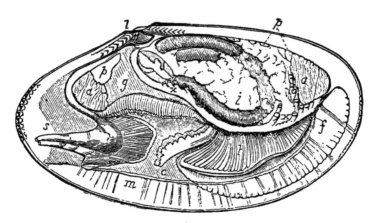

Fig. 182.   Yoldia n. sp. $\frac{3}{4}$.   Antarctic Expedition.

(From a drawing by Albany Hancock, Esq.) The internal organs are represented as seen, through the mantle, on the removal of the right valve.

a,a, adductors; *p,p*, pedal muscles; *l*, ligament; *g*, gills; *s*, siphons (much contracted); *t. c*, labial palpi and appendages; *i*, intestine; *f*, foot; *x,x*, lateral muscles of the foot; *m*, pallial line.

Sub-genus, *Yoldia*, Möller (dedicated to the Countess Yoldi).   Y. myalis, Pl. XVII. fig. 21.   *Shell* oblong, slightly attenuated behind, compressed, smooth or obliquely sculptured, with dark olive shining epidermis; external ligament slight; cartilage as in Leda; pallial sinus deep.   *Animal* with the branchial and anal siphons united, retractile; palpi very large, appendiculate; gills narrow, posterior; foot slightly heeled, deeply grooved, its margins crenulated; intestine lying partly close to the right side of the body, and producing an impression in the shell; mantle-margin plain in front, fringed behind; destitute of ventral lobes.   *Distr.* Arctic and Antarctic Seas; Greenland, Mass. Brazil; Norway, Kamtschatka.   *Fossil, Miocene* —.   (Crag and Glacial deposits.)   England, Belgium.

SOLENELLA, Sowerby.

Type, S. Norrisii, Pl. XVII. fig. 22.   S. ornata, fig. 23.

Syn. Malletia, Desm.   Ctenoconcha, Gray.   Neilo, Adams.

Shell oval or ark-shaped, compressed, smooth or concentrically furrowed, epidermis olive; ligament external, elongated, prominent: hinge with an anterior and posterior series of fine sharp teeth; interior sub-nacreous; pallial sinus large and deep; anterior adductor giving off a long oblique pedal line.

Animal like *Yoldia;* mantle-margins slightly fringed and furnished with ventral lobes; siphonal tubes united, long and slender, completely retractile; palpi appendiculate, convoluted, as long as the shell; gills narrow, posterior; foot deeply cleft, forming an oval disk, even-margined and striated across.

Distr. 2 sp. Valparaiso; New Zealand (shell like *S. ornata*).

Fossil, 1 sp. Miocene. Pt. Desire, Patagonia.

? Solemya, Lamarck.

Type, S. togata, Pl. XXII. fig. 17. *Syn.* Solenomya, Menke.

Shell elongated, cylindrical, gaping at each end; epidermis dark, horny, extending beyond the margins; umbones posterior; hinge edentulous; ligament concealed; pallial line obscure. Outer layer of long prismatic cells, nearly parallel with the surface, and mingled with dark cells, as in *Pinna;* inner layer also cellular.

Animal with the mantle lobes united behind, with a single siphonal orifice, hour-glass shaped, and cirrated; foot proboscidiform, truncated and fringed at the end; gills forming a single plume on each side, with the laminæ free to the base; palpi long and narrow, nearly free.

The shell resembles *Glycimeris* in the shortness of its posterior side, and the extraordinary development of its epidermis; the animal most resembles *Leda* in the structure of its foot and gills.

Distr. 4 sp. U. States, Canaries, W. Africa (Gaboon R.), Medit. Australia, New Zealand. Burrowing in mud; 2 fms.

Fossil, 4 sp. Carb. —. Brit. Belgium.

FAMILY V. Trigoniadæ.

Shell equivalve, close, trigonal, with the umbones directed posteriorly; ligament external; interior nacreous; hinge-teeth few, diverging; pallial line simple.

Animal with the mantle open; foot long and bent; gills two on each side, recumbent; palpi simple.

Trigonia, Bruguiere (not Aublet.)

Etym. Trigonos, three-angled. *Syn.* Lyriodon, G. Sby.

Ex. T. costata, Pl. XVII. fig. 24. T. pectinata, fig. 183.

Shell thick, tuberculated, or ornamented with radiating or concentric ribs; posterior side angular; ligament small and prominent; hinge-teeth 2.3, diverging, transversely striated; centre tooth of left valve divided; pedal impressions in front of the posterior adductor, and one in the umbo of the left valve; anterior adductor impression close to the umbo.

Animal with a long and pointed foot, bent sharply, heel prominent, sole bordered by two crenulated ridges; palpi small and pointed; gills ample, the outer smallest, united behind the body to each other and to the mantle.

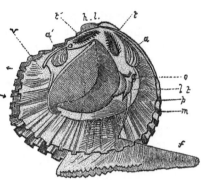

Fig. 183. *Trigonia pectinata.*‡

The shell of Trigonia is almost entirely nacreous, and usually wanting or metamorphic in limestone strata; casts of the interior are called "horse-heads" by the Portland quarry-men;* they spoil the stone. Silicified casts have been found at Tisbury, in which the animal itself, with its gills, was preserved.† The species with the posterior angle of the shell elongated, have a siphonal ridge inside. The epidermal layer of the recent shell consists of nucleated cells, forming a beautiful microscopic object. A Trigonia placed by Mr. S. Stutchbury on the gunwale of his boat leapt overboard, clearing a ledge of four inches; they are supposed to be migratory, as dredging for them is very uncertain, though they abound in some parts of Sydney Harbour.

Distr. 3 sp. (or varieties ?) Australia.

Fossil, 100 sp. Trias — Chalk; (not known in Tertiaries). Europe, U. S. Chile, Algeria, Cape, S. India.

MYOPHORIA, Bronn, 1830.

Type, M. vulgaris, Schl. *Syn.* Cryptina (Kefersteinii) Boue.

Shell trigonal, umbones turned forwards; obliquely keeled; smooth or sculptured; teeth 2.3, striated obscurely, centre tooth of left valve simple, anterior of right valve prominent; mould like *Trigonia*. *M. decussata*, Pl. XVII. fig. 25, has a lateral tooth at the dorsal angle of the left valve.

Fossil, 13 sp. Trias: Germany, Tyrol.

AXINUS, Sowerby, 1821.

Type, A. obscurus, Sby. *Syn.* Schizodus, King (not Waterhouse).

Shell trigonal, rounded in front, attenuated behind; rather thin, smooth, with an obscure oblique ridge; ligament external; hinge-teeth 2.3, smooth, rather small; anterior adductor slightly impressed, removed from the hinge, with a pedal scar close to it; pallial line simple.

Fossil, 20 sp. U. Silurian — Muschelkalk. U. States, Europe. *Mactra* tri-

---

* See Plott's Oxfordshire, T. vii. fig. 1.

† In the collection of the late Miss Benett of Warminster, now in Philadelphia.

‡ Fig. 183. From a specimen in alcohol; the gills slightly curled and contracted, they should terminate near the margin, between the arrows which indicate the inhalent and exhalent currents: *a,a'*, adductors; *hl*, ligament; *t.t'*, dental sockets; *o*, mouth; *lt*, labial tentacles or palpi; *p*, pallial line; *m*, margin; *f*, foot; *v*, cloaca.

gona, Goldf. *Isocardia* axiniformis. Ph. *Anodontopsis* securiformis, *Anatina* attenuata and *Dolabra* securiformis, M'Coy, probably belong to this genus. *Dolabra* equilateralis, *Amphidesma* subtruncatum, *Anodontopsis* angustifrons, M'Coy, with many others from the Palæozoic rocks, may constitute a distinct genus, but their generic character has yet to be discovered.

LYRODESMA, Conrad, 1841.

Type, L. plana, New York. *Syn.* Actinodonta, Phil.

Shell trigonia-shaped, rather elongated, with a striated posterior area; hinge with several (5—9) radiating teeth, striated across; ligament external.

Fossil, 3 sp. L. Silurian: Canada, U. States, Brit.

FAMILY VI. UNIONIDÆ. Naïdes.

Shell usually regular, equivalve, closed; structure nacreous, with a very thin prismatic-cellular layer beneath the epidermis; epidermis thick and dark; ligament external, large and prominent; margins even; anterior hinge-teeth thick and striated, posterior laminar, sometimes wanting; adductor scars deeply impressed; pedal scars 3, distinct, 2 behind the anterior adductor, one in front of the posterior.

Animal with the mantle-margins united between the siphonal orifices and, rarely, in front of the branchial opening; anal orifice plain, branchial fringed; foot very large, tongue-shaped, compressed, byssiferous in the fry; gills elongated, sub-equal, united posteriorly to each other and to the mantle, but not to the body; palpi moderate, laterally attached, striated inside: lips plain. Sexes distinct.

The river-mussels are found in the ponds and streams of all parts of the world. In Europe the species are few, though specimens are abundant; in N. America both species and individuals abound. All the remarkable generic forms are peculiar to S. America and Africa. Two of these are fixed, and irregular when adult, and have been placed with the chamas and oysters by the admirers of artificial systems; fortunately, however, M. D'Orbigny has ascertained that the *Mulleria*, which is fixed and *mono-myary* when adult, is locomotive and di-myary when young!*

Like other fresh-water shells, the naïds are often extensively eroded by the carbonic acid dissolved in the water they inhabit (p. 41).† This condition of the umbones is conspicuous in the great fossil *Uniones* of the Wealden,

---

* In the synopsis at p. 252 it will be seen that each of the principal groups of bivalves contains members which are fixed and irregular, and others which are byssiferous, or burrowing, or locomotive.

† Probably many of the organic acids, produced by the decay of vegetable matter, assist in the process. It has been suggested that sulphuric acid may sometimes be set free in river-water, by the decomposition of iron-pyrites in the banks: but Prof. Boye of Philadelphia states that it has not been detected in any river of the United States, where the phenomenon of erosion is most notorious.

but cannot be detected in the *Cardiniæ*, and some other fossils formerly referred to this family.

The outer gills of the female unionidæ are filled with spawn in the winter and early spring; the fry spins a delicate, ravelled byssus, and flaps its triangular valves with the posterior shell-muscle, which is largely developed, whilst the other is yet inconspicuous. The shells of the female river-mussels are rather shorter and more ventricose than the others. (See pp. 18, 34.)

UNIO, Retz. River-mussel.

Etym. Unio a pearl (Pliny). *Ex.* U. litoralis, Pl. XVIII. fig. 1.

Shell oval or elongated, smooth, corrugated, or spiny, becoming very solid with age; anterior teeth 1.2 or 2.2, short, irregular; posterior teeth 1.2, elongated, laminar.

Animal with the mantle-margins only united between the siphonal openings; palpi long, pointed, laterally attached. (Fig. 172, p. 246.)

U. plicatus (Symphynota, Sw. Dipsas, Leach) has the valves produced into a thin, elastic dorsal wing, as in *Hyria.** In the Pearl-mussel, *U. margaritiferus* (Margaritana, Schum. Alasmodon, Say) the posterior teeth become obsolete with age. This species, which afforded the once famous British pearls, is found in the mountain streams of Britain, Lapland, and Canada; it is used for bait in the Aberdeen Cod-fishery. The Scotch pearl-fishery continued till the end of the last century, especially in the R. Tay, where the mussels were collected by the peasantry before harvest-time. The pearls were usually found in old and deformed specimens; round pearls about the size of a pea, perfect in every respect, were worth £3 or £4. (Dr. Knapp.) An account of the Irish pearl-fishery was given by Sir R. Redding in the Phil. Trans. 1693. The mussels were found set up in the sand of the river-beds with their open side turned from the torrent; about one in 100 might contain a pearl, and one pearl in 100 might be tolerably clear. (See p. 38.)

Distr. 250 sp. N. America, S. America, Europe, Africa, Asia, Australia.

Fossil, 50 sp. Wealden —. Europe, India.

Sub-genera, Monocondylæa, D'Orb. M. Paraguayana, Pl. XVIII. fig. 2. *Shell* with a single large, round, obtuse cardinal tooth in each valve; no lateral teeth. *Distr.* 6 sp. S. America.

Hyria, Lam. H. syrmatophora, Pl. XVIII. fig. 3. *Syn.* Pachyodon and Prisodon, Schum. *Shell* Arca-shaped, hinge-line straight, with a dorsal wing on the posterior side; teeth elongated, transversely striated. *Distr.* 4 sp. S. America.

---

* This is the species in which the Chinese produce artificial pearls by the introduction of shot, &c., between the mantle of the animal and its shell (p. 38); Mr. Gaskoin has an example containing two strings of pearls, and another in the Brit. Mus. has a number of little josses made of bell-metal, now completely coated with pearl, in its interior.

CASTALIA, Lamarck.

Type, C. ambigua, Pl. XVIII. fig. 4. *Syn.* Tetraplodon, Spix.

Shell ventricose; trigonal; umbones prominent, furrowed; hinge-teeth striated; anterior 2.1, short; posterior 1.2, elongated.

Animal with mantle-lobes united behind, forming two distinct siphonal orifices, the branchial cirrated.

Distr. Rivers of S. America, Guiana, Brazil.

ANODON, Cuvier. Swan-mussel.

Type, A. cygneus, fig. 171. p. 245. *Etym. Anodontos*, edentulous.

Shell like *unio*, but edentulous; oval, smooth, rather thin, compressed when young, becoming ventricose with age.

Animal like unio: the outer gills of a female have been computed to contain 600,000 young shells (*Lea*). See p. 19.

Distr. 50 sp. N. America, Europe, Siberia. *Fossil*, 5 sp. Eocene —. Europe.

M. D'Orbigny relates that he found great quantities of small Anodons (*Bysso-anodonta Paraniensis*, D'Orb.) 4 lines in length, *attached by a byssus*, in the R. Parana, above Corrientes.

IRIDINA, Lamarck.

Syn. Mutela, Scop. Spatha, Lea (including *Mycetopus*).

Type. I. exotica, Pl. XVIII. fig. 5. *Etym. Iris*, the rainbow.

Shell oblong; umbones depressed; hinge-line long, straight, attenuated towards the umbones, crenated by numerous unequal teeth; ligament long and narrow.

Animal with mantle-lobes united posteriorly, forming two short siphons; mouth and lips small; palpi immense, oval; gills united to the body.

Iridina ovata (Pleiodon, Conrad), has a broader hinge-line.

Distr. 6 sp. Rivers of Africa, Nile, Senegal.

MYCETOPUS, D'Orbigny.

Etym. Mukes a mushroom, *pous* the foot.

Type, M. soleniformis, Pl. XVIII. fig. 6.

Shell elongated, sub-cylindrical, gaping in front; margins sub-parallel, hinge edentulous.

Animal with an elongated, cylindrical foot, expanded into a disk at the end; mantle open; gills equal; palpi short.

Distr. 3 sp. R. Parana, Corrientes; R. Amazon, Bolivia.

ÆTHERIA, Lamarck.

Type, Æ. semilunata, Pl. XVIII. fig. 7. (*aitherios*, aërial.)

Shell irregular, inequivalve; attached by the umbo, and tubular processes of one of the valves, usually the left; epidermis thick, olive; interior pearly, blistered (as if with air-bubbles); hinge edentulous; ligament external, with a conspicuous area and groove in the fixed valve; two adductor impressions, the anterior very long and irregular; pallial line simple.

Animal with the mantle-lobes open; body large, oblong, projecting backwards; no trace of a foot; palpi large, semi-oval; gills sub-equal, plaited, united posteriorly, and to the body and mantle.

Distr. R. Nile, from 1st Cataracts to Fazool;* R. Senegal.

MULLERIA, Férussac.

Dedicated to Otto Frid. Müller, author of the "Zoologia Danica."

Type, M. lobata, Fér. *Syn.* Acostæa (Guaduasana) D'Orb.

Shell when *young* free, equivalve, Anodon-shaped, with a long and prominent ligament, and two adductor impressions: *adult* irregular, inequivalve, attached by the right valve; umbones elongated, progressively filled up with shell, and forming an irregular " talon" in front of the fixed valve; epidermis thick; ligament in a marginal groove; interior pearly, muscular impression single, posterior.

Distr. R. Magdalena, near Bogota, New Granada.

Mr. Isaac Lea has determined the identity of *Mülleria* and *Acostæa* by examination of Férussac's type, and the suite of specimens, of different ages, in the collection of M. D'Orbigny.†

SECTION B. SIPHONIDA.

Animal with respiratory siphons; mantle-lobes more or less united.

a. Siphons short, pallial line simple; Integro-pallialia,

FAMILY VII. CHAMIDÆ.

Shell inequivalve, thick, attached; beaks sub-spiral; ligament external; hinge-teeth 2 in one valve, 1 in the other; adductor impressions large, reticulated; pallial line simple.

Animal with the mantle closed; pedal and siphonal orifices small, sub-equal; foot very small; gills two on each side, very unequal, united posteriorly.

CHAMA (Pliny) L.

Ex. C. macrophylla, Pl. XVIII. figs. 8, 9. *Syn.* Arcinella, Schum.

Shell attached usually by the *left* umbo; valves foliaceous, the upper smallest; hinge-tooth of free valve thick, curved, received between two teeth; in the other; adductor impressions large, oblong, the anterior encroaching on the hinge-tooth.

Animal with the mantle-margins united by a curtain, with two rows of tentacular filaments; siphonal orifices wide apart, branchial slightly prominent, fringed, anal with a simple valve; foot bent, or heeled; liver occupying the umbo of the attached valve only; ovary extending into both mantle-lobes, as far as the pallial line; lips simple, palpi small and curled; gills

---

* The " fresh-water oysters" discovered by BRUCE.

† The only specimen of Mülleria in England was purchased many years ago by Mr. Thos. Norris, of Bury, for £20.

deeply plaited, the outer pair much shorter and very narrow, furnished with a free dorsal border, and united behind to each other, and to the mantle; adductors each composed of two elements.

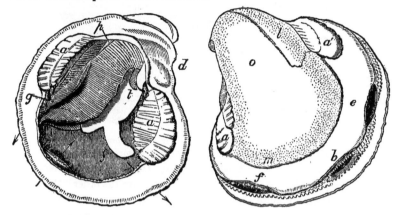

Fig. 184. *Right Side.* Fig. 185. *Left side.*
Animal of Chama (from Torres Str. Mr. Jukes.)
Fig. 184. Right side, with the umbonal portion of the mantle removed.
Fig. 185. Left side, showing the relative extent of the liver and ovarium.

a, a, adductors; *m,* pallial line; *e,* excurrent orifice; *b,* branchial; *f,* foot and pedal orifice; *p,* posterior pedal muscle; *t,* palpi; *g,* gills (contracted); *l,* liver; *o,* ovarium; *d,* dental lobes.

The shell of *Chama* consists of three layers; the external, coloured layer is laminated by oblique lines of growth, with corrugations at right angles to the laminæ; the foliaceous spines contain reticulated tubuli: the middle layer is opaque white and consists of ill-defined vertical prisms or corrugated structure; the inner layer, which is translucent and membranous, is penetrated by scattered vertical tubuli; the minute processes that occupy the tubuli give to the mantle (and to the casts of the shell) a granular appearance (fig. 185, *l, m.*)

Some Chamas are attached indifferently by either valve; when fixed by the right valve the dentition is reversed, the left valve having the single tooth. *Chama arcinella,* which is always attached by the right umbo, has the normal dentition 1 : 2; it is nearly regular and equivalve, and has a distinct lunule.

Distr. 50 sp. Tropical seas, especially amongst coral-reefs; — 50 fms. W. Indies, Canaries, Medit. India, China.

Fossil, 30 sp. Green-sand —. U. States, Europe.

Sub-genus ? Monopleura; Matheron (= Dipilidia, Math) *M. imbricata,* Math. Fig. 187. Neocomian, S. France. *Shell* attached by the *dextral* umbo; valves alike in structure and sculpturing; fixed valve straight, inversely conical, with a long, straight ligamental groove, and obscure hinge-area; opercular valve flat or convex, with an oblique, sub-marginal umbo.

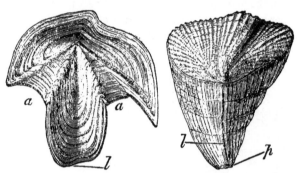

Fig. 186. *Bi-radiolites*, $\frac{3}{5}$.   Fig. 187. *Monopleura*, $\frac{1}{2}$.

p, point of attachment; *l*, ligamental groove; *a, a*, corresponding areas.

Fossil, 9 sp. Neocomian — Chalk. France, Texas. They are commonly found in groups, adhering laterally, or rising one above the other; the casts of such as are known are quite simple and chama-like.

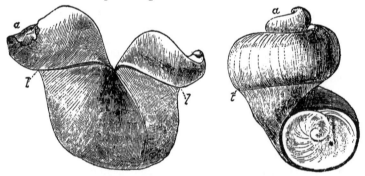

Fig. 188. *Diceras arietinum*, $\frac{1}{2}$.   Fig. 189. *Requienia ammonia*, $\frac{1}{4}$.

a, point of attachment; *l, l*, ligamental grooves; *t*, posterior adductor inflection.

DICERAS, Lamarck.

Type, D. arietinum, Pl. XVIII. figs. 10, 11, and fig. 188, 190.

Shell sub-equivalve, attached by either umbo; beaks very prominent, spiral, furrowed externally by ligamental grooves; hinge very thick, teeth 2.1, prominent; muscular impressions bounded by long spiral ridges, sometimes obsolete.

Distr. 5 sp. Middle oolite. Germany, Switz. France, Algeria.

Diceras differs from *Chama* in the great prominence of both its *umbones*, in having constantly two hinge-teeth in the right valve and one in the left, and in the prominent ridges bordering the muscular impressions. Similar ridges exist in *Cucullæa, Megalodon, Cardilia* and the Hippurite; they produce deep spiral furrows on the casts, which are of common occurrence in the Coral-oolite of the Alps. One or both the anterior furrows (fig. 190, *t, t*) are frequently obsolete. The dental pits are much deeper than the teeth which they receive, and are sub-spiral, giving rise to bifid projections (*c, c*) on the casts; the single tooth in the left valve consists of two elements, and the cavity (*fosset*) which receives it is divided at the bottom.

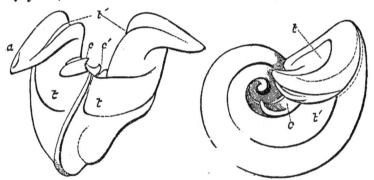

Fig. 190. *Diceras*, ¼.   Fig. 191. *Requienia*, ½.

Internal casts: *a*, point of attachment; *c, c'*, casts of dental pits; *t, t'*, furrows produced by spiral ridges. (Mus. Brit.)

REQUIENIA, Matheron,

Dedicated to M. Requien, author of a Catalogue of Corsican Mollusca.

Ex. R. Lonsdalii, Pl. XVIII. fig. 12 and fig. 191. R. Ammonia, fig. 189.

Shell thick, very inequivalve, attached by the *left* umbo; ligament external; teeth 2:1; left valve spiral, its cavity deep, not camerated; free valve smaller, sub-spiral; posterior adductor bordered by a prominent subspiral ridge in each valve.

The shell-structure of *Requienia* is like that of *Chama*. The relative size of the valves is subject to much variation; in *R. Favri* (Sharpe) they are nearly equal. The hinge-teeth are like those of *Diceras;* the cavity for the posterior tooth of the right valve is very deep and sub-spiral (fig. 191, *c'*). The internal muscular ridges are produced by duplicatures of the shell-wall, and are indicated outside by grooves (fig. 189, *t'*). In *R. sub-æqualis* and *Toucasiana* there is a second parallel ridge, as in Hippurites and *Caprotina*.

Fossil, 7 sp. Neocomian — L. Chalk. Brit. France, Spain, Algeria, Texas.

FAMILY VIII. HIPPURITIDÆ.

(Order *Rudistes*, Lamarck.)

Shell inequivalve, unsymmetrical, thick, attached by the *right* umbo; umbones frequently camerated; structure and sculpturing of valves dissimilar; ligament internal; hinge-teeth 1 : 2; adductor impressions 2, large, those of the left valve on prominent apophyses; pallial line simple, submarginal.

The shells of this extinct family are characteristic of the cretaceous

O

strata, and abound in many parts of the Peninsula, the Alps and E. Europe, where the equivalent of the Lower Chalk has received the name of "Hippurite limestone." They occur also in Turkey and in Egypt, and Dr. F. Rœmer has found them in Texas and Guadaloupe.

They are the most problematic of all fossils: there are no recent shells which can be supposed to belong to the same family; and the condition in which they usually occur has involved them in greater obscurity.* The characters which determine their position amongst the ordinary bivalves are the following:—

1. The shell is composed of two distinct layers.
2. They are essentially unsymmetrical, and right-and-left valved.
3. The sculpturing of the valves is dissimilar.
4. There is evidence of a large internal ligament.
5. The hinge-teeth are developed from the free valve.
6. The muscular impressions are 2 only.
7. There is a distinct pallial line.

The outer layer of shell in the Hippurite and Radiolite consists of prismatic cellular structure (fig. 123); the prisms are perpendicular to the shell-laminæ, and subdivided often minutely. The cells appear to have been empty, like those of *Ostrea* (p. 254).† The inner layer, which forms the hinge and lines the umbones is sub-nacreous, and very rarely preserved. It is usually replaced by calcareous spar (fig. 200), sometimes by mud or chalk, and very often it is only indicated by a vacuity between the outer shell and the internal mould (fig. 205). The inner shell-layer is seldom compact, its lamellæ are extremely thin, and separated by intervals like the water-chambers of *Spondylus;* similar spaces occur in the deposit, filling the umbonal cavity of the long-beaked oysters.‡

* 1. Buch regarded them as Corals. 1840, Leonh. and Bronn Jahrb. p. 573.
  2. Desmoulins, as a combination of the Tunicary and Sessile Cirripede.
  3. Dr. Carpenter, as a "group intermediate between the *Conchifera* and *Cirripeda*." An. Nat. Hist. XII. 390.
  4. Prof. Steenstrup, of Copenhagen, as Anellides.
  5. Mr. D. Sharpe refers *Hippurites* to the Balani; *Caprinella* to the Chamaceæ.
  6. Lapeirouse considered the Hippurites *Orthocerata;* the Radiolites, *Ostracea.*
  7. Goldfuss and D'Orbigny place them both with the *Brachiopoda.*
  8. Lamarck and Rang, between the *Brachiopoda* and *Ostraceæ.*
  9. Cuvier and Owen, with the Lamellibranchiate bivalves.
  10. Deshayes, in the same group with *Ætheria.*
  11. Quenstedt, between the *Chamaceæ* and *Cardiaceæ.*

† This is very conspicuous in *Radiolites* from the Chalk; a formation in which other prismatic-cellular fossils are solid.

‡ The water-chambers in some of the cylindrical Hippurites are large and regular, like those of the fossil corals *Amplexus* and *Cyathophyllum.* A section of *Hippurites bi-oculatus* passing through only one of the dental sockets, resembles an *Orthoceras* with a lateral siphuncle; whilst a *Caprinella* (fig. 207), which has lost its outer layer, might be mistaken for a sort of *Ammonite.*

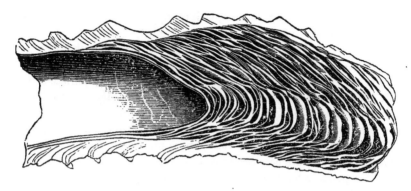

Fig. 192. Section of a fragment of *Ostrea cornucopiæ*.

The inner layer ceases at the pallial line, beyond which, on the rim of the shell, the cellular structure is often apparent; obscure bifurcating impressions radiate from the pallial line to the outer margin, (fig. 193, *v, v*.)

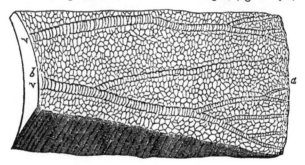

Fig. 123, Part of the rim of *Radiolites Mortoni*, Mantell, *

These have been compared to the vascular impressions of *Crania*. (figs. 157, 8) and constitute the only argument for supposing the *Rudistes* to have been *palliobranchiate;* but they occur on the *rim* of the shell, and not on the disk, as in *Crania*.† The chief peculiarity of the *Hippuritidæ* is the dissimilarity in the structure of the valves, but even this is deprived of much significance by its inconstancy.‡ The free valve of *Hippurites* is perforated by radiating canals which open round its inner margin, and communicate with

---

* Traced from the original specimen in the Museum of the School of Mines. *b*, is the inner edge: *a*, the outer edge; *v, v*, the dichotomous impressions; the horizontal laminæ are seen on the shaded side. Lower Chalk; Sussex.

† M. D'Orbigny considers they were produced by peculiar appendages to the mantle-margin, which, in *Hippurites*, were prolonged into the canals of the upper valve.

‡ The lower valves of some *Spondyli* are squamous or spiny, the upper plain; those of many oysters Pectens and some Tellens are diversely sculptured; but in no instance is the internal structure of the two valves different ? The inconstancy of the shell-structure in the *Rudista* has a parallel in *Rhynchonella* and *Terebratula* (p. 213), and in the condition of the hepatic organ in *Tritonia* and *Dendronotus*.

the upper surface by numerous pores, as if to supply the interior with filtered water; possibly, they were closed by the epidermis.*

In the closely allied genus *Radiolites* there is no trace of such canals, nor in *Caprotina*. Those which exist in the upper valve of *Caprina*, and in both valves of *Caprinella*, have no communication with the outer surface of the shell; they appear to be only of the same character with the tubular ribs of *Cardium costatum* (Pl. XIX. fig. 1), and it is highly improbable that they were permanently occupied by processes from the margin of the mantle.

The teeth of the left, or upper valve, are so prominent and straight, that its movement must have been nearly vertical, for which purpose the internal ligament appears to have been exactly suited by its position and magnitude; but it is probable that, like other bi-valves, they opened to a very small extent.

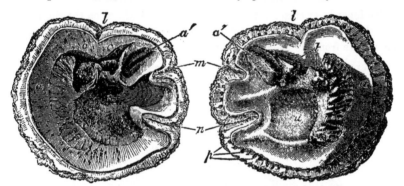

Fig. 194. *Interior of lower valve*, ½.   Fig. 195. *Upper valve* (restored).

Hippurites radiosus, Desm.   Lower Chalk, St. Mamest, Dordogne.†

a, a, adductor impressions and processes; *c, c,* cartilage pits; *t, t',* teeth and dental sockets; *u,* umbonal cavity; *p,* orifices of canals; *l,* ligamental inflection; *m,* muscular; *n,* siphonal inflection.

HIPPURITES, Lamarck.

Name, adopted from old writers, "fossil *Hippuris*" or Horse-tail.

Types, H. bi-oculatus, Lam. and *H. cornu-vaccinum,* fig. 198.

Shell very inequivalve, inversely conical, or elongated and cylindrical; *fixed valve* striated or smooth, with three parallel furrows (*l, m, n,*) on the cardinal side, indicating duplicatures of the outer shell layer: internal margin slightly plaited; pallial line continuous; umbonal cavity moderately deep, ligamental inflection (*l*) with a small cartilage-pit on each side (*c, c*); dental sockets sub-central, divided by an obsolete tooth; anterior muscular impression (*a*) elongated, double; posterior (*a'*) small, very deep, bounded by the second duplicature (*m*); third duplicature (*n*) projecting into the um-

---

* The valves of *Crania* are perforated by branching tubuli, but in that case they pass *vertically* through every part of the shell, and all its layers (p. 214.)

† From the original in the Brit. M. The inner layer of shell in this species has an irregularly cellular structure, to which its preservation is due.

bonal cavity: *free valve* depressed, with a central umbo, and two grooves or pits corresponding to the posterior ridges·in the lower valve; surface porous, the pores leading to canals in the outer shell-layer, which open round the pallial line upon the inner margin; anterior cartilage-pit deep and conical,

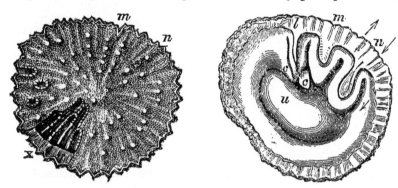

Fig. 196. *H. Toucasianus, upper valve,* ½.*   Fig. 197. *Lower valve, with mould,* ⅔.
l, ligamental; *m,* muscular; *n,* siphonal inflections; *x,* fracture, showing canals; *c,* cartilage: *u,* left umbo; the arrows indicate the probable direction of the branchial currents.

posterior shallow; umbonal cavity turned to the front (*u*); teeth 2, straight, sub-central, the anterior largest, each supporting a crooked muscular apophysis, the first broad, the hinder prominent, tooth-like; inflections (*m, n*) surrounded by deep channels.

H. cornu-vaccinum attains a length of more than a foot, and is curved like a cow's-horn; the outer layer separates readily from the core, which is furrowed longitudinally. The ligamental inflection (*l*) is very deep and narrow, and the anterior tooth further removed from the side than in *H. bi-oculatus* and *radiosus* (figs. 194, 5); the posterior apophysis (*a'*) does not nearly fill the corresponding cavity in the lower valve. In *H. bi-oculatus* and some other species there is no ligamental ridge inside; these, when they have lost their inner layer, present a cylindrical cavity with two parallel ridges, extending down one side. The third inflection (*n*) is possibly a siphonal fold, such as exists in the tube of *Teredo,* and sometimes in the valves of *Pholas, Clavagella,* and the caudate species of *Trigonia.*

The development of processes from the upper valve, for the attachment of the adductor muscles harmonizes with the other peculiarities of the Hippurite. The equal growth of the margins of the valves produces central umbones, and necessitates an internal cartilage; this again causes the removal

* This internal mould, representing the form of the animal, was obtained by removing the upper valve ·piecemeal with the chisel; a plaster-cast taken from it represents the interior of the upper valve, with the bases of the teeth and apophyses. See originals in Brit. Mus.

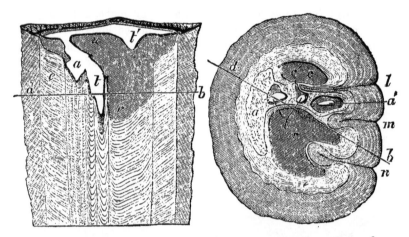

Fig. 198. Longitudinal section; upper half, ½. Fig. 199. Transverse section, ⅓.
Hippurites cornu-vaccinum, Bronn. Salzburg.

l, m, n, duplicatures; *u,* umbonal cavity of left valve; *r,* of right valve; *c, c',* cartilage-pits; *t, t',* teeth; *a, a',* muscular apophyses; *d,* outer shell-layer; Fig. 198 is taken in the line *d, b,* of fig. 199, cutting only the base of the posterior tooth (*t'*). Fig. 199, is from a larger specimen, at about the level *d, b* of fig. 198, cutting the point of the posterior apophysis (*a'*), and shewing the peculiar shell-texture deposited by the anterior adductor (*a*).

of the teeth and adductors further from the hinge-margin, to a position in which the muscles must have been unusually long, unless supported in the manner described. Supposing the animal to have had a small foot,* like

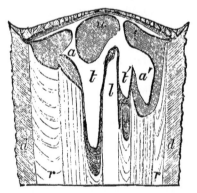 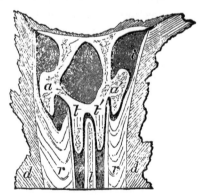

Fig. 200. *Hippurites cornu-vaccinum.*　　Fig. 201. *Radiolites cylindraceus,* ½.
Longitudinal sections taken through the teeth (*t, t'*) and apophyses (*a, a'*).
d, outer, *r,* inner shell-layer; *l,* dental plate of lower valve; *u,* umbonal cavity of upper valves; *i,* intestinal channel. Originals in Brit. M.

---

* This is extremely doubtful; since p. 253 was printed, we have examined an authentic specimen of *Aetheria,* and find that Rang and Cailliaud's account is incorrect : *it has no foot.*

Chama, the mantle-opening for that organ would have been completely ob-
structed by the adductor, but that the muscular support was hook-shaped
(fig. 200, *a*). The posterior adductor-process is similarly under-cut for the
passage of the rectum, which in all bivalves emerges between the hinge and
posterior adductor, winds round outside that muscle and terminates in the
line of the exhalent current. There is a groove (sometimes an inch deep)
round the second and third duplicatures in the upper valve, which seems intended to facilitate the passage of the alimentary canal, and the flow of water
from the gills into the exhalent channel. The smallness of the space for the
branchiæ may have been compensated by deep plication of those organs, as
in *Chama* and *Tridacna.*

Fossil, 16 sp. Chalk. Bohemia, Tyrol, France, Spain, Turkey, Syria
Algeria, Egypt.

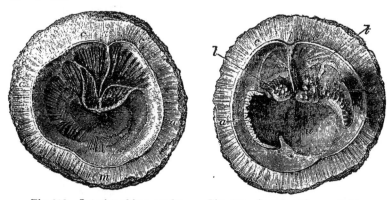

Fig. 202. Interior of lower valve.   Fig. 203. Interior of upper valve.
Radiolites mammillaris, Math. ½ L. Chalk. S. Mamest, Dordogne.
l, ligamental inflection ; *m,* pallial line ; *c, c,* cartilage pits ; *a, a,* adductor impressions
and processes; *t,* teeth and dental sockets.

RADIOLITES, Lamarck, 1801.

Etym. Radius, a ray. *Syn.* Sphærulites, De la Metherie, 1805.

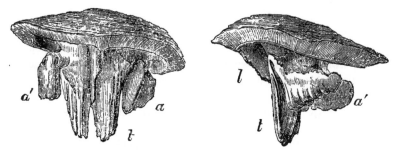

Fig. 204. Side views of the upper valve of *R. mammillaris;*' *l,* ligamental inflection
t, teeth ; *a, a',* muscular processes.

Shell inversely conical, bi-conic, or cylindrical ; valves dissimilar in structure; internal margins smooth or finely striated, simple, continuous; ligamental inflection very narrow, dividing the deep and rugose cartilage pits: *lower valve* with a thick outer layer, often foliaceous ; its cavity deep and straight, with two dental sockets and lateral muscular impressions; *upper valve* flat or conical, with a central umbo ; outer layer thin, radiated; umbonal cavity inclined towards the ligament; teeth angular, striated, supporting curved and sub-equal muscular processes.

The upper valve of *R. fleuriausus* has an oblique umbo, with a distinct ligamental groove. The foliations of the lower valve are frequently undulated; they are sometimes as thin as paper and several inches wide.

The umbonal cavity of the lower valve is partitioned off by very delicate funnel-shaped laminæ. Specimens frequently occur in which the outer shell layer is preserved, whilst the inner is wanting, and the mould ("birostrites") remains loose in the centre. The interior of the outer shell layer is deeply grooved with lines of growth, and exhibits a distinct ligamental ridge in each valve.

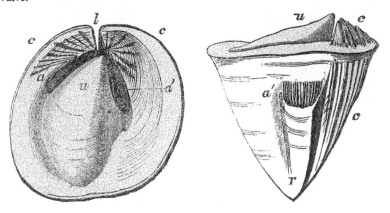

Fig. 205. Upper view.   Fig. 206. Side view.
Internal mould of *R. Hæninghausii*, Desm. ½. Chalk.
u, umbo of left valve; *r*, right umbo; *l*, ligamental groove; *c, c*, cartilage; *a*, anterior adductor muscle; *a'*, posterior.

In aged examples of *R. calceoloides* the ligamental inflection is concealed, the cartilage pits partially filled up and smoothed, and the teeth and apophyses so firmly wedged into their respective cavities, as to suggest the notion that the valves had become fixed about ¼ inch apart, and ceased to open and close at the will of the animal.

Fossil, 42 sp.   Neocomian — Chalk.   Texas ; Brit. France, Bohemia, Saxony, Portugal, Algeria, Egypt.

Sub-genus ? Bi-radiolites, D'Orb.   R. canaliculatus, (Fig. 186, upper valve).   Ligamental groove visible in one or both valves, sometimes occupying the crest of a ridge, and bordered by two similar areas, (*a, a.*)   *Fossil,* 5 sp.   Chalk, France.

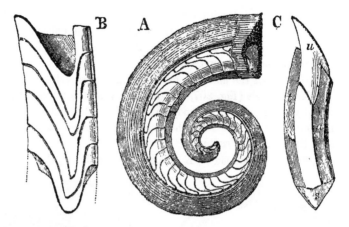

Fig, 207. *Caprinella triangularis*, Desm. U. Green-sand, Rochelle. $\frac{2}{5}$

A, portion of the left valve, after D'Orbigny,* the shell-wall is removed byweathering, exposing the camerated interior. B, mould of five of the water-chambers. C, mould of the body-chamber; *u*, umbo of right valve; *s*, of left valve; *t*, dental groove: *a*, surface from which the posterior lobe has been detached. From the originals in the Brit. M. presented by S. P. Pratt, Esq.

CAPRINELLA, D'Orbigny.

Type, C. triangularis, Desm. (Fig 207). *Syn*. Caprinula (Boissii) D'Orb.

Shell fixed by the apex of the right valve, or free; composed of a thick layer of open tubes, with a thin compact superficial lamina; cartilage internal, contained in several deep pits; umbones more or less camerated; right

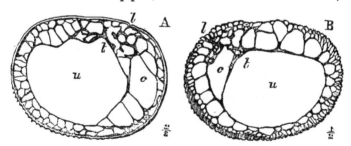

Fig. 208. Straight valve.　　　　Fig. 209. Spiral valve.

Transverse sections of *C. Boissii*, L. Chalk, Lisbon (Mr. Sharpe).

l, position of ligamental inflection; *t*, teeth; *c*, cartilage pits; *u*, umbonal cavity. Fig. 209 is from a weathered specimen, which has lost the outer layer. The tubes of the shell-wall are filled with limestone containing small shells.

* In M. D'Orbigny's figure the smaller valve has been added from another specimen, and is turned *towards* the spire of the large valve, (Pal. Franc. pl. 542, fig. 1.) In Mr. Pratt's specimens, and those collected by M. Sharpe in Portugal, the umbo of the smaller valve is turned *away* with a sigmoid flexure. (Geol. Journ. VI. pl. 18.)

valve conical or elongated, with a ligamental furrow on its convex side, and furnished with one strong hinge-tooth supported by an oblique plate: left valve oblique or spiral, with 2 hinge-teeth, the anterior supported by a plate which divides the umbonal cavity lengthwise.

In *C. triangularis* the umbonal cavity of the spiral valve is partitioned off at regular intervals (Fig. 207, A); the length of the water chambers is sometimes 3½ inches, and of the body-chamber from 2 to 7 diameters; specimens measuring a yard across may be seen on the cavernous shores of the islets near Rochelle.* (*Pratt.*)

Fossil, 6 sp.   Neocomian — L. Chalk.   France, Portugal, Texas.

 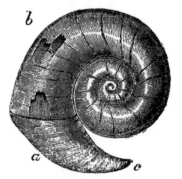

Fig. 210.   *C. Aguilloni*, left valve.   Fig 211.   *C. adversa* (after D'Orb.)
a. a, position of adductors; *l*, ligament; *u*, umbonal cavity; *t*, tooth of fixed valve, broken off and remaining in its socket; *c*, original point of attachment.

CAPRINA, C. D'Orb.

Etym. Caprina, pertaining to a goat.   *Syn.* Plagioptychus, Matheron.

Type, C. Aguilloni, C. D'Orb.   L. Chalk, Tyrol,   (= C. Partschii, Hauer.)

Shell with dissimilar valves, cartilage internal; fixed valve conical, marked only by lines of growth and a ligamental groove; hinge-margin with several deep cartilage-pits; and one large and prominent tooth on the posterior side; free valve oblique or spiral, thick, perforated by one or more rows of flattened canals, radiating from the umbo and opening around the inner margin; anterior tooth supported by a plate which divides the umbonal cavity lengthwise, posterior tooth obscure; hinge-margin much thickened, grooved for the cartilage.

In *C. adversa* (fig. 211) the free valve is (*b*) sinistrally spiral; its cavity is partitioned off by numerous septa, and divided longitudinally by the dental plate.   When young it is attached by the apex of the straight valve (*c*), but afterwards becomes detached, as the large specimens are found imbedded with

* These singular fossils were called *ichthyosarcolites* by Desmarest, from their resemblance to the flaky muscles of fishes.

the spire downwards. (Saemann). The lower valve of *C. Coquandiana* is sub-spiral.

Fossil, 5 sp. U. Green-sand and L. Chalk. Bohemia, France, Texas.

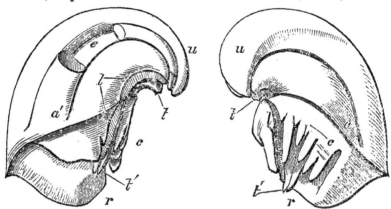

Fig. 212. Internal mould of *Caprotina quadripartita*, D'Orb. ½.
u, left umbo; *r*, right umbo; *l*, ligamental inflection; *c*, cartilage; *t, t'*, dental sockets; *a, a'*, position of adductors; at *e*, a portion of the third lobe is broken away.*
From a specimen collected by Mr. Pratt.

CAPROTINA, D'Orbigny.

Type, C. semistriata, Pl. XIX. fig. 13, 14. Le Mans, Sarthe.

Shell composed of two distinct layers; valves alike in structure, dissimilar in sculpturing; ligamental groove slight; cartilage internal; *right valve* fixed, striated, or ribbed, with one narrow tooth between two deep pits, cartilage pits several on each side of the ligamental inflection, posterior adductor supported by a plate: *free valve* flat or convex, with a marginal umbo; teeth 2, very prominent, supported by ridges (*apophyses*) of the adductor muscles (*a, a'*), the anterior tooth connected with a third plate (*n*), which divides the umbonal cavity.

The smaller *Caprotinæ* occur in groups, attached to oyster-shells; their muscular ridges are much less developed than in the large species (fig. 212). *C. costata* is like a little Radiolite.

Fossil, 4 sp. U. Green-sand, France. (The rest are Chamas, &c.)

FAMILY IX. TRIDACNIDÆ.

Shell regular, equivalve, truncated in front; ligament external; valves strongly ribbed, margins toothed; muscular impressions blended, sub-central, obscure.

Animal attached by a byssus, or free; mantle-lobe extensively united;

---

* The first and fourth lobes, those on each side of the ligamental inflection, appear to be the two divisions of a great internal cartilage, like that of the Radiolite. (Fig. 205, 206, *c, c*.)

pedal opening, large, anterior; siphonal orifices surrounded by a thickened pallial border; branchial plain; anal remote, with a tubular valve: shell-muscle single, large and round, with a smaller pedal muscle close to it behind; foot finger-like, with a byssal groove; gills 2 on each side, narrow, strongly plaited, the outer pair composed of a single lamina, the inner thick, with margins conspicuously grooved; palpi very slender, pointed.

The shell of *Tridacna* is extremely hard, being calcified until almost every trace of organic structure is obliterated. (*Carpenter.*)

TRIDACNA, Bruguière.   Clam-shell.

Etym. Tri- three, *dakno,* to bite; a kind of oyster. (*Pliny.*)

Ex. T. squamosa, Pl. XVIII. fig. 15.

Shell massive, trigonal, ornamented with radiating ribs and imbricating foliations; margins deeply indented; byssal sinus in each valve large, close to the umbo in front; hinge teeth 1.1, posterior laterals 2.1.

A pair of valves of *T. gigas,* weighing upwards ot 500 lbs. and measuring above 2 feet across, are used as *benitiers* in the Church of St. Sulpice, Paris. (Dillwyn.)   Capt. Cook states that the animal of this species sometimes weighs 20 lbs. and is good eating.*

Distr. 6 sp.   Indian Ocean, China Seas, Pacific.

Fossil, T. media.   *Miocene,* Poland (Pusch).   *Tridacna* and *Hippopus* are found in the raised coral-reefs of Torres Straits.   (Macgillivray.)

Sub-genus. Hippopus, Lamarck.   H. maculatus, Pl. XVIII. fig. 16. The "bear's-paw clam" has close valves with 2 hinge-teeth in each.   It is found on the reefs in the Coral Sea.   The animal spins a small *byssus.*

FAMILY X.   CARDIADÆ.

Shell regular, equivalve, free, cordate, ornamented with radiating ribs; posterior slope sculptured differently from the front and sides; cardinal teeth 2, laterals 1.1 in each valve; ligament external, short and prominent; pallial line simple or slightly sinuated behind; muscular impressions sub-quadrate.

Animal with mantle open in front; siphons usually very short, cirrated externally; gills 2 on each side, thick, united posteriorly; palpi narrow and pointed; foot large, sickle-shaped.

CARDIUM, L.   Cockle.

Etym. Kardia, the heart.   *Syn.* Papyridea, Sw.

Types, C. costatum, Pl. XIX. fig. 1.   C. lyratum, fig. 2.

Shell ventricose, close or gaping posteriorly; umbones prominent, sub-central; margins crenulated; pallial line more or less sinuated.

---

* " We staid a long time in the lagoon (of Keeling Id.), examining the fields of coral and the gigantic clam-shells, into which if a man were to put his hand, he would not, as long as the animal lived, be able to withdraw it."—Darwin's Journal, p. 460.

Animal with the mantle-margins plaited; siphons clothed with tentacular filaments, anal orifice with a tubular valve: branchial fringed; foot long, cylin rical, sickle-shaped, heeled.

The cockle (*C. edule*) frequents sandy bays, near low-water; a small variety lives in the brackish waters of the R. Thames, as high as Gravesend; it ranges to the Baltic, and is found in the Black Sea and Caspian. *C. rusticum* extends from the Icy Sea to the Medit. Black Sea, Caspian, and Aral. On the coast of Devon the large prickly cockle (*C. aculeatum*) is eaten.

Sub-genera. Hemicardium (Cardissa) Cuvier. C. hemicardium, Pl. XIX. fig. 3. *Shell* depressed, posterior slope flat, valves prominently keeled.

Lithocardium aviculare, Pl. XVIII. fig. 17. *Shell* triangular, keeled; anterior side very short; hinge-teeth 1.2, directed backwards; posterior laterals 2.1; anterior muscular pit minute, posterior impression large, remote from the hinge. *L. cymbulare, Lam.* exhibits slight indications of a byssal sinus in the front margins of the valves. *Fossil, Eocene,* France. These shells present considerable resemblance to *Tridacna.*

Serripes (grœnlandicus) Beck. Hinge edentulous. Arctic Seas, from C. Parry to Sea of Kara; fossil in the Norwich Crag.

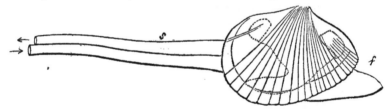

Fig. 213. C. læviusculum, Eichw. (after Middendorff.)

Adacna, Eichwald. C. edentulum, Pl. XIX. fig. 4. (Acardo, Sw. not Brug. Pholadomya, Ag. and Mid. not Sby.) *Shell* compressed, gaping behind, thin, nearly edentulous; pallial line sinuated. *Animal* with the foot (*f*) compressed; siphons (*s*) elongated, united nearly to the end, plain. *Distr.* 8 sp. Aral, Caspian, Azof, Black Sea, and the embouchures of the Wolga, Dnjestr, Dnjepr, and Don; burrowing in mud. *C. Caspicum* (Monodacna, Eichw.) has a single hinge-tooth, and *C. trigonoides* (Didacna, E.) rudiments of two teeth. The siphonal inflection varies in amount.

Distr. 200 sp: World-wide; from the sea-shore to 140 fathoms. Gregarious on sands and sandy mud.

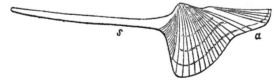

Fig. 214. Conocardium aliforme, Sby. Carb: Ireland. (Mus. Tennant.)

Fossil, 270 sp. U. Silurian —. Patagonia — S. India.

C. Hillanum, Sby. (Protocardium, Beyr.) is the type of a small group in which the sides are concentrically furrowed, the posterior slope radiately striated; the pallial line is slightly sinuated. Jura — Chalk; Europe; India.

CONOCARDIUM, Bronn.

Syn. Lychas, Stein. Pleurorhynchus, Ph. Lunulo-cardium, Münster.
Type, C. Hibernicum, Pl. XIX. fig. 5. C. aliforme, fig. 214.

Shell, equivalve trigonal, conical and gaping in front, truncated behind, with a long siphonal tube near the umbones; anterior slope radiately, posterior obliquely striated; margins strongly crenulated within; hinge with anterior and posterior laminar teeth: ligament external.

The truncated end has usually been considered *anterior*, a conclusion which seems incompatible with the vertical position and burrowing habits of most free and equivalve shells: if compared with *Adacna* (fig. 213) the large gape (*a*) will be for the foot, and the long tube (*s*) siphonal. *C. Hibernicum* has an expanded keel, like *Hemicardium inversum*. The shell-structure is prismatic-cellular, as first pointed out by Sowerby; but the cells are cubical, and much larger than in any of the *Aviculadæ*. In *Cardium* the outer layer is only corrugated or obscurely prismatic-cellular.

Fossil, 30 sp. U. Silurian — Carb. N. America, Europe.

FAMILY X. LUCINIDÆ.

Shell orbicular, free, closed; hinge-teeth 1 or 2, laterals 1—1 or obsolete; interior dull, obliquely furrowed; pallial line simple; muscular impressions 2, elongated, rugose; ligament inconspicuous or sub-internal.|

Animal with mantle-lobes open below, and having one or two siphonal orifices behind; foot elongated, cylindrical, or strap-shaped (*ligulate*), protruded at the base of the shell; gills one (or two) on each side, large and thick, oval; mouth and palpi usually minute.

The *Lucinidæ* are distributed chiefly in the tropical and temperate seas, upon sandy and muddy bottoms, from the sea-shore to the greatest habitable depths. The shell consists of two distinct layers.

LUCINA, Bruguière.

Etym. Lucina, a name of Juno.
Type, L. Pennsylvanica, Pl. XIX. fig. 6.

Shell orbicular, white; umbones depressed; lunule distinct; margins smooth or minutely crenulated; ligament oblique, semi-internal; hinge-teeth 2.2, laterals 1—1 and 2—2, or obsolete; muscular impressions rugose, anterior elongated within the pallial line, posterior oblong; umbonal area with an oblique furrow.

Animal with the mantle freely open below; siphonal orifices simple;

mouth minute, lips thin; gills single on each side, very large and thick; foot cylindrical, pointed, slightly heeled at the base.

The foot of *Lucina* is often twice as long as the animal, but is usually folded back on itself and concealed between the gills; it is hollow throughout. *L. lactea* (Loripes, Poli.) has a long, contractile anal tube. *L. tigrina* (Codakia, Scop.) has the ligament concealed between the valves, its lateral teeth are obsolete.

Distr. 70 sp. W. Indies, Norway, Black Sea, N. Zealand;—120 fms.

Fossil, 200 sp. U. Silurian —. U. States — T. del. Fuego; Europe — S. India.

Sub-genus, Cryptodon, Turton. L. flexuosa, Pl. XIX. fig. 7. *Syn.* Ptychina, Phil. Thyatira, Leach. Clausina (ferruginosa) Jeffr. *Shell* thin, edentulous; ligament quite internal, oblique, *Animal* with a long anal tube. *Distr.* Norway — N. Zealand. *Fossil,* Eocene —. U. S. Europe.

CORBIS, Cuvier.

Etym. Corbis, a basket. *Type,* C. elegans. Pl. XIX. fig. 8.

Syn. Fimbria, Muhl. not Bohadsch. " Idotæa," Schum.

Shell oval, ventricose, sub-equilateral, concentrically sculptured; margins denticulated within; hinge-teeth 2, laterals 2, in each valve; pallial line simple; umbonal area with an oblique furrow, muscular impressions round and polished; pedal scars close to adductors.

Animal with the mantle open below, doubly fringed; foot long pointed; siphonal opening single, with a long retractile tubular valve; lips narrow; palpi rudimentary; gills single on each side, thick, quadrangular, plaited, united behind.

Distr. 2 sp. India, China, N. Australia, Pacific.

Fossil, 80 sp. (including sub-genera). Lias —. U. States, Europe.

In *C. dubia* (Semi-corbis) Desh. *Eocene,* Paris, the lateral teeth are obsolete.

Sub-genera. Sphæra (corrugata) Sby. *Shell* globular, concentrically furrowed and obscurely radiated; ligament prominent; margins crenulated; hinge-teeth 2.2, obscure; laterals obsolete. *Fossil,* Trias — Chalk. Europe.

? *Unicardium,* D'Orb. (Mactromya, Ag. part) = Corbula cardioïdes, Sby. *Shell* thin, oval, ventricose, concentrically striated; ligamental plates elongated; pallial line simple; hinge with an obscure tooth, or edentulous *Fossil,* 40 sp. ? Lias — Portlandian. Europe.

? TANCREDIA, Lycett, 1850.

Dedicated to Sir Thos. Tancred, Bt. founder of the Cotteswolde Naturalists Club.

Ex. T. extensa, L. Pl. XXI. fig. 22. *Syn.* Hettangia, Turquem.

Shell trigonal, smooth; anterior side usually longest; cardinal teeth

2.2, one of them small; a posterior lateral tooth in each valve; ligament external; muscular impressions oval; pallial line simple.

Fossil, 11 sp. Lias — Bath Oolite. Brit. France.

DIPLODONTA. Bronn.

Etym. Diplos, twin, *odonta*, teeth.   *Syn.* Sphærella, Conrad.

Type, D. lupinus (Venus) Brocchi.   Pl. XIX. fig. 9.

Shell sub-orbicular, smooth ; ligament double, rather long, sub-marginal; hinge-teeth 2.2, of which the anterior in the left valve, and posterior in the right, are bifid; muscular impressions polished, anterior elongated.

Animal with the mantle-margins nearly plain, united; pedal opening large, ventral; foot pointed, hollow; palpi large, free ; gills 2 on each side, distinct, the outer oval, inner broadest in front, united behind; branchial orifice small, simple; anal larger, with a plain valve.

Distr. 12 sp. W. Indies, Rio, Brit. Medit. Red Sea, W. Africa, India, Corea, Australia, California.   D. diaphana (Felania Recluz) burrows in sand.

Fossil, Eocene —.   U. States, Europe.

? *Scacchia*, Philippi, 1844 ; Tellina elliptica, Sc.   *Shell* minute, ovate, posterior side shortest; hinge-teeth 1 or 2, laterals obsolete; ligament minute; cartilage internal, in an oblong pit.   *Animal* with mantle widely open; siphonal orifice single; foot compressed, linguiform; palpi moderate, oblong.   *Distr.* 2 sp. Medit.   *Fossil*, 1 sp.   Pliocene, Sicily.

? *Cyamium*, Philippi, 1845,   C. antarcticum, Pl. XIX. fig. 16.   *Shell* oblong, hinge-teeth 2.2; ligament double; cartilage in a triangular groove behind the teeth in each valve.   *Distr.* Patagonia.

UNGULINA, Daudin.

Etym. Ungulina, like a hoof.   *Type*, U. oblonga.   Pl. XIX. fig. 10.

Shell sub-orbicular; ligament very short ; epidermis thick, wrinkled, sometimes black; hinge-teeth 2.2 ; muscular impressions long, rugose.

Animal with the mantle open below, fringed; siphonal orifice single; foot vermi-form, thickened at the end and perforated, projecting from the base of the shell or folded up between the gills ; palpi pointed ; gills 2 on each side, unequal, the external narrower, with a free dorsal border, inner widest in front.

Distr. 4 sp.   Senegal, Philippines, excavating winding galleries in coral.

KELLIA, Turton, 1822.

Etym. Named after Mr. O'Kelly of Dublin.

Syn. Lasea (Leach) Br. 1827.   Cycladina (Adansonii) Cantr.   Bornia (sub-orbicularis) *Phil.* Poronia (rubra) Recluz (not Willd.)   Erycina (cycladiformis) Desh. (not Lam.)

Types, K. sub-orbicularis. Mont.   K. rubra.   Pl. XIX. fig. 12.

Shell small, thin, sub-orbicular, closed; beaks small ; margins smooth ; ligament internal, interrupting the margin (in *K. suborbicularis*), or on he thickened margins (in *K. rubra*); cardinal teeth 1 or 2, laterals 1—1 in ach valve.

Animal with the mantle prolonged in front into a respiratory canal, either mplete (in *K. suborbicularis*) or opening into the pedal slit (in *K. rubra*); ot strap-shaped, grooved; gills large, two on each side, united posteriorly, e external pair narrower and prolonged dorsally; palpi triangular; pos- rior siphonal orifice single, exhalent.

The hinges of these little shells are subject to variations, which are not nstantly associated with the modifications of the mantle-openings. They eep about freely, and fix themselves by a *byssus* at pleasure. *K. rubra* ; found in crevices of rocks at high-water mark, and often in situations nly reached by the spray, except at spring-tides; other species range as eep as 200 fms. *K. Laperousii* (Chironia) Desh. Pl. XIX. fig. 11, was btained, burrowing in sandstone, from deep-water, at Monterey, Cali- ornia.

Distr. 20 sp. Norway — New Zealaud — California.

Fossil, 20 sp. Eocene —. U. States, Europe.

Sub-genera. Turtonia (minuta) Hanley. *Shell* oblong, inequilateral, nterior side very short; ligament concealed between the valves; hinge-teeth .2. *Animal* with the mantle open in front; foot large, heeled; siphon ngle, slender, elongated, protruded from the long end of the shell. *Distr.* Greenland, Norway, Brit. In pools and crevices of rocks between tide-marks, and in the roots of sea-weeds and corallines. Mr. Thompson obtained them from the stomachs of mullets taken on the N.E. coast of Ireland.

Pythina (Deshayesiana) Hinds. (Myllita, D'Orb. and Recl.) *Shell* tri- gonal, divaricately sculptured; [ligament internal; right valve with 2 lateral teeth, left with 1 cardinal and 2 laterals. *Distr.* 2 sp. New Ireland, Australia, Philippines. *Fossil,* Eocene —. France.

MONTACUTA, Turton.

Dedicated to Col. George Montagu, the most distinguished of the earlier English malacologists.

Type, M. substriata. Pl. XIX. fig. 13.

Shell minute, thin, oblong, anterior side longest; hinge-line notched; ligament internal, between 2 laminar, diverging teeth (with a minute ossicle. Lovén).

Animal with the mantle open in front; margins simple; siphonal orifice single; foot large and broad, grooved.

The *Montacutæ* moor themselves by a byssus, or walk freely; *M. sub- striata* has only been found attached to the spines of the purple heart-urchin (*Spatangus purpureus*) in 5—90 fms. *M. bidentata* burrows in the valves of dead oyster-shells.

Distr. 3 sp. U. S. Norway, Brit. Ægean. *Fossil,* 2 sp. Miocene —. Brit.

LEPTON, Turton.

Etym. Lepton, a minute piece of money (from *leptos,* thin).

Syn. ? Solecardia (eburnea) Conrad, L. California.

Type, L. squamosum. Pl. XIX. fig. 14. Fig. 215.

Shell sub-orbicular, compressed, smooth, or shagreened, a little opened at the ends and longest behind; hinge-teeth 0.1 or 1.1. in front of an angular cartilage notch; lateral teeth 2.2 and 1.1.

Animal with the mantle (*m*) open in front, extending beyond the shell, and bearing a fringe of filaments, of which one in front (*t*) is very large; siphon (*s*) single; gills 2 on each side, separate; foot (*f*) thick, tapering, heeled and grooved, forming a sole or creeping disk. (*Alder.*)

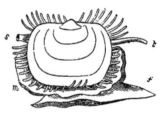

Fig. 215. *Lepton.*

Distr. 3 sp. U. S. Brit. Spain. Laminarian and Coralline Zones.

Fossil, Miocene —. U. S. Brit.

GALEOMMA, Turton.

Syn. Hiatella, Costa (not Daud.); Parthenopea, Scacchi (not Fabr.)

Type, G. Turtoni, Pl. XIX. fig. 15. (*Galee,* weasel, *omma,* eye.)

Shell thin, oval, equilateral, gaping widely below; invested with a thick, fibrous epidermis; beaks minute; ligament internal; teeth 0.1.

Animal with the mantle-lobes united behind and pierced with 1 siphonal orifice, margins double, the inner with a row of eye-like tubercles; gills large, sub-equal, united behind; lips large, palpi lanceolate, plaited; foot long compressed, with a narrow flat sole.

The *Galeomma* spins a byssus, but breaks from its mooring at will and creeps about like a snail, spreading out its valves nearly flat. (*Clarke.*)

Distr. 3 sp. Brit. Medit. Mauritius, Pacific.

Fossil, Pliocene —. Sicily.

FAMILY XI. CYCLADIDÆ.

Shell sub-orbicular, closed; ligament external; epidermis thick, horny; umbones of aged shells eroded; hinge with cardinal and lateral teeth; pallial line simple, or with a very small inflection.

Animal with mantle open in front, margins plain; siphons (1 or 2) more or less united, orifices usually plain; gills 2 on each side, large unequal, united posteriorly; palpi lanceolate: foot large, tongue-shaped.

All the shells of this family were formerly included in the genus *Cyclas,* a name now retained for the small species inhabiting the rivers of the north temperate zone; the *Cyrenæ* are found in warmer regions, on the shores of creeks and in brackish water, where they are gregarious, burying vertically in the mud, and often associated with members of marine genera.

CYCLAS, Bruguière.

Etym. Kuklas, orbicular. *Type,* C. Cornea. Pl. XIX. fig. 17.

Syn. Sphærium, Scop. Pisum, Muhlf. (not L.) Musculium, Link.

Shell thin, ventricose, nearly equilateral; cardinal teeth 2.1, minute, terals 1—1:2—2, elongated, compressed.

Animal ovo-viviparous; siphons partly united, anal shortest, orifices lain; gills very large, the outer smallest, with a dorsal flap; palpi small nd pointed.

The fry of *Cyclas* are hatched in the *internal* branchiæ, they are few in umber and very unequal in size; a full-grown *C. cornea* has about 6 in ach gill; the largest being ⅛ to ¼ the length of the parent. The young *yclades* and *Pisidia* are very active, climbing about submerged plants and ten suspending themselves by byssal threads; the striated gills and pulsat- g heart are easily seen through the shell.

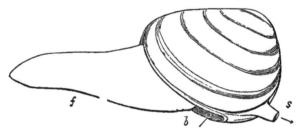

Fig 216. *Pisidium amnicum,* ³⁄₁. with its foot protruded.

Sub-genus, Pisidium, Pfr. P. amnicum, Pl. XIX. fig. 18. *Shell* ine- ⎡uilateral, anterior side longest; teeth stronger than in *Cyclas. Animal* ⱱith a single, small, excurrent siphon; branchial and pedal orifices confluent,

Distr. 30 sp. U. States, S. America, Greenland, Norway, Sicily, Algeria, ⎤ape, India, Caspian.

Fossil, 35 sp. Wealden —. Europe.

CYRENA, Lamarck.

Etym. Cyrene, a nymph. *Type,* C. cyprinoides, Pl. XIX. fig. 20.

Shell oval, strong, covered with thick, rough epidermis; ligament thick nd prominent; hinge-teeth 3.3, laterals 1—1 in each valve; pallial line lightly sinuated.

Animal (of type) with the mantle open in front and below, margins lain; siphons short, orifices fringed; gills unequal, square in front, plaited, nner lamina free at base; palpi lanceolate; foot strong, tongue-shaped.

Section, Corbicula, Muhlf. C. consobrina, Pl. XIX. fig. 21. *Shell* rbicular, concentrically furrowed, epidermis polished; lateral teeth elon- ;ated, striated across.

Distr. 25 sp. Tropical America (eastern); Egypt, India, China, Australia,

Pacific Ids. In the mud of rivers, and in mangrove swamps, usually near the coast. *C. consobrina* ranges from Egypt to Cashmere and China, and is found fossil in the Pliocene formations of England,* Belgium and Sicily.

Fossil, 70 sp. Wealden —. Europe, U. States.

? CYRENOIDES, Joannis.

Syn. Cyrenella, Desh. *Type,* C. Dupontii, Pl. XIX. fig. 19.

Shell orbicular, ventricose, thin, eroded at the beaks; epidermis dark olive; ligament external, prominent, elongated; cardinal teeth 3:2, the central tooth of the right valve bifid; muscular impressions long, narrow; pallial line simple.

Animal with the mantle open in front and below, margin simple, siphons short, united; palpi moderate, narrow; gills very unequal, narrow, united behind; foot cylindrical elongated.

Distr. 1 sp. R. Senegal. The marine sp. are *Diplodontæ.*

FAMILY XII. CYPRINIDÆ.

Shell regular, equivalve, oval or elongated; valves close, solid; epidermis thick and dark; ligament external, conspicuous; cardinal teeth 1—3 in each valve, and usually a posterior lateral tooth; pedal scars close to, or confluent with the adductors; pallial line simple.

Animal with the mantle-lobes united posteriorly by a curtain, pierced with two siphonal orifices; foot thick, tongue-shaped; gills 2 on each side, large, unequal, united behind, forming a complete partition; palpi moderate, lanceolate.

One half the genera of this family are extinct, and the rest (excepting *Circe*) were more abundant in former periods than at the present time. *Cyprina* and *Astarte* are boreal forms; *Circe and Cardita* abound in the Southern seas.

CYPRINA, Lamarck.

Etym. Kuprinos (from *Kupris*) related to Venus.

Type, C. Islandica, Pl. XIX. fig. 22. *Syn.* Arctica, Schum.

Shell oval, large and strong, with usually an oblique line or angle on the posterior side of each valve; epidermis thick and dark; ligament prominent; umbones oblique; no lunule; cardinal teeth 2:2, laterals 0—1, 1—0; muscular impressions oval, polished; pallial sinus obsolete.

Animal with the mantle open in front and below, margins plain; siphonal orifices close together, fringed, slightly projecting; outer gills semilunar, inner truncated in front.

The principal hinge-tooth in the right valve of *Cyprina* represents the

* Associated with the bones of *Elephas meridionalis, Rhinoceros leptorhinus, Mastodon Arvernensis, Hippopotamus major,* &c.

sccoud and third in *Venus* and Cytherea; the second tooth of the left valve is consequently obsolete.

Distr. C. Islandica ranges from Greenland and the U. S. to the Icy Sea, Norway, and England; in 5—80 fm. water. It occurs fossil in Sicily and Piedmont, but not alive in the Medit.

Fossil, 90 sp. (D'Orb.) Muschelkalk —. Europe.

CIRCE, Schumacher.

Etym. In Greek myth. a celebrated enchantress.

Ex. C. corrugata, Pl. XX. fig. 2. *Syn.* Paphia (undulata) Lam.*

Shell sub-orbicular, compressed, thick, often sculptured with diverging striæ; umbones flat; lunule distinct; ligament nearly concealed; margins smooth; hinge-teeth 3:3; laterals obscure; pallial line entire.

Animal (of *C. minima*) with the mantle open, margins denticulate, siphonal orifices close together, scarcely projecting, fringed; foot large, heeled; palpi long and narrow. Ranges from 8—50 fms. (Forbes.)

Distr. 37 sp. Australia, India, Red Sea, Canaries, Brit.

ASTARTE, Sowerby, 1816.

Syn. Crassina, Lam. Tridonta, Schum. Goodallia, Turton.

Ex. A. sulcata, Pl. XX. fig. 1. (*Astarte,* the Syrian Venus.)

Shell sub-orbicular, compressed, thick, smooth or concentrically furrowed; lunule impressed; ligament external; epidermis dark: hinge-teeth 2:2, the anterior tooth of the right valve large and thick; anterior pedal scar distinct; pallial line simple.

Animal with mantle open; margins plain or slightly fringed; siphonal orifices simple; foot moderate, tongue-shaped; lips large, palpi lanceolate; gills nearly equal, united behind, and attached to the siphonal band.

Distr. 14 sp. Behrings Sea, Wellington Channel, Kara Sea, Ochotsk, U. S. Norway, Brit. Canaries, Ægean (30—112 fms.)

Fossil, 200 sp. (D'Orb.) Lias —. N. and S. America, Europe, Thibet.

? *Digitaria,* Wood; Tellina digitaria, L. Medit. *Fossil,* Crag, Brit.

CRASSATELLA, Lamarck.

Syn. Ptychomya, Ag. Paphia (Lam. part) Roissy.

Type, C. ponderosa, Pl. XXI. fig. 4. *Etym. Crassus* thick.

Shell solid, ventricose, attenuated behind, smooth or concentrically furrowed; lunule distinct; ligament internal; margin smooth or denticulated;

---

* This name was employed by Bolten, in 1798, for sp. of *Veneridæ,* and by Lamarck, in 1801, for *Venus divaricata,* Chemn. (= Circe divaricata *and* Crassatella contraria) and *Mesodesma glabratum.* In 1808, Fabricius adopted the name for a group of butterflies, in which sense it is now widely employed, having been abandoned by Lamarck in his later works, and by all succeeding malacologists.

pallial line simple; hinge-teeth 1:2, striated, in front of cartilage pit; lateral teeth 0—1, 1—0; adductor impressions deep, rounded; pedal small, distinct.

Animal with mantle-lobes united only by the branchial septum ; inhalent margins cirrated; foot moderate, compressed, triangular grooved; gills smooth, unequal, outer semi-lunar, inner widest in front; palpi triangular.

Distr. 30 sp.   Australia, N. Zealand, Philippines, India, W. Africa, Canaries, Brazil.

Fossil, 50 sp.   Neocomian —.   Patagonia, U. S. Europe.

ISOCARDIA, Lam.   Heart-cockle.

Etym. Isos, like, *cardia,* the heart.   *Type,* I. cor. Pl. XX. fig. 3.

Syn. Glossus, Poli; Bucardium, Muhlfeldt; Pecchiolia, Meneghini.

Shell cordate, ventricose; umbones distant, sub-spiral; ligament exter-nal; hinge-teeth 2:2; laterals 1—1 in each valve, the anterior sometimes obsolete.

Animal with the mantle open in front; foot triangular, pointed, com-pressed; siphonal orifices close together, fringed; palpi long and narrow; gills very large, nearly equal.

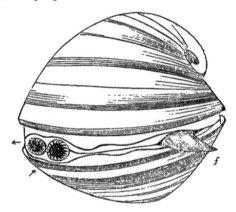

Fig. 217. Isocardia cor.

The heart-cockle burrows in sand, by means of its foot (*f*), leaving only the siphonal openings exposed.   (*Bulwer.*)

Distr. 5 sp.   Brit. Medit. China, Japan.

Fossil, 70 sp. Trias —.   U. S. Europe, S. India.

The Isocardia-shaped fossils of the old rocks belong to the genera *Car-diomorpha* and *Iso-arca;* many of those in the Oolites to *Ceromya.*   Casts of true *Isocardiæ* have only two transverse dental folds between the beaks, and no longitudinal furrows.

CYPRICARDIA, Lam.

Ex. C. obesa, Pl. XX. fig. 4.   *Syn.* Trapezium, Humph.   Libitina, Sch.

Shell oblong, with an oblique posterior ridge;. umbones anterior depressed; ligament external, in deep and narrow grooves; cardinal teeth 2:2, laterals 1—1 in each valve, sometimes obscure; muscular impressions oval, (of two elements); pallial line simple.

Animal (of *C. solenoides*) with mantle-lobes united, cirrated behind; pedal opening moderate; foot small, compressed, with a large byssal pore near the heel; siphons short, conical, unequal, cirrated externally; orifices fringed; palpi small; gills unequal, the outer narrower and shorter, deeply lamellated, united posteriorly, the inner prolonged between the palpi.

Distr. 13 sp. Red Sea, India, Australia. In crevices of rock and coral.

Fossil, 60 sp. L. Silurian —. N. America, Europe.

? *Sub-genera. Coralliophaga*, Bl. C. coralliophaga, Lam. *Shell* long, cylindrical, thin, slightly gaping behind; hinge-teeth 2:2, and a laminar posterior tooth; pallial line with a wide and shallow sinus. *Distr.* 2 sp. Medit. in the burrows of the *Lithodomus;* sometimes two or three dead shells are found one within the other, besides the original owner of the cell.

? *Cypricardites*, Conrad (part). An. Geol. Rep. 1841. (Sanguinolites, M'Coy). Employed for Cypricardia-shaped shells of the palæozoic rocks; some of them are more nearly related to *Modiola* (v. Modiolopsis, p. 266) but they bear no resemblance to *Sanguinolaria*.

PLEUROPHORUS, King, 1848.

Type, P. costatus, Brown. Permian, England, (Pal. Trans. 1850. Pl. XV. fig. 13—20.)

Syn. ? Cleidophorus, Hall (cast only). Unionites, Wissm. ? Mæonia, Dana.

Shell oblong; dorsal area defined by a line, or keel; umbones anterior, depressed; hinge-teeth 2.2; laterals 1.1; elongated posterior; anterior adductor impression deep, with a small pedal scar close to it, and bounded posteriorly by a strong rib from the hinge; pallial line simple.

Fossil, L. Silurian — Trias. U. States; Europe, N. S. Wales, Tasmania.

? CARDILIA, Deshayes.

Type, C. semisulcata, Pl. XVIII. fig. 18. *Syn.* Hemicyclonosta, Desh.

Shell oblong, ventricose, cordate; beaks prominent, sub-spiral; hinge with a small tooth and dental pit in each valve; ligament partly internal contained in a spoon-shaped inflection; anterior muscular scar long, with a pedal scar above; posterior adductor impression on a prominent sub-spiral plate; pallial line simple.

Distr. 2 sp. Chinese Sea; Moluccas.

Fossil, 2 sp. Eocene —. France, Piedmont.

MEGALODON, J. Sowerby.

Type, M. cucullatus, Pl. XIX. fig. 19. (*Megas*, large, *odous*, tooth.)

Shell oblong, smooth or keeled; ligament external; hinge-teeth 1:2, thick;

laterals 1.1, posterior; anterior adductor impression deep, with a raised margin; and a small pedal scar behind it.

In the typical species the beaks are sub-spiral, the lateral teeth obscure, and the posterior adductors bounded by prominent ridges.

Fossil, 14 sp.   U. Silurian — Devonian; U. States, Europe.

Sub-genera. ? *Goldfussia* (nautiloides) Castlenau.   Umbones spiral; anterior side concentrically furrowed; posterior side with two oblique ridges. *Fossil*, Silurian, U. States.

Megaloma (Canadensis) Hall, 1852.   U. Silurian, Canada.   Umbones very thick, hinge-teeth rugged, almost obliterated with age; posterior lateral teeth 1.1; no muscular ridges.

PACHYDOMUS (Morris) J. Sowerby.

Etym. Pachus, thick, *domos*, house.   *Syn.* Astartila, Dana. ? Cleobis (grandis) Dana. ? Pyramus (ellipticus) D. = Notomya, M'Coy. *Type*, P. globosus (Megadesmus) J. Sby. in Mitchell's Australia.

Shell oval, ventricose, very thick; ligament large, external; lunette more or less distinct; hinge-line sunk; teeth 1 or 2 (?) in each valve; adductor impressions deep; anterior pedal scar distinct; pallial line broad and simple, or with a very shallow sinus.

Fossil, 5 sp.   Devonian ?   N. S. Wales, Tasmania.

PACHYRISMA, Morris and Lycett.

Etym. Pachus, thick, *ereisma*, support.

Type, P. grande, M. and L.   Great Oolite (Bathonian) Minchinhampton.

Shell cordate, with large sub-spiral beaks; valves very thick near the umbones, obliquely keeled; hinge with one thick conical tooth (behind the dental pit, in the right valve), a small lateral tooth close to the deep and oval anterior adductor, and a posterior lateral-tooth (or muscular lamina ?) ; ligamental plates short and deep.

OPIS, Defrance.

Ex. O. lunulata, Pl. XIX. fig. 24.   (*Opis*, a name of Artemis.)

Shell strong, ventricose, cordiform, obliquely keeled; beaks prominent, incurved or sub-spiral; cardinal teeth 1.1; lunule distinct.

Fossil, 42 sp.   Trias — Chalk.   Europe.

CARDINIA, Agassiz.

Etym. Cardo-inis, a hinge.   *Type*, C. Listeri, Pl. XIX. fig. 23.

Syn. Thalassides, Berger 1833 (no descr.) Sinemuria, Christol. Pachyodon, Stutch. (not Meyer nor Schum.) Pronoe, Ag.

Shell oval or oblong, attenuated posteriorly, compressed, strong, not pearly, marked by lines of growth; ligament external; cardinal teeth obscure, laterals 1—0, 0—1, remote, prominent; adductor impressions deep pallial line simple.

Fossil, 20 sp. Lias —. Inf. Oolite, Europe; along with *marine* shells.

Sub-genus ? *Anthracosia*, King, 1844; Unio sub-constrictus. Sby. U. Sil. — Carb. 40 sp. They occur in the valuable layers of clay-ironstone called "mussel-bands," associated with *Nautili, Discinæ*, &c. In Derbyshire the mussel-band is wrought, like marble, into vases.

? MYOCONCHA, J. Sowerby.

Type, M. crassa, Pl. XIX. fig. 25. (*Mya*, mussel, *concha*, shell.)

Shell oblong, thick, with nearly terminal depressed umbones; ligament external, supported by long narrow appressed plates; hinge thick, with an oblique tooth in the right valve; anterior muscular impression round and deep, with a small pedal scar behind it; posterior impression large, single; pallial line simple.

This shell, which is not nacreous inside, is distinguished from any of the *Mytilidæ* by the form of its ligamental plates and muscular impressions; the hinge-tooth is usually overgrown and nearly obliterated by the hinge-margin as in aged examples of *Cardita orbicularis* and *Cypricardia vellicata*.

Fossil, 26 sp. Permian — Miocene. (D'Orb.) -Europe.

Sub-genus. ? *Hippopodium* (ponderosum, Sby.) Coneybeare. Lias, Europe. *Shell* oblong, thick, ventricose; umbones large; ligament external; ventral margin sinuated; hinge with one thick, oblique tooth in each valve, sometimes nearly obsolete; pallial line simple; anterior muscular scar deep. This shell appears to be a ponderous form of *Cypricardia* or *Cardita*; it is a characteristic fossil of the English Lias, but only very aged examples have been found.

CARDITA, Bruguière.

Syn. Mytilicardia and Cardiocardita, (ajar) Bl. Arcinella, Oken.

Type, C. calyculata, Pl. XX. fig. 5. *Etym. Cardia*, the heart.

Shell oblong, radiately ribbed; ligament external; margins toothed; hinge-teeth 1:2, and an elongated posterior tooth; pallial line simple; anterior pedal scar close to adductor.

Animal with the mantle lobes free, except between the siphonal orifices; branchial margin with conspicuous cirri; foot rounded and grooved, spinning a byssus; labial palpi short, triangular, plaited; gills rounded in front, tapering behind and united together, the outer pair narrowest.

C. pectunculus, Brug. (*Mytilicardia*, Bl.) has an anterior tooth; *C. concamerata*, Brug. found at the Cape, has a remarkable cup-like inflection of the ventral margin of each valve.

Sub-genus. Venericardia, Lam. V. ajar, Pl. XX. fig. 6. *Shell* cordate, ventricose; hinge without lateral teeth. *Animal* locomotive, with a sickle-shaped foot like the cockles.

P

Distr. 50 sp. Chiefly in tropical seas, on rocky bottoms and in shallow water; the *Venericardiæ* on coarse sand and sandy mud. W. Indies, U. S. W. Africa, Medit. Red Sea, India, China, Australia, New Zealand, Pacific, W. America. *C. borealis*, Conrad, inhabits the sea of Ochotsk; *C. abyssicola*, Hinds, ranges to 100 fms.; *C. squamosa*, to 150 fms.

Fossil, 100 sp. Trias —. U. S. Patagonia, Europe, S. India.

? VERTICORDIA, Searles Wood, 1844.

Syn. Hippagus, Philippi, not Lea. (*Verticordia*, a name of Venus.)

Type, V. cardiiformis (Wood, in Sby. Min. Con.) Pl. XVII. fig. 26.

Shell sub-orbicular, with radiating ribs; beaks sub-spiral; margins denticulated; interior brilliantly pearly; right valve with 1 prominent cardinal tooth; adductor scars 2, faint; pallial line simple; ligament internal, oblique; epidermis dark brown.

Distr. 2 sp. China Sea (Adams). Medit. ? (Forbes.)

Fossil, 2 sp. Miocene —. Brit. Sicily.

Hippagus isocardioides, Lea, 1833, Eocene, Alabama: is edentulous.

SECTION *b*. SINU-PALLIALIA.

Respiratory siphons long; pallial line sinuated.

FAMILY XIV. VENERIDÆ.

Shell regular, closed, sub-orbicular or oblong; ligament external; hinge with usually 3 diverging teeth in each valve; muscular impressions oval, polished; pallial line sinuated.

Animal free, locomotive, rarely byssiferous or burrowing; mantle with a rather large anterior opening; siphons unequal, united more or less; foot linguiform, compressed, sometimes grooved; palpi moderate, triangular, pointed; branchiæ large, sub-quadrate, united posteriorly.

The shells of this tribe are remarkable for the elegance of their forms and colours; they are frequently ornamented with chevron-shaped lines. Their texture is very hard, all traces of structure being usually obliterated. The *Veneridæ* appeared first in the Oolitic period, and have attained their greatest development at the present time; they are found in all seas, but most abundantly in the tropics.

VENUS, L.

Syn. Merceneria, Antigone and Anomalocardia (flexuosa) Schum. Chione, Megerle (not Scop,) Erycina (cardioides) Lam. 1818.

Type, V. paphia, L. Pl. XX. fig. 7.

Shell thick, ovate. smooth, sulcated or cancellated; margins minutely crenulated; cardinal teeth 3—3; pallial sinus small, angular; ligament prominent; lunule distinct.

Animal with mantle-margins fringed; siphons unequal, more or less separate; branchial- orifice sometimes doubly fringed, the outer pinnate; anal orifice with a simple fringe and tubular valve; foot tongue-shaped; palpi small, lanceolate.

V. textilis, and other elongated species, have a deep pallial sinus; *V. gemma* (Totten) has a very deep angular sinus, like *Artemis*; *V. reticulata* has bifid teeth, like *Tapes*; *V. tridacnoides*, a fossil of the U. States, has massive valves, ribbed like the clam-shell. The N. American Indians used to make coinage (*wampum*) of the sea-worn fragments of *Venus mercenaria*, by perforating and stringing them on leather thongs.

Distr. 176 sp. World-wide. Low-water — 140 fathoms. *V. astartoides*, Behrings' Sea. *V. verrucosa*, Brit. Medit. Senegal, Cape, Red Sea; Australia?

Fossil, 160 sp. Oolites —. Patagonia, U. S. Europe, India.

? *Volupia rugosa*, (Defrance, 1829.) Shell minute, Isocardia-shaped, concentrically ribbed, with a large lunule. *Eocene*, Hauteville.

Saxidomus (Nuttalli) Conrad. Oval, solid, with tumid umbones; lunule, 0; teeth 3—4, unequal, the central bifid; pallial sinus large. *Distr.* 8 sp. India, Australia, W. America.

CYTHEREA, Lam.

Etym. Cytherea, from Cythera, an Aegean Island.

Syn. Meretrix, Gray. Dione, Megerle.

Examples, C. dione, Pl. XX. fig. 8. C. chione, fig. 14, p. 26.

Shell like *Venus*; margins simple; hinge with 3 cardinal teeth and an anterior tooth beneath the lunule; pallial sinus moderate, angular.

Animal with plain mantle-margins; siphons united half-way.

Distr. Same as *Venus*. Recent 113 sp. *Fossil*, 80 sp.

MEROE, Schum.

Etym. Meroë, an island of the Nile.

Syn. Cuneus (part) Megerle (not Da Costa). Sunetta, Link.

Type, M. picta (= Venus Meroë, L. Donax, Desh.) Pl. XX. fig. 9.

Shell oval, compressed; anterior side rather longest; hinge with 3 cardinal teeth, and a long narrow anterior tooth; lunule lanceolate; ligament in a deep escutcheon.

Distr. 10 sp. Senegal, India, Japan, Australia.

TRIGONA, Mühlfeldt.

Etym. Trigonos, theee- cornered. *Type*, T. tripla, Pl. XX. fig. 10.

Shell trigonal, wedge-shaped, sub-equilateral; ligament short, prominent; cardinal teeth 3—4, anterior $\frac{2}{4}$ remote; pallial sinus rounded, horizontal.

Distr. 28 sp. W. Indies, Medit, Senegal, Cape, India, W. America.

Fossil, Miocene —. Bordeaux.

T. crassatelloides attains a diameter of 5 inches and is very ponderous.

P 2

Sub-genus, Grateloupia, Desm.　G. irregularis, Pl. XX. fig. 11.

Shell sub-equilateral, rounded in front, attenuated behind; hinge with I anterior tooth, 3 cardinal teeth and several small posterior teeth; pallial sinus deep, oblique. *Fossil,* 4 sp.　Eocene — Miocene.　U. States, France.

ARTEMIS, Poli.

Etym. Artemis, in Greek myth. Diana.

Type, A. exoleta, Pl. XX. fig. 12.　(*Syn.* Dosinia, Scopoli.)

Shell orbicular, compressed, concentrically striated, pale; ligament sunk; lunule deep; hinge like *Cytherea;* margins even; pallial sinus deep, angular, ascending.

Animal with a large hatchet-shaped foot, projecting from the ventral margin of the shell; mantle-margins slightly plaited; siphons united to their ends; orifices simple; palpi narrow.

Distr. 85 sp.　Boreal — Tropical seas; low-water—80 fms.

Fossil, 8 sp.　Miocene —.　U. States, Europe, S. India.

Sub-genera. Cyclina, Desh.　V. Sinensis, Chemn. Orbicular, ventricose, margins crenulated, no lunule, sinus deep and angular. *Distr.* 10 sp. Senegal, India, China, Japan. W. America. *Fossil,* 1 sp. *Miocene,* Bordeaux.

Clementia (papyracea) Gray. Thin, oval, white; ligament semi-internal; posterior teeth bifid, sinus deep and angular. *Animal* with long, united siphons, and a large crescentic foot, similar to *Artemis.* *Distr.* 3 sp. Australia, Philippines.

LUCINOPSIS, Forbes.

Syn. Dosinia, Gray, 1847 (not Scop.)　Mysia, Gray, 1851 (not Leach). Cyclina, Gray, 1853 (not Desh.)

Type, Venus undata, Pennant, Pl. XX. fig. 13.　(*Lucina,* and *opsis,* like.)

Shell lenticular, rather thin; right valve with 2 laminar, diverging teeth, left with 3 teeth, the central bifid: muscular impressions oval, polished; pallial sinus very deep, ascending.

Animal with mantle-margins plain; pedal opening contracted; foot pointed, basal; siphons longer than the shell, separate, divergent, with fringed orifices. (Clark.)

Distr. 1 sp.　Norway, Brit.　*Fossil,* 3 sp.　Miocene.　Brit. Belgium.

TAPES, Mühlfeldt.

Syn. Paphia, Bolten, 1798.　Pullastra, G. Sby.

Example, T. pullastra, Pl. XX. fig. 14.　(*Tapes,* tapestry.)

Shell oblong, umbones anterior, margins smooth; teeth 3 in each valve, more or less bifid; pallial sinus deep, rounded.

Animal spinning a byssus; foot thick, lanceolate, grooved; mantle plain or finely fringed; freely open in front; siphons moderate, separate half-way or throughout, orifices fringed, anal cirri simple, branchial ramose; palpi long, triangular.

Distr. 78 sp. Norway, Brit. Black Sea, Senegal, Brazil, India, China, New Zealand. Low-water—100 fms. (Beechey).

Fossil, Miocene —. Brit. France, Belgium, Italy.

The animal is eaten on the continental coasts; it buries in the sand at low-water or hides in the crevices of rocks, and roots of sea-weed.

Venerupis, Lamarck.

Etym. Venus, and *rupes,* a rock. *Syn. Gastrana,* Schum.

Example, V. exotica, Pl. XX. fig. 15.

Shell oblong, a little gaping posteriorly, radiately striated and ornamented with concentric lamellæ; three small teeth in each valve, one of them bifid; pallial sinus moderately deep, angular.

Animal with the mantle closed in front, pedal opening moderate; siphons united half-way, anal with a simple fringe and tubular valve, branchial siphon doubly fringed, inner cirri branching; palpi small and pointed.

Distr. 19 sp. Brit. — Crimea; Canaries; India, Tasmania; Kamtschatka. Behring's Sea — Peru. In crevices of rocks.

Fossil, Miocene —. U. States, Europe.

Petricola, Lamarck.

Etym. Petra, stone, *colo,* to inhabit.

Syn. Rupellaria, Bellevue; Choristodon, Jonas; Naranio, Gray.

Type, P. lithophaga, Pl. XX. fig. 16. P. pholadiformis, Pl. XX. fig. 17.

Shell oval or elongated, thin, tumid, anterior side short; hinge with 3 teeth in each valve, the external often obsolete; pallial sinus deep.

Animal with the mantle closed in front, much thickened and recurved over the edges of the shell; pedal opening small; foot small, pointed, lanceolate; siphons partially separate, orifices fringed, anal with a valve and simple cirri, branchial cirri pinnate; palpi small, triangular.

Distr. 30 sp. U. S. France, Red Sea, India, New Zealand, Pacific, W. America (Sitka—Peru). Burrows in limestone and mud.

Fossil, 12 sp. Eocene —. U. S. Europe.

Glaucomya, (Bronn) Gray.

Syn. Glauconome, Gray 1829 (not Goldfuss 1826).

Type, G. Sinensis, Pl. XX. fig. 18. (*Glaucos* sea-green, *mya* mussel.)

Shell oblong, thin; epidermis dark, greenish; ligament external; hinge with 3 teeth in each valve, one of them bifid; pallial sinus very deep and angular.

Animal with a rather small, linguiform foot; pedal opening moderate;

siphons very long, united, projecting far into the branchial cavity when retracted, their ends separate and diverging; palpi large, sickle-shaped; gills long, rounded in front, the outer shortest.

Distr. 11 sp. Embouchures of rivers; China, Philippines, Borneo, India.

FAMILY XV. Mactridæ.

Shell equivalve, trigonal, close, or slightly gaping; ligament (cartilage) internal, contained in a deep triangular pit; epidermis thick; hinge with 2 diverging cardinal teeth, and usually with anterior and posterior laterals; pallial sinus short, rounded.

Animal with the mantle more or less open in front; siphonal tubes united, orifices fringed; foot compressed; gills not prolonged into the branchial siphon.

Sections of the shell exhibit an indistinct cellular layer on the external surface and a distinct inner layer of elongated cells. (*Carpenter.*)

Mactra, L.

Etym. Mactra, a kneading trough. *Syn.* Trigonella, Da Costa (not L.) Schizodesma (Spengleri), Spisula (solida), Mulinia (lateralis) Gray.

Type, M. stultorum, Pl. XXI. fig. 1.

Shell nearly equilateral; anterior hinge tooth Λ-shaped, with sometimes a small laminar tooth close to it; lateral teeth doubled in the right valve.

Animal with the mantle open as far as the siphons, its margins fringed; siphons united, fringed with simple cirri, anal orifice with a tubular valve; foot large, linguiform, heeled; palpi triangular, long and pointed; outer gills shortest.

The Mactras inhabit sandy coasts, where they bury just beneath the surface; the foot can be stretched out considerably, and moved about like a finger, it is also used for leaping. They are eaten by the star-fishes and whelks. and in the I. of Arran *M. subtruncata* is collected at low-water to feed pigs. (*Alder.*)

Distr. 60 sp. All seas, especially within the tropics; — 35 fms.

Fossil, 30 sp. Lias —. U. States, Europe, India.

? *Sub-genus. Sowerbya,* D'Orb. S. crassa, Oxfordian, France. Cartilage-pit simply grooved; lateral teeth very large.

Gnathodon, Gray.

Etym. Gnathos a jaw-bone, *odous* a tooth. *Syn.* Rangia, Desm.

Type, G. cuneatus, Pl. XXI. fig. 2.

Shell oval, ventricose; valves thick, smooth, eroded; epidermis olive; cartilage-pit central; hinge teeth $\frac{2}{1}$; laterals doubled in the right valve, elongated, striated transversely; pallial sinus moderate.

Animal with the mantle freely open in front; margins plain; siphons short, partly united; foot very thick, tongue-shaped, pointed; gills unequal, the outer short and narrow; palpi large, triangular, pointed.

Distr. 1 sp. N. Orleans (3 other sp. ? Mazatlan, California; Moreton B. Australia. *Petit.*)

Fossil, 1 sp. Miocene —. Petersburg, Virginia.

G. cuneatus was formerly eaten by the Indians. At Mobile, on the Gulf of Mexico, it is found in colonies along with *Cyrena Carolinensis,* burying 2 inches deep in banks of mud; the water is only brackish, though there is a tide of 3 feet. Banks of dead shells, 3 or 4 feet thick, are found 20 miles inland: Mobile is built on one of these shell-banks. The road from New Orleans to Lake Pont-chartrain (6 miles) is made of Gnathodon shells procured from the east end of the lake, where there is a mound of them a mile long, 15, feet high, and 20—60 yards wide; in some places it is 20 feet above the level of the lake. (*Lyell.*)

LUTRARIA, Lamarck. Otter's-shell.

Type, L. oblonga, Gmel. Pl. XXI. fig. 3. (= L. solenoides, Lam.)

Shell oblong, gaping at both ends; cartilage-plate prominent, with 1 or 2 small teeth in front of it, in each valve; pallial sinus deep, horizontal.

Animal with closed mantle-lobes; pedal opening moderate; foot rather large, compressed; siphons united, elongated, invested with epidermis; palpi rather narrow, their margins plain; gills tapering to the mouth.

Distr. 18 sp. U. States, Brazil, Brit. Medit. Senegal, Cape, India, N. Zealand, Sitka.

Fossil, 10 sp. Miocene — U. States, Europe.

Resembles *Mya;* burying vertically in sand or mud, especially of estuaries; low-water, 12 fms. *L. rugosa* is found living on the coasts of Portugal and Mogador, fossil on the coast of Sussex. (Dixon.)

ANATINELLA, G. Sowerby.

Type, A. candida, (Mya) Chemn. Pl XXIII. fig. 6..

Shell ovate, rounded in front, attenuated and truncated behind; cartilage in a prominent spoon-shaped process, with 2 small teeth in front; muscular impressions irregular, the anterior elongated; pallial line slightly truncated behind.

Distr. 3 sp. Ceylon, Philippines; sands at low-water.

FAMILY XVI. TELLINIDÆ.

Shell free, compressed, usually closed and equivalve; cardinal teeth 2 at most, laterals 1—1, sometimes obsolete; muscular impressions rounded, polished; pallial sinus very large; ligament on shortest side of the shell, sometimes internal. Structure obscurely prismatic-cellular; prisms fusiform, nearly parallel with surface, radiating from the hinge in the outer layer, transverse in the inner.

Animal with the mantle widely open in front, its margins fringed; foot tongue-shaped, compressed; siphons separate, very long and slender; palpi large, triangular; gills united posteriorly, unequal, the outer pair sometimes directed dorsally.

The Tellens are found in all seas, chiefly in the littoral and laminarian zones; they frequent sandy bottoms, or sandy mud, burying beneath the surface; a few species inhabit estuaries and rivers. Their valves are often richly coloured and ornamented with finely sculptured lines.

TELLINA, L. Tellen.

Etym. Telline, the Greek name for a kind of mussel.

Syn. Peronæa (*part*) Poli. Phylloda (foliacea), Omala (planata) Schum. Psammotea (solidula) Turt. Arcopagia (crassa) Leach.

Examples, T. lingua-felis, Pl. XXI. fig. 5. , T. carnaria, fig. 6.

Shell slightly inequivalve, compressed, rounded in front, angular and slightly folded posteriorly, umbones sub-central; teeth 2.2, laterals 1—1, most distinct in the right valve; pallial sinus very wide and deep; ligament external, prominent.

Animal with slender, diverging siphons, twice as long as the shell, their orifices plain; foot broad, pointed, compressed; palpi very large, triangular; gills small, soft and very minutely striated, the outer rudimental and directed dorsally.

Tellinides, Lam. T. planissima, Pl. XXI. fig. 7. Valves with no posterior fold; lateral teeth wanting.

T. carnaria (*Strigilla*, Turt.) has the valves obliquely sculptured *T. fabula*, Gron. has the right valve striated, the other plain. *T. Burneti*, California, has the right valve flat; *T. lunulata, Pliocene*, S. Carolina, much resembling it in shape, has the left valve flat.

Distr. above 200 sp. In all seas, especially the Indian Ocean; most abundant and highly coloured in the tropics. Low-water — Coral zone, 50 fms. Wellington Channel; Kara Sea; Behrings' Sea; Baltic; Black Sea.

Fossil, 130 sp. Oolites —. U. States, S. America (Chiloe) Europe.

DIODONTA, Schumacher.

Etym. Di- two, *odonta* teeth. *Syn.* Fragilia, Desh.

Type, Tellina fragilis, L. Pl. XXI. fig. 8.

Shell equivalve, convex, with squamose lines of growth; cardinal teeth 2 in right valve, 1 bifid tooth in left; pallial sinus deep and rounded; umbonal area punctate; ligament external.

Animal with the mantle open in front, its margins fringed; siphons elongated, slender, separate, unequal, orifices with cirri; foot small, compressed, linguiform; palpi large, triangular; gills unequal, soft, finely striated.

Diodonta inhabits shallow water, boring in mud and clay, and not travelling about like the *Tellens.*

Distr. 3 sp. Greenland, Brit. Medit. Black Sea, Senegal, Cape. *Fossil,* Miocene —. Brit. France, Belgium.

CAPSULA, Schumacher.

Etym. Dimin. of *capsa,* a box.
Syn. Capsa (part) Brug. 1791. Sanguinolaria Lam. 1818, not 1801.
Type. C. rugosa, Pl. XX. fig. 19. (= Venus deflorata, Gmel.)
Shell oblong, ventricose, slightly gaping at each end; radiately striated; cardinal teeth 2 in each valve, one of them bifid; ligament external, large, prominent; siphonal inflection short.
Animal like *Psammobia;* foot moderate; gills deeply plaited, attenuated in front, outer small, dorsal border wide, fixed; siphons moderate.
Distr. W. Indies, Red Sea, India, China, Australia.
Fossil 4 sp. U. Green-sand —. U. States, Europe. (D'Orb.)

Fig. 218. *Psammobia vespertina,* Chemn, ½, Brit.

PSAMMOBIA, Lamarck. Sunset-shell.

Etym. Psammos sand, *bio* to live.
Syn. Psammotea (zonalis) Lam. Psammocola, Bl. Gari, Schum.
Ex. P. Ferroënsis, Pl. XXI. fig. 9. P. squamosa, Pl. XXI. fig. 10.
Shell oblong, compressed, slightly gaping at both ends; hinge-teeth $\frac{2}{1}$; ligament external, prominent; siphonal inflection deep, in contact with the pallial line; epidermis often dark.
Animal: mantle open, fringed; siphons very long, slender, nearly equal, longitudinally ciliated, orifices with 6—8 cirri; foot large, tongue-shaped; palpi long, tapering; gills unequal, recumbent, few plaited.
Distr. 40 sp. Norway, Brit. India, New Zealand, Pacific. Littoral — coralline zone, 100 fms. *P. gari* is eaten in India.
Fossil, 24 sp. Oolite? Eocene —. U. States, Europe.

SANGUINOLARIA, Lamarck.

Name, from the type, *Solen sanguinolentus,* Chemn.
Syn. Soletellina (*diphos*) Bl. Lobaria, Schum. Aulus, Oken.
Ex. S. livida, Pl. XXII. fig. 1. S. diphos, fig. 2, S. orbiculata, fig. 3.
Shell oval, compressed, rounded in front, attenuated and slightly gaping behind; hinge-teeth $\frac{2}{2}$, small; siphonal inflection very deep, connected with the pallial line; ligament external, on very prominent fulcra.
Animal: mantle open, fringed; siphons very long, branchial largest

orifices fringed; foot large, broadly tongue-shaped, compressed; palpi long pointed; gills recumbent, inner laminæ free, dorsal border wide.

Distr. 20 sp. W. Indies, Red Sea, India, Madagascar, Japan; Australia, Tasmania, Peru.

Fossil, 30 sp. Eocene —. U. States, Europe.

SÉMELE, Schumacher, 1817.

Etym. Semele, in Greek myth. the mother of Bacchus.

Syn. Amphidesma, Lam. 1818.* *Type,* S. reticulata, Pl. XXI. fig. 11.

Shell rounded, sub-equilateral, beaks turned forwards; posterior side slightly folded; hinge-teeth 2.2, laterals elongated, distinct in the right valve; external ligament short, cartilage internal, long, oblique; pallial sinus deep, rounded.

Distr. 40 sp. W. Indies, Brazil, India, China, Australia, Peru.

Fossil, 10 sp. Eocene —. U. States, Europe.

Sub-genera. Cumingia, G. Sowerby. C. lamellosa, Pl. XXI. fig. 12. Shell slightly attenuated and gaping behind, lamellated concentrically; cartilage-process prominent; pallial sinus very wide. *Distr.* 10 sp. In sponges, sand, and the fissures of rocks, — 7 fathoms. W. Indies, India, Australia, W. America. *Fossil,* Miocene —. Wilmington, N. Carolina.

Syndosmya, Recluz. *Syn.* Abra, Leach MS. Erycina (part) Lam. 1805.† *Type,* S. alba, Pl. XXI. fig. 13. *Shell* small, oval, white and shining; posterior side shortest; umbones directed backwards; cartilage-process oblique; hinge-teeth minute or obsolete, laterals distinct; pallial sinus wide and shallow. *Animal* with the mantle open, fringed; siphons long, slender, diverging, anal shortest, orifices plain; foot large, tongue-shaped, pointed; palpi triangular, nearly as large as the gills; branchiæ unequal, triangular. *Distr.* Norway, Brit. Medit. Black Sea, India. The sp. are few, and mostly boreal, ranging from the laminarian zone to 180 fms. (*Forbes.*) They live buried in sand and mud, but when confined are able to creep up the sides of the vessel with their foot. (*Bouchard.*) *Fossil,* 6 sp. Eocene —. Brit. France.

Scrobicularia, Schumacher. *Syn.* Trigonella (part) Da Costa (not L.) Ligula (part). Mont. " Le Lavignon" (Reaumur) Cuv. Listera, Turt. (not R. Brown.) Lutricola, Bl. Mactromya, D'Orb. (not Ag.) *Type,* S. piperata (Belon) Gmelin, Pl. XXI. fig. 14. (See p. 60.) *Shell* oval, compressed, thin; sub-equi-lateral; ligament external, slight; cartilage-pit shal-

---

* The name *Amphi-desma,* as employed by Lamarck, included species of *Semele, Loripes, Syndosmya, Mesodesma, Thracia, Lyonsia,* and *Kellia;* in addition to which it has since been applied to some Oolitic *Myacites.*

† The name *Erycina* was originally appplied by Lamarck to a number of minute fossil shells, including sp. of *Syndosmya, Venus, Lucina, Tellina, Astarte,* and *Kellia.* In 1808 Fabricius employed it for a well-known group of insects.

low, triangular; hinge-teeth small, 1 or 2 in each valve, laterals obsolete; pallial sinus wide and deep.

Animal with the mantle open, margins denticulated; siphons very long, slender, separate, orifices plain; foot large, tongue-shaped, compressed; palpi very large, triangular, gills minutely striated, the outer pair directed dorsally. Lives buried, vertically, in the mud of tidal estuaries, 5 or 6 inches deep. (*Montagu.*) The siphons can be extended to 5 or 6 times the length of the shell. (*Deshayes*). The animal has a peppery taste, but is sometimes eaten on the coasts of the Mediterranean. *Distr.* Norway, Brit. Medit. Senegal. *Fossil*, Pliocene, Brit.

MESODESMA, Deshayes.

Etym. Meso- middle, *desma* ligament. *Syn.* Eryx, Sw. (not Daud.) Paphia (part) Lam. 1799 (see p. 299, note). Erycina (part) Lam. 1818 (not Lam. 1805, nor Fabr. 1808). "Donacille," Lam. 1812 (not characterized).

Examples, M. glabratum, Pl. XXI. fig. 15. M. donacium. fig. 16.

Shell trigonal, thick, compressed, closed; ligament internal, in a deep central pit; a minute anterior hinge-tooth, and 1—1 lateral teeth in each valve; muscular scars deep, pallial sinus small.

Animal with mantle-margins plain; siphons short, thick, and separate, orifices cirrated, branchial cirri dendritic; foot compressed, broadly lanceolate: gills large, unequal; palpi small.

Sub-genus. Anapa, Gray. A. Smithii, Pl. XXI. fig. 17. Umbones anterior, siphonal inflection obsolete.

Distr. 20 sp. W. Indies, Medit. Crimea, India, New Zealand, Chili; sands at low-water.

Fossil, 7 sp. Neocomian —. U. S. Europe (*Donacilla*, D'Orb.)

ERVILIA, Turton. Lentil-shell.

Etym. Ervilia, diminutive of *ervum,* the bitter-vetch.

Type, E. nitens, P. XXI. fig. 18.

Shell minute, oval, close; cartilage in a central pit; right valve with a single prominent tooth in front and an obscure tooth behind; left valve with 2 obscure teeth; no lateral teeth; pallial sinus deep.

Distr. W. Indies, Brit. Canaries, Medit. Red Sea. — 50 fms.

DONAX, L. Wedge-shell.

Ex. D. denticulatus, Pl. XXI. fig. 19. *Etym. Donax,* a sea-fish, Pliny.

Syn. Chione, Scop. Cuneus, Da Costa. Capisterium, Meusch.* Latona and Hecuba, Schum. Egeria, Lea (not Roissy).

Shell trigonal, wedge-like, closed; front produced, rounded; posterior side short, straight; margins usually crenulated; hinge-teeth 2.2; laterals

---

* Meuschen was a Dutch auctioneer; the names occur in his " sale catalogues." *Idiotæ imposuere nomina absurda.* Linnæus.

1—1 in each valve; ligament external, prominent; pallial sinus deep, horizontal.

Animal with the mantle fringed, siphons short and thick, diverging, anal orifice denticulated, branchial with pinnate cirri; foot very large, pointed, sharp-edged, projected quite in front; gills ample, recumbent, outer shortest; palpi small, pointed.

Distr. 45 sp. Norway, Baltic, — Black Sea, all tropical seas. In sands near low-water mark (— 8 fms.) buried an inch or two beneath the surface.

Fossil, 30 sp. Eocene —. U. States, Europe.

Sub-genera. ? *Amphichæna,* Phil. A. Kindermanni, California. *Shell* oblong, nearly equilateral, gaping at each end; teeth $\frac{2}{3}$; ligament external, pallial line sinuated.

Iphigenia, Schum. (Capsa, Lam. 1818, not 1801. Donacina, Fér.) I. Brasiliensis, Pl. XXI. fig. 20. *Shell* nearly equilateral, smooth; hinge-teeth 2.2, one bifid, the other minute; laterals remote, obsolete in the left valve; margins smooth. *Distr.* 4 sp. W. Indies, Brazil, W. Africa, Pacific, Central America. Inhabits estuaries; *I. ventricosa,* Desh. is rayed like *Galatea,* and has its beaks eroded.

? *Isodonta* (Deshayesii) Buv. Bull. Soc. Geol. *Oxf.* France.

GALATEA, Bruguière.

Syn. Egeria, Roissy. Potamophila, Sby. Megadesma, Bowdich.

Type, G. reclusa, Pl. XXI. fig. 21.

Shell very thick, trigonal, wedge-shaped; epidermis smooth, olive; umbones eroded; hinge thick, teeth 1.2, laterals indistinct; ligament external, prominent; pallial sinus distinct.

Animal with the mantle open in front; siphons moderate, with 6—8 lines of cilia, orifices fringed; foot large, compressed; palpi long, triangular; gills unequal, united to the base of the siphons, the external pair divided into 2 nearly equal areas by a longitudinal furrow, indicating their line of attachment.

Distr. 2 or 7 sp.? Nile, and rivers of W. Africa.

FAMILY XVII. SOLENIDÆ.

Shell elongated. gaping at the ends; ligament external; hinge-teeth usually 2.3, compressed, the posterior bifid. External shell layer with definite cell-structure, consisting of long prisms, very oblique to the surface, and exhibiting nuclei; inner layer nearly homogeneous.

Animal with a very large and powerful foot, more or less cylindrical: siphons short and united (in the typical Solens, with long shells) or longer and partly separate (in the shorter and more compressed genera); gills narrow, prolonged into the branchial siphon.

Fig. 219. Solen siliqua, L. ⅓; the valves forcibly opened, and mantle divided as far as the ventral foramen, to show the foot.

SOLEN (Aristotle) L. Razor-fish.

Type, S. siliqua, Pl. XXII. fig. 4.

Syn. Hypogæa, Poli. Vagina, Megerle. Ensis, Schum. Ensatella, Sw.

Shell very long, sub-cylindrical, straight, or slightly recurved, margins parallel, ends gaping : beaks terminal, or sub-central; hinge-teeth $\frac{2}{2}$; ligament long, external; anterior muscular impression elongated; posterior oblong; pallial line extending beyond the adductors; sinus short and square.

Animal with the mantle closed except at the front end, and a minute ventral opening; siphons short, united, fringed; palpi broadly triangular; foot cylindrical, obtuse.

Distr. 25 sp. World-wide, except Arctic seas :—100 fms.

Fossil, 10 sp. Eocene —. U. States, Europe.

The Razor-fishes live buried vertically in the sand, at extreme low-water, their position being only indicated by an orifice like a key-hole; when the tide goes out they sink deeper, often penetrating to a depth of 1 or 2 feet. They never voluntarily leave their burrows, but if taken out soon bury themselves again. They may be caught with a bent wire, and are excellent articles of food, when cooked. (*Forbes.*)

CULTELLUS, Schumacher.

Type, C. lacteus, Pl. XXII. fig. 5. *Etym. Cultellus* a knife.

Shell elongated, compressed, rounded and gaping at the ends; hinge-teeth 2.3 ; beaks in front of the centre, supported internally by an oblique rib; pedal impression behind the umbonal rib; posterior adductor trigonal; pallial line not prolonged behind the posterior adductor; sinus short and square.

Animal (of C. Javanicus) with short, fringed siphons; gills narrow, half as long as the shell, transversely plaited; palpi large, angular, broadly attached; foot large, abruptly truncated.

Distr. 4 sp. Africa, India, Nicobar.

Sub-genera. Ceratisolen, Forbes. (Polia, D'Orb. Pharus, Leach, MS. Solecurtoides, Desm.) C. legumen, Pl. XXII. fig. 6. *Shell* narrow, subequilateral, anterior adductor impressions elongated, a second pedal scar near the pallial sinus. *Animal* with a long, truncated foot; siphons separate, diverging, fringed. *Distr.* 1 sp. Brit. Medit. Senegal, Red Sea. *Fossil,* 1 sp. Pliocene —. Italy.

Machæra, Gould. (Siliqua, Megerle. Leguminaria, Schum.) M. polita, Pl. XXII. fig. 7. *Shell* smooth, oblong; epidermis polished; umbonal rib extending across the interior of the valve; pallial sinus short. The animal, figured by Middendorff, is similar to *Solecurtus. Distr.* India, China, Ochotsk, Oregon, Sitka, Behring's Sea, Newfoundland. *M. costata,* Say, is often obtained from the maw of the cod-fish. *Fossil,* 4 sp. U. Greensand —. Brit. France.

SOLECURTUS, Blainville.

Etym. Solen and *curtus,* short.

Syn. Psammosolen Risso. Macha, Oken. Siliquaria, Schum.

Ex. S. strigilatus, Pl. XXII. fig. 8. S. Caribæus, Pl. XXII. fig. 9.

Shell elongated, rather ventricose, with sub-central beaks; margins subparallel; ends truncated, gaping; ligament prominent; hinge-teeth $\frac{2}{2}$; pallial sinus very deep, rounded; posterior adductor rounded.

Animal very large and thick, not entirely retractile within the shell; mantle closed below; pedal orifice and foot large; palpi triangular, narrow, lamellated inside; gills long and narrow, outer much shortest; siphons separate at the ends, united and forming a thick mass at their bases; anal orifices plain, branchial fringed.

The *Solecurti* bury deeply in sand or mud, usually beyond low-water, and are difficult to obtain alive. *P. Caribæus* occurs in countless myriads in the bars of American rivers, and on the coast of New Jersey in sand exposed at low-water; by removing 3 or 4 inches of sand its burrows may be discovered; they are vertical cylindrical cavities, 1¼ inches in diameter and 12 or more deep, the animal holds fast by the expanded end of its foot.

Distr. 25 sp. U. States, Brit. Medit. W. Africa, Madeira.

Fossil, 30 sp. Neocomian —. U. S. Europe.

Sub-genus, Novaculina, Benson. N. gangetica, Pl. XXII. fig. 10. *Shell,* oblong, plain; epidermis thick and dull; pallial sinus rather small; anterior pedal scar linear. *Distr.* India, China. In the mud of river-estuaries.

FAMILY XVIII. MYACIDÆ.

Shell thick, strong and opaque; gaping posteriorly; pallial line sinuated; epidermis wrinkled. Structure more or less distinctly cellular, with dark nuclei near outer surface; cartilage process composed of radiated cells.

Animal with the mantle almost entirely closed; pedal aperture and foot small; siphons united, partly or wholly retractile; branchiæ 2 on each side, elongated.

Fig. 220. *Mya truncata*, L. ½. Brit. (after Forbes.)

MYA, L. Gaper.

Etym. Myax (-acis) a mussel, Pliny. *Syn.* Platyodon, Conrad.

Types, M. truncata, Pl. XXIII. fig. 1. M. Arenaria, fig. 170, p. 244.

Shell oblong, inequivalve, gaping at the ends; left valve smallest, with a large flattened cartilage process; pallial sinus large.

Animal with a small straight linguiform foot; siphons combined, covered with epidermis, partially retractile; orifices fringed, the branchial opening with an inner series of large tentacular filaments; gills not prolonged into the siphon; palpi elongated, free.

M. anatina, Chemn. (Tugonia, Gray) W. coast of Africa; posterior side extremely truncated; similar cartilage-processes in each valve. *Fossil*, *Miocene*, Dax, and the Morea.

Distr. 10 sp. Northern Seas, W. Africa, Philippines, Australia, California. The Myas frequent soft bottoms, especially the sandy and gravelly mud of river-mouths; they range from low-water to 25 fathoms, rarely to 100 or 145 fms. *M. arenaria* burrows a foot deep; this species and *M. truncata* are found throughout the northern and Arctic seas, from Ochotsk and Sitka to the Russian Ice-meer, the Baltic, and British coast; in the Mediterranean they are only found fossil. They are eaten in Zetland and N. America, and are excellent articles of food. In Greenland they are sought after by the walrus, the Arctic fox, and birds. (*O. Fabricius.*)

Fossil, Miocene —. U. States, Brit. Sicily. Most of the fossil "Myas" have an external ligament, and are related either to *Panopæa* or *Pholadomya*.

CORBULA, Bruguière.

Etym. Corbula, a little basket. *Type*, C. sulcata, Pl. XXIII. fig. 2.

Syn. Erodona, Daud. (= Pacyodon, Beck.) Agina, Turt.

Shell thick, inequivalve, gibbose, closed, produced posteriorly; right valve with a prominent tooth in front of the cartilage pit; left valve smaller, with a projecting cartilage process; pallial sinus slight: pedal scars distinct from the adductor impressions.

Animal with very short, united siphons; orifices fringed; anal valve tubular; foot thick and pointed; palpi moderate; gills 2 on each side, obscurely striated.

Distr. 50 sp.   U. S. Norway, Brit. Medit. W. Africa, China.   Inhabits sandy bottoms; Lower laminarian zone—80 fms.

Fossil. 90 sp.   Inf. Oolite —.   U. States, Europe, India.   The external shell-layer consists of fusiform cells; the inner is homogeneous and adheres so slightly to the outer layer, that it is very frequently detached in fossil specimens.   Corbulomya, Nyst (*C. complanata*, Sby.) Crag. Brit.

Sub-genera. *Potamomya*, J. Sby.   P. gregaria, Eocene, I. Wight.   Cartilage process broad and spatulate, received between two obscure teeth in the right valve.   The estuary *Corbulæ* differ very little from the marine species. *P. labiata* (Azara, D'Orb.) Pl. XXIII. fig. 3, lives buried in the mud of the R. Plata, but not above Buenos Ayres, and consequently in water which is little influenced by the superficial ebb of the river.   The same species is found in banks widely dispersed over the Pampas near S. Pedro, and many places in the Argentine Republic, 5 yards above the R. Parana.   (*Darwin.*)

Sphenia, Turt.   S. Binghami, Pl. XXIII. fig. 4.   *Shell* oblong; right valve with a curved, conic tooth in front of the oblique, sub-trigonal cartilage-pit.   *Animal* with thick united siphons, fringed at the end, anal valve conspicuous; foot finger-like, with a byssal groove.   *Distr.* Brit. France. Burrowing in oyster-shells and limestone, in 10—25 fms.   *Fossil,* Miocene —. Brit.

NEÆRA, Gray.

Etym. *Neæra,* a Roman lady's name.

Type, N. cuspidata, Pl. XXIII. fig. 5.   *Syn.* Cuspidaria, Nardo.

Shell globular, attenuated and gaping behind; right valve a little the smallest; umbones strengthened internally by a rib on the posterior side; cartilage process spatulate, in each valve, (furnish.d with a moveable ossicle, *Deshayes*) with an obsolete tooth in front, and a posterior lateral tooth; pallial sinus very shallow.

Animal with the mantle closed; foot lanceolate; siphons short, united, branchial largest, anal with a membranous valve, both with a few long, lateral cirri.

Distr. 20 sp. Norway, Brit. Medit, Canaries, Madeira, China, Moluccas, New Guinea, Chile.   From 12—200 fms.

Fossil, 6 sp.   Oolite —.   Brit. Belgium, Italy.

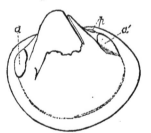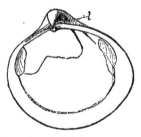

Fig. 221. *Thetis, minor,* Sby. *Neocomian,* I. Wight.

THETIS, Sowerby.

Etym. Thetis, in Greek myth. a sea-nymph.

Syn. Poromya (anatinoides) Forbes.   Embla (Korenii) Lovén ?.   Inoceramus (impressus) D'Orb. ?   Corbula (gigantea) Sby.

Type, T. minor, fig. 221.   T. hyalina, Pl. XXII. fig. 11.

Shell sub-orbicular, ventricose, thin, translucent, surface regularly granulated, interior slightly nacreous; ligament (*l*) external; hinge-teeth 1 or 2; umbones strengthened inside by a posterior lamina; adductor (*a, a'*) and pedal impressions (*p*) separate, slightly impressed, posterior adductor bordered by a ridge; pallial line nearly simple, sub-marginal.

Animal with short siphons, the branchial largest, surrounded at their base by 18—20 tentacles, generally reflected on the shell; mantle open in front; foot long, narrow and slender. (*M'Andrew.*)

Distr. 5 sp. Norway, Brit. Medit. Madeira, Borneo, China. 40—150 fms.

Fossil, 7 sp. Neocomian —.   Brit. Belgium, France, S. India.

Sub-genus ? *Eucharis,* Recluz; Corbula quadrata, Hinds, Guadaloupe. *Shell* equivalve, obliquely keeled, gaping; beaks anterior; hinge-teeth 1—1; ligament external; pallial line simple; surface granulated.

PANOPÆA, Menard de la Groye.

Etym. Panopè, a Nereid.   *Ex.* P. Americana, Pl. XXII. fig. 12.

Syn. ? Pachymya (gigas) Sby.   U. Greensand.   Brit. France.

Shell equivalve, thick, oblong, gaping at each end; ligament external, on prominent ridges; 1 prominent tooth in each valve; pallial sinus deep.

Animal with very long, united siphons, invested with thick, wrinkled epidermis; pedal orifice small, foot short, thick and grooved below; gills long and narrow, extending far into the branchial siphon, the outer pair much narrower, faintly pectinated; palpi long, pointed and striated.

In *P. Norvegica* the pallial line is broken up into a few scattered spots, as in *Saxicava;* the animal itself is like a gigantic Saxicava. (*Hancock.*) This species ranges from Ochotsk to the White Sea, Norway and N. Britain; it was formerly an inhabitant of the Medit. where it now occurs fossil. (= P. *Bivonæ,* Phil.) The British specimens have been caught, accidentally, by the deep-water fishing-hooks. *P. australis* is found at Port Natal, buried in the sand at low-water; the projecting siphons first attracted attention (doubtless by the strong jets of water they sent up when molested) but the shells were only obtained by digging to the depth of several feet. The Medit. sp. *P. glycimeris* attains a length of 6 or 8 inches.

Distr. 6 sp. Northern Seas, Medit. Cape, Australia, New Zealand, Patagonia. Low-water—90 fms.

Fossil, 140 sp. Inf. Oolite —. U. States, Europe, India.

SAXICAVA, Bellevue.

Etym. Saxum, stone, *cavo,* to excavate.　S. rugosa, Pl. XXII; fig. 13.

Syn. Byssomya, Cuv.　Rhomboides, Bl.　Hiatella (minuta) Daud.
Biapholius, Leach.　Arcinella (carinata) Phil.

Shell when young symmetrical, with 2 minute teeth in each valve; adult
rugose, toothless; oblong, equivalve, gaping, ligament external; pallial line
sinuated, not continuous.

Animal with mantle-lobes united and thickened in front; siphons large,
united nearly to their ends, orifices fringed; pedal opening small, foot finger-
like, with a byssal groove; palpi small, free; gills narrow, unequal, united
behind and prolonged into the branchial siphon.

Five genera and 15 species have been manufactured out of varieties and
conditions of this Protean shell.　It is found in crevices of rocks and corals,
and amongst the roots of sea-weed, or burrowing in limestone and shells; at
Harwich it bores in the cement stone (clay iron-stone), at Folkestone in the
Kentish-rag, and the Portland stone employed in the Plymouth Breakwater
has been much wasted by it.　Its crypts are sometimes 6 inches deep (*Couch*);
they are not quite symmetrical, and like those of the *Lithodomus* are inclined at various angles, so as to invade one another, the last comers cutting
quite through their neighbours; they are usually fixed by the byssus to a
small projection from the side of the cell.　The Saxicava ranges from low-
water to 140 fathoms; it is found in the Arctic Seas, where it attains its
largest size; in the Medit, at the Canaries, and the Cape.　It occurs fossil
in the Miocene tertiary of Europe and in the U. States, and in all the
Glacial deposits.

GLYCIMERIS, Lamarck.

Etym. Glukus, sweet, *meris,* bit.

Type, G. siliqua, Pl. XXII. fig. 14.　*Syn.* Cyrtodaria, Daud.

Shell oblong, gaping at each end; posterior side shortest; ligament large
and prominent; epidermis black, extending beyond the margins; anterior
muscular scar long, pallial impression irregular, slightly sinuated.

Animal larger than its shell, sub-cylindrical; mantle closed, siphons
united, protected by a thick envelope; orifices small; pedal opening small
anterior; foot conical; palpi large, striated inside, the posterior border plain;
gills large, extending into branchial siphon.

Distr. Arctic Seas, Cape Parry, N. W. America, Newfoundland.

Fossil, Miocene —.　Brit. Belgium.

FAMILY XIX.　ANATINIDÆ.

Shell often inequivalve, thin; interior nacreous; surface granular; ligament external, thin; cartilage internal, placed in corresponding pits and

furnished with a free ossicle; muscular impressions faint, the anterior elongated; pallial line usually sinuated.

Animal with mantle margins united; siphons long, more or less united, fringed; gills single on each side, the outer lamina prolonged dorsally beyond the line of attachment.

Pholadomya and its fossil allies have an external ligament only; *Cochlodesma* and *Pandora* have no ossicle. The external surface of these shells is often rough with large calcarious cells, sometimes ranged in lines, and covered by the epidermis; the outer layer consists of polygonal cells, more or less sharply defined; the inner layer is nacreous.

ANATINA, Lamarck. Lantern-shell.

Type, A. rostrata, Pl. XXIII. fig. 7. (*Anatinus*, pertaining to a duck.)
Syn. Laternula, Bolten M. S. Auriscalpium, Muhlf. Osteodesma, Bl. Cyathodonta (undulata) Conrad ? W. America.

Shell oblong, ventricose, sub-equivalve, thin and translucent, posterior side attenuated and gaping; umbones fissured, directed backwards, supported internally by an oblique plate; hinge with a spoon-shaped cartilage-process in each valve, furnished in front with a transverse ossicle; pallial sinus wide and shallow.

Animal with a closed mantle and long united siphons, clothed with wrinkled epidermis; gills one on each side, thick, deeply plaited; palpi very long and narrow; pedal opening minute, foot very small, compressed.

Distr. 20 sp. India, Philippines, New Zealand, W. America.
Fossil, 50 sp. Devonian ? — Oolite —. U. States, Europe.

Sub-genera. Periploma (inequivalvis) Schum. " Spoon-hinge" of Petiver; oval, inequivalve, left valve deepest; posterior side very short and contracted. *Distr.* W. Indies, S. America.

Cochlodesma, Couthouy, C. prætenue, Pl. XXIII. fig. 8. (Bontia, Leach MS. Ligula, Mont. part.) Oblong, compressed, thin, slightly inequivalve; umbones fissured; cartilage processes prominent, without an ossicle: pallial sinus deep. *Animal* with a broad, compressed foot; siphons long, slender, divided throughout; gills one on each side, deeply plaited, divided by an oblique furrow into two parts, the dorsal portion being narrower, composed of a single lamina only, and attached by its whole inner surface. (*Hancock.*) *Distr.* 2 sp. U. States, Brit. Medit. *Fossil*, Pliocene, Sicily.

Cercomya, Agassiz. C. undulata, Sby. (= Rhynchomya, Ag.) *Shell* very thin, elongated, compressed, attenuated posteriorly; sides concentrically furrowed, umbones fissured, posterior (cardinal) area more or less defined. *Fossil*, 12 sp. Oolite — Neocomian; Europe.

THRACIA (Leach) Bl.

Syn. Odoncinetus, Costa. Corimya, Ag. Rupicola (concentrica) Bellevue.

Type, T. pubescens, Pl. XXIII. fig. 9.

Shell oblong, nearly equivalve, slightly compressed, attenuated and gaping posteriorly, smooth or minutely scabrous; cartilage processes thick, not prominent, with a crescentic ossicle; pallial sinus shallow. Outer shell layer composed of distinct, nucleated cells.

Animal with the mantle closed; foot linguiform; siphons rather long, separate, with fringed orifices; gills single, thick, plaited; palpi narrow, pointed.

T. concentrica and *T. distorta,* Mont. are found in the crevices of rocks, and burrows of *Saxicava;* they have been mistaken for boring-shells.

Distr. 10 sp. Greenland, U. States, Norway, Brit. Medit. Canaries, China, Sooloo: 4—110 fms.

Fossil, 30 sp. (Trias ?) L. Oolite —. U. States, Europe.

PHOLADOMYA, G. Sowerby.

Recent Type, P. candida. Pl. XXII. fig. 15. I. Tortola.

Shell oblong, equivalve, ventricose, gaping behind; thin and translucent, ornamented with radiating ribs on the sides; ligament external; hinge with one obscure tooth in each valve; pallial sinus large.

Animal with a single gill on each side, thick, finely plaited, grooved along its free border, the outer lamina prolonged dorsally; mantle with a fourth (ventral) orifice. (*Owen.*)

Fossil, 150 sp. Lias —. U. S. Europe, Algeria, Thibet.

Homomya (hortulana) Ag. *Shell* thick, concentrically furrowed, without radiating ribs; 6 sp. Oolites, Europe.

MYACITES (Schlotheim) Bronn.

Syn. Myopsis (Jurassi) Ag. Pleuromya, Ag. Arcomya (Helvetica) Ag. Mactromya (mactroides) Ag. Anoplomya (lutraria) Krauss.

Ex. M. sulcatus, Flem. (Allorisma, King, Pal. Tr. 1850, Pl. XX. fig. 5.)

Shell oblong, ventricose, gaping, thin, often concentrically furrowed; umbones anterior; surface granulated; ligament external; hinge with an obscure tooth or edentulous; muscular impressions faint; pallial line deeply sinuated.

Fossil, 50 sp. L. Silurian — L. Chalk. U. S. Europe, S. Africa.

Sub-genera ? Goniomya, Ag. Mya literata, Pl. XXII. fig. 16. (Lysianassa, Münster, not M. Edw.) *Shell* equivalve, thin, granulated; ligament external, short, prominent. *Fossil,* 30 sp. U. Lias — Chalk. Europe.

Tellinomya (nasuta) Hall; *Silurian,* U. S. Europe. Not characterised.

? *Grammysia,* Verneuil. Nucula cingulata, His. *U. Silurian,* Europe. Valves with a strong transverse fold extending from the umbones to the middle of the ventral margin.

? *Sedgwickia* (corrugata) M'Coy. = ? Leptodomus (senilis) M'Coy.

Shell thin, ventricose, concentrically furrowed in front; escutcheon long and flat. Silurian — Carb. Europe.

CEROMYA, Agassiz.

Etym. Keraos horned, *mya*, mussel.

Type, C. concentrica (Isocardia) Sowerby, Min. Con. 491, fig. 1.

Shell Isocardia-shaped, slightly inequivalve ? very thin, granulated, often eccentrically furrowed; ligament · external; hinge edentulous; right valve with an internal lamina behind the umbo; pallial line scarcely sinuated?

Fossil, 14 sp. Inf. Oolite —. Green-sand ? Europe.

Sub-genus ? *Gresslya* (sulcosa) Ag. (Amphidesma and Unio. sp. Phil.) *Shell* oval, rather compressed; umbones anterior, incurved, not prominent; valves thin, close, smooth or concentrically furrowed; pallial sinus deep. *Fossil,* 17 sp. Lias — Portlandian. Europe. The lamina within the posterior hinge-margin o the right valve produces a furrow in the casts, which are more common than specimens retaining the shell.

? CARDIOMORPHA, Koninck. ,

Type, C. oblonga (Isocardia) Sby. (not Kon.) Carb. lime.

Shell Isocardia-shaped, smooth or concentrically furrowed, umbones prominent, hinge edentulous; hinge-margin with a narrow ligamental furrow, and an obscure internal cartilage-groove.

Fossil, 38 sp. L. Silurian — Carb. N. America, Europe.

EDMONDIA, Koninck.

Ex. E. sulcata, Ph. (T. Pal. Soc. 1850, Pl. XX. fig. 5.) Carb. Brit.

Syn. Allorisma, King (part). Sanguinolites, M'Coy (part).

Shell oblong, equivalve, thin, concentrically striated, close; umbones anterior; ligamental grooves narrow, external; hinge-line thin, edentulous, furnished with large oblique cartilage-plates, placed beneath the umbones, and leaving space for an ossicle ? pallial line simple ?

Fossil, 4 sp. Carb. — Permian. Europe.

LYONSIA, Turton, 1822 (not R. Brown).

Syn. Magdala, Leach, 1827. Myatella, Brown. Pandorina, Scacchi.

Type, L. Norvegica, Pl. XXIII. fig. 10.

Shell nearly equivalve, left valve largest, thin, sub-nacreous, close, truncated posteriorly; cartilage plates oblique, covered by an oblong ossicle; pallial sinus obscure, angular. Structure intermediate between *Pandora* and *Anatina ;* outer layer composed of definite polygonal cells.

Animal with the mantle closed; foot tongue-shaped, grooved, byssiferous ; siphons very short, united nearly throughout, fringed; lips large, palpi narrow, triangular.

Distr. 9 sp. Greenland, N. Sea, Norway, W. Indies, Madeira, India, Borneo, Philippines, Peru.

L. Norvegica ranges from Norway to the sea of Ochotsk; in 15—80 fms. *Fossil?* Miocene —. Europe. (100 sp. L. Sil. —. D'Orb.)

? *Entodesma* (Chilensis) Phil. *Shell* thin, saxicava-shaped, slightly inequivalve and gaping, covered with thick epidermis; hinge edentulous; each valve with a semi-circular process containing the cartilage.

PANDORA (Solander) Brug.

Type, P. rostrata, Pl. XXIII. fig. 11. (*Pandora*, the Grecian Eve.)

Shell inequivalve, thin, pearly inside; valves close, attenuated behind; right valve flat, with a diverging ridge and cartilage furrows; left valve convex, with two diverging grooves at the hinge; pallial line slightly sinuated. Outer layer of regular, vertical, prismatic cells, 250 times smaller than those of *Pinna* (fig. 260). (*Carpenter.*)

Animal with mantle closed, except a small opening for the narrow, tongue-shaped foot; siphons very short, united nearly throughout, ends diverging, fringed; palpi triangular, narrow; gills plaited, one on each side, with a narrow dorsal border.

Distr. 13 sp. U. States, Spitzbergen, Jersey, Canaries, India, N. Zealand, Panama: 4—110 fms. burrowing in sand and mud.

Fossil, 4 sp. Eocene —. U. States, Brit.

MYADORA, Gray.

Type, M. brevis, Pl. XXIII, fig. 12.

Shell trigonal. rounded in front, attenuated and truncated behind; right valve convex, left flat; interior pearly; cartilage narrow, triangular, between 2 tooth-like ridges in the left valve, with a free sickle-shaped ossicle; pallial line sinuated: structure like *Anatina*; outer cells large, rather prismatic.

Distr. 10 sp. N. Zealand, N. S. Wales, Philippines.

MYOCHAMA, Stutchbury.

Type, M. anomioides, Pl. XXIII. fig. 13.

Shell inequivalve, attached by the dextral valve and modified by form of surface of attachment; posterior side attenuated; left valve gibbose; cartilage internal, between 2 tooth-like projections in each valve, and furnished with a moveable ossicle; anterior muscular impression curved, posterior rounded, pallial sinus small.

Animal with mantle-lobes united; pedal opening and siphons surrounded by separate areas; siphons distinct, unequal, small, slightly fringed; a minute fourth orifice close to the base of the branchial siphon; visceral mass large, foot small and conical; mouth rather large, upper lip hood-like; palpi tapering, few-plaited; gills one on each side, triangular, plaited, divided by an oblique line into two portions; excurrent channels 4, 2 at the base of the gills and two below the dorsal laminæ. (Hancock, An. Nat. Hist. 1853.)

Distr. 3 sp. New South Wales; attached to Crassatella and Trigonia, in 8 fm. water; the fry (as indicated by the umbones) is free, regular, and Myadora-shaped.

CHAMOSTREA, Roissy.

Type, C. albida, Pl. XXIII. fig. 14. *Syn.* Cleidothærus, Stutch.

Shell inequivalve, chama-shaped, solid, attached by the anterior side of the deep and strongly-keeled dextral valve; umbones anterior, sub-spiral; left valve flat, with a conical tooth in front of the cartilage; cartilage internal, with an oblong, curved ossicle; muscular impressions large and rugose, the anterior very long and narrow; pallial line simple.

Animal with mantle-lobes united by their extreme edge between the pedal orifice and siphons; pedal opening small, with a minute ventral orifice behind it; siphons a little apart, very short, denticulated; body oval, terminating in a small, compressed foot; lips bilobed, palpi disunited, rather long and obtusely pointed; gills one on each side, large, oval, deeply plaited, prolonged in front between the palpi, united posteriorly; each gill traversed by an oblique furrow, the dorsal portion consisting of a single lamina with a free margin. (Hancock, An. Nat. Hist. Feb. 1853.)

Distr. 1 sp. New South Wales.

FAMILY XX. GASTROCHÆNIDÆ.

Shell equivalve, gaping; valves thin, edentulous, united by a ligament, sometimes cemented to a shelly tube when adult; adductor impressions 2, pallial line sinuated.

Animal elongated, truncated in front, produced behind into two very long, united, contractile siphons, with cirrated orifices; mantle-margins very thick in front, united, leaving a small opening for the finger-like foot; gills narrow, prolonged into the branchial siphon.

The shell-fish of this family, the *tubicolidæ* of Lamarck, are burrowers in mud or stone. They are often gregarious, living in myriads near low-water line, but are extracted from their abodes with difficulty.

GASTROCHÆNA, Spengler, 1783.

Type, G. modiolina, Pl. XXIII. fig. 15. (*Gaster,* ventral, *chæna,* gape.)

Shell regular, wedge-shaped, umbones anterior; gaping widely in front, close behind; ligament narrow, external; pallial sinus deep.

Animal with mantle closed, and thickened in front; foot finger-like, grooved, sometimes byssiferous, siphons long, separate only at their extromities; lips simple, palpi sickle-shaped, gills unequal, prolonged freely into the branchial siphon.

G. modiolina perforates shells and limestone; its holes are regular, about

2 inches deep and ½ inch diameter; the external orifice is hour-glass shaped, and lined with a shelly layer which projects slightly. When burrowing in oyster-shells it often passes quite through into the ground below, and then completes its abode by cementing such loose material as it finds into a flask-shaped case, having its neck fixed in the oyster-shell; in some fossil species the siphons were more separated, and the flasks have two diverging necks. The siphonal orifices are rarely 4-lobed ; Pl. XXIII. fig. 15 a.

Distr. 10 sp. W. Indies, Brit. Canaries, Medit. Red Sea, India, Mauritius, Pacific Ids. Gallapagos, Panama :—30 fms.

Fossil, 20 sp. Inf. Oolite —. U. States, Europe.

Sub-genus. Chæna, Retz. 1788. C. mumia. Pl. XXIII. fig. 16. (= Fistulana clava, Lam.) *Shell* elongated, contained within a shelly tube; posterior adductor nearly central, with a pedal scar in front; siphonal inflection angular, with its apex joining the pallial line. Tube round, straight, tapering upwards, transversely striated, closed at the lower end when complete, and furnished with a perforated diaphragm behind the valves. *Distr.* Madagascar, India, Philippines, Australia; burrowing in sand or mud.

Fossil, Inf. Oolite —. U. S. Europe, S. India.

CLAVAGELLA, Lamarck.

Ex. C. bacillaris, Pl. XXIII. fig. 17.

Shell oblong, valves flat, often irregular or rudimentary, the left cemented to the side of the burrow, when adult, the right always free; anterior muscular impression small, posterior large, pallial line deeply sinuated. Tube cylindrical, more or less elongated, sometimes divided by a longitudinal partition; often furnished with a succession of siphonal fringes above, and terminating below in a disk, with a minute central fissure, and bordered with branching tubuli.

Animal with the mantle closed in front, except a minute slit for the foot, and furnished with tentacular processes; palpi long and slender; gills 2 on each side, elongated, narrow (floating freely in the branchial siphon ?)

Some specimens of the recent *C. aperta* have 3 frills to their tubes, and *C. bacillaris* has twice that number occasionally. They are formed by the siphonal orifices when the animal continues elongating, after having fixed its valve and ceased to burrow ; or perhaps, in some instances, when it is compelled to lengthen its tubes upwards by the accumulation of sediment. Brocchi mentions that on breaking the tube of the fossil *C. echinata*, he sometimes found the shell of a *Saxicava* or *Petricola* beside the loose valve of the *Clavagella*, into whose tube they must have entered after its death. *C. elongata* is found in coral; *C. australis* lives at low tide, and spirts out water when alarmed. *Distr.* 6 sp. Medit. Australia, Pacific :—11 fms.

Fossil, 13 sp. U. Green-sand —. Brit. Sicily, S. India.

ASPERGILLUM,· Lam. Watering-pot shell.

Type, A. vaginiferum, Pl. XXIII. fig. 18. *Syn.* Clepsydra, Schum.

Shell small, equilateral, cemented to the lower end of a shelly tube, the umbones alone visible externally; tube elongated, closed below by a perforated disk with a minute central fissure; siphonal end plain or ornamented with (1—8) ruffles.

Animal elongated; mantle closed, thickened and fringed with filaments in front; foot conical, anterior, opposed to a minute slit in the mantle; palpi lanceolate; gills long, narrow, united posteriorly, continued into and attached to the branchial siphon.

Distr. 4 sp. Red Sea, Java, Australia, N. Zealand; in sand.

Fossil, 1 sp. (A ? Leognanum, Hœning. *Miocene,* Bordeaux.)

FAMILY XXI. PHOLADIDÆ.

Shell gaping at both ends; thin, white, brittle and exceedingly hard; armed in front with rasp-like imbrications; without hinge or ligament, but often strengthened externally by accessory valves; hinge-plate reflected over the umbones, and a long curved muscular process beneath each; anterior muscular impression on the hinge-plate; pallial sinus very deep.

Animal club-shaped, or worm-like; foot short and truncated; mantle closed in front, except the pedal orifice; siphons large, elongated, united nearly to their ends; orifices fringed; gills narrow, prolonged into the exhalent siphon, attached throughout, closing the branchial chamber; palpi long; anterior shell-muscle acting as substitute for a ligament.

The *Pholadidæ* perforate all substances that are softer than their own valves (p. 242);* the burrows of Pholas are vertical, quite symmetrical, and seldom in contact. The ship-worms (*Teredines*) also make symmetrical perforations, and however tortuous and crowded never invade each other, guided either by the sense of hearing or by the yielding of the wood. The burrow

---

* M. Cailliaud has proved that these valves are quite equal to the work of boring in limestone, by imitating the natural conditions as nearly as possible, and *making such a hole with them.* Mr. Robertson also, has kept the living Pholades in blocks of chalk, by the sea-side at Brighton, and has watched the progress of the work. They turn from side to side never going more than half-round in their burrow, and cease to work as soon as the hole is deep enough to shelter them; the chalk powder is ejected at intervals by spasmodic contractions from the *branchial* siphon, the space between the shell and burrow being filled with this mud. (Journ. Conch. 1853, p. 311.) It is to be remarked that the condition of the Pholades is always related to the nature of the material in which they are found burrowing; in soft sea beds they attain the largest size and greatest perfection, whilst in hard, and especially gritty rock, they are dwarfed in size and all prominent points and ridges appear worn by friction. No notice has been taken of the hypothesis which ascribes the perforation of rocks, &c., to *ciliary action,* because, in fact, there is no current between the shell or siphons and the wall of the tube.

Q

has frequently a calcarious lining, within which the shell remains free; *Teredina* cements its valves to this tube when full-grown.. The. opening of the burrow, at first very minute, may become enlarged progressively by the friction of the siphons, which are furnished with a rough epithelium; but it usually widens with much more rapidity by the *wasting of the surface*. As the timber decomposes the shelly tubes of the *Teredo* project, and as the beach wears away the *pholas* burrows deeper.

PHOLAS, L.   Piddock.

Etym. Pholas, a burrowing shell-fish, from *pholeo*, to bore.

Type, P. dactylus, fig. 222.   *Ex.* P. Bakeri, Pl. XXIII. fig. 19.

Shell elongated, cylindrical; dorsal margin protected by accessory valves; pallial sinus reaching the centre of the shell.

Animal with a large truncated foot, filling the pedal opening; body with a fin-like termination; combined siphons large, cylindrical, with fringed orifices.

The common piddock is used for bait on the Devon coast; its foot is white and translucent when fresh, like a piece of ice; the *hyaline stylet* (p. 29) lodged in it, is large and curious.   *P. costata* is sold in the market of Havannah, where it is an article of food.

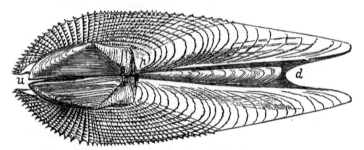

Fig 22. *Pholas dactylus*.   Chalk, Sussex Coast.

u, umbonal valves; *p*, post-umbonal valve; *d*, dorsal valve.

P. dactylus has two accessory valves to protect the umbonal muscle, with a small transverse plate behind; a long unsymmetrical plate fills up the space between the valves in the dorsal region.   *P. candida* and *parva* have a single umbonal shield, and no dorsal plate; these differences are only of *specific* value.   In *P. crispata*, L. (*Zirfæa*, Leach) the umbonal shield is not distinctly calcified, but there is a small posterior plate; the surface of the valves is divided into two areas by a transverse furrow.

Distr. 25 sp.   U. S. Norway, Brit. W. Africa, Medit. Crimea, India, Australia, N. Zealand, W. America :—25 fms.

Fossil, 25 sp. (U. Lias —) Eocene —. U. States, Europe. The secondary species belong to the next group.

PHOLADIDEA, Turton, 1819.

Type, P. papyracea, Pl. XXIII. fig. 20.

Shell globose-oblong, with a transverse furrow; anterior gape large, closed in the adult by a callous plate; 2 minute accessory valves in front of the beaks.

Animal with a fringed disk at the end of the combined siphons, and a horny cup at their base.

Distr. 6 sp. Brit. N. Zealand, Ecuador. Low-tides—10 fms.

Sub-genera. Martesia (Leach) Bl. 1825. M. striata, Pl. XXIII. fig. 21 Valves lengthened behind, when full grown, by a plain border; umbonal valves 1 or 2; dorsal and ventral margins often with narrow accessory valves. 10 sp. W. Indies, Africa, India. *M. striata* burrows in hard timber. *M. terediniformis* was found in cakes of floating wax on the coast of Cuba. (G. B. Sby.) *M. australis* in (fossil?) resin, on the coast of Australia. *M. rivicola* in timber 12 miles from the sea, in Borneo. *M. scutata,* Eocene, Paris, lines its burrow with shell.

Jouannetia (semicaudata) Desm. (Pholadopsis, Conrad; Triomphalia, Sby.) *Shell* very short, sub-globose; right valve longest behind: anterior opening closed by a callous plate developed from the left valve overlapping the margin of the right valve, and fixed to the single unsymmetrical umbonal plate. *Distr.* 3 sp. Philippines, W. America. *Fossil,* Miocene —. France.

Parapholas, Conrad, P. bisulcata, Pl. XXIII. fig. 22. Valves with 2 radiating furrows. *Distr.* 4 sp. California, Panama, Torres Strts.

XYLOPHAGA, Turton.

Etym. Xulon, wood, *phago,* to eat.

Types, X. dorsalis, Pl. XXIII. fig. 23; X. globosa, Sby. Valparaiso.

Shell globular, with a transverse furrow; gaping in front. closed behind; pedal processes short and curved; anterior margins reflected, covered by 2 small accessory valves; burrow oval, lined with shell.

Animal included within the valves, except the slender contractile siphons, which are furnished with pectinated ridges, and divided at the end; foot thick, very extensile.

Distr. 2 sp. Norway, Brit. S. America. Bores an inch deep, and across the grain, in floating wood, and timbers which are always covered by the sea.

TEREDO (Pliny) Adanson.

Type, T. Norvegica, Pl. XXIII. figs. 26, 27. *Syn.* Septaria, Lam.

Shell globular, open in front and behind, lodged at the inner extremity of a burrow partly or entirely lined with shell; valves 3 lobed, concentrically striated, and with one transverse furrow; hinge-margins reflected in front marked by the anterior muscular impressions; umbonal cavity with a long curved muscular process.

Fig. 223. Ship-worm, *Teredo Norvegica*, removed from its burrow.

Animal worm-like; mantle-lobes united, thickened in front, with a minute pedal opening; foot sucker-like, with a foliaceous border; viscera included in the valves, heart not pierced by the intestine; mouth with palpi; gills long, cord-like, extending into the siphonal tube; siphons very long, united nearly to the end, attached at the bifurcation and furnished with 2 shelly pallets or styles; orifices fringed.

T. navalis is ordinarily a foot long, sometimes 2½ feet; it destroys soft wood rapidly, and teak and oak do not escape; it always bores in the direction of the grain unless it meets the tube of another *Teredo*, or a knot in the timber.* In 1731-2 it did great damage to the piles in Holland, and caused still more alarm; metal sheathing, and broad-headed iron nails have been found most effectual in protecting piers and ship-timbers. The *Teredo* was first recognised as a bivalve mollusc by Sellius, who wrote an elaborate treatise on the subject, in 1733. (*Forbes.*)

T. corniformis, Lam. is found burrowing in the husks of cocoa-nuts and other woody fruits floating in the tropical seas; its tubes are extremely crooked and contorted, for want of space. The fossil wood and palm-fruits (*Nipadites*) of Sheppy and Brabant are mined in the same way. The tube of the giant Teredo (*T. arenaria*, Rumph. Furcella, Lam.) is often a yard long and 2 inches in its greatest diameter; when broken across it presents a radiating prismatic structure. The siphonal end is divided lengthwise, and sometimes prolonged into two diverging tubes. *T. Norvegica* and *T. denticulata* are divided longitudinally and also concamerated by numerous, incomplete transverse partitions, at the posterior extremity.

T. bipalmulata (Xylotrya, Leach) has the siphonal pallets elongated and penniform (Pl. XXIII. fig. 28); a species with similar styles occurs in the fossil wood of the Green-sand of Blackdown.

Distr. 14 sp. Norway, Brit. Black Sea; Tropics :—119 fms.

Fossil, 24 sp. Lias —. U. States, Europe.

Sub-genus, Teredina, Lam. T. personata, Pl. XXIII. figs. 24, 25. *Eocene*, Brit. France. *Valves* with an accessory plate in front of the umbones; free when young, united by their margins to the shelly tube when adult. The tube is sometimes concamerated; its siphonal end is often truncated; and the opening contracted by a lining which makes it hour-glass shaped, or six-lobed (fig. 25a.).

---

* The operations of the *Teredo* suggested to Mr. Brunel his method of tunnelling the Thames.

A

MANUAL OF THE MOLLUSCA;

OR,

RUDIMENTARY TREATISE

OF

RECENT AND FOSSIL SHELLS.

BY

S. P. WOODWARD, F.G.S.

ASSOCIATE OF THE LINNEAN SOCIETY;
ASSISTANT IN THE DEPARTMENT OF MINERALOGY AND GEOLOGY
IN THE BRITISH MUSEUM; AND
MEMBER OF THE COTTESWOLDE NATURALISTS' CLUB.

ILLUSTRATED WITH
NUMEROUS ENGRAVINGS AND WOODCUTS.

PART III.
CONTAINING THE TUNICATA;
GEOGRAPHICAL DISTRIBUTION, ETC.; SUPPLEMENT,
AND INDEX.

LONDON:
JOHN WEALE, 59, HIGH HOLBORN.
MDCCCLVI.

LONDON:
PRINTED BY WILLIAM OSTELL,
HART STREET, BLOOMSBURY.

A

MANUAL OF THE MOLLUSCA.

PART III.

CLASS VI. TUNICATA, LAMARCK.

(Order *Hetero-branchiata*, Blainville.)

The lowest order of Acephalous Mollusca are called *Tunicaries*, being protected by an elastic tunic in place of a shell. They are extremely unlike shell-fish in appearance, and are denied a place in most works on conchology; having no hard skeleton they neither furnish objects for the cabinet of the collector, nor materials for the speculations of the geologist.*

Many of the Tunicaries are curious objects when seen fresh from the sea; or still better when living in those miniature *aquaria*, which—thanks to MR. GOSSE—are now so popular.† The transparent sorts are beautiful even when preserved in spirits. To the naturalist they present many points of interest unknown amongst the other mollusca, for here he meets with compound animals, and the phenomenon of alternate generation; they afford excellent illustrations of the structure of the breathing-organ and mechanism of aquatic respiration; and they also exhibit the simplest form and condition of the vascular system, in which the blood no longer circulates in one unvarying direction, but ebbs and flows like the tides.‡ (pp. 31, 49.)

The principal forms of tunicated mollusca are given in plate 24, and the woodcut (fig. 224) represents one of the largest and simplest kind, which is drawn as if it were transparent, so as to shew the whole of its internal structure. These large solitary tunicaries are termed *Ascidians*, from their

---

* König supposed the *Sphaeronites* to be tunicaries allied to *Boltenia;* they are globular bodies, with a tessellated surface and two orifices, found in the Silurian strata, and belong to the order *Cystideae* amongst the *Echinodermata.* The genus *Eschadites* of König was also supposed to be a fossil tunicary; its nature is still problematical. See Murchison's "Siluria."

† At the gardens of the London Zoological Society there are examples of *Ascidium* and *Cynthia*, the compound and starlike *Botryllus* (pl. 24. fig. 8) and a delicate little pearly *Clavellina*, whose presence was first detected by Mr. Tennent the intelligent and obliging keeper of the aquarium.

‡ In *Appendicularia* Mr. Huxley finds no reversal of the current.

Q

resemblance to a water-skin, or small leather bottle (*ascidium*). They attain a length of several inches, and are fixed to rocks or shingle, or sea-weed, but sometimes so slightly that they are brought up detached, and yet uninjured, by the dredge. Their appearance is sufficiently unpromising; their surface often rugged or concealed by adhering sand and fragments of shell; sea-weeds grow upon them, and small bivalves (*crenella*) burrow in their tunic. They are hollow and elastic, and have two orifices, from which (especially the terminal opening), they squirt water, as the bivalve shell-fish do when molested.

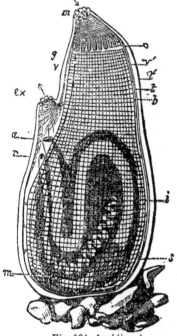

If the soft outer shell (*t'*) is opened there will be found inside a second tunic (*t*) which is compared to the *mantle* of the bivalves; it is extremely muscular, the fibres circling round it closely, especially near the orifices, whilst some others are oblique and longitudinal. The mantle lines the tunic, but is only slightly attached to it at the two orifices, and at those points where the blood-vessels pass through.*

During life the outer tunic follows the contractions of the muscular mantle; and when the latter relaxes, the tunic returns to its original shape by virtue of its elasticity. But when preserved in spirit the mantle contracts to such an extent as to tear itself away from the tunic, and if such a specimen is opened the muscular sac looks like a little tunicary quite loose within the large one. Within this a

Fig, 224, Ascidian.†

third and more delicate tunic is formed by the respiratory or *branchial* sac (*b*) having only one external orifice by which it is suspended, a little within the terminal (or exhalent) opening of the outer tunics; as its texture is porous the water passes through it readily into the mantle cavity, and thence by the second

* In the thick pellucid test of *Ascidium mamillatum* the eye can discern an extensive network of vascular ramifications. The blood-vessels enter the test near the base. In the closely allied genus *Cynthia* there is no such vascular connexion, but the mantle is more strongly united to the test at the orifices; in Chelysoma the tunics are extensively united by muscular fibres. (Rupert Jones) The relation between the Ascidian test and mantle is that of the *epidermis* to the *cutis vera*, precisely as in the lamellibranchiate bivalves; the union of the two in the majority of Ascidians is exceedingly intimate in the fresh state. (*Huxley.*)

† Fig. 224, *Ascidium monachus;* '*in.* incurrent; *ex.* excurrent orifice; *t'.* outer tunic; *t.* muscular tunic; *b.* branchial sac; *o.* tentacular fringe; *g.* nervous ganglion;

outlet (*ex.*) At the bottom of the *branchial sac* is the animal's mouth (*m*) or commencement of the digestive canal, which ends (at *a*), near the second external orifice. This digestive system is accompanied by other organs, forming the body of the animal, but it appears only like a thickening of one side of the muscular tunic.

If the animal presenting this organization be compared with the mussel (represented in fig. 30* p. 53,) or the *mya* (fig. 170, p. 244), it will be seen that each has a *test* lined by a mantle and furnished with an inhalent and an exhalent orifice; in each the respiratory cavity is separated from the channel of the out-going current by a sieve-like breathing organ, and in each the currents are produced and food brought to the mouth by microscopic *cilia* fringing the pores of the gill. The inhalent orifice of each is guarded by tentacles developed from the *mantle*,* and the exhalent opening is often furnished with a valve to prevent a reversal of the current when the animal expands after one of its occasional spasmodic contractions.

These points of *analogy* are so obvious and striking, as to have induced many naturalists to believe in a very close relationship between the Ascidians and bivalve shell-fish. We must, however, hesitate before we assume that the organs which perform identical functions, are themselves identical, ("homologous.") Mr. Hancock has pointed out (in the excellent memoir just referred to,) that the branchial sac of the Ascidian is not the anatomical equivalent of the gills of *mya*, but a portion of the alimentary canal;† and that the peculiarities of their circulation and mode of reproduction are more in harmony with what obtains amongst the higher zoophytes (*bryozoa*). A similar view is expressed by M. Milne Edwards in his memoir on the Composite Ascidians.‡

These statements are referred to more particularly, since of late years an

v. v'. referring to the space between the mantle and the branchial sac, indicate the dorsal and ventral sinuses of Milne-Edwards; *m.* mouth, at the bottom of the branchial sac ; *s.* stomach, plaited lengthways; *i*, intestine, lying between the brachial sac and muscular tunic, on the further side ; *a*, termination of the intestine ; *r*, reproductive organ, ending in the cloaca.

* These tentacular filaments are not anatomically connected with the branchial sac as supposed by Farre and Owen. See Hancock on the Anatomy of the Freshwater *Bryozoa*. An. Nat. Hist. vol. V. p. 196.

† Dr. Farre compared the Ascidian gill to the *pharynx* of the bryozoa; but M. Van Beneden and Mr. Hancock consider it homologous with the circle of oral tentacles in the retracted or undeveloped bryozoon.

‡ The Ascidians have less intimate analogies with the Mollusca, properly so called than is usually believed. They resemble, it is true, these animals in the arrangement of their digestive apparatus, and in some peculiarities of the respiratory system; but they depart from the Molluscan type in mode of circulation, in the metamorphosis which the fry undergo, and above all, in the singular power which most of them possess, of multiplying by gemmation. In these latter characters, so very important in a physiological point of view, they closely approach the polypes. (Milne-Edwards, Mem. Inst., France, 1842.)

opinion has been gaining ground with anatomists that not only the tunicaries, but the *bryozoa*, (or Ascidian Zoophytes of Dr. Johnston) should be regarded as *mollusca*; this view was recommended by Prof. Forbes, though not adopted by him, and is advocated by Prof. Allman and Mr. Huxley.

Those who have only seen the horn-coloured sea-weeds such as *Flustra* and *Notamia*, drifted by the wind on the sea-beach, may have admired their minute lace-work or chain-like cells, without once dreaming they were ex-amining compound animals—shell-fish, anatomically considered. But the minute polypes which studded these zoophytes when alive, were undoubtedly as active, and in some respects as highly organized as the lower mollusca. The question is whether their organization is of the same kind, or *type*, as the molluscan, and in this respect their claims are nearly on a parallel with those of the *Tunicata*. The relation of the *bryozoa* is to the *Terebratulae*, as shown in their oral apparatus and muscular system (*Hancock*), but they have neither heart, arteries or veins, and the nutrient fluid is contained in the com-mon visceral cavity. The ciliated gemmules of the bryozoa are not, however, more unlike molluscan larvæ* than are the tadpole-shaped fry of the tunicaries.

Before proceeding further with the description of the tunicaries, we are glad to avail ourselves of a diagram by Mr. Huxley, which will make it more intelligible.

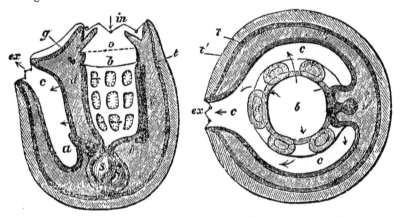

225, *Longitudinal*,   226, *Transverse section*.

in. inhalent orifice; *ex.* exhalent orifice; *b.* branchial sac; *c.* atrium ("thoracic chamber" of Milne Edwards); *o*, tentacular filaments; *g*, nerve ganglion and auditory vesicle; *d*, thoracic vessel, (hypo-pharyngeal band); *v v'*, great vascular sinuses; *t'*, test; muscular mantle; *e*, endostyle; *s*, stomach; *a*, intestine; *h*, position of heart. The shading is accidently omitted on a small portion of the test by the letter *g*; the branchial sac (*b*) is connected with the wall of the atrium by (*branchio-parietal*) vessels crossing the cavity *c, c.*

* The embryo of *antiopa* (p. 196) is bell-shaped at first, with a fringe of long cilia round the rim which afterwards becomes the two-lobed *velum.*

In these figures the outer circle represents the test (t') lined by the muscular mantle (t). The branchial sac in the centre (b) is perforated by a few large openings which are fringed with *cilia*; the arrows mark the direction of the respiratory currents which enter at the òral opening, passs through the branchial sac into the atrium or "thoracic chamber" ($c\ c$) and escape by the anal orifice (ex).

The *atrium* does not exist in the embryo; it is formed by an inflection of the tunics, and its ultimate extent varies in different genera. At first the whole space between the mantle and viscera is a common vascular *sinus*, as in the *bryozoa*, but the formation of the atrium divides it into two portions, one lining the mantle, the other investing the alimentary canal. The outer portion, or parietal sinus, is further subdivided by the union of its walls at definite points, leaving spaces and channels of various sizes and degrees of regularity. Of these, the principal are the dorsal and ventral sinuses ($v\ v'$) communicating by transverse channels.* The lower part of the alimentary canal continues surrounded by a vascular space termed the *peri-intestinal sinus*, whilst the pharyngeal portion with its vascular envelope becomes perforated to form the branchial sac.† It has been mentioned that the branchial openings are microscopic and innumerable in the solitary ascidians, whilst they are comparatively few and large in the social and compound species. In *Salpa* the branchial sac is so much reduced that the respiratory process must be exercised chiefly by the vascular lining of the mantle itself.

The heart is near the posterior or fixed end of the body; it is elongated, and slightly muscular, open at each end, and contracts progressively like the dorsal vessel of the anellides, the direction of its contractions being periodically reversed. The nervous system consists of filaments connected with a single ganglion placed in the sinus between the external orifices.‡ The organs of special sense are an auditory capsule sometimes containing an otolithe, (fig. 225. *g*) and coloured spots, supposed to be rudimentary eyes, placed between the segments of the outer openings.

The *neural* side, or that on which the nerve-ganglion is placed, should be considered *ventral* in these as in other invertebrate animals; and the *haemal* side, where the heart is situated, ought to be regarded as *dorsal*.§ The

---

* See the figure of *Salpa*, Pl. 24, fig. 22. The thick black lines represent the sinuses; the heart is near the lower end of the figure, outside the visceral nucleus. The sinuses have no visible lining membrane but resemble those already referred to (pp. 31, 198) as existing in all classes of mollusca.

† The resemblance of the pharyngeal sac of the tunicaries to the gills of fishes was pointed out by Mr. Goodsir in his memoir on the Lancelet (*amphioxus*).

‡ In Plate 24, the position of the nervous ganglion is indicated in several instances by a small star.

§ Milne-Edwards has employed these terms in an opposite sense, apparently

first flexure of the intestinal canal in the tunicaries is always to the haemal side, but it is usually turned again in the opposite direction.

The food of the ascidians, judging by the contents of their stomachs, consists chiefly of minute particles of the articulated sea-weeds and *diatomaceæ;* and it is a remarkable circumstance that the outer tunic of these animals contains *cellulose,* a ternary organic substance formerly supposed to be peculiar to vegetables.* They also contain radiated concretions, sometimes silicious, but more frequently calcarious, like the bodies found in *alcyonium* and *gorgonia.*

All the Tunicata appear to possess the power of reproduction by buds—or gemmation; but in one group the individuals, however produced, become entirely distict, in another they remain connected by a vascular canal, and in a third they become blended into a common mass. These three groups are the "·solitary," "social," and "compound ascidians" of Milne-Edwards; these are all fixed in their adult state, whilst the two remaining families swim freely in the open sea, *Pyrosoma* being compound, and *Salpa* alternately aggregated and solitary. The separate individuals of these composite masses are termed Zoïds.

The sexes are united in all the Tunicata but *Doliolum* and *Appendicularia.* The young produced from eggs undergo a metamorphosis, which has been observed in many genera. The larvae are shaped like the tadpole of the frog; the body is oval and furnished with black eye specks, short tentacular processes, and a long tail by the vibrations of which they swim (Pl. 24, fig. 18). Ultimately they fix themselves, the tail is absorbed, and the young ascidian, or first zoïd of a compound tunicary, is developed.

The *tunicata* are found in all seas, from low-water to a considerable depth. Four genera are pelagic, and several belong to the Arctic province viz., *Boltenia, Chelyosoma, Synœcium* and *Cystingia.*

Mr. Huxley divides the Tunicaries into three groups—

1. *Ascidia Branchiales.* Branchial sac occupying the whole, or nearly the whole, length of the body; intestine lying on one side of it. (*Ascidiadae —Perophora—Botryllus—Pyrosoma.*)

2. *Ascidia Intestinales.* Alimentary canal completely behind the branchial sac, which is comparatively small. (Other genera.)

3. *Ascidia Larvales.* Permament larval form. (*Appendicularia.*)†

---

guided by the analogy of the ganglionic side of the tunicata to the dorsal region of the *lamellibranchiata.* Still more confusion exists in the employment of the terms *anterior* and *posterior;* the *inhalent* orifice is anterior if compared with the mouth of a polype, but Milne-Edwards makes it *posterior.*

* Discovered by Dr. Schmidt, in 1845. The observation has been confirmed by M. M. Löwig and Kolliker, and by M. Payen, who gives the following as the chemical composition of the ascidian tunic;—Cellulose 60.34, azotised matter 27.00, inorganic 12.66. The cellulose portion is not acted upon by soda or hydrochloric acid.

† See Knight's "English Cyclopædia," article MOLLUSCA.

FAMILY I. Ascidiadae. · Simple Ascidians.

Animal simple, fixed; solitary or gregarious; oviparous; sexes united; branchial sac simple or disposed in (8—18) deep and regular folds.

The simple ascidians were called *tethya* and well described in Aristotle's History of Animals.* Many of them are esteemed as articles of food in Brazil, China and the Mediterranean; at Cette they are regularly taken to market; and *Cynthia microcosmus†* furnishes a delicate morsel, much sought after.

Ascidium, Baster 1764. Sea-squirt.

Etym. Diminutive of *askos*, a skin-bottle.

Syn. Alina, Risso: Phallusia, Pirena, Ciona, Savigny.

Ex. A monachus, Cuv. fig. 224, Tenby.

Body sessile, covered with a coriaceous or gelatinous tunic; branchial orifice 8-lobed, furnished inside with a circle of simple tentacular filaments; anal 6-lobed; branchial sac not plaited, its meshes papillated.

The ascidia vary in length from 1 inch to 5 or 6 inches. The test is pale and semitransparent, the inner tunic orange or crimson, or sometimes marbled with crimson and white; the ocelli are red, or yellow with a central red spot. The surface of *A. echinatum* is studded with conical papillae, each with 4—7 radiating bristles. The ascidia range from low-water to 20 fathoms, attached to rocks, shells, and fuci.

Distr. Greenland, Spitzbergen, U. States, Europe, (especially in the north), Brit. 19 sp. Medit. New Zealand.

Molgula, Forbes.

Etym. Diminutive of *molgos*, a bag of skin.

Ex. M. arenosa, A. and H. (not M. tubulosa Rathke), Pl. 24, fig. 1.

Body more or less globular, attached or free; test membranous, usually invested with extraneous matter; orifices on very contractile, naked tubes; oral opening 6-lobed, anal 4-lobed.

M. arenosa is found in the muddy lochs and bays of the west of Scotland; it comes up in the dredge like a little ball of sand. At Tenby it occurs between tide-marks, and in the laminarian zone.

M. oculata was dredged, adhering to a scallop, in 25 fathoms, off Plymouth; its orifices are like dark eyes in a spectacle-formed frame. (*Forbes*).

Distr. 3 sp. Denmark, Brit.

---

* Linnaeus used the name *Tethyum* for the Tunicaries in the earlier editions of his " Systema Naturae," and recognising their resemblance to the bivalves, called the animal of the latter " a tethys." Afterwards he adopted Baster's name *Ascidium*, and used Tethys for a nudibranche; *Tethya* (Lam.) is now e ployed for a genus of globular sponges.

† So called from the little world of parasites that ofter grow upon it.

CYNTHIA, Savigny, 1816.

Etyn. A name of Diana, from Mt. Cynthos, Delos.

Syn. Stycla (pomaria) Sav.   Caesira (quadridentata) Sav.

Ex. C. papillosa, Pl. 24, fig 2.

Body coriaceous, sessile, orifices 4-lobed, branchial sac plaited longitudinally, surmounted by a circle of tentacular filaments; ovaries two.

Sub-genera. Dendrodoa (glandaria) Mc Leay.   Sub-cylindrical, smooth; orifices terminal, minute; ovary single, on left side.

Pandocia (mytiligera) Sav.   Right ovary only developed.

Distr. Norway—Medit.   Sometimes on sand and very slightly attached; or on oysters, stones and sea-weed, from low-water to 30 fathoms.   Occasionally gregarious in vast numbers, forming large bunches in consequence of the interlacing of their root-fibres.   The test is often orange-coloured or crimson.   The branchial sac, in this and the following genera, is thrown into deep folds to increase its extent of surface.   Greenland, Brit. 14 sp.

PELONÆA, Forbes and Goodsir.

Etym. Pelos, mud, *naio* to inhabit.

Ex. P. glabra, Pl. 24, fig. 3.   Rothesay bay; 7 fms.

Body elongated, cylindrical, smooth or wrinkled; orifices terminal 4-cleft, on two small conical eminences; posterior end blunt pointed, villose with fine rootlets; mantle adherent to the test; no tentacles; ovaries 2, symmetrical.

Distr. 2 sp. N. Brit.   Norway (Mc Andrew and Barrett).

Pelonæa resembles *Sipunculus,* one of the worm-like Echinoderms, *in appearance.*   It is not free, but rooted in mud and quite as apathetic as the other ascidians.*

CHELYOSOMA, Broderip and Sby.

Etym. Chelyon tortoise-shell, *soma* body.

Type, C. Macleayanum, Pl. 24, fig. 4.   Greenland.

Body depressed, oblong; test coriaceous, its upper surface composed of 8 polygonal plates; orifices small, prominent, 6-valved; gills plaited; tentacles simple.

BOLTENIA, Sav.

Named after Dr. Bolten, a Hamburgh naturalist.

Syn. ? Bi-papillaria, Lam. 1816.   Australia.

Ex. B. pedunculata, Pl. 24, fig. 5.

Body globular, pedunculated; test coriaceous, orifices lateral, 4-cleft; branchial sac longitudinally plaited; tentacles compound.

---

* Pelonæa is not so extraordinary as at first supposed.   The very erroneous statement at p. 32, lines 27, 28, should be erased.

The young *Bolteniae* sometimes grow on the stem of the parent. The branchial orifice is nearest the stalk, but as the body is pendulous it becomes higher than the other opening, as usual amongst the ascidians. (Rupert Jones.) *B. reniformis*, Mc L. lives attached to stones in deep water; it is sometimes brought up by the fishing hooks. (*Gould.*) Elizabeth harbour, 70 fms. (*Ross.*)

Distr. N. Zealand; Greenland, (*B. ovifera*—Vorticella, L.) Mass. U. S.

Sub-genus ? Cystingia (Griffithi) Mc Leay, 1824. Arctic seas, Felix harbour and Fox's channel. *Test* sub-coriaceous, anal orifice irregular, terminal.

FAMILY II. Clavellinidae. Social Ascidians.

Animal compound, fixed; individuals connected by creeping tubular prolongations of the common tunic, through which the blood circulates, (or by a common gelatinous base).

These small or microscopic creatures are found on stones, shells and seaweed, adhering by numerous root-like projections of their outer tunic. They are so transparent and colourless that they may be examined without dissection (Pl. 24, figs. 6, 7). The position of the stomach is indicated by an orange-coloured spot; the œsophagus is long, and the intestine returns parallel to it. The heart and ovary are near the stomach. The gill, perforated by rows of holes, completely separates the branchial cavity from the cloaca; a series of membranous processes (*languettes*) project from its neural side. The creeping tube contains two channels through which the blood circulates in opposite directions.

Reproduction is effected by ova and by buds produced on filaments given off by the creeping tube. These off-shoots are hollow, and lined by a membrane continuous with the inner tunic of the ascidian; the circulation passes into them and they grow and branch and form buds containing little organized masses from which the internal organs are gradually developed. The branchial sac is perfectly outlined before it communicates with the interior, and the curved digestive tube is seen before the oral opening is formed. The new individual may continue united with the parent, or become completely free by the rupture of the connecting tube. (*Milne-Edwards.*)

Clavellina, Sav.

Etym. Clavella, a small staff, *Syn.* ? Rhopalaea, Phi.

Type, C. lepadiformis, Pl. 24, fig. 6.

Body elongated, erect, more or less pedunculated; test smooth and transparent; orifices without rays; thoracic region usually marked with coloured lines.

Distr. Greenland, Brit. Medit. On rocks and stones at low-water.

PEROPHORA, (Wiegm.) Lister, 1834.

Etym. Pera, a sac, and *phoros* bearing.
Type, P. Listeri, Wiegman, Pl. 24, fig. 7.
Body pedunculated, suborbicular, compressed; thoracic region plain.

This curious little species was discovered by Mr. J. Lister at Brighton, growing on *Conferva elongata*. It occurs in groups consisting of several individuals, each having its own heart, respiration, and system of nutrition, but fixed on a peduncle that branches from a common creeping stem, and all being connected by a circulation that extends throughout. (Lister).

Mr. Forbes has dredged it adhering to weed on the coast of Anglesey; he remarks "it is beautifully transparent, appearing on the weed like little specks of jelly dotted with orange and brown. When dried, as it may often be met with on sea-weed cast on shore, these bodies appear like the minute ova of some mollusk." According to Mr. Huxley's view this genus differs widely from the last, being a "branchial ascidian" whilst *Clavellina* is an "abdominal" one.

SYNTETHYS, Forbes and Goodsir.

Type, S. Hebridicus, F and G. Croulin Id. near Applecross.
Animals compound, gelatinous, orbicular, sessile; *individuals* very prominent, arranged sub-concentrically in the common mass; branchial and anal orifices simple, not cut into rays.

Syntethys is a Clavellina with the habit of a *Diazona*. The only known species forms compact greenish translucent gelatinous masses of half a foot in diameter, and nearly equal height, affixed to rocks or stones by a short base. The individual ascidians are when full grown 2 inches in length. Their inner tunics are remarkably irritable, withdrawing themselves into the common mass when pinched. (*Forbes, Brit. Moll.* iv., 244).

FAMILY III. BOTRYLLIDAE, Compound Ascidians.

Animals compound, fixed, their tests fused, forming a common mass in which they are imbedded in one or more groups; individuals not connected by any internal union; oviparous and gemmiparous.

Milne-Edwards divides the compound ascidians into three tribes:—

1. *Botryllina.* Individuals united in systems around common excretory cavities (*cloacae*). Thorax and abdomen not distinct.

2. *Didemnina.* Thorax and abdomen distinct.

3. *Polyclinina.* Body divided into three distinct portions—1, thorax, with the branchial apparatus;—2, superior abdomen with the digestive organs; —3, post-abdomen, containing the heart and reproductive organs.

Tribe 1, Botryllina—Botryllians.

BOTRYLLUS, Gaertner, 1774.

Etym. Botrys, a cluster of grapes.

Syn. ? Pyura, Bl. Polycyclus, Lam.

Ex. B. violaceus, Pl. 24, fig. 8, two stars from a group.

Test gelatinous or cartilaginous, incrusting; systems numerous, prominent, round or star-shaped, with central cavities; individuals 6—20 in each system, lying horizontally, with the vent far from the simple branchial orifice.

Distr. 10 sp. U. States, Europe. Brit. 6 sp. On stones and sea-weed near low-water mark. *B. violaceus* is greenish grey, with dark blue stars, yellow in the centre round the common orifice. *B. racemosus, N. Zealand.*

BOTRYLLOIDES, M. Edw., 1841.

Ex. B. rotifera, Pl. 24, fig 9, a zoïd detached, with a cluster of reproductive germs.

Animals nearly vertical, in star-like groups irregular and ramifying; cloacae prolonged into the common mass, forming irregular channels, along each side of which the individuals are placed in linear series; orifices closely approximate.

Distr. European coasts, on roots of sea-weed and under sides of stones between tide marks. Brit. 4 sp.

Tribe 2, Didemnina. "Didemnians."

Division *a*, unistellate, (oral orifice rayed.)

DIDEMNIUM, Sav.

Etym. Di-demnium double-couch (or cavity).

Ex. D. gelatinosum, Pl. 24, fig. 10, zoïd detached.

Test coriaceous, polymorphous, incrusting; systems numerous, compressed, without central cavities or distinct circumscription; individuals scattered; abdomen pedunculate; ovary by the intestinal loop, increasing in length when the ova are fully developed.

Distr. Europe.

EUCŒLIUM, Sav.

Etym. Eu-koilios much excavated.

Ex. E. hospitiolum, Pl. 24, fig. 11.

Test gelatinous, incrusting; systems numerous, without central cavities or distinct circumscription: animals scattered or arranged quincuncially branchial orifice circular; anal minute; abdominal viscera beside the thorax.

Distr. Europe.

LEPTOCLINUM, M. Edw.

Etym. Leptos thin, *kline* tunic.

Type, L. maculosum Edw. (L. gelatinosum, F. and H. Pl. A, B. fig. 5.)

Test coriaceous or gelatinous, thin, incrusting; systems few; individuals grouped irregularly round common cloacal cavities; abdomen pedunculate, short, smaller than the thorax.

Distr. Brit. 6 sp.   On roots of *laminariae ;* in colour white, yellowish, or variegated with blue.

Division *b.*   Bi-stellate Didemnians.

DISTOMUS, Gaertner.

Etym. Distomos two-mouthed.   *Syn.*   Polyzona, Flem.

Ex. D. fuscus Pl. 24, fig. 12, a detached zoïd.

Test semi-cartilaginous, polymorphous, sessile: systems numerous, usually circular; individuals 1 or 2 ranked at unequal distances from their common centre; both orifices 6-rayed.

Distr. Europe, S. Africa, Australia.   Brit. 2 sp.

DIAZONA, Sav.

Etym. Dia-zonai in circles.

Ex. D. violacea, Pl. 24, fig. 13   Medit ..

Test gelatinous, orbicular, sessile or somewhat pedunculate; tunicaries very prominent, arranged in concentric circles on an expanded disk, forming a single flower-like system; orifices 6-rayed; abdomen pedunculate; ovary inclosed in the intestinal loop.

Tribe 3. Polyclinina.

Division *a,* unistellate Polyclinians.

POLYCLINUM, Sav.

Etym. Polys many, *kline* cavities.

Ex. P. constellatum, Pl. 24, fig. 15.

Test gelatinous or cartilaginous, polymorphous, sessile or slightly pedunculate; systems numerous, convex, somewhat stellate, with central cloacal cavities; tunicaries 10—150, at very unequal distances from centres; abdomen much smaller than thorax, post-abdomen pedunculate.

Distr. 6 sp. Brit., Medit., Red Sea, India.

APLYDIUM, Sav.   Sea-fig.

Etym. Aploos simple.   *Ex.* A. lobatum, Pl. 24, fig. 14.

Test gelatinous or cartilaginous, sessile; systems very numerous, slightly prominent, annular or sub-elliptical, without central cavities; tunicaries (3—25) in single rows, equidistant from centres; branchial orifice 6-rayed; division of thorax and abdomen not always distinctly marked.

Distr. 6 sp. Europe, Red Sea.   Attached to shells, &c., in deep water.

SIDNYUM, Sav.

Type, S. turbinatum, Sav. British coast. (F. and H. Pl. A, B. fig. 2.)

Test gelatinous, incrusting; systems numerous, conical, truncated and starred at the summit; tunicaries 5 or 6 to 10 or 12, forming a margin round a depressed centre; branchial orifice 8-toothed; vent simple, tubular; ovary pedunculate.

Found on the under surfaces of shelving rocks, at low-water spring tides, forming translucent amber-coloured masses.

AMOROECIUM, M. Edw.

Etym. Amoiros incomplete, *oikos* house.

Ex. A. argus, Pl. 24, fig. 17. A proliferum, (larva) fig. 18.

Test fleshy or coriaceous, polymorphous, incrusting or slightly pedunculate; systems numerous; tunicaries grouped round common apertures; abdominal divisions indistinct.

Distr. 4 sp. British Channel, Medit., Aegean.

Sub-genus Parascidium, M. Edw. P. flavum, 24, fig. 16. Oral openings 8-lobed, each accompanied by 2 oculiform points.

SYNOECIUM, Phipps, 1773.

Etym. Synoikos united house. *Type*, S. turgens, Pl. 24, fig. 19.

Test semicartilaginous, cylindrical, pedunculate, isolated or gregarious; systems single, circular, terminal, tunicaries 6—9 ; branchial orifice 6-rayed, anal of 6 unequal rays; post-abdomen sessile.

Division b. Bistellate Polyclinians.

SIGILLINA, Sav.

Etym. Sigillum, a seal, *Ex.* S. Australis, Pl. 24, fig. 20.

Test gelatinous, solid, conical, elongated, pedunculate, solitary or gregarious; systems single, of many individuals, in irregular circles one above another; orifices both 6-rayed ; abdomen larger than thorax; post-abdomen long and slender.

FAMILY IV. PYROSOMIDÆ.

Animal compound, free, pelagic.

PYROSOMA, Péron, 1804.

Etym. Pyr (pyros) fire, *soma* body.

Ex. P. giganteum, Pl. 24, fig. 21.

Body cartilaginous, non-contractile, cylindrical, hollow, open at one end only; exterior covered by the numerous pointed zoïds, grouped in whirls interior mamillated and pierced by the exhalent orifices of the tunicaries.

The Pyrosomes are 2—14 inches long and ½—3 inches in circumference; they are composed of innumerable tunicaries united side by side, with their orifices so arranged that the inhalent openings are external, the exhalent inside the tube, and the result of so many little currents discharged into the cavity is to produce one general outflow, which impels the floating cylinder with its closed end foremost.

The ganglionic side of each zoïd is turned towards the open end of the tube; the respiratory cavity is large, and completely inclosed by a quadrangular net-work; the test and mantle are united and lined by a vascular sinus-system. There is an "endostyle" on the haemal side, as long as the branchial sac. The ventral column (hypo-pharyngeal band) supports a series of *languet*

The sexes are combined; reproduction takes place by *buds* developed amongst the adult zoïds, and by solitary *ova* connected with the inner tunic by a pedicle near the posterior termination of the endostyle; 2 or 3 ova are perceptible in the young zoïd at a very early period.

The Pyrosomes are often gregarious in vast numbers; in the Mediterranean they sometimes abound to such an extent as to clog the nets of the fishermen. They are phosphorescent at night. The light of *P. atlanticum* is very vivid and of a greenish blue colour; when touched the light appears in very minute sparks, issuing from each of the separate individuals, it first appears at the part touched, and gradually spreads over the body; it disappears after death. (*Müller*). Placed in a vessel of salt-water, and at rest, they emit no light, and the light excited by touching them gradually fades after the removal of irritation; but immersed in *fresh-water* they continue glowing with their brightest refulgence for several hours—as long as life remains. Péron first noticed them as "a phosphorescent band, stretched across the waves and occupying an immense tract in advance of the ship. Those most distinctly seen resembled incandescent cylinders of iron." Humboldt speaks of the *Pyrosomæ* as forming a light 1½ feet in diameter, by which the fishes were visible !

FAMILY V. Salpidae.

Animals free, oceanic; alternately solitary and aggregated.

Salpa, Forskahl, 1775.

Etym. Salpe a luminous fish.   *Syn.* Dagysa, Banks and Solander. Thalia, Brown. Biphora, Brug. Pegea and Jasis, Sav.

Ex. S. maxima, Pl. 24, fig. 22, solitary form.

Animal oblong, sub-cylindrical, truncated in front by the oral orifice, pointed posteriorly; anal orifice sub-terminal; *test* thin, transparent; muscular mantle incomplete, forming a set of transverse or oblique bands; mantle cavity lined by a system of vascular sinuses; gill rudimentary, forming an oblique· band across the interior; visceral nucleus posterior. *Sexes* combined; young produced by gemmation in chains, consisting of individuals unlike the parent and becoming oviparous, the alternate generations only being alike.

Distr. North sea, Brit. Medit. Australia, N. Zealand.

The individual Salpians are from ½ an inch to 10 inches in length; the chains vary from a few inches to many feet, but are often broken up, indeed the *adults* appear to be always separate. They swim with either end foremost, although the pointed end would seem the normal one, as the motion is produced by the forcibly expulsion of the water from the mantle. Each orifice is furnished with a valve, and there is no division between the atrium and respiratory cavity except the rudimentary gill, or " hypopharyngeal band." ·The Salpa-chains also swim, with a regular serpentine movement.

The solitary and aggregate forms differ. so much that they were always named and described as distinct species before the remarkable discovery made by *Chamisso*,* that each form always produced the other. The free form of *S. democratica*, Forsk. is a four-sided prism, with a rough surface, and 8 prominent spines at the posterior end; it has 7 muscular bands which completely encircle the body. The aggregate form (*S. mucronata*, Forsk.) is ovoid, pointed behind, smooth, and has only 5 muscular bands, whose dorsal ends are separate. (*Huxley*.)†

The *solitary* Salpae always contain a chain of embryos winding spirally round the visceral nucleus; the embryos are attached in pairs to a double tube (or " proliferous stolon ") connected with the sinus to the right of the heart. Sometimes they increase in size gradually from the heart outwards to the free end of the stolon, but usually the embryos are developed in groups, and each portion of the series when it is detached consists of young Salpas of the same size. These portions are liberated in succession through an aperture produced in the tunic opposite the extremity of the stolon.

The *aggregate* Salpae produce a single ovum at a time, which is attached by a pedicle to the posterior part of the respiratory cavity. It remains there until it has attained a considerable size, and exhibits the proliferous stolon already partly developed, and those external characters which permanently distinguish it from its parent.

It was in *Salpa* that Hasselt first observed the periodic change in the direction of the circulating currents. The heart itself is a muscular membrane not forming a complete tube, but open on one side. The dorsal sinus contains the long tubular filament (fig. 225, e) called the *endostyle*. In the ventral sinus is the ganglion, and the auditory vesicle containing 4 otolithes. The gill is a hollow column, or band, representing only the thoracic vessel (" hypo-pharyngeal band ") of the Ascidians (fig. 226, d) and the respi-

* Chiefly known in England as the author of PETER SCHLEMIHL.
† Phil. Trans, 1851, Part II. p. 567.

ratory function is performed by the entire pallial cavity. The muscles of the Salpae consist of single layers of tranversely striped fibre.

DOLIOLUM, Quoy and Gaimard.

Etym. Diminutive of *dolium* a cask.   *Syn.* ? Anchinaea, Esch.

Type, D. denticulatum, Pl. 24, fig. 23.

Body transparent, cask-shaped, open at the ends, 2—10 lines in length; oral extremity a little prominent, with about 12 rounded denticulations; posterior end fringed; muscular bands 6, equidistant, besides the sphincters of the orifices; branchiae consisting of two bands stretched across the interior, one above (*epi*) and one below (*hypopharyngeal*), connected by transverse bars with one another and the parietes; mouth on the dorsal side, in front of the fourth band; heart above and in front of the mouth. (*Huxley.*)

Distr. 2 sp. Amboina, Vanicoro, N. Zealand.

APPENDICULARIA, Chamisso.

Etym. Appendiculus, a small appendage.

Syn. Vexillaria, Müll. 1846.   Oikopleura, Mertens, 1831.

Type, A. flabellum, Pl. 24, fig. 24.

Body ovoid, $\frac{1}{8}$—$\frac{1}{4}$ inch long, with a long curved tail or swimming-organ; smaller end perforated, leading into a large cavity lined by a sinus-system; gill represented by the ciliated pharynx, which communicates with the exterior by two funnel-shaped canals opening on the hæmal surface beside the rectum; œsophagus short, slightly curved, leading into a wide stomach; intestine turned forwards, ending on *dorsal* side in front of appendage; heart between lobes of the stomach; tail lanceolate, horizontally compressed. All the examples hitherto observed have been males. (*Huxley.*)

These minute creatures appear to be the lowest forms of the *Tunicata;* typifying in their adult age the larval state of the higher ascidians.

Distr. Behring's Straits, N. Brit. Tenby, Cape, New Guinea, S. Pacific.

Prof. Forbes relates that "when cruising off the north coast of Scotland in 1845, with Mr. Mc Andrew, their attention was attracted by the appearance of cloudy patches of red colouring matter in the water, and on procuring some and submitting it to microscopic examination, it was found to consist entirely of the curious and anomalous creatures called *Appendiculariæ.*"*

---

* The most complete and accurate history of the class Tunicata is contained in the Article TUNICATA of Todd's *Cyclopædia of Anatomy*, by Mr. T. Rupert Jones.

CONCLUSION.

CHAPTER I.

NUMERICAL ESTIMATE.

The number of living and fossil species of each genus of mollusca has been stated in the preceding pages, so far as they could be ascertained. With some modifications derived from recent data, these numbers give the following totals, by which the relative numerical development of the orders and families will be seen.

	Recent.	Fossil.
CEPHALOPODA. *Dibranchiata.*		
Argonautidæ	4	1
Octopodidae........	58	—
Teuthidæ	91	31
Belemnitidæ	—	67
Sepiadæ	30	1
Spirulidæ	3	—
	186	100
Tetrabranchiata.		
Nautilidæ	4	174
Orthoceratidæ....	—	281
Ammonitidæ	—	904
	4	1,359
GASTEROPODA. *Prosobranchiata.*		
Strombidæ*	83	195
Muricidæ	870	697
Buccinidæ	1,048	352
Conidæ	856	390
Volutidæ..........	686	210
Cypraeidæ	225	97
Naticidæ	245	340
Pyramidellidæ	216	322
Cerithiadæ........	192	610
Melaniadæ........	424	50
Turritellidæ†......	196	290
Litorinidæ	315	220
Paludinidæ........	132	110
Calyptræidæ	160	100
Turbinidæ	855	906
Haliotidæ	99	136
Fissurellidæ	194	72
Neritidæ	300	100
Patellidæ..........	130	100
Dentaliadæ........	50	70
Chitonidæ	230	24
	7,506	5,391

	Recent.	Fossil
Pulmonifera.		
Helicidæ	3,900	280
Limacidæ	72	4
Limnæidæ	160	155
(Marine)	86	28
(Ditto, shell-less) ..	16	0
	4,234	467
Operculated. Pulmonifera.		
Cyclostomidæ	700	23
Aciculidæ	26	1
	726	24
Tecti-branchiata.		
Tornatellidæ........	50	152
Bullidæ	158	78
Aplysiadæ	79	4
Pleurobranchidæ....	9	—
Phyllidiadæ	10	—
	326	234
Nudibranchiata.		
British	90	--
Foreign	220	—
	310	—
Nucleobranchiata.		
Shell-less	14	—
Shell-bearing	30	100
	44	100
PTEROPODA.		
Hyaleidæ	50	32
Limacinidæ	16	—
Cliusidæ	13	—
	79	32

* Including *Aporrhais*.   + With *Scalaria*.

	Recent.	Fossil.		Recent.	Fossil.
BRACHIOPODA.			(CONCHIFERA.)		
Terebratulidæ......	50	300	Tridacnidæ	7	3
Spiriferidæ	—	254	Cardiadæ.........:..	200	300
Rhynchonellidæ....	3	300	Lucinidæ	120	351
Orthidæ............	—	200	Cycladidæ	200	105
Productidæ	—	100	Astartidæ	46	373
Craniadæ	5	30	Cyprinidæ	108	356
Discinidæ :........	7	50	Veneridæ	573	260
Lingulidæ	7	38	Mactridæ..........	82	41
			Donacidæ	73 ,	40
	75	1,272	Tellinidæ	315	200
CONCHIFERA.			Solenidæ	55	45
Ostreidæ	270	1,062	Myacidæ	90	250
Aviculidæ	85	570	Anatinidæ	66	400
Mytilidæ	112	242	Gastrochænidæ....	23	34
Arcadæ	288	616	Pholadidæ	64	50
Trigoniadæ	3	136			
Unionidæ	320	50		3,150	5,612
Chamidæ	50	50	TUNICATA (about)....	150	
Hippuritidæ	—	78			

Of the recent marine shell-fish some are in great measure animal feeders, while the rest live on algæ and infusoria.

Animal feeders.		*Vegetable feeders.*	
Cephalopoda.....................	190	Gasteropoda rostrifera	3,127
Proboscidean Gasteropoda	4,329	Opistho-branchiata (part)........	128
Dentaliadæ	50	*Infusorial feeders.*	
Opistho-branchiata (part)........	508	Bivalve shellfish	3,226
Nucleobranchiata	44	Tunicaries.....................	150
Pteropoda	79		6,631
	5,200	Pulmonifera	4,960

	Recent.		Fossil.
Fresh-water shells	1,504	....	800
Marine shells	10,002	....	13,300
Land snails	4,626	....	491
Total of *Shell-bearing Mollusca*	16,132	.... Total	14,591
Naked Mollusks	660	....	—
Total of *Recent Mollusca**	16,792	......(British	4,590)

* The total number of living *Vertebrate* animals amounts to about 16,000; the number of Plants is estimated at 100,000, and the Insect class is supposed to include not less then 300,000 species.

CHAPTER II.

GEOGRAPHICAL DISTRIBUTION OF THE MOLLUSCA.

It is one of the most familiar facts in Natural History, that many countries possess a distinct Fauna and Flora, or assemblages of animals and plants peculiar to themselves ; and it is equally true, though less generally understood, that the sea also has its provinces of animal and vegetable life.

The most important, or best known of these provinces are indicated on the accompanying map ; different names, in some instances, and different letters and numbers being employed to distinguish the marine from the terrestrial regions.*

The division of the surface of the globe into natural history provinces ought to be framed upon the widest possible basis. The geographical distribution of every class of animals and plants should be considered, in order to arrive at a theory of universal application.

The *Land Provinces* hitherto proposed have been chiefly founded on botanical grounds, but the evidence afforded by insects, and the higher classes of animals, confirms the existence of these divisions.

The *Marine Provinces* have also been investigated by botanists; and the striking peculiarities of the fisheries have been taken into account as well as the distribution of shell-fish and corals.

In order to constitute a distinct province it is considered necessary that at least *one-half* the species should be *peculiar,* a rule which applies equally to plants and animals. Some genera, and sub-genera are limited to each province, but the proportion is different in each class of animals and in plants.†

Specific areas.—Species vary extremely in their range, some being

---

* The author regrets that, on account of the expense, this map appears without the advantage of colours. He would recommend those who are sufficiently interested in the subject, to colour their own copies, distinguishing the shores of the marine provinces by the following tints :—

Blue 1. Arctic province; 15. Magellanic.

Green. 2. Boreal; 11, Aleutian, 5. Aralo-Caspian.

Orange. 3. Celtic.

Purple. 4. Lusitanian ; 10. Japonic; 12. Californian ; 18, Trans-Atlantic.

Yellow. 6. W. African; 8. Indo-Pacific; 13. Panamic; 17. Caribbean.

Lake. 7. S. African ; 9. Australo-Zealandic; 14. Peruvian; 16. Patagonian.

† The genera of plants amount to 20,000, and consist on an average of only 4 species apiece ! The genera of shells commonly admitted are only 400 in number, and average 40 species each. It follows that the areas of the molluscan genera (*cæteris paribus*) ought to be 10 times as great as those of plants.

limited to small areas, while others, more widely diffused, unite the local populations into fewer and larger groups. Those species which characterise particular regions are termed "endemic;" they mostly require peculiar circumstances, or possess small means of migrating. The others, sometimes called "sporadic," possess great facilities for diffusion, like the lower orders of plants propagated by *spores*, and more easily meet with suitable conditions. The space over which a species is distributed is called a "centre," or more properly specific *area*. The areas of one-half the species are smaller (usually much smaller) than a single province.

In each specific area there is frequently one spot where individuals are more abundant than elsewhere; this has been called the "metropolis" of the species. Some species which appear to be no-where common can be shown to have abounded formerly; and many probably seem rare only because their head-quarters are at present unknown. (*Forbes*.)

Specific centres are the points at which the particular species are supposed to have been created, according to those who believe that each has originated from a common stock (p. 56); these can only be known approximately in any case. The doctrine that each species originated from a single individual, or pair, created once only, and at one place, derives strong confirmation from the fact that so "many animals and plants are indigenous only in determinate spots, while a thousand others might have supported them as well."[*]

Generic areas.—Natural groups of species, whether called genera, families or orders, are distributed much in the same manner as species;[†] not for the

[*] Mrs Somerville's Physical Geography, II. 95.

[†] "What we call class, order, family, genus, are all only so many names for *genera* of various degrees of extent. Technically, a *genus* is a group to which a *name* (as *Ribes*) is applied: but essentially, *Exogens, Ranunculaceæ; Ranunculus*, are genera of different degrees.

One of the chief arguments in favour of the *naturalness* of genera (or *groups*), is that derived from the fact that many genera can be shown to be *centralized* in definite geographical areas (*Erica*, for example); *i.e.* we find the species gathered all, or mostly, within an area, which has some one point where the *maximum* number of species is developed.

But, in *geographical space*, we not unfrequently find that the same genus may have two or more areas, within each of which this phenomenon of a point of *maximum* number of species is seen, with fewer and fewer species radiating, as it were, from it.

In *time*, however (or, in other words, in *geological distribution*), so far as we know, each generic type has had an unique and continuous range. When once a generic type has ceased, it never re-appears.

A genus is an abstraction, a divine idea. The very fact of the centralization of groups of allied species, *i.e.* of genera, in space and time, is sufficient proof of this. Doubtless we make many so-called genera that are artificial; but a true genus is natural; and, as such, is not dependent on man's will." *E. Forbes*. (See An. Nat. Hist. July, 1852, and Jan. 1855, p. 45.)

same reason, since their constituents are not related by descent, but apparently from the intention of the Creator.

Sub-generic areas are usually smaller than generic; and the areas of orders and families are as a matter of course larger than those of the included genera. But it is necessary to remember that groups of the same denomination are not always of equal value; and since species vary in range, it often happens that specific areas of one class or family are larger than generic areas of another. The smallest areas are usually those of the forms termed *aberrant* (p. 61) ; the *typical* groups and species are most widely distributed. (*Waterhouse*.)

" When a generic area includes a considerable number of species, there may be found within it a point of maximum, (*metropolis*) around which the number of species becomes less and less. A genus may have more centres than one.—It may have had unbroken extension at one period, and yet in the course of time and change, may have its centre so broken up that there shall appear to be out-lying points. When, however, the history of a natural genus shall have been traced equally through its extension in *time* and *space*, it is not impossible that the area, considered in the abstract, will be found to be necessarily unique." (*Forbes*.)

To illustrate the doctrine of the *unity of generic areas* Prof. Forbes has given several examples, showing that some of the most exceptional cases admit of explanation and confirm the rule. One of these relates to the genus *Mitra* of which there are 400 species; it has its metropolis in the Philippine Islands and extends by the Red Sea to the Mediterranean and West Africa, the species becoming few, small, and obscure. Far away from the rest a single species is found on the coast of Greenland! But this very shell occurs fossil in Ireland along with another *mitra* now living in the Mediterranean. Another case is presented by the genus *Panopæa*, of which the six living species are widely separated,—*a*, in the Mediterranean ; *b*, in Patagonia ; *c*, at the Cape; *d*, Tasmania; *e*, New Zealand; *f*, Japan. Of this genus above 100 fossil species are known, distributed over many places within the wide area, on whose margin the relics of this ancient form of life seem to linger, like the last ripple of a circling wave.[*]

According to this view the specific centres are scattered thickly over the whole surface of the globe; those of the genera more thinly distributed; and the points of origin of the large groups become fewer in succession, until we have to estimate the probable position or scene of creation of the primary divisions themselves; and are led to speculate whether there may not have been some common focus—the centre of centres—from which the first and greatest types of life have emanated.

Boundaries of Natural History Provinces. The land provinces are sepa-

[*] The most striking and conclusive instances may be met with in the distribution of the higest classes of vertebrate animals.

rated by lofty mountains, deserts, seas, and climates; whilst the seas are divided by continents and influenced by the physical character of coast-lines, by climates and currents. These "natural barriers" as they were called by Buffon, retard or altogether prevent the migrations of species in particular directions.

Influence of Climate.—Diversity of climate has been the popular explanation of most of the phenomena of geographical distribution, because it is so' well-known that some species require a tropical amount of warmth, whilst others can endure a great variety of temperature, and some only thrive amidst the rigours of the arctic regions. The character of the vegetation of the zones of latitude has been sketched by Baron Humboldt; Fabricius and Latreille have divided the world into climatal Insect-provinces; and Prof. E. Forbes has constructed a map of the *homoiozoic belts* or zones of marine life. To all these the remark of Mr. Kirby is applicable—that any division of the globe into provinces, by means of *equivalent* parallels and meridians, wears the appearance of an artificial and arbitrary system, rather than of one according to nature. Prof. Forbes has been careful to point out that although the "Faunas of regions under similar physical conditions bear a striking resemblance to each other"—this resemblance is produced, "not by identity of species, or even of genera, but by *representation.*" (p. 56).

Origin of the Natural History Provinces.—Mr. Kirby appears to have been the first to recognize the truth that physical conditions were not the primary causes of the zoological provinces, which he "regarded as fixed by the will of the Creator, rather than as regulated by isothermal lines."* Mr. Swainson also has shown that the "circumstances connected with temperature, food, situation and foes, are totally insufficient to account for the phenomena of animal geography," which he attributes to the operation of unknown laws.†

The most important contribution towards a knowledge of these "unknown laws" has been made by Prof. E. Forbes, who was perhaps the first naturalist ever in a position to avail himself of the great storehouse of facts accumulated by geologists, respecting the distribution of organic life in "the former world." This subject will be referred to again in connection with the subject of Fossil Shells; meanwhile it may be stated, that according to this evidence, the Faunas of the Provinces are of various ages, and that their origin is connected with former (often very remote) geological changes, and a different distribution of land and water over the surface of the globe.

MARINE PROVINCES.

Amongst the genera of marine shells, there are some which have been considered particularly indicative of climate. From the Arctic list the follow-

---

* Introduction to Entomology.

† Treatise on Geography and Classification of Animals, Lardner's Cabinet Cyclopædia.

ing may be taken as examples of the shells of high latitudes; those marked *
being found in the southern, as well as in the northern hemisphere:—

Buccinum.	Velutina.	*Crenella.
*Chrysodomus.	Lacuna.	*Yoldia.
*Trophon.	*Margarita.	*Astarte.
Admete.	——	Cyprina.
*Trichotropis.	*Rhynchonella.	Glycimeris.

The following have been thought peculiar to the warmer regions of the
sea:

Nautilus.	Conus.	Columbella.	Perna.
Rostellaria.	Harpa.	Cypræa.	Vulsella.
Triton.	Oliva.	Nerita.	Tridacna.
Cancellaria.	Voluta.	Spondylus.	Crassatella.
Terebra.	Marginella.	Plicatula	Sanguinolaria.

But it must not be inferred that these genera were always characteristic
of extreme climates. On the contrary, the whole of them have existed in the
British seas at no very remote geological period. *Rhynchonella* and *Astarte*
were formerly "tropical shells;" and since the period of the English chalk-
formation there have been living *Nautili* in the North Sea, and Cones and
Olives in the "London basin." It is not true that the same *species* have
been at one time tropical, at another temperate, but the *genera* have in many
instances enjoyed a much wider range than they exhibit now. Some of the
"tropical" forms are more abundant and extend farther in the Southern
hemisphere; several large Volutes range to the extremity of South America,
and the largest of all inhabits New Zealand.

The tropical and sub-tropical provinces might be naturally grouped in
three principal divisions, viz., the Atlantic, the Indo-Pacific, and the West-
American,—divisions which are bounded by meridians of longitude, not by
parallels of latitude. The Arctic province is comparatively small and exceptional; and the three most southern Faunas of America, Africa, and Australia
differ extremely, but not on account of climate.

If only a small extent of sea-coast is examined, the character of its
mollusca will be found to depend very much upon the nature of the shore,
the tides, depth, and local circumstances, which will be referred to again in
another page. But these peculiarities will disappear when the survey is
extended to a region sufficiently large to include every ordinary variety of
condition.

It has been stated that each Fauna consists of a number of peculiar
species, properly, more than half; and of a smaller number which are common to some other provinces. By ascertaining the direction of the tides and
currents, and the circumstances under which the species occur, it may be
possible to determine to which province these more widely diffused mollusca
originally belonged. And when species occur both recent and fossil it is easy
to perceive the direction in which their migrations have taken place.

The Fauna of the Mediterranean has been critically examined by Prof. Forbes and M. Philippi, with this result,—that a large proportion of its population has migrated into it from the Atlantic, and a smaller number from the Red Sea, and that the supposed peculiar species are diminishing so rapidly with every new research in the Atlantic, that it can no longer rank as a province distinct from the Lusitanian.

When the Faunas of the other regions have been tested in the same manner, and disentangled, the result will probably be the establishment of a much greater number of provinces than we have ventured at present to indicate on the map.

It may be desirable to notice here the extraordinary range attributed to some of the marine species. These statements must be received with great hesitation; for when sufficiently investigated, it has usually proved that some of the localities were false, or that more than one species was included. The following are given by Dr. Krauss in his excellent monograph of the South African Mollusca :—

Ranella granifera: Red Sea, Natal, India, China, Philippines, New Zealand.

Triton olearius: Brazil, Mediterranean, Natal, Pacific.

Purpura lapillus: Greenland (Senegal, Cape).

Venus verrucosa: (W. Indies) Brit. Senegal, Canaries, Mediterranean, Red Sea, Cape (Australia).

Octopus vulgaris: Antilles, Brazil, Europe, Natal, Mauritius, India.

Argonauta argo: (Antilles), Medit. Red Sea, Cape.

Lucina divaricata is said to be "found on the shores of Europe, India, Africa, America, and Australia," (*Gray*.) In this case several species are confounded. The rock-boring *Saxicava* has been carried to all parts of the world *in ballast*, and it remains yet to be ascertained whether *the same species* occurs in a living state beyond the Arctic Seas and North Atlantic.

Lastly, the *money cowry* is always catalogued as a shell of the Mediterranean and Cape, although its home is in the Pacific, and it has no other origin in the Atlantic than the occasional wreck of one of the ships in which such vast quantities of the little shell are annually brought to this country to be exported again to Africa.

I. Arctic Province.

The North Polar Seas contain but one assemblage of *Mollusca*, whose Southern limit is formed by the Aleutian Islands in the Pacific, but in the North Atlantic is determined chiefly by the boundary of floating ice, descending as low as Newfoundland on the West, and thence rising rapidly to Iceland and the North Cape. A very complete general account of the Arctic Mollusca is given by Dr. Middendorff;* those of Greenland have been catalogued and

* Malaco-zoologia Rossica; Mem. del'Acad. Imp. des Sc. Petersb. T. 6, pt. 2, 1849.

described by Otho Fabricius and Möller ;* and scattered notices occur in the Annals of Natural History,† and the Supplements to the Narratives of the Arctic Voyagers,—Phipps, Scoresby, Franklin, Back, Ross, Parry, and Richardson. The existence of the same marine animals in Behring's Sea and Baffin's Bay, was long since held to prove at least a former North-West passage; but the occurrence of recent sea shells in banks far inland, rendered it probable that even recent elevation of the land in Arctic America might have much reduced the passage. During the "Glacial period," this Arctic Sea, with the same fauna, extended over Britain; over Northern Europe, as far as the Alps and Carpathians; and over Siberia, and a considerable part of North America. The shells now living in the Arctic Seas, are found fossil in the deposits of " Northern Drift," over all these countries ; and a few of the species yet linger within the bounds of the two next provinces, especially in tracts of unusual depth. The Arctic shells have mostly a thick greenish epidermis (p. 40.) they occur in very great abundance, and are remarkably subject to variation of form, a circumstance attributed by Professor E. Forbes to the influence of the mixture of fresh water produced by the melting of great bodies of snow and ice.

ARCTIC SHELL-FISH.

R. Russian Lapland. F. Finmark. I. Iceland. G. Greenland. D. Davis Straits (west coast). B. Behring's Straits. O. Ochotsk. * British species. ** Brit. fossils.

Octopus granulatus. G.	Buccinum angulosum. N. Zemla,
Cirroteuthis Mülleri. G.	Icy C. Spitz.
Rossia palpebrosa. G. P. Regent Inlet.	,, tenue. N. Zemla. G.
Onychoteuthis Bergii. F. B.	,, Groenlandicum. D.
- ,, Fabricii. G.	,, undulatum G.
,, amœna. G.	,, scalariforme. G.
*Ommastrephes todarus. F. Newf.	** ,, ciliatum. G.
————	,, boreale (Leach). Baffin's B
Limacina arctica G. O.	,, sericatum. D. P. Refuge.
Spirialis stenogyra. F.	,, Hollböllii (Mangelia, Mol.)
,, balea. G.	G. F.
*Clio borealis N. Zemla. G.	* ,, Dalei. R. B.
————	*Fusus antiquus. N. Zemla. B.
*Nassa incrassata. F.	** ,, carinatus. G.
*Buccinum undatum, var. Kara. O.	* ,, contrarius. R. O.
,, hydrophanum. D. Prince	,, deformis. R. Spitz.
Regent Inlet.	** ,, despectus. G. Spitz.
,, tenebrosum. R. G. B.	,, heros. C. Parry.
* ,, Humphreysianum. R. G.	,, latericeus. G.
** ,, cyaneum. F. D. G. Icy	** ,, Sabini. D. Mass.
C. St. Lawrence.	,, pellucidus. D.
,, glaciale. Kara. O. C.Parry.	,, Kroyeri. G. Spitz.
G. Spitzbergen.	,, decemcostatus. B. Newf.

* Index Molluscorum Grœnlandiæ. Hafn. 1842.

† Hancock, An. Nat. Hist. vol. 18, p. 323, pl. 5.

*Fusus Berniciensis. R. B.
　　„　Spitzbergensis. Spitz.
* 　„　Islandicus. F.
* 　„　gracilis. F. R. G. B.
*Trophon clathratus. R. G. B.
** 　　„　scalariformis. Spitz. Newf. B.
** 　　„　Gunneri. F. G.
** 　　„　craticulatus. R. I. G.
* 　　„　Barvicensis. F.
　　„　harpularius. F. U.S.
*Purpura lapillus. R. G. B.
Mangelia, 9 sp. G.
　　„　decussata. D.
*Bela turricula. F. G.
* 　„　rufa. F. G.
**Mitra Grœnlandica. G.
**Admete viridula. R. Spitz. G. B.
*Trichotropis borealis. F. G. B.
　　　　　　P. Regent Inlet.
　　„　conica. G.
　　„　insignis. B.
　　„　bicarinata. B.
*Natica helicoides R. G. B.
** 　„　clausa, F. N. Zemla. G. Mel-
　　　ville Id. P. Regent Inlet. B.
　　„　pallida. R. O.
　　„　flava. N. Zemla. B. Newf.
* 　„　pusilla (grœnlandica). G.
　　　Norway. Spitz.
　　„　nana. G.
*Velutina lævigata. R. B.
* 　„　flexilis. F.
　　„　zonata. R. G.
　　„　lanigera. G.
Lamellaria prodita. F.
　　„　Grœnlandica. G. B.
**Scalaria Grœnlandica. F. G. B.
　　„　borealis, (Eschrichti). G.
Amaura candida. G.
Chemnitzia albula. G.
Mesalia lactea. G.
Turritella polaris. G.
Aporrhais occidentalis. Labrador.
*Litorina obtusata. R.
* 　„　tenebrosa. N. Zemla. D.
　　„　Grœnlandica. G. F.
　　„　palliata (arctica). G.
　　„　limata. F.
*Lacuna vincta. R. Newf. G.
　　„　labiosa. F. P. Refuge.
* 　„　crassior. R.
　　„　glacialis. G.
* 　„　pallidula. G.
* 　„　puteolus. F. Newf.

Lacuna frigida. F.
　　„　solidula. F.
Hydrobia castanea. R. G.
Rissoa scrobiculata. G.
　　„　globulus. G.
　　„　saxatilis. G.
*Skenea planorbis. G. F.
Margarita cinerea. F. U.S.
* 　„　undulata. R. G.
* 　„　alabastrum. F.
* 　„　helicina. G. White Sea.
　　　Spitz.
　　„　sordida. R. Spitz. G. B.
　　„　umbilicalis. D. B.
　　„　Harrisoni. D.
　　„　glauca. G.
　　„　Vahlii. G.
* 　„　costulata. G.
*Puncturella Noachina. F. G.
*Acmæa testudinalis. R. Iceland. G.
*Lepeta cæca. G. F. Spitz. C.Eden.
Pilidium rubellum. F. G. D.
*Chiton ruber. F. G. Spitz.
* 　„　albus. F. G.
Dentalium, entale. Spitz.

Bulla Reinhardi. G.
　　„　subangulata. G.
*Cylichna alba. G. F. Spitz.
　　„　turrita. G.
*Philine scabra. Norway. G.
　　„　punctata (Müll.) G.
Doris liturata. G
　　„　acutiuscula. G.
　　„　obvelata. G.
*Dendronotus arborescens. F. G.
Æolis bodocensis. G.
Tergipes rupium. G.
Euplocamus Holböllii. G.

*Terebratulina caput-serpentis. Spitz.
　　　F. Mass. Medit.
*Waldheimia cranium. F.
　　„　septigera. F.
Terebratella Spitzbergensis. Sp.
　　„　Labradorensis. Labr.
**Rhynchonella psittacea. R. Baffin's
Bay, 76 deg. N. Melville, I. B.
*Crania anomala. Spitz.

*Anomia squamula. R.
* 　„　aculeata. R.

**Pecten Islandicus. F. N. Zemla.
Spitz. G. B. St. Lawrence.
„ vitreus. F. Arctic America.
„ Grœnlandicus. R. Spitz. D.
Limatula sulcata. G. F.
*Mytilus edulis. R. G. B.
*Modiola modiolus. R. B.
*Crenella discors (lævigata). G. D.
N. Zemla.
„ decussata. R. G.
„ nigra. N. Zemla. R. G. D.
„ faba. G.
„ vitrea G.
Arca glacialis. P. Regent Inlet.
Nucula corticata. G.
„ inflata. G. D.
Leda buccata. G.
„ macilenta. G.
** „ rostrata (pernula). F. Spitz.
Arctic America.
** „ minuta (Fabr.) F Spitz. G. D.
„ lucida. F. (=navicularis? Spitz.)
* „ pygmæa. G. F. Siberia.
**Yoldia arctica Gr. (myalis). G. U.S.
Spitzbergen.
** „ lanceolata (arctica B. & S.)
Icy Cape.
„ limatula. F. U S. Kamts.
„ hyperborea. Spitz.
** „ thraciæformis (angularis). G.
Mass.
** „ truncata, Br. (Portlandica, Hit.)
P. Refuge. Arctic America.
**Astarte borealis (arctica). F. Ice-
land. G.
** „ semisulcata (corrugata). Kara
Sea. N. Zemla. Spitz. P.
Regent Inlet. C. Parry.
Icy Cape.
⌐ „ elliptica. F. G. Spitz.
* „ sulcata. R. N. Zemla. O.
** „ crebricosta. F. Spitz. Newf.
„ crenata. P. Regen nlet.

Astarte Warhami. Davis Str.
„ globosa. G.
* „ compressa. N. Zemla. G.
„ Banksii. Spitz. Baffin's B.
*Cardium edule var. rusticum. R.
„ Islandicum. N. Zemla. G.
** „ Grœnlandicum. Kara. Spitz.
C. Parry. St. Lawrence.
„ elegantulum. G.
*Cryptodon flexuosus. G. F.
*Turtonia minuta. G. F.
*Cyprina Islandica. R. Labrador.
**Cardita borealis. Mass. O.
*Tellina calcaria. F. G. B.
** „ Grœnlandica. (=Balthica, L.)
N. Zemla. Spitz. F. G. B.
** „ edentula. B.
*Mya truncata. R. Spitz. G. C. Parry. B
** „ Uddevallensis. St. Lawrence. D.
P. Regent Inlet. Melville I.
* „ arenaria N. Zemla. G. O.
Saxicava rugosa (arctica). N. Zemla.
Spitz. G. C. Parry. B.
* „ (Panopæa) Norvegica. White
Sea. O.
Machæra costata. Labrador. O.
Glycimeris siliqua. C. Parry. Newf.
*Lyonsia Norvegica. F. O.
„ arenosa. G. D. P. Refug
Thracia myopsis. G.
Pandora glacialis. Spitz. Baff. (Leach.

Chelyosoma Macleayanum. G.
Cynthia glutinosa. G.
Ascidium, 9 sp. including :
* „ echinatum. G.
* „ conchilegum. G.
., „ rusticum. G. Spitz.
Clavellina crystallina G.
Boltenia reniformis. G.
„ ciliata. G.
Synœcium turgens. Spitz.
Cystingia Griffithi. Felix H.

II. Boreal Province.

The Boreal Province extends across the Atlantic from Nova Scotia and Massachusetts to Iceland, the Faeroe and Shetland Islands, and along the coast of Norway from North Cape to the Naze.

Of the 289 Scandinavian shells catalogued by Dr. Lovén,* 217, or 75 per

* Index Molluscorum Scandinaviæ ; extracted from the " Ofversigt af K. Vet. Akad. Forh." 1846. The climate of Finmark is much less severe than Russian Lapland ; Hammerfest has an open harbour all the year.

cent. are common to Britain, and 137 range as far as the North coast of Spain.

The boreal shells of America are described by Dr. Gould.* From these lists it appears that out of 140 sea-shells found on the coast of Massachusetts north of Cape Cod, more than half are common to Northern Europe.

Many of the species, it is believed, could only have extended their range so distantly, by means of continuous lines of connecting coast, now no longer in existence.†

Boreal Shells common to Europe and North America.

* British species.

*Teredo navalis.
*Pholas crispata.
*Solen ensis.
* (Panopæa) Norvegica.
*Mya arenaria.
* „ truncata.
*Thracia phaseolina (Conradi, Couth).
Mactra ponderosa (ovalis, G.)
? Montacuta bidentata.
*Turtonia minuta.
? Kellia rubra.
? Lepton nitidum (fabagella, Conr.?)
*Saxicava rugosa (arctica).
Tellina solidula, var. (fusca, Say).
* „ calcaria (sordida, Couth).
*Lucina borealis.
? „ divaricata.
*Cryptodon flexuosus.
*Astarte borealis.
* „ triangularis? (quadrans, G.)]
*Cyprina Islandica.
?(Cardium Islandicum, U.S.—N. Zemla.)
Yoldia limatula.
„ arctica, Gr. (=myalis).
*Leda pygmæa.
* „ caudata.
? „ navicularis (lucida, Lovén?)

*Nucula tenuis.
*Mytilus edulis.
*Modiola modiolus.
*Crenella nigra.
* „ discors, L.
* „ decussata, (glandula, Tot.)
Pecten Islandicus.
? Ostrea edulis (borealis, Lam.?)
*Anomia ephippium.
* „ aculeata.
„ squamula?

*Terebratulina caput-serpentis.
*Rhynchonella psittacea.

*Dendronotus arborescens.
Polycera Lessonii?
? Amphisphyra hyalina (debilis?)
Cylichna alba (triticea, C.)
* „ obtusa (pertenuis).
*Philine quadrata (formosa, St.)

*Chiton cinereus.
* „ marmoreus.
* „ ruber.
* „ lævis.
* „ asellus.

* Report on the Invertebrata of Massachusetts. 1841.

† Forbes, Memoirs of the Geol. Survey, I. p. 379. Sir John Richardson, when speaking of the cod-tribe and turbot-tribe, says : " Most of the fish of this order feed on or near the bottom, and a very considerable number of the species are common to both sides of the Atlantic, particularly in the higher latitudes where they abound. It does not appear that their general diffusion ought to be attributed to migration from their native haunts, but rather that in this respect they are analogous to the owls, which, though mostly stationary birds, yet include a greater proportion of species common to the old and new worlds than even the most migratory families. Several of the *Scomberoideæ* (Mackerel-tribe) which feed on the surface, have been previously noted as traversing many degrees of longitude in the Atlantic : but the existence of the ground-feeding *Gadoideæ* in very distant localities must be attributed to a different cause, as it is not probable that any of them wander out of soundings or ever approach the mid-seas."—Report Zool. N. America, p. 218.

*Chiton albus.
*Dentalium (entale, L.?)
? Lepeta cæca (candida, C.)
*Acmæa testudinalis (amoena, S.)
*Puncturella Noachina.
*Adeorbis divisus (= Skenea serpuloides).
Margarita cinerea.
*   „   costulata? (Skenea).
*   „,   helicina.
*   „   undulata.
*   „   alabastrum (= occidentalis,?)
Litorina grœnlandica.
*   „   tenebrosa (vestita).
      „   palliata?
*Lacuna vincta (divaricata).
*   „   puteolus (Montagui).
*Skenea planorbis.
*Velutina lævigata.
      „,   zonata.
*Lamellaria perspicua.
*Natica helicoides.

Natica clausa.
*   „   pusilla.
*Scalaria grœnlandica.
(Ianthina communis).
Odostomia producta.
Cancellaria (admete) viridula.
*Trichotropis borealis.
*Fusus antiquus (tornatus).
*   „   islandicus.
*   „   propinquus.
      „   ? rosaceus.
*Trophon muricatus.
*   „   clathratus.
      „   scalariformis.
      „   harpularius.
*Purpura lapillus.
*Buccinum undatum.
*   „   (Cominella) Dalei.
*Bela turricula.
*   „   Trevelyana.
*   „   rufa (Vahlii)?

Ommastrephes sagittatus and *Cynthia microcosmus* are also common to both sides of the North Atlantic. The genera,

Machæra,   Glycimeris,   Cardita, and
Solemya,   Mesodesma (deauratum).   Crepidula,

are peculiar to the American side of the Boreal Province.

Several other species now living on the coast of the U. States occur fossil in England: *e.g. Trophon cinereus*, Say., is believed to be the *Fusus Forbesi*, Strickland, of the Isle of Man; others are marked in the Arctic list.

III. CELTIC PROVINCE.

The Celtic province, as described by Prof. E. Forbes, includes the British island coasts, Denmark, Southern Sweden, and the Baltic.* The fauna of this region (which includes the principal herring-fisheries) is essentially Atlantic; many of the species are of ancient origin, being known fossil in the Pliocene Tertiaries.

The British mollusca described by Forbes and Hanley amount to 682, viz. :—

14 Cephalopoda.   100 Pulmonifera.   175 Acephala.
220 Marine Univalves.   4 Pteropoda.   73 Tunicata.
91 Nudibranchiata.   5 Brachiopoda.

Of this number two-thirds of the *Nudibranches*, 55 marine univalves, and

* The great work of Messrs. Forbes and Hanley contains all that is known respecting British *Testacea* up to the present time. The *Nudibranchiata* alone have been more fully described, in the publications of the Ray Society, by Messrs. Alder and Hancock. For the marine zoology of the coasts of Denmark the "Zoologia Danica" of O. F. Müller is still the most important work.

7 bivalve shell-fish, are, at present only known in British seas; but as most of these are minute or "critical" species it is considered they will yet be met with elsewhere.

A few of the species belong to the Lusitanian province, whose northern limits include the Channel Islands, and just impinge upon our coast.

Phasianella pullus.	Murex corallinus.	Cytherea chione.
Haliotis tuberculata.	Avicula Tarentina.	Petricola lithophaga.
Truncatella Montagui.	Galeomma Turtoni.	Venerupis irus.
Oncidium celticum.	Pandora rostrata.	Cardium rusticum, L. (tu-
Bulla hydatis	Ervilia castanea.	berculatum).
Volva patula.	Mactra helvacea.	

Of the *Gasteropoda* 54 are common to the seas both north and south of Britain; 52 range further south, but are not found northward of these islands; and 34 which find here their southern limit occur not only in Northern Europe, but most of them in Boreal America. Nearly half of the bivalves range both north and south of Britain; 40 extend southward only, and about as many more are found in Scandinavia, 27 of them being common to N. America. (*Forbes*.)

In the lists of Arctic and Boreal shells the British species are distinguished by an asterisk.

According to Mr. M'Andrew's estimate there are 406 British shell-bearing mollusca, of which

217	or 53 per cent.	are common to	Scandinavia.
246	or 61	„	North of Spain.
227	or 56	„	S. Spain and Medit.
97	or 24	„	Canary Islands.

The following are at present peculiar to Britain:—

Assiminea, sp.	Odostomia, 19 sp.?	Montacuta ferruginosa.
Jeffreysia, sp.	Buccinum fusiforme.	Argiope cistellula.
Otina otis.	Fusus Berniciensis.	Pecten niveus.
Rissoa, sp.	„ Turtoni.	Syndosmya tenuis.
Stylifer turtoni.	Natica Kingii.	Thracia villosiuscula.

The most common edible species are:—

Ostrea edulis.	Mytilus edulis.	Fusus antiquus.
Pecten maximus.	Cardium edule.	Litorina litorea.
„ opercularis.	Buccinum undatum.	

Amongst the species characteristic of the Celtic province—or most abundant in it—are the following:—

Trophon muricatus.	Litorina litoralis.	Venus striatula.
Nassa reticulata.	Trochus Montagui.	„ casina.
Natica Montagui.	„ millegranus.	Donax anatinus.
„ monilifera.	„ tumidus.	Solen ensis.
„ nitida.	Patella vulgata.	Pholas candida.
Velutina lævigata.	„ pellucida.	Mactra elliptica.
Turritella communis.	Acmæa virginea.	„ solida.
Aporrhais pes-pelecani.	Chiton cinereus.	Periploma prætenuis.
Rissoa cingillus.	Scaphander lignarius.	Thracia distorta.
Scalaria Trevelyana.	Tellina crassa.	Syndosmya prismatica.

The wide expanse of the Baltic affords no shell-fish unknown to the coasts of Britain and Sweden. The water is brackish, becoming less salt northward, till only estuary shells are met with, and the Litorinæ and Limnæans are found living together, as in many of our own marshes. This scanty list is taken from the Memoirs of Dr. Middendorff and M. Boll.

Buccinum undatum.	Neritina fluviatilis.	Tellina Balthica.
Purpura lapillus.	Limnæa auricularia.	„ tenuis.
Nassa reticulata.	„ ovata.	Scrobicularia piperata.
Litorina litorea.	Mytilus edulis.	Mya arenaria.
Patella (tarentina).	Donax (trunculus).	„ truncata.
Hydrobia muriatica.	Cardium edule var.	

IV. LUSITANIAN PROVINCE.

The shores of the Bay of Biscay, Portugal, the Mediterranean, and N. W. Africa, as far as Cape Juby, form one important province, extending westward in the Atlantic as far as the Gulf weed bank, so as to include Madeira, the Azores, and Canary Islands.*

In the Atlantic portion of the province occur the following genera, not met with in the Celtic and Boreal seas, although two of them, *Mitra* and *Mesalia*, occur on the coast of Greenland.

Argonauta.	Pisania.	Litiopa.	Umbrella.
Philonexis.	Dolium.	Truncatella.	Glaucus.
Chiroteuthis.	Cassis.	Solarium.	——
	Triton.	Bifrontia.	Carinaria.
——	Ranella.	Turbo.	Firola.
Conus.	Cancellaria.	Monodonta.	Atlanta.
Pleurotoma.	Sigaretus.	Haliotis.	Oxygyrus.
Marginella.	Crepidula.	Gadinia.	——
Cymba.	Mesalia.	Siphonaria.	Cleodora.
Mitra.	Vermetus.	Auricula.	Cuvieria.
Terebra.	Fossarus.	Pedipes.	Creseis.
Columbella.	Planaxis.	Ringicula	——

* In the northern part of the Lusitanian province are the Pilchard fisheries; in the Mediterranean, the Tunny, Coral, and Sponge fisheries.

The Gulf-weed banks (represented in the map) extend from 19° to 47° in the middle of the North Atlantic, covering a space almost seven times greater than the area of France. Columbus, who first met with the *sargasso* about one hundred miles west of the Azores, was apprehensive that his ships would run upon a shoal. (*Humboldt.*) The banks are supposed by Prof. E. Forbes to indicate an ancient coast-line of the Lusitanian land-province, on which the weed originated. Dr. Harvey states that species of *Sargassum* abound along the shores of tropical countries, but none exactly correspond with the Gulf-weed (*S. bacciferum*). It never produces fructification—the "berries" being air-vesicles, not fruit—but yet continues to grow and flourish in its present situation, being propagated by breakage. It may be an abnormal condition of *S. vulgare*, similar to the varieties of *Fucus nodosus* (Mackayi) and *F. vesiculosus* which often occur in immense strata; the one on muddy sea-shores, the other in salt marshes, in which situations they have never been found in fructification. (*Manual of British Algæ, Intr.* 16, 17.)

Megerlia.	Chama.	Cardita.	Ervilia.
———	Crassatella.	Cytherea.	Panopæa.
Spondylus.	Lithodomus.	Petricola.	
Avicula.	Ungulina.	Venerupis.	
Solemya.	Galeomma.	Mesodesma.	

Spain and Portugal.

The coast of Spain and Portugal is less known than any other part of the province, but the facilities for exploration are in some respects greater than in the Mediterranean, on account of the tides. Shell-fish are more in demand as an article of food here than with us, and the Lisbon market afforded to Mr. M'Andrew the first indication that the genus *Cymba* ranged so far north.

On the coasts of the Asturias and Gallicia, especially in Vigo Bay, Mr. M'Andrew obtained, by dredging, 212 species, of a somewhat northern character, 50 per cent. of them being common to Norway, and 86 per cent. common to the south of Spain.

On the southern coast of the Peninsula 353 species were obtained, of which only 28 per cent. are common to Norway and 51 per cent. to Britain.

The identical species are chiefly amongst the shells dredged from a considerable depth (35—50 fathoms); the litoral species have a much more distinct aspect.

The shells of the coast of Mogador are generally identical with those of the Mediterranean and Southern Peninsula.

Canary Islands. The shells of the Canaries collected by MM. Webb and Berthelot,[*] and described by M. D'Orbigny, amount to 124, to which Mr. M'Andrew has added above 170. Of the 300 species 17 per cent. are common to Norway, 32 per cent. to Britain, and 63 per cent. to the coasts of Spain and the Mediterranean. Two only are W. Indian shells, *Neritina viridis* and *Columbella cribaria.* Of the African shells found here, and not met with in more northern localities, the most remarkable are :—

Crassatella divaricata.	Ranella lævigata.	Cymba proboscidalis.
Cardium costatum.	Cassis flammea.	Conus betulinus.
Lucina Adansoni.	„ testiculus.	„ Prometheus.
Cerithium nodulosum.	Cymba Neptuni.	,, Guinaicus.
Murex saxatilis.	„ porcina.	„ papilionaceus.

Madeira. Mr. M'Andrew obtained 156 species at Madeira, of which 44 per cent. are British, 70 per cent. common to the Mediterranean, and 83 to the Canaries. Amongst the latter are the two W. Indian shells before mentioned, and the following African shells :—

Pedipes.	Mitra fusca.	Patella crenata.
Litorina striata.	„ zebrina.	„ guttata.
Solarium.	Marginella guancha.	„ Lowei.
Scalaria cochlea.	Cancellaria.	„ Candei.
Natica porcellana.	Monodonta Bertheloti.	Pecten corallinoides.

---

[*] Hist. Naturelle des Iles Canaries ; the list of shells is reprinted with the additions made by Mr. M'Andrew, as one of the Catalogues of the British Museum.

· *Azores.* Amongst the litoral shells which range to the Azores, are Pedi-pes, Litorina striata, Mitra fusca, and Ervilia castanea; the other species obtained there are Lusitanian. (*M'Andrew.*)

The Mediterranean fauna is known by the researches of Poli, Delle Chiaje, Philippi, Verany, Milne-Edwards, Prof. E. Forbes, and Deshayes. In its western part it is identical with that of the adjacent Atlantic coasts; the number of species diminishes eastward, although reinforced by a consider-able number of new forms as yet only known in the Mediterranean; and a few accessions (about 30) of a different character from the Red Sea. The total number of shell-bearing species is estimated at 600, viz.:—

Cephalopoda 1	Nucleobranchiata .. 6	Lamellibranchiata ...200
Pteropoda............ 13	Gasteropoda 370	Brachiopoda 10

On the coast of Sicily, M. Philippi has found altogether 619 marine mollusca, viz.:—

Bivalves 188	Pteropoda..,....... 13	Gasteropoda 319
Brachiopoda 10	Nudibranches...... 54	Cephalopoda 15

Of the 522 which are provided with shells, 162 have not been found fossil, and are presumed to be of post-tertiary origin, so far as concerns their presence in the Medit. The remaining 360 occur fossil in the newer tertiary strata, along with nearly 200 others which are either extinct or not known living on those coasts; a few of them are living in the warmer regions of Senegal, the Red Sea, and the West Indies:—

Senegal.	*Antilles.*	*Red Sea.*
Lucina columbella.	Lucina pennsylvanica.	Argonauta hians.
Cardium hians.	Vermetus intortus.	Dentalium elephantinum.
Terebra fusca.		Terebra duplicata.
	Morocco.	Phorus agglutinans.
	Trochus strigosus.	Niso terebellum.
		Pecten medius.
		Diplodonta apicalis.

Most of them, however, are of northern origin, such as:—

Saxicava rugosa.	Tellina crassa.	Rhynchonella psittacea.
(Panopæa) Norvegica.	Cyprina Islandica.	Patella vulgata.
Mya truncata.	Leda pygmæa.	Eulimella Scillæ.
Periploma prætenuis.	Limopsis pygmæa.	Buccinum undatum
Lutraria solenoides.	Ostrea edulis.	Fusus contrarius.

Of the 522 Sicilian testacea about 35 (including 10 oceanic species) are common to the West Indies—if the species have been correctly determined; 28 are stated, with more probability, to be common to West Africa, including *Murex Brandaris* and other common species; 74, including *Murex trun-culus*, are common to the Red Sea; *Crania ringens* cannot be distinguished from the species found in New South Wales (*Davidson*); and *Columbella corniculum* ranges from the north coast of Spain to Australia, the specimens from these distant localities being only distinguishable as geographical varieties. (*Gaskoin*.) Six other species are included in Menke's Australian Catalogue, but require verification.

The following genera, nine of which are naked molluscs, are supposed to be now peculiar to the Mediterranean ; the small number of species show they are aberrant or expiring forms. *Cassidaria, Terebratula*, and *Thecidium* are ancient, widely-distributed genera, and the Mediterranean *Thecidium* occurs fossil in Brittany and the Canaries.

Histioteuthis, 2 sp.	Lobiger, 1.	Pedicularia, 1.
Verania, 1.	Pleurobranchæa, 1.	Terebratula, 1.
Gastropteron, 1.	Tethys, 1.	Morrisia, 2.
Doridium, 1.	Tiedemannia, 1.	Thecidium, 1.
Icarus, 1.	Cassidaria, 4 ?	Scacchia, 2.

The genera *Fasciolaria, Siliquaria, Tylodina, Notarchus, Verticordia?* *Clavagella*, and *Crania*, occur only in this portion of the Lusitanian province.

Amongst the peculiar species are :—

Nassa semistriata.	Argiope cuneata.	Artemis lupinus.
Fusus crispus.	Clavigella angulata.	Trigona nitidula.
Tylodina Rafinesquii.	Spondylus Gussonii.	Lucinopsis decussata.
Crania rostrata.	Astarte bipartita.	

Ægean Sea. Prof. E. Forbes obtained 450 species of mollusca in the Ægean, belonging to the following orders :—

Cephalopoda	4	Nudibranches	15	Brachiopoda	8
Pteropoda	8	Opisthobranches	28	Lamellibranches	143
Nucleobranches	7	Prosobranches	217	Tunicata	22

Of these 71 were new species, but several have since been found in the Atlantic. and even in Scotland.* The only marine air-breather met with was *Auricula myosotis*.

Black Sea. In the northern part a few Aralo-Caspian shells are found, otherwise the Black Sea only differs from the Mediterranean in the paucity of its species; Dr. Middendorff enumerates 68 only. The water is less salt, and there is no tide, but a current flows constantly through the Dardanelles to the Mediterranean.†

V. ARALO-CASPIAN PROVINCE.

The only inland salt-seas that contain peculiar shell-fish are the Aral and Caspian. The shells chiefly consist of a remarkable group of Cockles which burrow in the mud (see fig. 213, p. 291). No explorations have been made with the dredge, but other species, probably still existing in these seas, have been found in the beds of horizontal limestone which form their banks and extend in all directions far over the *steppes*. This limestone is of brackish-

---

* Trans. Brit. Assoc. (for 1843) 1844, p, 130.

+ A current from the Atlantic sets in perpetually through the Straits of Gibraltar, and there is scarcely any tide; it only amounts to 1 foot at Naples and the Euripus, 2 feet at Messina, and 5 at Venice and the Bay of Tunis.

water origin, being sometimes composed of myriads of *Cyclades*, or the shells of *Dreissena* and *Cardium*, as in the islets near Astrakhan. It is believed to indicate the former existence of a great inland sea, of which the Aral and Caspian are remnants, but which was larger than the present Mediterranean at an age previous to that of the Mammoth and Siberian Rhinoceros. The present level of the Caspian is 83 feet below that of the Black Sea; that of the Aral has been stated to be 117 feet higher than the Caspian, but is probably not very different; their waters are only brackish, and in some parts drinkable. The steppe limestone rises to a level of 200—300 feet above the Caspian; it spreads eastward to the mountains of the Hindoo Kush and Chinese Tartary, southward over Daghestan and the low region E. of Tiflis, and westward to the northern shores of the Black Sea. The extent to which it has been traced is represented by oblique lines on the map.* Some of the Caspian shells still exist in the Sea of Azof and the estuaries of the Dnieper and Dniester. Our information upon this seldom-visited region is derived from the works of Pallas, Eichwald,† Krynicki,‡ Middendorff, and Sir Roderick Murchison.

Aralo-Caspian Shells.

A, Aral; C, Caspian; B, Black Sea.

The Species marked * are found also in the steppe limestone.

*Cardium edule, L. C. (very small) B. Baltic.
　　„　edule, var. (rusticum, Chemn.) A. C. B. Icy Sea.
*Didacna trigonoides, Pal. C (Azof. M. Hommaire).
　　„　Eichwaldi, Kryn. (crassa, Eich.) C. B. (Nikolaieff).
Monodacna Caspia, Eich. C.
　　„　pseudo-cardium, Desh. (pontica, Eich.) B.
Adacna læviuscula, Eich. C.
　　„　vitrea, Eich. C. A.
*　„　edentula, Pallas. C.
　　„　plicata, Eich. C. B. (Dniester, Akerman, Odessa).
　　„　colorata, Eich. C. B. (Azof, Dnieper).
*Mytilus edulis, L. C. B. (not in Middendorff's list.)
　　„　latus, Chemn. B.
*Dreissena polymorpha, Pal. C. B.

Paludinella stagnalis, L. (pusilla Eich.) C. B. (Odessa). Ochotsk.
*　„　variabilis, Eich. C.
*Neritina liturata, Eich. C. on sea weed.
*Rissoa Caspia, Eich. C.
　　„　oblonga, Desm. B.
　　„　cylindracea, Kryn. B.§

* From a sketch kindly prepared by Professor Ramsay.

† Geogr. des Kaspischen Meeres, des Kaukasus und des Südlichen Russlands Berlin, 1838. Fauna Caspio-Caucasica, 1841.

‡ Bull. des Nat. Moscow, 1837.

§ The *Velutina (Limneria) Caspiensis*, A. Ad. was founded on a specimen of *Limnæa Gebleri, Midd.* (1851) from Bernaoul, Siberia.

The following species are described by Eichwald, from the steppe lime-stone. (Murchison, Russia, p. 297.)

"Paludina" Triton.	————	Donax priscus.
„ exigua.	Mactra Caspia.	Monodacna propinqua.
Rissoa conus.	„ Karagana.	„ intermedia.
„ dimidiatus.	Cyclas Ustuertensis.	„ Catillus.
Bullina Ustuertensis.	Mytilus rostriformis.	Adacna prostrata.

No other inland bodies of salt water are known to have peculiar marine shells; those of the modern deposits, in Mesopotamia (at Sinkra and Warka), collected by Mr. W. K. Loftus, are species still abounding in the Persian Gulf. *

VI. West African Province.

The tropical coast of Western Africa is rich in conchological treasures, and far from being wholly explored. The researches of Adanson,† Cranch (the naturalist to the Congo expedition‡), and the officers of the Niger expedition, have left much to be done. Dr. Dunker has described 149 species in his *Index Moll. Guineæ, coll. Tams.* Cassel, 1853.

At *St. Helena*, Mr. Cuming collected 16 species of sea-shells, 7 of them new. *Litorina Helenæ* is found on the shore of St. Helena, and *L. miliaris* and *Nerita Ascensionis*, at Ascension.

West African Shells.

Onychoteuthis, 3 sp.	Lagena nassa.	Cypræa picta.
Cranchia, 2 sp.	Terebra striatula.	Vermetus lumbricalis.
Strombus rosaceus.	„ ferruginea.	Cerithium Adansonii.
Triton ficoides.	? Halia priamus.	Turritella torulosa.
Ranella quercina.	Mitra nigra.	Mesalia.
Dolium tessellatum.	Cymba.	Litorina punctata.
Harpa rosea.	Marginella.	Collonia.
Oliva hiatula.	Persicula.	Clanculus villanus.
Pusionella.	Pleurotoma mitriformis.	Haliotis virginea.
Nassa Pfeifferi.	Tomella lineata.	„ coccinea.
Desmoulinsia.	Clavatula mitra.	Nerita Senegalensis.
Purpura nodosa.	„ coronata.	„ Ascensionis.
Rapana bezoar.	„ bimarginata.	Pecten gibbus.
Murex vitulinus.	„ virginea.	Arca ventricosa.
„ angularis.	Conus papilionaceus.	„ senilis.
„ megaceros.	„ genuinus.	Cardium ringens.
„ rosarius.	„ testudinarius.	„ costatum.
„ duplex.	„ achatinus.	Lucina columbella.
„ cornutus.	„ monachus.	Ungulina rubra.
Clavella ? filosa.	Natica fulminea.	Diplodonta rosea.
„ afra.	Cypræa stercoraria.	Cardita ajar.

* A species of coral (*Porites elongata*, Lam.) now living at the Seychelles has been said to be found in the Dead Sea. (v. Humboldt's Views of Nature, Bohn, ed. p. 260.)

† Hist. Nat. de Senegal, 4to. Paris, 1757. This able but eccentric naturalist destroyed the utility of his own writings by refusing to adopt the bi-nomial nomenclature of LINNÆUS, and employing instead the most barbarous chance-combinations of letters he could invent.

‡ Appendix to Capt. Tuckey's Narrative (1818), by Dr. Leach.

Artemis africana.	Cytherea africana.	Mactra rugosa.
,, torrida.	Venus plicata.	,, nitida.
Cyclina Adansonii.	Tellina.	Pholas clausa.
Trigona bicolor.	Strigilla Senegalensis.	Tugonia anatina.
,, tripla.	Gastrana polygona.	———
Cytherea tumens.	Mactra depressa.	Discina radiosa.

VII. SOUTH AFRICAN PROVINCE.

The fauna of South Africa, beyond the tropic, possesses few characters in common with that of the western coast, and is more like the Indian Ocean fauna, as might be expected from the direction of the currents. But, together with these it has a large assemblage of marine animals found nowhere else, and the "Cape of Storms" forms a barrier between the populations of the two great oceans, scarcely less complete than the far-projecting promontory of South America. The coast is generally rocky, and there are no coral-reefs; accumulations of sand are frequent, and sometimes very extensive, like the Agulhas Bank. The few deep sea-shells which have been obtained off these banks possess considerable interest, but explorations in boats are said to be difficult, and often impossible on account of the surf. Shells from the Cape are too frequently dead and water-worn specimens picked up on the beach. The shell-fish of South Africa have been collected and described by Owen Stanley, Hinds, A. Adams, and, especially, by Dr Krauss, who has published a very complete monograph.* Of 400 sea-shells recorded in this work, above 200 are peculiar, and most of these belong to a few litoral genera. Only 11 species are common to the coast of Senegal, whilst 18 are found in the Red Sea.

South African Shells.

Panopæa natalensis.	Chiton, 16 sp.	Pleurotoma, 6 sp.
Solen marginatus.	Patella, 20 sp.	Clionella (sinuata).
Mactra spengleri.	,, cochlea.	Typhis arcuatus.
Gastrana ventricosa.	,, compressa.	Triton dolarius
Nucula pulchra, Hinds.	,, apicina.	,, fictilis, 50-60 fm.
(L'Agulhas bank, 70 fm.)	,, longicosta, &c.	Harpa crassa.
Pectunculus Belcheri, 120	Helcion pectinata.	Cominella ligata.
fm.	Siphonaria, 5 sp.	,, lagenaria.
Modiola Capensis.	Pupillia (aperta).	,, limbosa.
,, pelagica, Forbes.	Fissurella, 10 sp.	,, tigrina.
Septifer Kraussi.	Crepidula, 4 sp.	Bullia lævissima.
———	Haliotis sanguinea.	,, achatina.
Terebratulina abyssicola,	Delphinula granulosa.	,, natalensis.
132 fm.	,, cancellata.	Nassa plicosa.
Terebratella (Kraussia),	Trochus, 22 sp.	,, capensis.
,, rubra.	Turbo sarmaticus.	Cyclonassa Kraussi.
,, cognata.	Litorina Africana (7 sp.)	Eburna papillaris.
,, pisum.	Phasianella, 6 sp.	Columbella, 5 sp.
,, Deshayesii, 120 fm.	Bankivia varians.	Ancillaria obtusa.
———	Turritella, 4 sp.	Mitra, 5 sp.

* Die Südafrikanischen Mollusken, 4to. Stutt. 1848.

Imbricaria carbonacea.	Trivia ovulata.	————
Voluta armata.	Cypræa, 22 sp.	Octopus argus.
„ scapha.	Luponia algoënsis.	Sepia, 4 sp.
„ abyssicola, 132 fm.	Cyprovulum (capense).	
Marginella rosea.	Conus, 8 sp.	

The following are stated to be common to the Cape and European seas.*

Saxicava (arctica?) Greenland, Medit.
. Tellina fabula, Brit. Medit.
Lucina lactea, Medit. Red Sea.
„ fragilis, Medit.
Venus verrucosa, W. Indies? Brit. Se-
negal, Canaries, Red Sea, Australia?
Tapes pullastra, North Sea.
„ geographica, Medit.
Arca lactea, Medit.

Chama gryphoides, Medit. Red Sea.
Pecten pusio, Brit.
Diphyllidia (lineata?) N. Brit. Medit.
Eulima nitida, Medit.
Purpura lapillus ?? (not in Medit.).
Nassa marginulata.
Octopus vulgaris? Brit.
Argonauta argo, Medit.

VIII. Indo-Pacific Province.

This is by far the most extensive area over which similar shell-fish and other marine animals are distributed. It extends from Australia to Japan, and from the Red Sea and east coast of Africa to Easter Island in the Pacific, embracing three-fifths of the circumference of the globe and 45° of latitude. This great region might indeed be subdivided into a number of smaller provinces, each having a particular association of species, and some peculiar shells; such as the Red Sea, the Persian Gulf, Madagascar, &c.; but a considerable number of species are found throughout the province, and their general character is the same.† Mr. Cuming obtained more than 100 species of shells from the eastern coast of Africa, identical with those collected by himself at the Philippines, and in the eastern coral islands of the Pacific.‡ This is pre-eminently the region of coral reefs, and of such shell-fish as affect their shelter. The number of species inhabiting it must amount to several thousands. The Philippine Islands have afforded the greatest variety, but their apparent superiority is due, in a measure, to the researches of Mr. Cuming; no other portion of the province has been so thoroughly explored.§

Amongst the genera most characteristic of the Indo-Pacific, those marked (*) are wholly wanting on the coasts of the Atlantic, but half of them occur fossil in the older tertiaries of Europe. Those in italics are also found on the west coast of America.

---

* Marks of doubt are added to some of the species, and other are quite omitted.

† See Mrs. Somerville's Physical Geography, II. p. 233.

‡ Journal Geol. Soc. 1846, vol. II. p. 268.

§ Mr. Cuming collected 2500 species of sea-shells at the Philippines, and estimates the total number at a thousand more. The genera most developed are *Conus* 120 sp., *Pleurotoma* 100, *Mitra* 250, *Columbella* 40, *Cypræa* 50, *Natica* 50, *Chiton* 30, *Tellina* 50.

*Nautilus.	*Magilus.	Stomatella.	Hemicardium.
*Pterocera	*Melo.	Gena.	*Cypricardia.
*Rimella	Mitra.	*Broderipia.	*Cardilia.
*Rostellaria.	*Cylindra.	*Rimula.	*Verticordia.
*Seraphs.	*Imbricaria.	*Neritopsis.	*Pythina.
Conus.	Ovulum.	*Scutellina.	Circe.
Pleurotoma.	*Pyrula (type).	*Linteria.	*Clementia.
*Cithara.	*Monoptygma.	*Dolabella.	*Glaucomya.
*Clavella.	Phorus,	*Hemipecten.	*Meröe.
*Turbinella (typ.)	Siliquaria.	*Placuna.	Anatinella.
Cyllene.	*Quoyia.	*Malleus.	Cultellus.
Eburna.	*Tectaria.	*Vulsella.	*Anatina.
Phos.	Imperator.	*Pedum.	*Chæna.
Dolium.	Monodonta.	*Septifer.	*Aspergillum.
Harpa.	Delphinula.	*Cucullæa,	*Jouannetia.
*Ancillaria.	Liotia.	*Hippopus.	*Lingula.
*Ricinula.	*Stomatia.	*Tridacna.	Discina.

The strictly litoral species vary on each great line of coast : for example, *Litorina intermedia* and *Tectaria pagodus* occur on the east coast of Africa; *Litorina conica* and *melanostoma*, in the Bay of Bengal; *Litorina sinensis* and *castanea*, and *Haliotis venusta*, on the coast of China; *Litorina scabra* and *H. squamata*, in N. Australia; *H. asinina*, New Guinea; and *L. picta*. at the Sandwich Islands.

Red Sea (Erythræan).

Of the 408 mollusca of the Red Sea, collected by Ehrenberg and Hemprich, 74 are common to the Medit. from which it would seem that these seas have communicated since the first appearance of some existing shells. Of the species common to the two seas 40 are Atlantic shells which have migrated into the Red Sea by way of the Medit. probably during the newer pliocene period ; the others are Indo-Pacific shells which extended their range to the Mediterranean at an earlier age.

The genera wanting in the Medit. but existing in the Red Sea, show most strikingly their diversity of character, and the affinity of the latter to the Indian fauna.

Pterocera.	Ancillaria.	Siphonaria.	Limopsis.
Strombus, 8 sp.	Harpa.	Placuna.	Tridacna
Rostellaria.	Ricinula.	Plicatula.	Crassatella.
Turbinella.	Magilus.	Pedum.	Trigona.
Terebra.	Pyramidella.	Malleus.	Sanguinolaria.
Eburna.	Parmophorus.	Vulsella.	Anatina.
Oliva.	Nerita.	Perna.	Aspergillum.

Other genera become abundant, such as *Conus*, of which there are 19 species in the Red Sea, *Cypræa* 16, *Mitra* 10, *Cerithium* 17, *Pinna* 10, *Chama* 5, *Circe* 10.

Persian Gulf.

The marine zoology of the Persian Gulf and adjoining coast has not been yet explored, although the E. India Company maintains a squadron of five or six ships constantly cruising in the Gulf.* The following shells were picked up on the beach at Kurachee by Major Baker, with many others evidently new, but not in a satisfactory state for description. (1850.)

Rostellaria curta.
Murex tenuispina var.
Pisania spiralis.
Ranella tuberculata.
,,   spinosa.
,,   crumena.
Triton lampas.
Bullia, n. sp.
Eburna spirata.
Purpura persica.
,,   carinifera.
Columbella blanda.
Oliva subulata.
,,   Indusica.
,,   ancillaroides.
Cypræa Lamarckii.
,,   ocellata.
Natica pellis-tigrina.
Sigaretus sp.
Odostomia sp.
Phorus corrugatus.
Planaxis sulcata.
Imperator Sauliæ.
Monodonta sp,
Haliotis sp.
Stomatella imbricata.
,,   sulcifera.
Fissurella Ruppellii.
,,   Indusica.
,,   salebrosa.
,,   dactylosa.

Fissurella funiculata.
Pileopsis tricarinatus.
Nerita ustulata.
Dentalium octangulatum.
Ringicula sp.
Bulla ampulla.
Anomia achæus.
,,   enigmatica.
Pecten sp.
Spondylus sp.
Plicatula depressa.
Mytilus canaliculatus.
Arca obliqua.
,,   sculptilis, &c.
Chama sp.
Lucina sp.
Cardium fimbriatum.
,,   latum.
,,   impolitum.
,,   pallidum.
,,   assimile.
Venus pinguis.
,,   cor.
,,   purpurata.
Meroë Solandri.
,,   effossa.
Trigona trigonella?
Artemis angulosa. |
,,   exasperata.
,,   subrosea ?
Venerupis sp.

Petricola sp.
Tapes sulcosa.
,,   Malabarica.
Cypricardia vellicata.
Cardita crassicostata ?
,,   calyculata.
,,   Tankervillii.
Mactra Ægyptica, &c.
Tellina angulata.
,,   capsoides.
Mesodesma Horsfieldii.
Psammobia sp.
Syndosmya sp.
Semele sp.
Solen sp.
Solecurtus politus.
Donax scortum.
,,   scalpellum.
Sanguinolaria diphos.
,,   violacea.
,,   sinuata.
Corbula sp.
Diplodonta sp.
Anatina rostrata.
Pandora sp.
Martesia sp.
Pholas australis.
,,   Bakeri, Desh.
,,   orientalis.
(Meleagrina v. p. 261.)

At the *Cargados* or St. Brandon shoals, north of Mauritius, *Voluta costata, Conus verrucosus, Pleurotoma virgo,* and *Turbinella Belcheri* have been obtained by dredging.

IX. Australo-Zelandic Province.

Most remote from the Celtic seas, this province is also most unlike them in its fauna, containing many genera wholly unknown in Europe, either living or fossil, and some which occur fossil in rocks of a remote period. The province includes New Zealand, Tasmania, and extra-tropical Australia, from

---

* The "Brindled Cowry," (*Cypræa princeps*) from the Persian Gulf, was valued at £50; the only known specimen is in the British Museum.

Sandy Cape on the east, to the Swan River. The shells, which are nearly all peculiar, have been catalogued by Gray,[*] Menke,[†] and Forbes.[‡] Of the following genera some are peculiar (*), others attain here their greatest development:—

*Pinnoctopus.	*Macgillivraia.	Cypricardia.	Imperator.
*Struthiolaria.	*Amphibola.	Mesodesma.	Monoptygma.
Phasianella	*Trigonia.	Terebratella.	Siphonaria.
Elenchus.	*Chamostrea.	Spirula.	Pandora.
Bankivia.	*Myadora.	Oliva.	Anatinella.
Rotella.	*Myochama.	Conus.	Clavagella.
*Macroschisma.	Crassatella.	Voluta.	Placunomia.
Parmophorus.	Cardita.	Terebra.	Waldheimia.
Risella.	Circe.	Fasciolaria.	Crania.

Some of the genera of this province are only met with elsewhere at a considerable distance:—

Solenella—Chile.	Bankivia—Cape.	Rhynchonella—Arctic seas.
Panopæa—Japan.	Kraussia—Cape.	Trophon—Fuegia; ,,
Monoceros—Patagonia.	Solemya—Medit.	Assiminea—India; Brit.

Amongst the litoral shells of South Australia are *Haliotis elegans, H. rubicunda*, and *Litorina rugosa. Haliotis iris* and *Litorina squalida* are found on the shores of N. Zealand; and *Cyprovula umbilicata* in Tasmania.

Mr. Gray's New Zealand list amounts to 104 marine species, among which are three volutes, including *V. magnifica*, the largest of its genus; *Strombus troglodytes, Ranella argus*, the great *Triton variegatus;* 6 Cones, (all doubtful), *Oliva erythrostoma, Cypræa caput-serpentis, Ancillaria australis, Imperator heliotropium, Chiton monticularis*, &c.

Venus Stutchburyi and *Modiolarca trapezina* have been found at Kerguelen's Id. and *Patella illuminata* at the Auckland Ids.

X. Japonic Province.

The Japanese Islands and Corea represent the Lusitanian province. A few shells were collected here by Mr. A. Adams, but they are chiefly known through the Dutch dealers.[§] *The Astarte Japonica* of the Catalogues is nothing more than *A. borealis*, and is stated to have come from Lapland by Jay and Cuming. *Panopæa Japonica* belongs to the same type with *P. intermedia* of the London Clay.

---

* Travels in New Zealand, by Dr. E. Dieffenbach. 8vo, London, 1843.

† Moll. Nov. Hollandiæ, 1843.

‡ Narrative of the Voyage of H.M.S. Rattlesnake, 1846-50, by J. Macgillivray. Supplement by Prof. E. Forbes.

§ For many years the Dutch have been allowed to send one ship annually to Japan for trade, whilst all other nations have been excluded; a state of things which the Americans will perhaps alter. The work of Siebold, on the Natural History of Japan, does not contain any account of the shells.

Octopus areolatus.
Sepia chrysopthalma.
Sepiola Japonica.
———
Conus Sieboldi.
Pleurotoma Coreanica.
Terebra serotina.
„ stylata.
Eburna Japonica.
Cassis Japonica.
Murex eurypterus.
,: rorifluus.
„ plorator.
„ Burneti.
Cancellaria nodulifera.
Mitra.
Strombus corrugatus.
Cypræa fimbriata.

Cypræa miliaris.
Radius birostris.
Cerithium longicaudatum.
Imperator Guilfordiæ.
Haliotis Japonica.
„ discus.
„ gigantea.
Bulla Coreanica.
Siphonaria Coreanica.
Pecten asperulatus
„ Japonicus.
Spondylus Cumingii.
Nucula mirabilis.
„ Japonica.
Cardium Bechei.
Crassatella compressa.
Diplodonta alata.
„ Coreanica.

Isocardia Moltkiana.
Venus Japonica.
Cyclina orientalis.
Cytherea petechialis.
Artemis sericea.
„ bilunata.
„ Sieboldi.
„ Japonica.
Circe Stutzeri.
Tapes Japonica.
Petricola radiata.
Solen albidus.
Panopæa Japonica.
Terebratulina Japonica.
„ angusta.
Waldheimia Grayi.
Terebratella Coreanica.
„ rubella.

XI. ALEUTIAN PROVINCE.

The Boreal province is represented on the northern coasts of the Pacific, where, according to Dr. Middendorff, the same genera and many identical species are found. In addition to those indicated in the Arctic list (p. 355), the following species occur at the Shantar Ids. in the Sea of Ochotsk (O), Saghalien, the Kuriles (K), Aleutians and Sitka (S).

Patella (scurra). S.
Acmæa, 3 sp. S.
Pilidium commodum. O.
Paludinella, 3 sp. O.
Litorina, 6 sp. O. K. S.
Turritella Eschrichtii. S.
Margarita sulcata. A.
Trochus, 6 sp. S.
Scalaria Ochotensis.
Crepidula Sitchana.
„ minuta. S.
„ grandis. A.
Fissurella violacea. S.
„ aspera. S.
Haliotis Kamtschatica.
„ aquatilis. K.
Velutina coriacea. K.
„ cryptospira. O.
Trichotropis inermis. S.
Purpura decemcostata. (Mid.) S.
„ Freycineti. O. S.
„ septentrionalis. S.
Pleurotoma Schantarica.
„ simplex. O.
Murex monodon. S.
„ lactuca. S.

Fusus (Chrysodomus) Sitchensis.
„ decemcostatus. A.
„ Schantaricus.
„ Behringii.
„ Baerii. A.
„ luridus. S.
Buccinum undatum var. Schantaricum.
„ simplex. O.
„ Ochotense.
„ cancellatum. A
„ ovoides. O.
Pisania scabra. A.
Bullia ampullacea. O.
Onychoteuthis Kamtschatica.

Terebratella frontalis. O.
Placunomia macroschisma. O.
Pecten rubidus. S.
Crenella vernicosa. O.
„ cultellus. Kamts.
Nucula castrensis. S.
Pectunculus septentrionalis. A.
Cardita borealis O.
Cardium Nuttalli. S.
„ Californicum. S.
Saxidomus Petiti. S.

Saxidomus giganteus. S.	Tellina lutea. A. nasuta. S.
Petricola cylindracea. S.	„ edentula. A.
„ gibba. S.	Lutraria maxima. S.

The influence of the Asiatic coast-current is shewn in the presence of two species of *Haliotis*, whilst affinity with the fauna of W. America is strongly indicated by the occurrence of *Patella* (*scurra*), three species of *Crepidula*, two of *Fissurella*, and species of *Bullia, Placunomia, Cardita, Saxidomus*, and *Petricola*, which are more abundant, and range farther north than their allies in the Atlantic.

Provinces on the Western coast of America.

The mollusca of the Western coast of America are equally distinct from those of the Atlantic and those inhabiting the central parts of the Pacific.

Mr. Darwin states in his Journal (p. 391) that "not one single sea-shell is known to be common to the Islands of the Pacific and to the west coast of America," and he adds that "after the comparison by Messrs. Cuming and Hinds of about 2000 shells from the Eastern and Western coasts of America, only one single shell was found in common, namely the *Purpura patula*, which inhabits the West Indies, the coast of Panama, and the Gallapagos." Even this single identification has since been doubted. Mr. Cuming, who resided many years at Valparaiso, did not discover any West India species on that coast, and M. D'Orbigny makes the same observation. On the other hand M. Mörch of Copenhagen says he has received *Tellina operculata* and *Mactra alata* from the west coast and also from Brazil; and M. Deshayes gives the following extraordinary ranges in his "Catalogue of *Veneridæ* in the British Museum:"

> Artemis angulosa, Philippines—Chile.
> Cytherea umbonella, Red Sea—Brazil.
> „ maculata, W. Indies—Philippines, Sandwich.
> „ circinata, West Indies—West coast America.

In these instances there is doubtless some mistake, either about the locality or the shell. As regards the last, Mr. Carrick Moore has shown that the error has arisen from confounding the *Cytherea alternata* of Broderip with *C. circinata* of Born. M. D'Orbigny collected 628 species on the coast of S. America,—180 from the eastern side, and 447 from the Pacific coast, besides the *Siphonaria Lessonii* which ranges from Valparaiso in Chile to Maldonado on the coast of Uruguay.* These shells belong to 110 genera, of which 55 are common to both coasts, while 34 are peculiar to the Pacific, and 21 to the Atlantic side of S. America; an extraordinary amount of diversity, attributable partly to the different character of the two coasts—the

---

* The dispersion of this coast shell may perhaps have taken place at the time when the channel of the river S. Cruz formed a strait, joining the Atlantic and Pacific oceans, like that of Magellan. (Darwin, p. 181.) Mr. Couthouy makes 3 sp. *S. Lessonii*, nearly smooth, Atlantic coast; *S. antarctica*, ribbed. Pacific coast; and *S. lateralis*, thin, oblique, Fuegia.

eastern low, sandy or muddy, the western rocky, with deep water near the shore.*

The comparison of the shells of Eastern and Western America is of considerable interest to geologists; for if is true that any number of living species are common to the Pacific and Atlantic shores, it becomes probable that some portion of the Isthmus of Darien has been submerged *since* the Eocene Tertiary period. Any opening in this barrier would allow the Equatorial current to pass through into the Pacific—there would be no more Gulf Stream—and the climate of Britain might from this cause alone, become like that of Newfoundland at the present day.

XII. CALIFORNIAN PROVINCE.

The shells of Oregon and California have been collected and described by Mr. Hinds,† Mr. Nuttall‡, and Mr. Couthouy, naturalist of the American Exploring Expedition.§

Shells common to U. California and Sitka. (Middendorff.)

Tritonium scabrum.	Fissurella aspera.	Trochus euryomphalus.
Litorina modesta.	Trochus ater.	Petricola cylindracea.
,, aspera.	,, mœstus.	Lutraria maxima.
Fissurella violacea.	,, Fokkesii.	

Scarcely any species are common to this province (extending from Puget Sound to the peninsula) and the Bay of California, which belongs to the Panamic province. The following list probably contains some shells which should be referred to the latter.

Fusus Oregonensis.	Dentalium politum.	Cardita ventricosa.
Murex Nuttalli.	Patella, 15 sp.	Cardium, 4. Lucina, 3.
Monoceros unicarinatus.	Acmæa scabra.	Cypricardia Californica.
,, punctatus.	,, pintadina.	Chironia Laperousii.
Cancellaria urceolata.	Chiton Mertensii.	Solecardia eburnea.
Trivia Californica.	,, scrobiculatus, &c.	Venus Californiensis.
Natica herculea.	Cleodora exacuta.	,, callosa.
,, Lewisii.	————	Artemis ponderosa
Calyptræa fastigiata.	Waldheimia Californica.	Saxidomus Petiti.
Crepidula exuviata.	Discina Evansii.	,, Nuttalli.
,, navicelloides.	————	,, giganteus.
,, solida, &c.	Anomia pernoides.	Venerupis cordieri.
Imperator Buschii.	Placunomia cepa.	Petricola mirabilis.
Haliotis Cracherodii.	Hinnites giganteus.	Mactra, 2. Donax, 1.
,, fulgens.	Perna, 1. Pinna, 2.	Tellina Bodegensis.
,, corrugata.	Mytilus, 1, Pecten 2.	,, secta, &c.
Fissurella crenulata.	Mytilimeria Nuttalli.	Semele decisa.
,, cucullata.	Modiola capax.	Cumingia californica.
Puncturella, 2 sp.	Chama exogyra.	Sanguinolaria Nuttalli.

* Voyage dans l'Amérique Méridionale. 1847, t. v. p. v.

† Voyge of H. M. S. Sulphur; Zoology by R. B. Hinds, 4to. 1844.

‡ Described by T. A. Conrad, Journ. Acad. N. S. Philadelphia, 1834.

§ Gould in Bost. Nat. Hist. Soc. Proceedings, 1846 ; and U. S. Exploring Exped. (Commander Wilkes) vol. xii. Mollusca, with Atlas. 4to. Philad. 1852.

Lutraria Nuttalli.	Cyathodonta undulata.	Machaera maxima.
Platyodon cancellatus.	Sphenia californica.	Mya præcisa.
AmphichænaKindermanni.	Periploma argentaria.	Panopæa generosa.
Lyonsia, 1. Thracia, 1.	Solecurtus subteres.	Pholas Californica.
Pandora, 1. Saxicava, 2.	Machaera lucida.	,, concamerata.

XIII. PANAMIC PROVINCE.

The Western coast of America, from the Gulf of California to Payta in Peru, forms one of the largest and most distinct provinces. The shells of Mazatlan and the Gulf have been imperfectly catalogued by Menke and are now under examination by Mr. P. Carpenter, who states that they amount to about 500 species, of which perhaps half are common to Panama and Peru ; a *very few* are common to the west coast of the Promontory and very few (including *Purpura patula* and *Mactra similis*) to the West Indies; still fewer to the Pacific coasts and islands, and one or two identical or closely analogous with Senegambian and British species, (e. g. *Kelliä suborbicularis*.)

The late Prof. C. B. Adams of Amherst published, in 1852, a very valuable work on the shells of Panama, in which the total number of species found in the province is estimated at 1500, of which " perhaps none exist beyond— all of the few examples which are supposed to have a wider range, are more or less doubtful." He remarks that " in general there is a great dissimilarity between the shells of this and the Caribbean Province" in which he had himself collected extensively; the number of large species was much greater in Panama.*

The river-openings of this coast are bordered by mangroves, amongst which are found *Potamides*, Arcas, Cyrenas, Potamomyas, Auriculas and Purpuras, whilst *Litorinæ* climb the trees and are found upon their leaves. The ordinary tide at Panama amounts to 16 or 20 feet, the extreme to 28 feet, so that once a fortnight a lower zone of beach may be examined and other shells collected; the beach is of fine sand, with reefs of rocks in the bay.

Gallapagos Islands.—Out of 90 sea-shells collected here by Mr. Cuming 47 are unknown elsewhere ; 25 inhabit Western America, and of these 8 are distinguishable as varieties ; the remaining 18 (including one variety) were found by Mr. Cuming in the Low Archipelago, and some of them also at the Philippines. (*Darwin*, p. 391.)

Litoral shells common to Panama and the Gallapagos (C. B. Adams.)

Cypræa rubescens	Columbella atramentaria.	Ricinula reeviana.
Mitra tristis.	,, bicanalifera.	Cassis coarctata.
Planaxis planicostatus.	,, hæmastoma.	Oniscia tuberculosa.
Purpura carolinensis.	Columbella nigricans.	Conus brunneus.

* Mr. Adams found but one shell common to the two sides of the Isthmus— *Crepidula unguiformis*—wich is said to be found throughout the warmer latitudes, but is really an abnormal form of many distinct species of Crepidula, caused by growing in the interior of other shells.

Conus nux.
Strombus granulatus.
Turbinella cerata.

Pleurotoma eccentrica.
Hipponyx radiata.
Fissurella macrotrema.

Fissurella nigro-punctata.
Siphonaria gigas.

Panama shells.

Strombus gracilior.
Murex erythrostomus.
  ,, regius.
  ,, imperialis.
     radix.
  ,, brassica.
  ,, monoceros, &c.
Rapana muricata.
  ,, Kiosquiformis.
Myristica patula.
Ricinula clathrata.
Purpura, many sp.
Monoceros, many sp.
  ,, brevidentatus.
  ,, cingulatus.
Clavella ? distorta.
Oliva porphyria.
  ,, splendidula, &c.
Northia pristis.
Harpa crenata.
Malea ringens.
Mitra Inca, &c.
Terebra luctuosa, &c.
Conus regularis, &c.
Pleurotoma, many sp.
Cancellaria goniostoma.
  ,, cassidiformis.
  ,, chrysostoma.
Columbella, many sp.

Columbella strombiformis.
Marginella curta.
Cypræa nigro-punctata.
Trivia.
Pyrula ventricosa.
Natica glauca.
Pileopsis hungaricoides.
Crucibulum auriculatum,
    &c.
Trochita mamillaris.
Crepidula arcuata &c.
Litorina pulchra.
Turritella Californica.
Truncatella, 2 sp.
Cœcum, 8 sp.
Imperator unguis, &c.
Trochus pellis serpentis.
Vitrinella, 12 sp.
Nerita ornata.
Patella maxima.
——
Discina strigata.
  ,, Cumingii.
Lingula semen.
  ,, albida.
  ,, audebardi.
——
Placunomia foliacea.
Ostrea æquatorialis.

Spondylus princeps.
Pecten magnificus.
Arca lithodomus, &c.
Pectunculus tessellatus, &c.
Nucula exigua.
Leda, 5 sp.
Cardium senticosum,
  ,, maculosum.
Cardita laticosta.
Gouldia Pacifica.
Cytherea, many sp.
Venus gnidia.
  ,, histrionica.
Artemis Dunkeri.
Trigona crassatelloides.
Cyclina subquadrata.]
Venerupis foliacea.
Petricola californica, &c.
Tellina Burneti.
Cumingia coarctata.
Semele, 7 sp.
Saxicava purpurascens.
Gastrochæna.
Solecurtus lucidus.
Lyonsia brevifrons.
Pandora arcuata, &c.
Pholas melanura, &c.
Parapholas.
Jouannetia pectinata.

XIV. PERUVIAN PROVINCE.

The coast of Peru and Chile, from Callao to Valparaiso, affords a large and characteristic assemblage of shells, of which only a small part have been catalogued, although the district has been well explored, especially by D'Orbigny, Cuming and Philippi. M. D'Orbigny collected 160 species, one half of which are common to Peru and Chile, whilst only one species found at Callao was also met with at Payta, a little beyond the boundary of the region. Mr. Cuming obtained 222 species on the coast of Peru, and 172 in Chile. The Island of Juan Fernandez is included within this province. Only a few of the Peruvian shell-fish can be here enumerated.

Onychoteuthis peratop-
  tera.
——
Æolis Inca.
Doris Peruviana.
Diphyllidia Cuvieri.

Posterobranchæa.
Aplysia Inca.
Tornatella venusta.
Chiton, many species.
Patella scurra.
Acmæa scutum.

Crucibulum lignarium.
Trochita radians.
Crepidula dilatata.
Fissurella, many sp.
Liotia Cobijensis.
Gadinia Peruviana.

Litorina Peruviana.
„ araucana.
Rissoina Inca.
Cancellaria buccinoides.
Sigaretus cymba.
Fusus Fontainei.
Murex horridus.
Ranella ventricosa.
Triton scaber.
Nassa dentifera.
Columbella sordida.
Oliva Peruviana.
Rapana labiosa.
Monoceros giganteus.
„ crassilabris.

Monoceros acuminatus.
Purpura chocolata.
Concholepas.
Mitra maura.

———

Terebratella Fontainei.
„ Chilensis.
Discina lamellosa.
„ lævis.

———

Pholas subtruncata, &c.
Lyonsia cuneata.
Solen gladiolus.
Solecurtus Dombeyi.
Mactra Byronensis.

Mesodesma Chilensis.
Cumingia lamellosa.
Semele rosea, &c.
Petricola, many sp.
Saxidomus opacus, &c.
Cyclina Kroyeri.
Venus thaca.
Crassatella gibbosa.
Nucula, many sp.
Leda, many sp.
Solenella Norrisii.
Lithodomus Peruvianus.
Saxicava solida.

XV. MAGELLANIC PROVINCE.

This region includes the coasts of Tierra del Fuego, the Falkland Ids. (Malvinas) and the Mainland of South America, from P. Melo, on the east coast, to Concepcion, on the west. It is described by M. D'Orbigny and Mr. Darwin (Journal, p. 177 et seq.). The southern and western coasts are amongst the wildest and stormiest in the world; glaciers in many places descend into the sea, and the passage round Cape Horn has often to be made amidst icebergs floating from the south polar continent.* The greatest tides in the straits amount to 50 feet. "In T. del Fuego the giant sea-weed (*Macrocystis pyrifera*), grows on every rock from low-water mark to 45 fathoms, both on the outer coast and within the channels; it not only reaches up to the surface, but spreads over many fathoms and shelters multitudes of marine animals, including beautiful compound Ascidians, various patelliform shells, Trochi, naked mollusca, cuttle-fish and attached bivalves. The rocks, at low-water, also abound with shell-fish, which are very different in their character from those of corresponding northern latitudes, and even when the genera are identical the species are of much larger size and more vigorous growth."†

Shells of the Magellanic province (* Falkland Islands).

Buccinum antarcticum.
„ Donovani?
Bullia cochlidium.
Monoceros imbricatus.
„ glabratus.
„ calcar.
Trophon Magellanicus.
Voluta Magellanica.
„ ancilla.

Natica limbata.
Lamellaria antarctica.
Litorina caliginosa.
Chemnitzia Americana.
*Scalaria brevis.
*Trochita pileolus.
Crepidula Patagonica.
Trochus Patagonicus.
*Margarita Malvinæ.

*Scissurella conica.
*Fissurella radiosa.
Puncturella conica.
Nacella cymbularia.
*Patella deaurata.
* „ barbara.
* „ zebrina.
Siphonaria lateralis.
Chiton setiger.

* Familiar to the admirers of Coleridge's "Ancient Mariner," and graphically described in Dana's "Two Years before the Mast."

+ Shell-fish are here the chief support of the natives as well as of the wild animals. At Low's harbour a sea-otter was killed in the act of carrying to its hole a large Volute, and, in T. del Fuego, one was seen eating a cuttle-fish. (*Darwin.*)

Doris luteola.
Æolis Patagonica.
*Spongiobranchæa.
Spirialis? cucullata, 66° S.
———
Terebratella crenulata.
* „ Magellanica, many
varieties.

Waldheimia dilatata.
Pecten Patagonicus.
„ corneus.
Mytilus Magellanicus.
*Modiolarca trapezina.
Leda sulculata.
*Cardita Thouarsii.
*Astarte longirostris.

*Venus exalbida.
*Cyamium antarcticum.
Mactra edulis.
*Lyonsia Malvinensis.
Pandora cistula.
Saxicava antarctica.
Boltenia coacta.
Octopus megalocyathus.

XVI. PATAGONIAN PROVINCE.

From S. Catharina, south of the Tropic, to P. Melo. This coast-line has shifted considerably since the era of its present fauna. M. D'Orbigny and Mr. Darwin observed banks of recent shells, especially *Potamomya labiata*, in the valley of the La Plata and the Pampas around Bahia Blanca. Mr. Cuming also met with *Voluta Brasiliana*, and other living shells, in banks 50 miles inland. Of 79 shells obtained by M. D'Orbigny on the coast of N. Patagonia, 51 were peculiar, 1 common to the Falkland Ids. and 27 to Maldonado and Brazil. At Maldonado 37 species were found, 8 being special, 10 common to N. Patagonia, 2 to Rio, and 17 to Brazil. Of the latter 8 range as far as the Antilles; viz.:

Crepidula aculeata.
„ protea.
Pholas costata.

Mactra fragilis.
Venus flexuosa*.
Lucina semi-reticulata.

Modiola viator.
Plicatula Barbadensis.

At Bahia Blanca, in lat. 39° S., the most abundant shells observed by Mr. Darwin (p. 243) were

Oliva auricularia.
„ puelchana.

Oliva tehuelchana.
Voluta Brasiliana.

Voluta angulata.
Terebra Patagonica.

M. D'Orbigny's list also includes the following genera and species:

Octopus tehuelchus
Columbella sertularium.
Bullia globulosa.
Pleurotoma Patagonica.
Fissurellidæa megatrema.
Panopæa abbreviata.
Periploma compressa.
Lyonsia Patagonica.
Solecurtus Platensis.

Æolis.
Paludestrina.
Scalaria.
Natica.
Chiton.
Solen.
Lutraria.
Donacilla.
Nucula.

Leda.
Cytherea.
Petricola.
Corbula.
Pinna.
Mytilus.
Lithodomus.
Pecten.
Ostrea.

XVII. CARIBBEAN PROVINCE.

The Gulf of Mexico, the West Indian Islands, and the eastern coast of South America, as far as Rio, form the fourth great tropical region of marine life. The number of shells is estimated by Prof. C. B. Adams at not less than 1500 species. Of these 500 are described by M. D'Orbigny in Ramon de la Sagra's History of Cuba, and a small number of the Brazilian species in the same author's Travels in South America.

---

* The variety of *Venus flexuosa* found at Rio, can be distinguished from the West Indian shell, which is the *Venus punctifera* of Gray.

The coasts of the Antilles, Bermuda, and Brazil, are fringed with coral reefs, and there are considerable banks of gulf-weed at some distance from the coast of the Antilles.

West India Shells.

Argonauta.	Ommastrephes.	Cleodora.	Cheletropis.
Octopus.	Sepioteuthis.	Creseis.	Ianthina.
Philonexis.	Sepia.	Cuvieria.	Glaucus.
Loligo.	Spirula.	Atlanta.	Notarchus Plei.
Cranchia.	Hyalea.	Oxygyrus.	Aplysia.
Onychoteuthis			

Strombus gigas.	Natica canrena.	Arca Americana.
„ pugilis.	Pyramidella dolabrata.	Yoldia tellinoides.
Murex calcitrapa.	Planaxis nucleus.	Chœma arcinella.
Pisania articulata.	Litorina zic-zac.	„ macrophylla.
Enzina turbinella.	„ flava.	Cardium lævigatum.
Triton pilearis.	„ lineolata.	Lucina tigrina.
„ cutaceus.	Tectaria muricata.	„ Pennsylvanica.
Fusus morio.	Modulus lenticularis.	„ Jamaicensis.
Fasciolaria tulipa.	Fossarus	Corbis fimbriata.
Lagena ocellata.	Truncatella caribbæa.	Coralliophaga.
Cancellaria reticulata.	Torinia cylindracea.	Crassatella.
Fulgur aruanum.	Turritella exoleta.	Gouldia parva.
Terebra acicularis.	„ imbricata.	Venus paphia.
Myristica melongena.	Trochus pica.	„ dysera.
Purpura patula.	Imperator tuber.	„ crenulata.
„ deltoidea.	„ calcar.	„ cancellata.
Oniscia oniscus.	Fissurella Listeri.	„ violacea.
Cassis tuberosa.	„ nodosa.	Cytherea dione.
„ flammea.	„ Barbadensis.	„ circinata.
„ Madagascariensis.	Nerita.	„ maculata.
Columbella mercatoria.	Neritina.	„ gigantea.
„ nitida, &c.	Hemitoma 8 radiata.	„ flexuosa.
Voluta vespertilio.	Hipponyx mitrula.	Artemis concentrica.
„ musica.	Pileopsis militaris.	„ lucinalis.
Oliva brasiliensis.	Calyptræa equestris.	Cyclina saccata.
„ angulata.	Crepidula aculeata.	Trigona mactroides.
„ jaspidea.	Patella leucopleura.	Petricola lapicida.
„ oryza, &c.	Chiton squamosus.	Capsula coccinea.
Ancillaria glabrata.	Hydatina physis.	Tellina Braziliana.
Conus varius, &c.		„ bimaculata.
Clavatula zebra.	Bouchardia tulipa.	Strigilla carnaria.
Marginella.	Discina antillarum.	Semele reticulata.
Erato Maugeriæ.		„ variegata.
Cypræa mus.	Placunomia foliata.	Cumingia.
„ exanthema.	Plicatula cristata.	Iphigenia Brasiliensis.
„ spurca, &c.	Lima scabra.	Lutraria lineata.
Trivia pediculus.	Mytilus exustus.	Periploma inæquivalvis.
Ovulum gibbosum.	Lithodomus dactylus.	Pholadomya candida.

XVIII. TRANS-ATLANTIC PROVINCE.

The Atlantic coast of the United States was supposed by Prof. E. Forbes to consist of two provinces, 1. the *Virginian*, from C. Cod to C. Hatteras,

S

and 2. the *Carolinian*, extending to Florida; but no data were supplied for such a division. The total number of mollusca is only 230, and 60 of these range further north, 15 being moreover common to Europe.

Dr. Gould describes 110 shells from the coast of Massachusetts south of Cape Cod, of which 50 are not found to the northward, but form the commencement of the proper American type. The shells of New York and the southern Atlantic States are described by De Kay, in the State Natural History of New York; this list supplies 120 additional species, of which at least a few are stragglers from the Caribbean province; *e.g. Chama arcinella, Iphigenia lævigata, Capsula deflorata.**

M. Massachusetts. Y. New York. SC. South Carolina. F. Florida.

Conus mus. F.
Fusus cinereus. M. SC.
Nassa obsoleta. M. F. (Mex.)
„ trivittata. M. SC.
„ vibex. M. F. (Mexico).
Purpura Floridana. (Mex.)
Terebra dislocata. V. SC.
Pyrula? papyracea. F.
Fulgur carica. M. SC.
„ canaliculatum. M. SC.
Oliva literata. SC.
Marginella carnea. F.
Fasciolaria distans. SC. (Mex.)
Columbella avara. M. Y.
Ranella caudata. M. Y.
Natica duplicata. Y. SC.
Sigaretus perspectivus. Y. SC.
Scalaria lineata. M. SC.
„ multistriata. M. Y.
„ turbinata. NC.
Cerithium ferrugineum. F.
„ 4 sp. M.
Triforis nigro-cinctus. M.
Odostomia, 6 sp. M. Y.
Turritella interrupta. M. Y.
„ concava. SC.
(Vermetus lumbricalis. M. ?)
Calyptræa striata. Y.
Crepidula convexa. M. Y.
„ fornicata. M. F. (Mex.)
Litorina irrorata. Y.
Fissurella alternata. (Say) ?
Chiton apiculatus. M. SC.
Tornatella puncto-striata. M. Y.
Bulla insculpta. M. Y.

Ostrea equestris. SC. F.
Pecten irradians (*scallop*).
Avicula Atlantica. F.
Mytilus leucophantus. SC.
Modiola Carolinensis.
„ plicatula. M. Y.
Pinna muricata. SC.
Arca ponderosa. SC.
„ pexata. M. F.
„ incongrua. SC.
„ transversa. M. Y.
Solemya velum. M. Y.
„ borealis. M.
Cardium ventricosum. SC.
„ Mortoni. M. Y.
Lucina contracta. Y.
Astarte Mortoni. Y.
„ bilunulata. F.
Cardita incrassata. F.
Venus mercenaria. M. SC.
„ Mortoni. SC. F.
„ gemma. M. Y.
Artemis discus. SC.
Petricola dactylus. M. SC.
„ pholadiformis. Y
Mactra similis. SC. M.
„ solidissima. M. Y.
„ lateralis. M. Y.
Lutraria lineata. F.
„ canaliculata. V. F.
Mesodesma arctata. M. Y.
Tellina tenta. M. SC.
„ 8 sp. SC. F.
Semele æqualis. SC.
Cumingia tellinoides. M.
Donax fossar. Y.

* The sea-shells of the United States have also been collected and described by Say, Le Sueur, Conrad, and Couthouy.

Donax variabilis. G. F.	Periploma papyracea. M. Y.
Solecurtus fragilis. M. SC.	Lyonsia hyalina. Y.
„ caribbæus. M. F.	Pandora trilineata. M. F.
Corbula contracta. M. F.	Pholas costata SC. F.
Periploma Leana. M. Y.	„ semicostata. SC.

LAND REGIONS.

Distribution of Land and Fresh-water Shells.

The boundaries of the Natural-history land-regions are more distinctly marked, and have been more fully investigated, than their counterparts in the sea. Almost every large island has its own fauna and flora ; almost every river-system its peculiar fresh-water fish and shells; and mountain-chains like the Andes appear to present impassable barriers to the " nations" of animals and plants of either side. Exceptions, however, occur which shew that beyond this first generalisation there exists a higher law. The British Channel is not a barrier between two provinces, nor is the Mediterranean ; and the desert of Sahara separates only two portions of the same zoological region. In these and other similar instances the " barrier " is of later date than the surrounding fauna and flora.

It has been often remarked that the northern part of the map of the world presents the appearance of vastly-extended, continental plains, much of which is, geologically speaking, new land. In the southern hemisphere the continents taper off into promontories and peninsulas, or have long since broken up into islands. Connected with this is the remarkable fact that only around the shores of the Arctic Sea are the same animals and plants found through every meridian ; and that in passing southward, along the three principal lines of land, specific identities give way to mere identity of genera, these are replaced by family resemblances, and at last even the families of animals and plants become in great measure distinct—not only on the great continents, but on the islands—till every little rock in the ocean has its peculiar inhabitants—the survivors, seemingly, of tribes which the sea has swallowed up.(*Waterhouse.*)

The two largest genera, or principal types of the land and fresh-water shells, *Helix* and *Unio*, have an almost universal range, but admit of many geographical subdivisions.* Amongst the land-snails are several species to which a nearly world-wide range has been assigned, sometimes erroneously as when *Helix cicatricosa* is attributed to Senegal and China, or *Helix similaris* Fér. to Brazil and India ; and often correctly, but only because they have been carried to distant localities by human agency. Land-snails are in

* In cataloguing *Unionidæ* the river and country of each species should be stated. American authors are too often contented with recording such localities as " Nashville " and " Smithville," which are quite unintelligible. Almost as uncertain in their meaning are S. Vincent, S. Cruz, S. Thomas, Prince's Id. ; whilst the latinized names of places often defy all attempts at re-translation.

favour with Portuguese sailors, as "live sea-stock;" and they have naturalized the common garden-snail of Europe (*Helix aspersa*) in Algeria, the
Azores, and Brazil; and *Helix lactea* at Teneriffe and Mte. Video. *Achatina
fulica* has been taken from Africa to the Mauritius, and thence to Calcutta,
where it has been established by a living naturalist; and *Helix hortensis* has
been carried from the old country to America, and naturalized on the coast
of New England and the banks of the St. Lawrence. *Bulimus Goodalli*,
indigenous to the West Indies and S. America, has been introduced into
English pineries and to Mauritius. *Helix pulchella*, one of the small species
found in moss and decayed leaves, inhabits Europe, the Caucasus, Madeira,
the Cape (introduced), and N. America as far as the Missouri. *Helix cellaria*
inhabits Europe and the Northern States of America, and has been carried
abroad with the roots of plants, or attached to water-casks, and naturalized
at the Cape and New Zealand.

The fresh-water *Pulmonifera—Limnæa, Physa, Planorbis, Ancylus—*
and the amphibious *Succinea*, have a nearly world-wide range; and like
aquatic plants and insects often re-appear, even at the antipodes, under
familiar forms. The range of the gill-breathing fresh-water shells is more
restricted.

The Old World and America may be regarded as provinces of paramount
importance, having no species in common (except a few in the extreme
north), and each possessing many characteristic genera.

America.	Old World.	America.	Old World.
Anastoma.	Zonites.	Choanopoma.	Pomatias.
Tridopsis.	Nanina.	Chondropoma.	Otopoma.
Sagda.	Vitrina.	Cistula.	Craspedopoma.
Stenopus.	Helicolimax.	Trochatella.	Diplommatina.
Proserpina.	Daudebardia.	Alcadia	Aulopoma.
Bulimus.	Achatina.	Stoastoma.	Pupina.
Odontostomus.	Achatinella.	Geomelania	Acicula.
Lignus.	Clausilia.	——	——
Glandina.	Paxillus.	Hemisinus.	Vibex.
Cylindrella.	Pupa.	Melafusus.	Pirena.
Megaspira.	——	Ceriphasia.	Melanopsis.
Simpulopsis.	Testacella.	Anculotus.	Paludomus.
Amphibulima.	Parmacella.	Melatoma.	Lithoglyphus.
Omalonyx.	Limax.	Amnicola.	Navicella.
——	Arion.	——	——
Philomycus.	Phosphorax.	Mülleria.	Ætheria.
Peltella.	Incilaria.	Mycetopus.	Iridina.
——	Oncidium.	Castalia.	Galatea.
Chilinia.	——	Monocondylæa.	Cyrenoïdes.
Gundlachia.	Latia.	Gnathodon.	Glaucomya.

The Land Provinces represented on the map are the principal Botanical
Regions of Prof. Schouw, as given in the Physical Atlas of Berghaus; and it
is proposed to inquire how far these divisions are confirmed by the land and
fresh-water shells, more especially by the land-snails, (*Helicidæ, Limacidæ*,

and *Cyclostomidæ*), which have been so elaborately catalogued by Dr. L. Pfeiffer.*

The first Botanical region—that of Saxifrages and Mosses—has not been numbered on the map, although its boundary is given by the line of northern limit of trees. This line nearly coincides with the Isotherm of 32°, or permanent ground-frost; but in Siberia the pine-forests extend 15° further, owing to the absence of winter rains and the bright clear air.

In this region shells are very rare; Dr. Middendorff found *Physa hypnorum* in Arctic Siberia, and *Limnæa geisericola* (Beck) inhabits the warm springs of Iceland. The few species discovered by Möller in Greenland are supposed to be peculiar:—

Helix Fabricii.	Succinea Grœnlandica.	Limnæa Holböllii.
Pupa Hoppii.	Limnæa Vahlii.	Planorbis arcticus.
Vitrina angelicæ.	,, Pingelii.	Cyclas Steenbuchii.

1. Germanic Region.

The whole of Northern Europe and Asia, bounded by the Pyrenees, Alps, Carpathians, Caucasus, and Altai, constitutes but one province, with a fauna by no means proportioned in richness to its extent.†

The land-snails amount to more than 200, but nearly all (or at least five-sixths) are common to the Lusitanian region.‡

Helix............... 90	Pupa 44	Cyclostoma 1
Bulimulus 10	Clausilia 52	Acicula 1
Zua .... } 5	Vitrina 5	Limax 9
Azeca.............	Succinea 5	Arion 4
Cionella	Balea 1	Carychium 1

The fresh-water shells belong to these genera and sub-genera :—

Limnæa............. 20	Velletia.............. 1	Unio, sp. and vars.. 20
Amphipeplea 2	Neritina, vars......... 3	Anodon, vars....... 20
Physa.............. 5	Paludina and Bithynia 23	Alasmodon 3
Aplexa 1	Valvata 5	Cyclas 6
Planorbis 16	Conovulus (Alexia) .. 3	Pisidium 11
Ancylus 7	Dreissena 1	

The British land-shells amount to 74, fresh-water *pulmonifera* 24, fresh-water *pectinibranchiata* 7, marine *pulmonifera* 4; fresh-water bivalves 15. Of the species formerly thought peculiar, *Pupa anglica* and *Helix fusca* have been found in France, and *Helix lamellata* in Holsace. *Helix excavata* (Bean) is still unknown upon the Continent; and *Geomalacus maculosus* and

* The distribution of the *Cycladidæ* is taken from the British Museum Catalogue, by M. Deshayes.

† The mean temperature of the winter and summer months averages 36°—57°; in Western Europe autumn rains prevail, and summer rains in Eastern Europe and Siberia.

‡ It was the opinion of Prof. E. Forbes that *all* the species of the Post-pliocene land of Northern Europe and Asia had *originated* beyond the bounds of that region.

Limnæa involuta have only been met with in the south-west of Ireland, but are possibly Lusitanian species. *Dreissena polymorpha* has been permanently naturalized in canals (p. 267), and *Testacella Maugei* and *haliotidea* in gardens; *Bulimus decollatus* and *Goodalli* have been often established in greenhouses. Some species are now very scarce in England that were formerly abundant, as :—

Clausilia plicatula.	Vertigo Venetzii.	Succinea oblonga.
Vertigo minutissima.	Helix lamellata.	Acicula fusca.

Others which occur in the newer tertiary deposits have become quite extinct in England, such as :—

Helix fruticum,   living in France and Sweden.
    ,,   ruderata .... Germany.
    ,,   labyrinthica (Eocene) New England.
Paludina marginata........ France.
Corbicula consobrina ...... Egypt and India.
Unio litoralis .... France and Spain.

On the other hand, some of the commonest living species have not been found fossil; *e.g. Helix aspersa, pomatia,* and *cantiana.* Several genera only occur fossil in the older tertiaries, viz. :—

Glandina.	Cyclotus.	Nematura.
Proserpina.	Megalomastoma.	Melania.
Cylindrella?	Craspedopoma.	Melanopsis.

The land and fresh-water shells of Scandinavia are 56, all common European species; *H. pomatia* has been naturalized at Stockholm.*

Dr. Middendorff gives the following list of Siberian shells in his *Sibirische Reise* (Band II. th. 1. Petersb. 1851) :—

Helix carthusiana. Irkutsk.	Planorbis complanatus, Altai.
,, Schrenkii, M. Tunguska, 58°.	,, albus, Barnaul, ,,
,, hispida, Beresov. Bernaul.	,, contortus, ,,
,, ruderata, Stanowoj Mtn.	,, vortex, ,,
,, pura, ,,	,, leucostoma, ,,
,, sub-personata, ,, ; Ochotsk.	,, nitidus, Irkutsk.
Pupa muscorum, Bernaul.	Bithinia tentaculata, Bernaul.
Zua lubrica, ,,	,, Kickxii, R. Ami, Altai.
Succinea putris, ,, ; Irkutsk.	Valvata cristata, var. Sibirica, Bernaul,
Limnæa Gebleri, M. Bernaul.	Beresov; Kamtschatka.
,, auricularia, Nertschinsk.	,, piscinalis, R. Ami.
,. ovata, Bernaul.	Unio complanatus, Kamtschatka.
,, Kamtschatica, Mid.	Unio Dahusicus, Mid. Schilka.
., peregra, Bernaul, Beresov.	,, Mongolicus, M. Gorbitza, Dauria.
,, stagnalis, ,, Irkutsk.	Anodon herculeus, M. Scharanai.
,, palustris, ,, ,,	,, anatinus, Tunguska.
,, truncatula, ,, Tomsk.	,, cellensis var. Beringiana, Kamt-
,, leucostoma, Irkutsk.	schatka.
Physa hypnorum, Bernaul; Taimyr-lande.	Cyclas calyculata, Bernaul, R. Lena, R. Ami, S. Kamts.
Planorbis corneus, Bernaul; Beresov; Kirgisensteppe, Altai.	Pisidium fontinale, Beresov.
	,, obliquum, Bernaul, Tomsk.

* Norske Land- og Fersk-vands Mollusker, Joachim Friele, 1851.

2. LUSITANIAN REGION.

The countries bordering the Mediterranean, with Switzerland, Austria and Hungary, the Crimea (*Taurida*), and Caucasus, form a great province (or rather cluster of provinces) to which Prof. E. Forbes applied the term *Lusitanian*. The Canaries, Azores, and Madeira are outlying fragments of the same region.*

In Southern Europe about 600 land-snails are found, of which above 100 are also spread over the Germanic region and Siberia; and 20 or 30 are common to Northern Africa. Besides these 60 others are found in Algeria and Egypt, 100 in Asia Minor and Syria, and 135 in the Atlantic Islands, making a total of nearly 900 species of *Helicidæ*.†

Of the 12 species of *Zonites* (proper) 10 are peculiar to Lusitania.

The species of *Bulimus*, *Achatina*, and *Pupa* are small and minute, belonging to the sub-genera *Bulimulus*, *Cionella*, *Zua*, *Azeca*, *Vertigo*, &c.; 4 (of which two are Algerian) have been referred to *Glandina*.

In this region are also found 22 species of *Cyclostomidæ* and 44 *Limacidæ*:—

Helix 392	Vitrina 11	Cryptella 1
Bulimus 80	Daudebardia 3	Cyclostoma 5
Succinea 8	Helicolimax 3	Craspedopoma 3
Achatina 25	Limax 28	Pomatias 10
Tornatellina 3	Arion 7	Acicula 4
Balea.............. 4	Phosphorax 1	———
Pupa 120	Testacella.......... 2	Carychium 3
Clausilia‡ 247	Parmacella 5	

The fresh-water shells are of the same genera as in the Germanic province, and their numbers about the same; with the addition of several species of *Melania, Melanopsis, Lithoglyphus*, and *Cyrena*. *Melanopsis buccinoides* is found in Spain, Algeria, and Syria, having become extinct in the intervening countries. Two species of *Lithoglyphus* inhabit the Danube; *Cyrena* (*Corbicula*) *Panormitana* is found in Sicily, two others in the Euphrates, and *C. consobrina* in the Alexandrian Canal.

The Lusitanian province includes numerous minor regions, the islands and mountain tracts especially being centres or *foci* where a number of peculiar species are associated with those living around. Thus, of species not as yet recorded from other localities, Switzerland has 28, the Austrian Alps 46, Carpathians 28, N. Italy and Dalmatia 100, Roumelia 20, Greece and its

---

* In the South of Europe rain seldom falls in summer, but is frequent at other seasons, especially in winter. The mean temp. is 54°-72°.

† The writer is greatly indebted to W. H. Benson, Esq. for information respecting the land-shells of the Lusitanian province, Africa, and the remote islands.

‡ Many of these cannot be considered *species*, in the sense here understood, but only as *races*, or geographical varieties.

Archipelago 90, Anatolia 50, Caucasia 20, Syria 30, Lower Egypt and Algeria 60, Spain 26, and Portugal 15 *Helicidæ* and 9 *Limacidæ.*

Mediterranean Islands.

Corfu, Cyprus, Rhodes, Syra, Candia, and Crete, have each a few peculiar land snails, amounting to 40 species altogether.

Balearic Isles. Helix Graellsiana, hispanica (var. balearica,) nyellii, minoricensis; and Cyclostoma ferrugineum, common to Spain and Algeria.

Corsica. Helix Raspaili, tristis, Clausilia 4 sp.

Sardinia. Helix Sardiensis, meda, tenui-costata, Pupa 2, Clausilia 1.

Malta has two peculiar species of Helix, and a Clausilia (*scalaris*).

Sicily has 40 peculiar Helices and 3 Limaces. This island is connected with N. Africa by a winding shoal with deep water on each side.

Madeira Group.

These ancient volcanic islands, 660 miles S. W. of Portugal, consist of Madeira, with Fora and 3 other islets called Dezertas, and Porto Santo, 26 miles to the N. E., with the rocky islets Ferro, Baxo and Cima.* The land-snails have been described by the Rev. R. T. Lowe,† and form the subject of a monograph by Dr. Albers;‡ the investigations of Mr. Vernon Wollaston have nearly doubled the number of known species, which now amount to 132. The *Vitrinæ* belong to the section *Helico-limax*; the Cyclostomas to the sub-genus *Craspedopoma,* and half the Pupas to *Vertigo.*

Arion 1	Bulimus 2	Cionella 3	Limnæa........ 1
Limax 4	Glandina...... 4	Pupa.......... 23	Ancylus........ 1
Testacella...... 2	Azeca 3	Balea.......... 1	Conovulus 3
Vitrina 3	Tornatellina .. 1	Clausilia 3	Pedipes (afra.).. 1·
Helix 76	Zua 2	Cyclostoma.... 2	

Of the 92 found in Madeira or the Dezertas, 70 are peculiar; 54, of which 39 are peculiar, inhabit Porto Santo and its islets; 11 others, of which 4 are widely diffused, are common to Madeira and Porto Santo. One species is peculiar to the Dezerta Grande; 1 species and 1 variety to the S. Dezerta (Bugio); 1 to the Northern (Cho); one variety to Ferro. Seven species are common to the Dezertas; 1 to the great and northern Dezertas; 5 to Madeira and Dezerta Grande; and 3 to Madeira, P. Santo, and the Dezertas. Of those species, which inhabit more than one island, the specimens from each locality are recognizable as distinct *races,*

---

* These islands, and also the Canaries and Azores, contain marine formations (volcanic grits and tufas) with Miocene Tertiary shells. The islet of Baxo is quarried for lime.

† Primitiæ et novitiæ Faunæ et Floræ Maderæ et Portus Sancti. 12mo. Lond., 1851. Descriptive list of all the species, by same author, Zool. Proc. *for* 1854, p.161. The statements and numbers given above are taken from this last monograph, corrected by Mr. Wollaston.

‡ Malacographia Maderensis, 4to. Berlin, 1854, with figures of all the species.

or geographical varieties. *Helix subplicata* and *papilio* are found on the Ilheo Baxo; *H. turricula* on Cima. Of the total 'number (132) 111 species are peculiar to the Madeira group; 5 are common to the Canaries; 4 to the Azores, and 1 to the Guinea coast; 11 are common to S. Europe, besides 2 *Limnæids*, and 7 slugs, which may have been recently introduced viz. :—

Arion empiricorum	Helix cellaria.	Zua lubrica, var.
Limax variegatus.	,, crystallina.	,, folliculus.
,, antiquorum.	., pisana.	Bulimus decollatus.
,, agrestis.	,, pulchella.	,, ventrosus, Fer.
,, gagates.	,, lenticula.	Balæa perversa (p. 166).
Testacella Maugei.	(,, lapicida, *fossil*).	Limnæa truncatula
,, haliotidea.	Cionella acicula.	Ancylus fluviatilis.

Great quantities of *dead shells* of the land-snails are found in ancient sand-dunes near Caniçal, at the eastern extremity of Madeira, and in Porto Santo, including 64 of the living species and 13 which have not been found alive. As the fossil examples of several species are larger than their living descendants, it is possible that some of those reputed to be extinct have only degenerated. It is a remarkable fact that some of the commonest living species are not found fossil, whilst others, now extremely scarce, occur abundantly as fossils.*

Extinct land-snails of Madeira.

Helix delphinula, Lowe. M.
,, arcinella, Lowe, P.
,, coronula, Lowe, S. Deserta.
,, vermetiformis, Lowe, P.
,, Lowei, Fer. (porto-sanctana, var.?). P.
,, fluctuosa, Lowe (= chrysomela, Lowe). P.
,, psammophora, Lowe (phlebophora var. ?). P.
,, Bowdichiana, Fer. (punctulata, major ?). M. P.
Glandina cylichna, Lowe. P. Santo.
Cionella eulima, Lowe, P.
Pupa linearis, Lowe. M. (= minutissima, Hartm ?)
,, abbreviata, Lowe. M.

The problem of the colonization of these islands receives additional light from the circumstances noticed at other oceanic islands, especially the Canaries and St. Helena. There is evidence that this mountain group has not arisen newly from the sea, and great probability that it has become insulated by the subsidence of the surrounding land.† The character and arrangement of its fauna is probably nearly the same now as when it formed part of a continent, and the diminution of its land-shells in variety and size .

* *Helix tiarella*, W. and B. was supposed to be extinct, but in the last summer, (1855) Mr. Wollaston detected it alive in two almost inaccessible spots on the north coast of Madeira: it is not a native of the Canaries.

† See the Observations of Mr. James Smith, and of Sir C. Lyell and Mr. Hartung (Geol. Journ. 1854).

may be the result of a modern change of physical conditions brought about by human agency, as at St. Helena.   The annual fall of rain is now 29.82 inches, whereas it was remarked by Columbus, 350 years ago, "that, formerly, the quantity of rain was as great in Madeira, the Canaries and the Azores, *as in Jamaica*, but since the trees, which shaded the ground, had been cut down, *rain had become much more rare.*" *

The *Azores* are a group of 9 volcanic islands, 800 miles W. of Lisbon, the loftiest being Pico, 7,613 ft. Only 13 land-shells have been found, of which 3 are common to the Canaries, 1 to the Canaries and Madeira, 3 to Madeira, 1 to the Canaries and C. de Verdes, and 2 are peculiar, viz.: *Helix Azorica* and *Bulimus cyaneus.   Helix barbula* is also found in Portugal, *H. pisana* and *cellaria* are common to Madeira and Europe, and *H. aspersa* has been introduced recently.

The *Canary islands* are 60 miles W. of Africa, with a temperature of 60°—66° in the coolest half-year, and 78°—87° in the hottest. The landsnails are about 80 in number, including *Helix* 50, *Nanina* 1, *Vitrina* 3, *Bulimus* 16, *Achatina* 3, *Pupa* 5, *Limax* 1, *Phosphorax* 1, *Testacella* 2, *Cryptella* 1, and 4 *Cyclostomidæ*. Of these, 60 are peculiar, 12 are common to S. Europe, and 4 to the West Indies? 1 to Morocco, 1 to Algeria (also European), and 1 to Egypt.   The fresh-water shells are *Physa* 2, *Ancylus* 1.

Helix ustulata and *McAndrei* are peculiar to the rocky islets known as the "Salvages" north of the Canaries.

The absence of W. African land-shells and the presence of W. Indian species may be explained by the currents, which come from the Antilles, as shown on the map. † Some of the European species may have been introduced (*e. g. Helix lactea, pisana, cellaria*); but the presence of 20 Lusitanian species, in a total of 80, is too remarkable to be accidental.

The *Cape de Verde Islands*, although much further to the south, are also much farther from the continent, being 320 miles West of C. de Verde; the mean temperature is 65°—70°, and the vegetation, as Dr. Christian Smith remarked, is more like that of the Mediterranean coast than W. Africa.   Of the 12 land-shells, two are common to the Canaries and Azores.

Lusitanian species of wide distribution.

Helix amanda, Sicily — Palma.
 „   planata, Morocco — Canaries.
 „   lenticula, S. Europe — Madeira — Canaries.
 „   rozeti, Sicily, Morea — Algeria — C. de Verde — Canaries.
 „   lanuginosa, Majorca — Algeria — Palma.

---

* Cosmos, II. 660, Bohn ed.   It seems likely that Jamaica itself has since undergone a similar change; the fall of rain is stated to be 49.12, whilst in the neighbouring islands it exceeds 100 inches.

† Long before the discovery of America it was observed that the westerly gales washed ashore stems of bamboos, trunks of pines, and even *living men in canoes*. =Humboldt, II. p. 462.

Helix simulata, Syria — Egypt — Lancerotte.
„ Michaudi, summit of Porto Santo — Teneriffe?
„ cyclodon, Azores — Canaries — C. de Verdes.
„ advena, (= *erubescens* Lowe,) Madeira—Azores — St. Vincent.
„ plicaria and planorbella, Canaries—Porto Rico?
Bulimus subdiaphanus, Canaries — Azores — C. de Verdes.
„ bœticatus and badiosus, Canaries—St. Thomas?

Ascension. This barren volcanic island, in the midst of the Atlantic Ocean, is not known to possess any terrestrial *Pulmonifera* beside a slug, the *Limax Ascensionis.* Mr. Benson thinks that some *Helicidæ* might possibly be found on the Green Mountain, 2840 feet high, where the garrison have their gardens. Mr. Darwin remarks "we may feel sure that at some former epoch, the climate and productions of Ascension were very different from what they now are."

St. Helena. (No. 28 of Map).

The Island of St. Helena is 800 miles S. E. of Ascension, and 1200 from the nearest African coast of Benguela. It is entirely volcanic. The indigenous plants are all peculiar, and not more related to those of Western Africa than to Brazil.* The land shells are also peculiar; 13 species have been described; viz:—*Helix*, 3 sp. *Bulimus* 5, *Achatina* 2, *Pupa* 1, *Succinea* (*Helisiga*), 2. As many more have been met with only in the condition of dead shells, rarely retaining their colour and translucency. They are found beneath the surface-soil in the sides of ravines worn by the heavy rains, at a height of 1200 to 1700 feet; " their extinction has probably been caused by the entire destruction of the woods, and the consequent loss of food and shelter, which occurred during the early part of last century."—(Darwin's Journal, p. 488). A living *Bulimus*, related to the extinct *B. Blofieldi*, is found feeding on the cabbage-trees, only on the highest points of the Island.

Extinct land-shells of St. Helena.†

Bulimus auris vulpinus.	Bulimus relegatus.
„ Darwini.	Helix bilamellata.
„ Blofieldi.	„ polyodon.
„ Sealei.	„ spurca.
„ subplicatus.	„ biplicata.
„ terebellum.	„ Alexandri.
„ fossilis.	Succinea Bensoni.

The large *Bulimus*, (fig. 91, p. 164) has no living analogue in Africa,

* "It might perhaps have been expected that the examination of the vicinity of the Congo would have thrown some light on the origin, if I may so express myself, of the Flora of *St. Helena*. This, however, has not proved to be the case; for neither has a single indigenous species, nor have any of the principal genera characterising the vegetation of that Island, been found either on the banks of the Congo, or on any other part of this coast of Africa."—R. Brown, Appendix to Captain Tuckey's Narrative of the Congo Expedition, (p, 476.) 1818.

† G. Sowerby in Darwin's " Volcanic Islands," p. 73. Forbes, Journ. Geol. Soc' 1852, p. 197.—Benson, An. Nat. Hist. 1851, VII. 263.

but is a member of of a group characteristic of tropical America (to which the names *Plecochilus, Pachyotis* and *Caprella* have been given) including *B. signatus, B. bilabiatus, B. goniostomus,* and especially *B. sulcatus* (Chilonopsis, Fischer) of St. Iago.* The four next species belong to the same type, but are smaller and slenderer. "The marine mollusks of the coast of St. Helena would lead us to infer the very ancient isolation of that island, whilst at the same time a pre-existing closer geographical relationship between the African and the American continents than now maintains is dimly indicated. The information we have obtained respecting the extinct and existing terrestrial mollusks would seem to point in the same direction, and assuredly to indicate a closer geographical alliance between St. Helena and the east coast of S. America than now holds."—(Forbes).

Tristan d'Acunha. (No. 29 of Map).

Two peculiar species of *Balea* (Tristensis and ventricosus) are found on this remote and lofty island, which attains an elevation of 8,236 feet.

3. AFRICAN REGION.

Tropical Western Africa, with its hot and swampy coasts and river valleys is the region of the great *Achatinæ* and Achatina-like *Bulimi,* the largest of all living land-snails. Dr. Pfeiffer enumerates—*Vitrina* 3 sp. *Streptaxis* 7, *Helix* 8, *Pupa* 5, *Bulimus* 35, *Achatina* 39, *Succinea,* 3. *Streptaxis Recluziana* inhabits the Guinea Islands. *Helix Folini, Bulimus numidicus* and *fastigiatus, Pupa crystallum* and *sorghum, Achatina columna, striatella* and *lotophaga* are found ou Princes Island. *Pupa putilla* on Goree Island. *Bulimus (Pseudachatina) Downesi, Achatina iostoma* and *Glandina cerea* at Fernando Po. The reversed river-snail (*Lanistes*) is generally diffused in the fresh waters of Africa; several species of *Potamides* and *Vibex* are found in the embouchures of the western rivers and *Pedipes* on the sea-shore. The freshwater bivalves of Senegal are similar to those of the Nile;—

Pisidium parasiticum, Egypt.		Iridina exotica,	Senegal.
Cyrenoides Duponti, Senegal.		„ rubens,	„
Corbicula, 4 sp.	Egypt.	Pleiodon ovatus	„
Iridina nilotica	„	Ætheria semilunata „	Nile.
„ aegyptiaca	„	Galatea radiata	„

4. CAPE REGION.

Dr. Krauss describes 41 species of land-snail from South Africa, and Mr. Benson has furnished a list containing 22 others; these are all peculiar, except a *Succinea* which appears to be only a variety of the European

---

* As Dr. Pfeiffer includes this (with a sign of doubt) amongst the synonymes of *B. auris-vulpinus* he must have suspected that the specimens came from St. Helena and not from St. Iago. The only other group of Bulimi resembling the St Helena shells occurs in the Pacific Islands:—*Bulimus Caledonicus* at Mulgrave I.—*B. auris zovinæ* at the Solomons, and *B. shongi* in New Zealand.

S. putris, and two European Helices (H. *cellaria* and *pulchella*) probably imported to the environs of the Cape. There are also 3 slugs, 9 freshwater Pulmonifera, 7 marine Pulmonifera, 5 freshwater bivalves and 5 univalves. The species found at the Cape, Algoa Bay, Natal, &c., are for the most part different—*Potamides decollatus, Clionella sinuata* and an *Assiminea* inhabit brackish waters.

Limax 1	—	—
Arion 1	Limnæa 1	Paludina 3
—	Physa 4	Neritina 1
Vitrina 4	Physopsis 1	—
Helix 29	Ancylus 1	Corbicula.............. 1
Succinea 4	Planorbis............ 3	Cyclas 1
Bulimus 9	—	Pisidium 1
Pupa 6	Vaginulus 1	Unio 1
Achatina 5	Oncidium............ 1	Iridina 1
Cyclostoma 6	Auricula 6	

5. YEMEN—MADAGASCAR.

The S. W. Highlands of Arabia (Yemen) form a distinct Botanical province isolated by rainless deserts to the north. The land snails consist of a few species of *Helix* and *Bulimus, Cyclostoma lithidion*, and 3 species of the section *Otopoma*, a group also found in Madagascar. Two species are common to the island of *Socotra*, (No. 30) which also has a species (of *Pupa*) common to Madagascar. *Bulimus guillaini, Cyclostoma gratum, modestum* and *Souleyeti* are found on the island of Abd-el-Gouri.

Very few land shells have been collected on the mainland of Eastern Africa, although it is a rainy region, and well wooded in the southern part; 5 species only are recorded from Mogadoxa and Ibu, belonging to the genera *Helix, Bulimulus, Achatina, Pupa,* and *Otopoma*. On the Island of Zanzibar are found, *Achatina Rodatzi,* and *allisa, Cyclostoma Creplini,* and *Zanguebarica; Pupa cerea* is common to Zanzibar and Madagascar.

Madagascar itself is rich in land shells; Dr. Pfeiffer enumerates—*Helix* 28 *sp., Bulimus* 6, *Succinea* 14, *Pupa* 1, *Achatina* 4, (one of which, *eximia,* is allied to *A. Columna, of W. Africa*), and 32 Cyclostomidæ, chiefly of the section with spiral ridges (*Tropidophora*), 3 of the division *Otopoma. Cyclostoma cariniferum* and *Cuvieri* are found on the Island of Nosse Be; *Helix guillaini* on S. Maria I. Amongst the fresh-water shells are *Melania amarula, Melanatria fluminea* and *Neritina corona*.

The land shells of the *Mascarene Islands* are all peculiar; we are indebted to Mr. W. H. Benson for most of the information existing in respect to them.

Comoro Islands.

Helix russeola and *Achatina simpularia* are found in Mayotte; *Cyclostoma pyrostoma* in Mayotte and Madagascar.

Seychelles, (No. 31 of Map).

Parmacella Dussumieri
Helix unidentata
  ,, Studeri
  ,, Souleyeti
  ,, Tranquebarica
Streptaxis Souleyeti

Bulimus ornatus
  ,, fulvicans
Cyclostoma insulare
  ,, pulchrum
Cyclotus conoideus

Mauritius, (32).

Parmacella perlucida
  ,, Rangii
  ,, mauritii
Helix philyrina
  ,, inversicolor
  ,, stylodon
  ,, mauritiana
  ,, mauritianella
  ,, rawsoni
  ,, semicerina
  ,, mucronata
  ,, nitella
  ,, rufa
  ,, similaris
  ,, suffulta
  ,, albidens

Helix Barclayi
  ,, odontina
Vitrina angularis
Tornatellina cernica
Gibbus Antoni
  ,, Lyonneti
Succinea sp. .
Bulimus clavulinus
  ,, Mauritianus
Pupa pagoda
  ,, fusus
  ,, sulcata
  ,, clavulata
  ,, modiolus
  ,, funicula
  ,, versipolis

Pupa Largillierti
Cyclostoma Barclayi
  ,, Michaudi
  ,, carinatum
  ,, undulatum
  ,, insulare ?
Cyclotus conoideus ?
Otopoma Listeri
  ,, hæmastoma
Realia rubens
  ,, aurantiaca
  ,, multilirata
  ,, expansilabris
  ,, globosa
Megalomastoma croceum

Two large species of *Achatina (fulica* and *panthera)* abounding in the coffee plantations, are believed to have been introduced. The fall of rain in Mauritius is 35.25.

Bourbon, (No. 33).

Helix cælatura
  ,, detecta
  ,, delibata ?

Helix tortula
  ,, Brandiana
Pupa Largillierti—Mauritius.

Rodriguez.

Cyclostoma articulatum Madagascar?   Streptaxis—pyriformis.

No. 34. *Kerguelen's Land.* Helix Hookeri was collected at this island when visited by the Antarctic Expedition.

6. INDIAN REGION.

Proceeding eastward, in Asia, the species of *Achatina, Pupa, Clausilia, Physa, Limax,* and *Cyclostoma* rapidly diminish or quite disappear. Helices of the section *Nanina* become plentiful, amounting to 150 species, and *Bulimulus* and *Cyclophorus* attain their maximum. *Leptopoma* and *Pupina* are peculiar to the Asiatic islands.

Our catalogue of Indian land shells must be very imperfect, including only about 180 *Helicidæ* and 50 *Cyclostomidæ.* A very few of the Indian species are common to China and the Asiatic Islands, or even to Ceylon. The shells of northern India resemble those of the Lusitanian region : in the south they approximate more to the large and vividly coloured species of the Asiatic Islands. In the Himalaya land shells are numerous, and ascend as high as the region of Junipers and Rhododendrons, 4,000—10,000 feet above the sea.

Helix	37	Pupa	7	Cyclophorus	26
Nanina	46	Clausilia	7	Leptopoma	1
Ariophanta	8	Vitrina	9	Pterocyclus	10
Streptaxis	3	Succinea	7	Cyclotus	3
Bulimus	40	Parmacella	2	Megalomastoma	4
Achatina	13	Cyclostoma	3	Diplommatina	3

Parmacella and *Vaginulus* are found in India, and the typical fresh-water species of *Oncidium*. Ordinary forms of *Limnæa* and *Planorbis* are abundant, and there is one species of *Ancylus*. *Physa* occurs only in a fossil state or is represented by the singular *Camptoceras* of Benson. *Hypostoma Boysii*, *Auricula Judæ* and *Polydonta scarabæus* are also Indian forms.

The gill-breathing fresh-water shells of India are very numerous, especially the Melanias and Melanatrias, and species of *Pirena, Paludomus, Hemimitra* (retusa), *Ampullaria, Paludina, Bithynia, Nematura* (deltæ), *Assiminea* (fasciata), *Neritina* (particularly crepidularia and Smithii) *and Navicella* (tessellata).

The brackish-water species of *Cerithidium, Terebralia,* and *Pyrazus* are mostly common to India and North Australia.

The fresh-water bivalves are a few ordinary forms of *Unio*, 3 species of *Cyrena*, a *Corbicula* (of which 6 species have been made), *Cyclas Indica, Arca scaphula, Glaucomya cerea* and *Novaculina gangetica.*

Ceylon.—The land-shells of Ceylon have been investigated by Mr. Benson who has favoured us with a list of 112 species; they most resemble those of the Neilgherry hills, but are nearly all specifically distinct, and even some of the genera are peculiar. It seems entitled to rank as a province. *Helix Waltoni* and *Skinneri*, are examples of the most characteristic form of Helices, the Vitrini-form type (*Nanina*) is also common. *H. hæmastoma*, one of the most conspicuous species, found on trees at P. Galle, is common to the Nicobar Islands. The Achatinas belong to a distinct section (*Leptinaria*, Beck) also represented on the Continent. Some of the *Bulimi* approach the Philippine forms.

Helix	36	Succinea	1	Pterocyclus	5
Nanina	9	Pupa	3	Aulopoma	4
Vitrina	3	Achatina	7	Leptopoma	7
Streptaxis	2	Cyclophorus	12	Cataulus	10
Bulimus	11				

The fresh-water shells belong to the genera Limnæa, Physa, 2 species, (not found on the Continent); Planorbis, Melania, Tanalia 10 (peculiar), Paludomus, Bithynia, Ampullaria, Neritina, Navicella, Unio, and Cyrena.

At the Nicobar Islands are found—Cataulus tortuosus, Helicina Nicobarica and Pupina Nicobarica. *Helix castanea* is from Sumatra. (*Beck*).

7. China and Japan.

The few land-snails known from China are of Indian and Lusitanian types; viz.—Helix 12, Nanina 4, Streptaxis 1, (Cochin-China), Bulimus 5, Achatina 1, Pupa 1, *Clausilia* 11, Succinea 1, *Helicarion* 6, Cyclophorus, 1, Cyclotus 1, Otopoma 1. In the I. of Chusan Dr. Cantor discovered the genera *Lampania* and *Incilaria*. The most characteristic bivalves are *Glaucomya Sinensis* and *Symphynota plicata*; 3 species (or varieties) of Cyrena and 9 Corbiculas are described by Deshayes, and a *Planorbis* by Dunker.

In the Japanese and Loo-choo Islands only 9 species of Helix, 2 of Nanina, 2 of Clausilia and 2 of Helicarion have been hitherto obtained.

8. Philippine Islands.

The extraordinary richness of these islands has been developed mainly by the researches of Mr. Cuming. The *Helicidæ* (above 300) are inferior in number only to those of Lusitania and the Antilles, and vastly superior in size and beauty of colouring. The *Cyclostomidæ* (55) are not much fewer than in India. Nearly all the species are confined to particular islands, and the repetition of forms makes it probable that many of them are geographical varieties. The climate is equable, with a temperature like that of S. China (66° —84°) woods are prevalent, and the rains heavy—all circumstances favourable to the *individual* abundance of land snails.

Helix	152	Clausilia	1	Cyclotus	6
Nanina	32	Vitrina	14	Megalomastoma	1
Helicarion ?	3	Cyclophorus	15	Pupina	9
Bulimus	95	Leptopoma	16	Helicina	7

The Helices belong in great part to the section *Callicochlias* (Ag.) and *Helicostyla* (mirabilis) Fér. Some with sharply-keeled whirls have been called *Geotrochi* (Iberus of Albers.) The Bulimi are chiefly of the section *Orthostylus* (Beck), large and highly coloured, with a *hydrophanous* epidermis, the bands becoming translucent when wetted; others, like the well-known *B. perversus*, represent the typical Brazilian forms. To these islands belong most of the helicina-shaped *Cyclophori* (*Leptopoma*.)

The fresh-water shells are numerous; above 100 were obtained by Mr. Cuming, including many species of *Melania* (54 ?) *Navicella lineata* and *suborbicularis*, 5 sp. of *Glaucomya*, *Unio verecundus*, a *Corbicula*, and 11 sp. (?) of *Cyrena*.

Celebes and *Moluccas*. From these islands we have on record, at present, 16 sp. of Helix, Nanina 19, Bulimus 3, Vitrina 2, (viridis and flammulata, Quoy), Cyclophorus 1. In the fresh-water ponds and rivulets Mr. A. Adams found sp. of Melania, Assiminea, Ampullaria and Navicella; Auricula subulata, and Conovulus leucodon. Neritina sulcata was found on the foliage of trees several hundred yards from the water.

9. JAVA.

The Java group, including Floris and Timor, have been partially explored from the head-quarters of the Dutch settlement at Batavia. The land and fresh-water shells are nearly all peculiar, a few only being common to the Philippines and N. Australia; they have been described and figured by M. Albert Mousson (8vo. Zurich, 1849, 22 plates).

Helix	15	Platycloster?	3	Navicella	2
Nanina	8	ˋMeghimatium	2		
Ariophanta	1	—		Unio and	} 4
Bulimus	10	Limnæa	1	Symphynota	}
Clausilia	6	Auricula	2	Alasmodon	2
Cyclophorus	4	—		Anodon	1
Cyclotus	2	Melania	5	Cyrena	7
Leptopoma	1	Ampullaria	1	Corbicula	4
Parmacella	3	Neritina	2		

10. BORNEO.

The land shells of this great island are almost unknown, and the only reason for mentioning it separately is the doubt whether it should be considered part of the Javanese Province, or associated with the Moluccas and Philippines.

Helix	12	Paxillus	1	Leptopoma	3
Nanina	8	Succinea	2	Cyclotus	1
Bulimus	1	Cyclophorus	2	Pterocyclus	2

The freshwater bivalves are Glaucomya rostralis, Corbicula tumida and Cyrena triangularis. *Pholas rivicola* was found burrowing in floating logs used as landing places, 12 miles from the sea, up the Pantai river. The mangrove swamps abound with Cerithidium, Terebralia Telescopiúm, Potamides palustris and Quoyia; Auricula Midae and Polydonta scarabæus inhabit the damp woods.

11. PAPUA AND NEW IRELAND.

The landshells of New Guinea are nearly all distinct from those of the Philippines and Moluccas and include some related to the Polynesian types. The Louisiade Islands to the south-east and New Ireland on the North of New Guinea are included with it.

Helix	26	Partula	3	Leptopoma	1
Nanina	4	Pupina	3	Cyclotus	1
Bulimus	2	Otopoma	1	Helicina	2

Cyrenæ are numerous in this region. *Cyclostoma australe* is common to the Australian Islands and New Ireland; *C. Massenæ* to Australia and New Guinea, and *C. Vitreum* to New Ireland, New Guinea, the Philippines and India.

12. AUSTRALIAN REGION.

Both Fauna and Flora of Tropical Australia are distinct from those of New South Wales and Tasmania, the principal barrier being the desert character of the interior; but the localities of the landshells have not been defined with sufficient accuracy to shew whether they are equally distinct. The most complete list is given by Prof. E. Forbes, in the Appendix to Mc Gillivray's Narrative of the Voyage of H. M. S. Rattlesnake (1846-50); it specifies 48 Helices (of which *H. pomum* is the most conspicuous), 10 Bulimi, an Achatina, 6 Vitrinas (*Helicarion*) belonging to the main land, and one from the Lizard Islands, and a dextral *Balea* (australis). Pupa and Helicina (*Gouldiana*) are only found on the islets off the N. E. coast, and Pupina (*bilinguis*) at C. York and the adjacent islets; a portion of the province which is densely wooded, and lies within the *rain region* of the Asiatic Islands. *Cyclostoma bilabre* of Menke's Catalogue is probably West Indian. The fresh-water shells of Australia are *Planorbis Gilberti, Iridinae?* (Victoria R.) *Unio auratus, cucumoides, superbus, (Hyridella) australis, Corbicula* 4 sp. *Cyrena* 3, *Cyclas egregia* (Hunter R.)*Pisidium semen* and *australe*, the last common to Timor.

13. S. AUSTRALIA and TASMANIA.

From extra-tropical Australia we have the following:—Helix 9, Helicarion 2, Bulimus 2, Succinea 1 (common to Swan R. and Tasmania) Limax olivaceus, and one Ancylus. Two of the largest land snails, *Helix Cunninghami* and *Falconeri*, are found in N. South Wales. The coasts of this region are thinly wooded, but much of it is rendered desert by want of rain; in N. S. Wales droughts recur at intervals of twelve years, and sometimes last three years, during which time scarcely any rain falls.

14. NEW ZEALAND.

The moist and equable climate of these islands (which have a mean temp. of 61°—63°) is favourable to the existence of numerous land snails. Nearly 100 species of land and fresh-water shells are already determined, and are all peculiar; the genus *Helix* musters 60 species, some of which (including the great *H. Busbyi*) resemble in shape the European *Helicellae;* Bulimus 3; Balea (peregrina), Vitrina 2 of peculiar form, Tornatellina 1, Cyclophorus cytora and Omphalotropis egea. There are two slugs, Limax antipodarum and Janella bitentaculata; two fresh-water *pulmonifera*, Physa variabilis and Latia neritoides; several marine air-breathers,—Oncidium (*Peronia*) 2, Siphonaria 3, Amphibola 1 (*avellana*). The other fresh-water shells are Melanopsis trifasciatus (a Lusitanian type), Assiminea antipodarum and Zelandiæ, Amnicola? corolla, Cyclas Zelandiæ and Unio Menziesii and Aucklandicus.

Vitrina zebra is found at the Auckland Islands.

15. Polynesian Region.

The Pacific Islands are partly the volcanic summits of submerged mountain ranges, usually fringed or surrounded with coral reefs ; and partly *atolls* or lagoon islands, scarcely rising above the sea and presenting no vestige of the rock on which they are based. The low coral-islands form a long stream of archipelagos, commencing in the west with the Pelews, Carolines, Radack, Gilbert, and Ellice groups, then scattered over a wider space and ending eastwards in the Low Archipelago ; they are chiefly, perhaps entirely, colonized by drift from the other islands.

The volcanic groups are the Ladrones, Sandwich Islands and Marquesas, to the north of the low coral zone ; and to the south of it, the Salomons, New Hebrides, New Caledonia and Feejees,—the Friendly Islands, Navigator's and Cook's Islands,—Society and Austral Islands, ending with Pitcairn's and Elizabeth Island. Many of these are very lofty, and are perhaps the most ancient land in the world.* Their molluscan fauna is entirely peculiar, but it has most affinity with those of New Zealand and the Asiatic Islands, and great analogy with those of St. Helena, Brazil, and the W. Indies.

Salomons—New Hebrides—New Caledonia—Feejees.

The most remarkable land-shells of these islands are the great auriculoid, *Bulimi* (e. g. *B. auris-bovinæ* and *B. miltochilus* of the Salomons). *Acicula striata* and 2 sp. of *Cyrena* are found at Vanicoro ; and *Physa sinuata Peronia acinosa* and *corpulenta*, and several Neritinas and coronated Melanias have been obtained at the Feejees.†

Helix	18	Bulimus	10	Cyclophorus	2
Nanina	2	Partula	6	Omphalotropis	1
Vitrina	6	Acicula	1	Helicina	6

Friendly Islands—Navigator's—Society Islands.

The principal lofty and rocky islands of the southern Pacific, at which land-shells have been obtained, are Tonga, Samoa, Upolu and Manua ; Taheiti, Oheteroa, and Opara ; Pitcairn's Island and Elizabeth Island. Each appears to have some peculiar species and some common to other islands ; the little raised coral islet Aurora (*Metia*) N. E. of Taheiti, 250 feet in elevation, has four land snails which have been found nowhere else ;— *Helix pertenuis, dædalea, Partula pusilla, Helicina trochlea.* " Samoa and the Friendly Islands must have intimate geological relations ; the same forms, and many of the same species of land-shells occur on both groups ; not a single Feejeean species was collected on either."—(*Gould.*)

---

* Islands composed partly of stratified rocks must be *newer* than those rocks ; Volcanic Islands may be of any degree of antiquity.

† The Feejees (*Viti*) are more nearly allied to the westward islands, such as the New Hebrides, than the Friendly Islands. *Succinea* and *Partula*, so plentiful at the latter, are not found at the Feejees.—(Gould, U. S. Exploring Expedition,)

Helix 13	Tornatellina........ 6	Cyclophorus 5
Nanina 18	Pupa 3	Omphalotropis 6
Bulimus 1	Succinea 12	Helicina 13
Partula 15	Electrina 1	

The fluviatile shells are species of *Physa, Melania, Assiminea* (*Taheitana*), *Neritina,* and *Navicella;* the two last being often litoral, or even marine in their habit.

Low Coral-islands.

The Atolls, or lagoon-islands, are less prolific; 2 *Helices,* and 2 *Partulæ* are found at Oualan, in the Caroline Archipelago; and from Chain Island (*Annaa*), the centre of commerce in the eastern Archipelago, have been obtained.—*Helix* 2 sp., *Nanina* 1, *Partula* 1, *Tornatellina* 1, *Cyclophorus* 1, and *Melampus mucronatus.*

Sandwich Islands.

The land shells of these islands exceed 100, and are all, or nearly all, peculiar; there is one *Limax;* and in the fresh-waters are found *Limnæa volutatrix, Physa reticulata* (Gould), *Neritopsis?* *Neritina Nuttalli* and *undata,* and *Unio contradens* (Lea).

In the I. Kaui, two species of *Achatina* have been found; the Achatinellæ are elongated (*Leptachatina,* G.) and the Helices planorboid and multispiral. In Molokai the Achatinellæ are large and coloured. In Maui and Oahu the Helices are small and glabrous, or hispid, ribbed and toothed. In Hawaii, Succineas prevail, and Achatinellae are rare.—(*Gould*).

Helix 13	Achatina 3	Pupa 2
Nanina 4	Achatinella 56	Vitrina 2
Bulimus 5	Tornatellina........ 3	Succinea 10
Partula 4	Balea.............. 1	Helicina 6

The Island of Guam, Ladrones, has 3 sp. of *Partula,* 2 of *Achatinella,* and 1 *Omphalotropis.* At the Marquesas have been found 3 sp. of *Nanina,* 1 *Partula* and 1 *Helicina.*

NEW WORLD.
16. CANADIAN REGION.

The country drained by the Great Lakes and the river St. Lawrence possesses very few peculiar shells, and these mostly of fresh-water genera. It is chiefly remarkable for the presence of a few European species, which strengthen the evidence before alluded to (p. 358.) of a land-way across the north Atlantic having remained till after the epoch of the existing animals and plants.*

* For example, the common Heather (*Calluna vulgaris*), one of the most abundant *social* plants of Europe, characteristic of the moorland zone, and seldom rising above 3000 feet on the mountains of Scotland.—(*Watson.*) According to Pallas it abounds on the western flanks of the Ural Mountains, but disappears on their eastern side and is not found in Siberia. In the *Pliocene period* it appears to have spread itself nothward and westward to Iceland, Greenland and Newfoundland, where it still grows, *the only heath indigenous to the New World.*—(Humboldt.)

Helix hortensis (imported) coast of New England and banks of St. Lawrence.
„ pulchella (smooth var. only) Boston, Ohio, Missouri.
Helicella cellaria (glaphyra, Say ?) N. E. and middle States.
„ pura, nitida and fulva ?
Zua lubrica, North West Territory.
Succinea amphibia (= campestris, Say ?)
Limax agrestis (= tunicatus, G.) Mass.
„ flavus, New York, introduced.
Vitrina pellucida (= Americana ?)   Limnæa palustris (= elodes, Say ?)
Arion hortensis, New York (Dekay.)   „   truncatula (= desidiosa ?)
Aplexa hypnorum (= elongata, Say ?)
Auricula deticulata, Mont., New York Harbour.•
Alasmodon margaritiferus (= arcuatus, Barnes.)
Anodon cygneus (= fluviatilis, Lea ?)

The shells proper to Canada, or derived from the adjoining States, are only 6 sp. of Helix, 2 Succineas, and 1 Pupa; 8 sp. of *Cyclas* have been obtained from the region of Lake Superior.

The following species occur in New England :—

Helix	13	Physa	2	Unio	5
Succinea	2	Planorbis	11	Alasmodon	2
Pupa	7	Paludina	1	Anodon	2
Limnæa	7	Valvata	2	Cyclas	6
Ancylus	2	Auricula	1	Pisidium	1

Carychium exiguum, Say, is found in Vermont, and *Limnæa* (Acella) *gracilis* in Lake Champlain; *Valvata tricarinata* and *Paludina decisa* are characteristic forms.

The genera *Clausilia* and *Cyclostoma* are entirely wanting in Canada and the Northern states. The *Limacidæ* are represented by *Philomycus*, of which there are 9 reputed species, ranging from Mass. to Kentucky and South Carolina.

17. ATLANTIC STATES.

The parallel of 36° N. Lat. forms the boundary-line of two botanical regions in the U. States, but the evidence of the fresh-water shells, in which they are particularly rich, seems to favour a division into two hydrographical provinces,—the region of the Atlantic streams and the basin of the Mississippi. Above 50 fresh-water *Pulmonifera*, 150 *pectinibranchiata*, and 250 bivalves are reputed to be found in the States, and it is supposed that only a few species are common to both sides of the Alleghanies. *Cyclas mirabilis, Pisidium Virginicum, Cyrena Carolinensis*, and *Unio complanatus* and *radiatus* are characteristic of the eastern rivers; *Melania depygis* is said to be the only member of that large genus found eastward of the Hudson River. Of the American land-snails, 29 sp. of Helix, 6 Succineas, and 13 Pupas are enumerated from the Atlantic States. In Florida the propinquity of the West Indian Fauna is strongly indicated by the occurrence of the great *Glandina truncata*, by species of *Cylindrella*, and a *Helicina*. A Cuban species of

Chondropoma (C. dentatum), is also said to occur in Florida, and *Ampullaria depressa* in Florida and Georgia.

18. AMERICAN REGION.

The mass of American land and fresh-water shells are found in the central and southern states, the country drained by the Mississippi and its tributaries. The *Helicidæ* are not more remarkable for size and colour than those of northern Europe ; the most characteristic forms belong to the subgenus *Polygyra* (or *Tridopsis*, Raf.), such as *Helix tridentata, albolabris, hirsuta,* and *septemvolvis.* The truly North American forms all belong to three genera, viz.—Helix 43, Succinea 8, Pupa 3 species. In the Southern States are also found 5 sp. of Bulimus, 3 Cylindrellas, 2 Glandinas, and 5 Helicinæ, genera whose metropolis is in the Antilles or in tropical America.

The fresh-water univalves include above 100 species of *Melaniadæ* belonging to the genera *Ceriphasia, Melafusus, Anculotus, Melatoma,* and *Amnicola,* 15 *Paludinæ,* some keeled, and one muricated, (*P. magnifica*) ; and species of *Valvata, Limnæa, Physa,* (15) *Planorbis,* and *Ancylus,* (5).

The fresh-water bivalves are also extremely numerous ; the *Unionidæ* are unequalled for their ponderous solidity, the rich tinting of their interiors, and the variety of their external forms.* *Gnathodon cuneatus, Cyrena floridana,* 16 sp. of *Cyclas,* and *Pisidium altile,* belong to this region.

19. OREGON AND CALIFORNIA.

The Fauna of the region beyond the Rocky Mountains is believed to be almost entirely distinct from that of the United States. *Arion* (foliolatus) and *Limax* (Columbianus,) genera not indigenous to eastern America, were found near Puget Sound, (*Gould*). We have no information respecting the land and fresh-water shells of Russian America, but from analogy we may expect to find a few there identical with those already mentioned as occurring in Siberia.†

The shells of Oregon and California are as yet only imperfectly known by the researches of Mr. Nuttall and Mr. Conthouy.

Helix	22	Physa	1	Cyrena	2
Bulimus	1	Ancylus	2	Cyclas	1
Achatina	1	Planorbis	3	Unio	1
Succinea	4	Melania	2	Alasmodon	1
Limnæa	4	Potamides	2	Anodon	3

Limnæa fragilis, a Canadian species, is said to range westward to the Pacific ; and *L. jugularis* to be common to Michigan, the North-west terri-

---

* The private cabinet of Mr. Jay contains above 200 species of North American *Unionidæ,* and very many varieties.

† The affinity between the *Mammalia* of the Old and New Worlds is greatest in eastern Asia and north-west America, and diminishes with distance from those regions. —(*Waterhouse,* in Johnston's Physical Atlas, No. 28.)

tory and Oregon(De Kay.) *Limnæa umbrosa*, Say? and *Planorbis·corpulentus* Say, are found in the Columbia R.

20. MEXICAN REGION.

The lowlands of the northern half of Tropical America constitute only one botanical region, extending from the R. Grande del Norte to the Amazon; but on zoological grounds it may be divided into two smaller areas. The Mexican province, including Central America, itself comprises three physical regions; the comparatively rainless and treeless districts of the west; the mountains or high table-lands with their peculiar flora; and the rainy wooded region that borders the Caribbean Sea. The land snails of Central America resemble those of the Antilles in the prevalence of some characteristic genera—*Glandina, Cylindrella* and *Helicina*,—of which very few species are found on the northern Coast of the Gulf of Mexico. The *Bulimi* are nume-rous but chiefly thin, translucent species.

Helix 33	Glandina 25	Cistula 7
Proserpina 1	Tornatellina 1	Cyclophorus 3
Bulimus 30	Pupa 1	Chondropoma........ 3
Succinea 6	Cylindrella 11	Megaloma............ 2
Achatina (Spiraxis).. 12	Cyclotus 1	Helicina 22

Amongst the fresh-water shells are *Neritina picta, Cyclas macula'a, Cor-bicula convexa*, and 7 species of *Cyrena*. From Mazatlan, Mr. Carpenter describes *Cyrena olivacea* and *Mexicana, Gnathodon trigonus, Anodon ciconia* (allied to the Brazilian *A. anserina*), *Physa aurantia* and *elata, Pla-norbis* sp. *Melampus olivaceus*. Two brackish-water species, *Cerithidium varicosum* and *Montagnei*—are common to S. America.

21. ANTILLES.

The West Indian Islands have supplied nearly 500 species of *Helicidæ*, a larger number than any province except the Lusitanian; and above 260 *Cy-clostomidæ*, or nearly 3 times as many as India. They are also richest in generic forms, and the climate is highly favourable to the .multiplication of individuals. The mean temp. of the Antilles is 59°—78°, and the annual fall of rain exceeds 100 inches in most of the islands.

Helix 200	Pupa................ 26	Cyclophorus........ 1
Stenopus 2	Cylindrella.......... 73	Cyclotus............ 14
Sagda 20	Clausilia............ 1	Megaloma.......... 8
Proserpina 5	Balea 1	Helicina............ 43
Bulimus............ 53	Succinea 16	Alcadia 17
Achatina 27	Chondropoma 15	Trochatella 16
Glandina 46	Choanopoma........ 53	Lucidella 6
Spiraxis 9	Adamsiella.......... 10	Stoastoma 20
Tornatellina 1	Cistula.............. 36	Geomelania 21

Probably every island has some peculiar species, and those of the great islands, like Cuba and Jamaica are nearly all distinct. To Jamaica belong the species of *Stoastoma*, *Sagda* and *Geomelania*, the small subgenus *Lucidella*, the *Alcadias* and the mass of beautiful Cyclostomas with a decollated spire and fringed lip (*Choanopoma*, *Adamsiella*, *Jamaicia*, *Chondropoma*, part, and *Cistula*, part.* The solitary *Clausilia* is found in P. Rico, the *Balea* in Haiti, and the *Tornatellina* in Cuba; *Stenopus* is peculiar to St. Vincents. Bermuda has 4 Helices of which one is common to Texas and one to Cuba. The Chondropomas are found in Cuba and Haiti.

The West Indian *Achatinæ* belong to the subgenera *Glandina*, *Liguus*, and *Spiraxis*; the *Bulimi* are sharp-lipped and mostly small and slender (*Subulina*, *Orthalicus*). *Helix* (Sagda) *epistylium*, *H. Carocolla*, and *Succinea* (Amphibulima) *patula* are characteristic forms.

Although connected with Florida by the chain of the Bahamas, and with Trinidad by the Lesser Antilles, very few species are common to the mainland of either North or South America; the relation is generic chiefly.

The *Limacidæ* are represented by *Vaginulus* (Sloanei); and in the freshwaters there are species of *Physa* (3,) *Planorbis*,8, *Ancylus* and the peculiar *Gundlachia*, *Valvata pygmæa*, *Ampullaria* (fasciata), *Paludestrina* (minute sp.) *Hemisinus*, and 2 sp. of *Pisidium*.

In the brackish-waters are *Cerithidium*, *Neritina* (e. g. meleagris, pupa, virginea, viridis), *Melampus* (coniformis) and *Pedipes quadridens*.

22. COLUMBIAN REGION.†

The tract shaded in the map comprehends several minor regions; 1, the rainy and wooded states of New Granada and Ecuador; 2, the elevated and nearly rainless province of Venezuela, with a flora like that of the higher regions of the Andes; 3, the Guianas, including the Valley of the Amazon, where the forests are most luxuriant, and rain falls almost daily (amounting to 100 or even 200 inches in the year). Most of the low lands, like those of the Mexican Province, belong to the " Cactus Region" of botanists, and have a mean temp. of 68°—84°. Landshells are abundant in the forests and underwood of the lower zone of the mountains, where the temperature is 10° less and the rains more copious. *Bulimi* are the predominant forms, especially the succinea-shaped species, (e. g. *B. succinoides*).

Helix 37	Pupa 7	Cistula ....
Streptaxis............ 3	Clausilia 3	Bourciera ....
Bulimus .!............ 45	Cylindrella 1	Cyclotus ....
Succinea .:.......... 9	Vitrina 1	Adamsiella ....
Tornatellina.......... 1	Limax 1	Helicina ....
Achatina 10	Choanopoma 2	Trochatella ....
Glandina 5	Cyclophorus 2	

* A magnificent collection of Jamaica land shells has been presented to the Britisl Museum by the Hon. E. Chitty whose researches were conducted with the late Prof C. B. Adams.

† In 1821 the states of New Granada, Venezuela and Ecuador united to form th " Columbian Republic," but dissolved again in 1831.

The presence of several species of the old-world genera *Clausilia* and *Streptaxis*—both wanting in North America, becomes a significant fact when taken in connection with the affinities of the higher animals of South America and Africa. These imply a land-way across the Atlantic (at some *very remote* period,) more direct than would be afforded by the continent which is believed to have united the boreal regions at the close of the Miocene Age.*

Corbicula cuneata and 3 sp. of *Cyrena* are found in the Orinoco and smaller rivers; and the remarkable genus *Mülleria*, representing the African *Ætheria*, inhabits the Rio Magdalena. A sp. of *Ancylus* is recorded from Venezuela.

Galapagos Islands. No. 35.

The fauna and flora of these islands is peculiar, but related to tropical South America. The only known land-shells are 11 small and obscure species of *Bulimus*, of which the most remarkable is *B. achatinellinus*. Some of them are peculiar to particular islands, like the birds and reptiles, viz:— Chatham I. 2, Charles I. 3, Jacob I, 2, James I. 1. "The Archipelago is a little world within itself, or rather a satellite attached to America, whence it has derived a few stray colonists, and has received the general character of its indigenous productions."—(Darwin's Journal, p. 377.)

23. Brazilian Region.

The "region of Palms and Melastomas," extending from the Amazon to the southern tropic, is one of the richest zoological provinces. It includes Bolivia, and the largest portion of Peru, all that lies to the east of the Andes. The greater part of the region is mountainous and rainy and densely wooded, but intersected by extensive plains (*Llanos*), some grassy and fertile, others dry, rocky and rainless, especially in the south; it is watered by numerous streams—the affluents of the Amazon and Plata. The hydrographical areas of these two great rivers have been represented on the map, but the southern boundary of the Brazilian Province extends beyond the line of watershed to the tropic, including the head-waters of the Plata, in which the same remarkable fresh-water bivalves are found as in the Bolivian streams. (*D'Orbigny*). The mountains around the Lake Titicaca are the highest in the New World, and there M. D'Orbigny found several species of Helix up to the elevation of 14,000 feet; *Bulimus Tupaici* ranges to 9,000 feet. The large and typical species of *Bulimus* belong to this province; *B. ovatus* and *oblongus* are found near the coast, (p. 164,) and *B. maximus* farther inland. The auriculoid *Bulimi*, (*Otostomus*, and *Pachyotis*, Beck,) those with an

* In Lieut. Maury's physical map of the Atlantic, the contour of this former land is partly shewn by the 2000 fathom line, extending beyond the Canaries and Madeira, and sending out a promontory to the Azores. *Clausiliæ* are found in Eocene strata; perhaps even in the Coal-measures, (p. 160.)

T

angular mouth, (*Goniostomus*, Beck,) and the pupiform species, with a toothed aperture, (*Odontostomus*,) are characteristic of this region, and also some of the most elongated forms, (*Obeliscus*). The lamp snails (*Anastoma*) and *Megaspira*, genera inhabiting France during the Eocene period, are now peculiar to Brazil; *Simpulopsis* is also peculiar, and *Streptaxis* attains its maximum there. The *Cyclostomida* are few, and the other W. Indian forms have almost disappeared.

Helix	34	Glandina	1	Cyclophorus	2
Streptaxis	9	Tornatellina	1	Cyclotus	1
Anastoma	7	Vitrina	5	Cistula	1
Bulimus	172	Omalonyx	1	Helicina	12
Megaspira	2	Simpulopsis	5		

The land slugs are *Peltella palliolum, Vaginulus solea, and Limax andicolus.* The fresh-waters of the interior are rich in bivalves of peculiar genera;*—

Physa	1	Ampullaria	2	Unio	4
Ancylus	1	Corbicula	2	Iridina	1
Planorbis	4	Pisidium	1	Hyria	1
Paludestrina	2	Anodon	1	Castalia	2
Marisa	1	Monocondylæa	1	Mycetopus	3

24. Peruvian Region.

The long and narrow tract between the Andes and Pacific, extending from the equator to 25° S. lat. forms a distinct, though comparatively unproductive province, including the coast of Ecuador, Peru and Bolivia. It is warm and almost rainless; the clouds discharge themselves on the east side of the Andes, and rain is so rare on the west coast that in some parts it only falls two or three times in a century. In Peru, during great part of the year, a vapour rises in the morning, called the "garua;" it disappears soon after mid-day, and is followed by heavy dews at night.

Mr. Cuming collected 46 species of land snails in Peru; and Dr. Pfeiffer enumerates 100, but perhaps half the latter were from the eastern side of the Andes, belonging to the Brazilian Province. They are mostly *Bulimi*, and are smaller and less richly coloured than those of Bolivia and Brazil; *B. Denickei, solutus,* and *turritus* are peculiar forms. *Cistula Delatreana* is the only operculated land snail, and *Vaginulus limayanus* the only slug.

Helix	12	Pupa	1	Ancylus	1
Bulimus	79	Balea	1	Ampullaria	1
Succinea	5	Cistula	1	Paludestrina	2
Glandina	1	Physa	1	Cyrena	3
Tornatellina	1	Planorbis	3	Anodon	1

25. Argentine Region.

The "region of arborescent Compositæ" has afforded scarcely any land

---

* The American Expedition explored 40 Brazilian streams, and found only 1 *Ampullaria*, 1 *Melania*, and 1 *Planorbis.*—(Gould.)

snails, only 7 species of *Bulimus,* and 3 *Helices* are recorded, but some
others may have been included with those of Brazil and Chile. From Bolivia
this province is separated by the wide plains of the Great Desert, or northern
prolongation of the Pampas; and all the eastern part has been submerged at
a recent (geological) period; so that the only promising districts are
Paraguay, and the eastern declivities of the Chilian Andes. The fresh-water
shells of the La Plata and its tributaries are more remarkable; M. D'Orbigny
gives the following :—

Chilinia	1	Cyclas	1	Byssoanodon	1
Planorbis	1	Pisidium	1	Monocondylæa	6
Ancylus	2	Corbicula	2	Mycetopus	1
Ampullaria	7	Unio	7	Castalia	1
Asolene	1	Anodon	10	Iridina	1
Paludestrina	7				

Ampullaria (Marisa) *cornu-arietis* is a characteristic shell; *Paludestrina
lapidum* has a claw-like (non-spiral) operculum, and appears to belong to the
Melaniadæ.

26. CHILIAN REGION.

The northern part of Chile belongs to the same physical region with Peru,
consisting of dry and rainless plains. Here the land snails are few and small,
and only seen after the dews. At Valparaiso rain is abundant during the
three winter months, and the southern coasts are luxuriantly wooded, and
extremely wet. The characteristic pulmonifera are the fresh-water *Chilinias.*
The genus *Buchanania* is doubtful. There are 25 sp. of Bulimus (including *B. Chilensis,* Plectostylus) and 4 of Helix; *Succinea Chiloensis, Ancylus
Gayanus* (Valparaiso), *Planorbis fuscus, Paludestrina* sp. *Unio Chilensis,
Pisidium Chilense* (Valdivia). *Helix Binneyana* is found on the Island of Chiloë.

The Island of *Juan Fernandez* (36) has at least 20 species of land shells,
all peculiar to it :—

Helix quadrata	Omalonyx Gayana	Tornatellina minuta
„ arctispira	Achatina diaphana	„ trochiformis
, pusio	„ splendida	Succinea Cumingi
„ tessellata	„ bulimoides	„ mamillata
, ceroides	„ conifera	„ fragilis
„ marmorella	„ acuminata ?	Parmacella Cumingi.
„ helicophantoides	Spiraxis consimilis	

In the adjoining Island, Masafuera, are found—

Tornatellina Recluzii	Succinea semiglobosa
Succinea rubicunda	„ pinguis

27. PATAGONIAN REGION.

The Pampas, or great plains of Patagonia are dry and rainless nearly all
the year; the vegetation which springs up during the light summer rains
becomes converted into natural hay for the support of the wild animals. In

T 2

Fuegia the mean temperature is 33°—50°, and there is rain and snow throughout the year; yet the bases of the mountains are clothed with forests of evergreen beech.* *Bulimus sporadicus* is found on the banks of the River Negro, and *B. lutescens* at the Straits of Magellan; *Helix lyrata* (costellata, D'Orb. ?) and *H. saxatilis* inhabit Fuegia. *Succinea magellanica* is also found at the Straits, and *Chilinia fluminea, Limnæa viatrix,* a *Paludestrina, Anodon puelchanus,* and *Unio Patagonicus* in the River Negro. *Peronia marginata* and *Potamides cælatus* were discovered in Fuegia by Mr. Couthouy.

The *Falkland Islands* are 300 miles east of Patagonia, and the only recorded shells are two species of *Paludestrina*. There is zoological evidence that these islands were united to the mainland of S. America at no very distant geological period. The flora consists of characteristic plants of Fuegia and Patagonia, mingled, and overspreading the whole surface; few species are peculiar. (J. D. Hooker).†

Since the preceding pages were in type we have seen the following remark by Dr. Gould, referring to certain statements about the distribution of shells (p. 354). "The doctrine of *distinct zoological regions* is well illustrated by the mollusca. The many thousand localities carefully noted on the records of the American Exploring Expedition go to prove beyond dispute, that no such random or wide-spread distribution obtains."

* Humming-birds are seen fluttering about delicate flowers, and parrots feeding amidst the evergreen-woods. (*Darwin*, p. 251.)

† Dr. Hooker has suggested that not only the Falkland Islands, but the far distant Tristan d'Acunha (p 390) and Kerguelen's-land (p. 392) may be mountain-tops of a continent which has been submerged since the epoch of their existing flora. "There are five detached groups of islands between Fuegia and Kerguelen's land, (a region extending 5,000 miles,) all partaking of the botanical peculiarities of the southern extremity of the S. American continent. Some of these detached spots are much closer to the African and Australian continents, whose vegetation they do not assume, than to the American; and they are situated in latitudes and under circumstances eminently unfavourable to the migration of species."

"The botany of Tristan d'Acunha (which is only 1,000 miles distant from the Cape of Good Hope, but 3,000 from the Straits of Magellan) is far more intimately allied to that of Fuegia than Africa. Of 28 flowering plants, 7 are natives of Fuegia, or typical of S. American botany."

"The flora of Kerguelen's-land is similar to, and many of the species identical with, those of the American continent. (Its geological structure) would bespeak an antiquity for the flora of this isolated speck on the surface of our globe, far beyond our power of calculation. We may regard it as the remains of some far more extended body of land."—(Botany of Antarctic Voyage, I. Pt. 2, 1847.)

GEOGRAPHICAL DEVELOPMENT.

Rough estimate of known Species proper to each Province.

MARINE PROVINCES.		LAND REGIONS.	
I. Arctic	100	1. Germanic	100
II. Boreal (New England)	200	2. Lusitanian	900
		3. African	150
III. Celtic	250	4. Cape	60
IV. Lusitanian (Medit. Madeira, &c.)	450	5. Mascarene	150
		6. Indian	350
		7. Chinese	50
V. Aralo-Caspian (N. Euxine)	30	8. Philippine	350
		9. Javanese	80
VI. West African (St. Helena)	500	10. Bornean	30
		11. Papuan	80
VII. South African	350	12. Australian	80
VIII. Indo-Pacific	4000	13. Austro-Tasmanian	50
IX. Austro-Zelandic (Tasmania)	400	14. Zelandic	80
		15. Polynesian	300
X. Japonic	300	16. Canadian	30
XI. Aleutian (Ochotsk)	100	17. Atlantic States	60
		18. American	80
XII. Californian.	250	19. Californian	30
XIII. Panamic (Galapagos)	1000	20. Mexican	170
		21. Antillean	760
XIV. Peruvian	500	22. Equatorial	180
XV. Magellanic (Falklands)	100	23. Brazilian	260
		24. Peruvian	100
XVI. Patagonian	170	25. Argentine	50
XVII. Caribbean	1000	26. Chilian	60
XVIII. Trans-Atlantic	300	27. Patagonian	10

| Sea-Shells | 10,000 | Land-Shells | 4,600 |

The inequality of these provinces, in size and importance, is partly natural, and partly caused by the unequal facilities they present for sub-division. The "Indo-Pacific" is not of the same rank with the Japonic, but results from the fusion of several provinces. Mr. Waterhouse terms the great regions in which the large groups of animals are distinct, *ordinal* and *family* provinces; the smaller regions *generic* or *specific* provinces.—(Johnston's Physical Atlas, 28.)

CHAPTER III.

ON THE DISTRIBUTION OF THE MOLLUSCA IN TIME.

THE historian of modern geology, SIR CHAS. LYELL, has taught us to regard the stratified rocks as so many monuments, recording the physical condition and living inhabitants of the earth in past ages.

Each *formation* consists of a similar, and more or less complete series of limestones, sandstones, clay, coal, and other *strata,* representing the deep and shallow seas, the fresh-waters, and the terrestrial portions of the surface of the globe, at one particular period of time.*

The organic remains found in the strata exhibit no such repetitions, but are changed gradually and regularly, from the earliest to the latest formations; so that the *mass of species* in each period must have been peculiar and distinctive.

The important theory, that strata may be identified by fossils, was taught by WILLIAM SMITH, early in the present century, and is thus expressed in his *Stratigraphical System:* " Organized fossils are to the naturalist as coins to the antiquary; they are the antiquities of the earth; and very distinctly show its gradual, regular formation, with the various changes of inhabitants in the watery element."—"They are chiefly submarine, and as they vary generally from the present inhabitants of the sea, so at separate periods of the earth's formation they vary as much from each other; insomuch that each layer of these fossil organized bodies must be considered as a separate creation; or how could the earth be formed, *stratum super stratum,* and each abundantly stored with a different race of animals and plants."†

The "Prodrome" of M. D'Orbigny is a catalogue of the shells (and radiate animals) of each formation, from which it appears that the mass of the living population of the globe has been changed twenty times since the close of the First or Palæozoic Age; and although the fossils of the older rocks have not been generally classified with the same minuteness, yet enough is known to shew that at least ten great changes had taken place before the Secondary epoch.

In the following Table, the first column gives the names of the Formations, or Periods; the second contains those by which the principal strata are known.

---

* The coal-measures and chalk of England cannot indeed be called *similar,* but the Cretaceous formations of the whole *world* afford *mineral* types corresponding to perhaps every variety of Carboniferous rock.

† Stratigraphical System of Organized Fossils, 4to, Lond. 1817.

I. GEOLOGICAL TABLE.

FORMATIONS OR PERIODS.		NAMES OF STRATA.
PALÆOZOIC AGE.	I. { 1. Tremadocian.	{ Longmynd slate. (Bangor, Wicklow.) Lingula flags = Primordial group. Festiniog slate. Potsdam sandstone.
	2. Snowdonian.	{ Llandeilo flags } Bala or Coniston Caradoc sandstone } group.
	II. { 3. Wenlock.	{ May-hill sandstone = Clinton group. Woolhope and Dudley limestones.
	. Ludlow.	L. Ludlow, Aymestry lime.,.U. Ludlow.
	III. { ¯. Hercynian. ₄. Eifelian. 🜪. Clymenian.	Spirifer sandstone; Rhine. } Killas, or Plymouth limestone. } Old Red Petherwin limestone. } Sandstone.
	IV { 8. Bernician. 9. Demetian.	Carboniferous limestone (shale and coal.) Coal-measures. (Millstone grit, coal, &c.)
	V. 10. Permian.	Magnesian lime. = Zechstein. (Perm.)
SECONDARY AGE.	VI. { 11. Conchylian. 12. Saliferous.	{ New-red-sandstone = Bunter. (Muschel-kalk = Ceratite limestone.) Red marls = Keuper. Lias bone-bed.
	VII. { 13. Liassic. 14. Toarcian. 15. Bajocian. 16. Bathonian.	L. Lias = *Sinemurien* & *Liasien*. Marlstone, Alum-shale. (Thouars.) Inf. Oolite, Fuller's-earth. (Bayeux.) { Great Oolite. (Stonesfield slate; G. Ool. Bradford cl. Forest m. Cornbrash.)
	VIII. { 17. Oxfordian. 18. Corallian. 19. Kimmeridgian. 20. Portlandian.	{ Kelloway rock = *Callovien*, D'Orb. Oxford clay. (White Jura.) Coral-rag and Calcarious grit. Kimmeridge clay. (Dorsetshire.) Portland stone and Purbeck beds.
	IX. { 21. Wealden. 22. Neocomian.	Hastings sand and Weald clay. { Speeton clay? (Neuchatel.) Lower Green-sand, & *Aptien*, D'Orb.
	X. { 23. Albian. 24. Cenomanian. 25. Hippuritic. 26. Senonian.	Gault. (District of the Aube, or *Albe*.) Upper Green-sand. (Mans, *Cenomanum*.) Chalk-marl and L. Chalk = *Turonien*. { Chalk with flints = Baculite limestone. Maestricht chalk = *Danien*, D'Orb.
TERTIARY.	XI. { 27. Londinian. . 28. Nummulitic.	Thanet sands, Plastic clay, London clay. { Bracklesham; Barton; I. Wight; = *Parisien*. Hempstead; Fontainbleau; = *Tongrien*.
	XII. 29. Falunian.	Faluns of Touraine; Bordeaux, Vienna.
	XIII. 30. Icenian.	Crag of E. Co. = *Sub-apennin*, D'Orb.

It must be observed that the number and magnitude of the " Formations " was determined by accident in the first instance, and afterwards modified to suit the requirements of theory, and to make them more nearly equal in value.[*]

According to MM. Agassiz and D'Orbigny, all, or nearly all the fossils of each formation are peculiar; very few species being supposed to have survived from one period to another. Sudden and entire changes of this kind only take place when the nature of the deposit is completely altered,—as when sands or clays rest upon chalk;—and in these instances there is usually evidence (in the form of beds of shingle, or a change of dip) that an interval must have elapsed between the completion of the lower stratum and the commencement of the upper.

We have seen that distinct faunas may be separated by narrow barriers in existing seas; and differences almost as great may occur on the same coastline without the interposition of any barrier, merely in passing from a sea-bed of rock and weed to one of sand or mud, or to a zone of different depth. It would be unreasonable to expect the same fossils in a limestone as in a sandstone; and even in comparing similar strata we must consider the probability of their having been formed at different depths, or in distinct zoological provinces.

The most careful observations hitherto made, under the most favourable circumstances, tend to show that all sudden alterations have been *local*, and that the law of change over the whole globe, and through all time, has been gradual and uniform. The hypothesis of Sir C. Lyell—that species have been created, and have died out, *one by one*, agrees far better with facts, than the doctrine of periodic and general extinctions and creations.

As regards the Zoological value of the " formations," we shall be within the truth if we assume that those already established correspond in importance with geographical provinces; for at least half the species are peculiar, the remainder being common to the previous or succeeding strata. This will give to each Geological Period a length equal to three times the average duration of the species of marine shells.[†]

---

[*] The names of Formations are in great measure provisional, and open to criticism. Some of them were given by Brongniart and O. D'Halloy; others have been more recently applied by D'Orbigny, Sedgwick, Murchison, and Barrande; and some are adopted from popular usage. *Geographical* names, and those derived from characteristic fossils have been found the best, but no complete scheme of *zoological* nomenclature has been framed.

The epithet " Turonien " (25) is rejected, because it conveys the same meaning with " Falunian " (29), or Middle Tertiary, the type of which was taken from Touraine.

The term *Icenian* is proposed for the Pliocene strata, because their order of succession was first determined, by Mr. Charlesworth, in the eastern counties of England, the country of the ICENI.

[†] The exact value of these *periods* cannot be ascertained, but some notion of their length may be obtained by considering that the deposits in the valley of the Mississippi,

The distribution of species in the strata (or in time), is like their distribution in *space*. Each is most abundant in one horizon, and becomes gradually less frequent in the beds above and below; the locality of the newest rock in which it occurs being often far removed from that of the oldest.*

That species should be created at a single spot, and gradually multiply and diffuse themselves, is sufficiently intelligible. That, after attaining a certain climax of development, they should decline and disappear is a fact involved in mystery. But even if it depends on physical causes, and is not a law of all Being, its operation is equally certain, and does not appear to vary beyond moderate limits.

The deep-sea shells (such as *Rhynchonella, Terebratula,* and *Yoldia*), enjoy a longer range in time, as well as in space, than the litoral species; whilst the land and fresh-water shells are most remarkable for specific longevity.†

In each stratum there are some fossils which characterize small subdivisions of rock, just as there are living species of very limited range.

When species once die out they never reappear; one evidence of their having become extinct consisting in their replacement by other species, which fulfilled their functions, and are found in deposits formed under similar conditions. (*Forbes.*)

The total number of species is greater in the newest formations than in those of older date; but the ratio of increase has not been ascertained.‡

Distribution of Genera in Time. The doctrine of the Identification of strata by fossils derives its chief value from the fact that the development and distribution of Genera is as much subject to law as the distribution of species; and so far as we know, follows a similar law.

Groups of strata, like the Zoological provinces, may be of various magnitudes; and whilst the smaller divisions are characterized by peculiar *species,*

estimated to represent 100,000 years, have been accumulated since the era of many existing shells. The same may be said of the elevation of Mont Blanc, the formation of the Mediterranean Sea and other grand physical events. The great cities of antiquity—Rome, Corinth, and Egyptian Thebes, stand upon raised sea-beds, or alluvial deposits, containing recent shells.

* M. Agassiz and Prof. E. Forbes have represented, diagramatically, the distribution of genera in time, by making the horizontal lines (such as in p. 415) swell out in proportion to the development of the genera. Those whose commencement, climax, and end are ascertained may be represented by a line of this kind ——————
Genera which attain their *maxima* in the present seas are thus expressed ——————

† Land and Fresh-water shells of existing species are found with the fossil bones of the *Mastodon* and *Megalonyx*, in N. America. (*Lyell.*)

‡ The number in each formation depends on the extent to which it has been investigated, and on the opinions entertained as to the strata referable to it. Prof. Phillips has discussed this subject in his work on Devonian fossils (p. 165), and in the "Guide to Geology."

the larger groups have distinct sub-genera, genera, and families, according to their size and importance.

Wm. Smith himself observed that "Three principal families of organized fossils occupy nearly three equal parts of Britain."

"*Echini* are most common in the superior strata;

"*Ammonites* to those beneath;

"*Producti* with numerous *Encrini* to the lowest."

This kind of generalization has justly been considered, by Prof. E. Forbes, of higher importance than the identification of strata by *species ;* a method only applicable to moderate areas, and becoming less available with distance. Indeed it might be assumed that strata geographically distant, yet containing some identical species, must differ in age by the time required for the migration of those species from one locality to the other.

A table of the characteristic *species* of the English strata is of little use in America or India, except to shew how few and doubtful are the identical fossils. Whereas the characteristic genera, and order of succession of the larger groups are the same at the most distant localities; and whatever value there may be in the assumption that particular systems of rocks contain most workable coal, lead, or rock-salt, is not lessened by the circumstance that the species of fossils in those rocks are not everywhere the same, since the genera alone are sufficient to identify them.

Genera, like species, have a commencement, a climax, and a period of decline; the smallest usually range through several formations, and many of the typical genera equal the families in duration.

Groups of formations are called Systems, and these again are combined in three principal series—Palæozoic—Secondary—and Tertiary.

Thirteen geological systems, each having a number of peculiar genera are shewn in the accompanying table. (No. II.) Some of the genera cited have a wider range, like *Belemnites.* but are mentioned because of their abundance in one particular system. The names *in italics* are existing genera.*

The third table contains the names of some of the larger genera, arranged according to the order of their appearance. This diagram conveys the impression that the series of fossiliferous strata is not completely known; or that the beginning of many groups of fossils has been obliterated in the universal metamorphism of the oldest stratified rocks.†

---

* The *Pliocene strata* contain no extinct genera, and represent only the commencement of the present order of things. All the deposits now taking place will not constitute an additional " Formation," much less a " Quaternary System."

† It was on this account Prof. Sedgwick proposed the term " Palæozoic. rather than " Protozoic," for the oldest fossiliferous rocks.

II. TABLE OF CHARACTERISTIC GENERA.

SYSTEMS	GENERA AND SUB-GENERA.
1. CAMBRIAN, or Lower Silurian.	Camaroceras, Endoceras, Gonioceras, Pterotheca. Maclurea, Raphistoma, Holopea, Platyceras. Orthisina, Platystrophia, Porambonites, Pseudo-crania. Ambonychia, Modiolopsis, Lyrodesma.
2. SILURIAN.	Actinoceras, Phragmoceras, Trochoceras, Ascoceras. Theca, Holopella, Murchisonia, Atrypa, Retzia. Cardiola, Clidophorus, Goniophorus, Grammysia.
3 DEVONIAN.	Bactrites, Gyroceras, Clymenia, Apioceras, Serpularia. Spirifera, Uncites, Merista, Davidsonia, Calceola, Stringocephalus, Megalodon, Orthonota, Pterinea.
4. CARBONIFEROUS.	Nautiloceras, Discites, Goniatites, Porcellia. Naticopsis, Platyschisma, Metoptoma, Producta. Aviculo-pecten, Anthracosia, Conocardium, Sedgwickia.
5. PERMIAN.	Camarophoria, Aulosteges, Strophalosia. Myalina, Bakewellia, Axinus, Edmondia.
6. TRIAS.	Ceratites, Naticella, Platystoma, Koninckia, Cyrtia. Monotis, Myophoria, Pleurophorus, Opis.
7. L. JURASSIC.	Belemnites, Beloteuthis, Geoteuthis, Ammonites. Alaria, Trochotoma, Rimula, Pileolus, Cylindrites. Waldheimia, *Thecidium*, Spiriferina, Ceromya. Gryphæa, Hippopodium, Cardinia, Myoconcha.
8. U. JURASSIC.	Coccoteuthis, Acanthoteuthis, Leptoteuthis, Nautilus. Spinigera, Purpurina, Nerinæa, Neritoma. Pteroperna, Trichites, Hypotrema, Diceras. Trigonia, Pachyrisma, Sowerbia, Tancredia.
9. L. CRETACEOUS.	Crioceras, Toxoceras, Hamulina, Baculina. Requienia, Caprinella, Sphæra, Thetis.
10. U. CRETACEOUS.	Belemnitella, Conoteuthis, Turrilites, Ptychoceras. Hamites, Scaphites, Pterodonta, Cinulia, Tylostoma. Acteonella, Globiconcha, Trigonosemus, Magas, Lyra. Neithea, Inoceramus, Hippurites, Caprina, Caprotina
11. EOCENE.	Beloptera, Lychnus, *Megaspira. Glandina, Typhis.* *Volutilithes, Clavella, Pseudoliva, Seraphs, Rimella.* Conorbis, Strepsidura, Globulus, *Phorus*, Velates. Chilostoma, Volvaria, Lithocardium, Teredina.
12. MIOCENE.	Spirulirostra, Aturia, Vaginella, Ferussina, Halia, Proto, Deshayesia, *Niso, Cassidaria*, Carolia. Grateloupia, *Artemis, Tapes, Jouannetia.*
13. PLIOCENE.	*Argonauta, Strombus, Purpura, Trophon.* *Yoldia, Tridacna, Circe, Verticordia.*

III. RANGE OF GENERA IN TIME.

Genera, arranged in their Order of Appearance.	Cambrian.	Silurian.	Devonian.	Carbonif.	Permian.	Trias.	L. Jura.	U. Jura.	L. Cret.	U. Cret.	Eocene.	Miocene.	Pliocene.
Lituites, Raphistoma, Obolus	—	—											
Camaroceras, Atrypa, Pterinea..	—	—	—										
Gomphoceras, Bellerophon, Pentamerus	—	—	—	—									
Orthis, Conularia, Murchisonia	—	—	—	—	—								
Spirifera, Athyris, Posidonomya	—	—	—	—	—	—	—						
Isoarca	—	—	—	—	—	—	—	—	—	—	—	—	—
Conocardium, Megalodon, Chonetes ..	—	—	—										
Cardiomorpha..	—	—	—										
Orthoceras, Loxonema, Cyrtia	—	—	—	—	—	—							
Pleurotomaria..	—	—	—	—	—	—	—	—					—
Producta, Macrochilus, Streptorhynchus	—	—	—										
Goniatites, Porcellia, Pleurophorus ..	—	—	—			—							
Edmondia, Myalina		—	—										
Acteonina..				—	—	—	—	—					
Terebratula, Pinna, Cyprina					—	—	—	—	—	—			
Lima					—	—	—	—	—	—	—	—	—
Gervillia, Myoconcha					—	—	—	—	—	—			
Ammonites, Naticella, Opis..						—	—	—	—	—			
Trigonia, Isocardia, Thecidium						—	—	—	—	—	—	—	—
Cerithium, Plicatula, Cardita						—	—	—	—	—	—	—	—
Trochotoma, Tancredia, Gryphaea							—	—					
Ancyloceras, Inoceramus, Unicardium ..							—	—	—	—			
Astarte, Pholadomya, Corbis							—	—	—	—	—	—	—
Nerinaea, Goniomya, Exogyra..								—	—	—			
Terebratella, Limopsis, Neæra								—	—	—			
Baculites, Cinulia, Radiolites									—	—			
Physa, Paludina, Unio, Cyrena..									—	—			
Aporrhais, Tornatella, Pyrula									—	—			
Pectunculus, Thetis, Crassatella									—	—	—	—	—
Crenella, Chama, Argiope										—	—	—	—
Voluta, Conus, Mitra, &c. &c.										—	—	—	—
Aturia											—	—	
Helix, Auricula, Cyclostoma											—	—	—
Pseudoliva, Rostellaria, Seraphs											—	—	—
Purpura, Strombus, Haliotis												—	—
Argonauta, Tridacna													

The genera of the *older rocks* are believed to be nearly all extinct; for although the names of many recent forms appear in the catalogues of Palæozoic fossils, it must be understood that they are only employed in default of more exact information. *Buccinum*, *Melania*, and *Mya*, have been long since expunged; and *Modiola*, *Nucula*, and *Natica*, are only retained until the characters which distinguish them are better understood.

IV. RANGE OF FAMILIES IN TIME.

Systems of Strata.	Cambrian.	Silurian.	Devonian.	Carbonif.	Permian.	Trias.	L. Jura.	U. Jura.	L. Cret.	U. Cret.	Eocene.	Miocene.	Pliocene.	Recent.
Argonautidæ													—	—
Teuthidæ—Sepiadæ							—	—	—		—	—	—	—
Belemnitidæ						—	—	—	—	⊥				
Nautilidæ				—	—	—	—	—	—	—	—	—	—	—
Ammonitidæ		—	—	—	—	—	—	—	—					
Orthoceratidæ	—	—	—			—								
Atlantidæ—Hyaleidæ		—	—	—	—	—	—	—	—	—	—	—	—	—
Strombidæ—Buccinidæ											—	—	—	—
Conidæ—Volutidæ										—	—	—	—	—
Naticidæ—Calyptræidæ											—	—	—	—
Pyramidellidæ	—	—				—	—	—	—	—	—	—	—	—
Cerithiadæ—Litorinidæ						—	—	—	—	—	—	—	—	—
Turbinidæ—Janthinidæ						—	—	—	—	—	—	—	—	—
Fissurellidæ—Chitonidæ	—	—				—	—	—	—	—	—	—	—	—
Neritidæ—Patellidæ						—	—	—	—	—	—	—	—	—
Dentaliadæ	—	—				—	—	—	—	—	—	—	—	—
Tornatellidæ	—	—				—	—	—	—	—	—	—	—	—
Bullidæ						—	—	—	—	—	—	—	—	—
Helicidæ—Limacidæ											—	—	—	—
Limnæidæ—Melaniadæ						—	—	—		—	—	—	—	—
Auriculidæ—Cyclostomidæ											—	—	—	—
Terebratulidæ				—	—	—	—	—	—	—	—	—	—	—
Rhynchonellidæ	—	—	—	—	—	—	—	—	—	—	—	—	—	—
Spiriferidæ—Orthidæ	—	—	—	—	—	—	—							
Productidæ	—	—	—	—										
Craniadæ—Lingulidæ	—	—	—	—	—	—	—							
Pectinidæ						—	—	—	—	—	—	—	—	—
Aviculidæ—Mytilidæ	—	—	—	—	—	—	—	—	—	—	—	—	—	—
Arcadæ—Trigoniadæ	—	—	—	—	—	—	—	—	—	—	—	—	—	—
Unionidæ											—	—	—	—
Chamidæ—Myadæ									—	—	—	—	—	—
Hippuritidæ									—	—				
Tridacnidæ													—	—
Cardiadæ—Lucinidæ				—	—	—	—	—	—	—	—	—	—	—
Cycladidæ									—	—	—	—	—	—
Cyprinidæ—Anatinidæ	—	—	—	—	—	—	—	—	—	—	—	—	—	—
Astartidæ						—	—	—	—	—	—	—	—	—
Veneridæ—Tellinidæ									—	—	—	—	—	—
Mactridæ								—			—			
Solenidæ									—	—	—	—	—	—
Gastrochænidæ—Pholadidæ						—	—	—	—	—	—	—	—	—

Distribution of Families of Shells in Time.　Employing the term "families" for natural groups of genera, and adopting the smallest possible number of them, we find that sixteen, or nearly one-fifth, range through all the geological systems.　Only seven have become extinct, viz.—

Belemnitidæ.	Spiriferidæ.	Hippuritidæ.
Ammonitidæ.	Orthidæ.	
Orthoceratidæ.	Productidæ.	

Three others are nearly extinct:—

Nautilidæ.	Rhynchonellidæ.	Trigoniadæ.

And several have passed their maximum, and become less varied and abundant than formerly, *e.g.*—

Tornatellidæ.	Cyprinidæ.	Anatinidæ.

The extinct families and genera appear to have attained their *maxima* more rapidly than their *minima;* continuing to exist, under obscure forms, and in remote localities, long after the period in which they flourished.

The introduction of new forms, also, is more rapid than the process of extinction.　If four Palæozoic families disappear, twenty-six others replace them in the Secondary series ; and three of the latter are succeeded by fifteen shell-bearing families in the Tertiaries and existing seas.

In consequence of this circumstance, the number of types is three times greater in the newer tertiary than it was at the Silurian period ; and since there is no evidence or indication that the earth was ever destitute of life, either wholly or in part, it follows almost as a matter of necessity that the early types must have been more widely distributed and individually developed, than those of the present day.

From the following Table it will be seen that the number of Genera and Families increases with an amount of regularity, which cannot be accidental. Moreover the relation of these numbers is not liable to be much altered by the progress of discovery, or the caprice of opinion.　The *discovery* of new types, is not likely to be frequent ; the imposition of new names, in place of the old, will not increase the number of Palæozoic genera ; and the establishment of fresh and arbitrary distinctions will affect all the groups in due proportion.

If the number of groups called "Systems" were reduced to seven, (viz. three Palæozoic, three Secondary, and one Tertiary, as shewn in the following table,) then the *average* duration of a genus of shells would be equal to a System of Formations.

The duration of the smallest well-defined Families of shells is about equal to one of the three great Geological Divisions, or Ages.

DEVELOPMENT OF FAMILIES, GENERA, AND SPECIES, IN TIME.

			GEOLOGICAL SYSTEMS.	Total of Genera.	Cephalopoda.	Gasteropoda.	Brachiopoda.	Conchifera.	Total Number of Species. (D'Orbigny).	Families.	
PALÆOZOIC.	1	{	Cambrian	49	12	11	15	11	362	18	
			Silurian	53	13	11	16	13	317	20	
	2		Devonian	77	14	20	23	20	1035	24	} 32
	3	{	Carboniferous	79	11	26	19	23	835	30	
			Permian*	66	6	24	16	20	74	30	
SECONDARY.	4		Trias	81	9	25	16	31	713	35	
	5	{	L. Jurassic	107	12	35	12	48	1502	42	
			U. Jurassic	108	13	36	9	50	1266	49	} 57
	6	{	L. Cretaceous	123	20	41	9	53	784	52	
			U. Cretaceous	148	16	59	14	59	2147	56	
TERTIARY.	7	{	Eocene	172	4	85	11	72	2636	60	
			Miocene	187	3	97	11	76	2242	60	} 78
			Pliocene	192	1	100	12	79	437	62	
			Recent	400	21	251	13	115	16,000	78	
			Recent & Fossil	520	56	280	34	150	30,000	85	

Order of appearance of the groups of Shells. The first and most important point shewn in the preceding Tables, is the co-existence of the four principal classes of *testacea* from the earliest period. The highest and the lowest groups were most abundant in the palæozoic age; the ordinary bivalves and univalves attain their climax in existing seas. If there be any meaning in this order of appearance it is connected with the general scheme of creation, and cannot be inquired into separately; but it may be observed that the last-developed groups are also the most typical, or *characteristic of their class.* (p. 61.)

The *Cephalopoda* exhibit amongst themselves unmistakable evidence of order in their appearance and succession. The tetrabranchiate group comes earliest, and culminates about the period of the first appearance of the more highly-organized cuttle-fishes.[†] The families of each division which are least unlike (*Orthoceratidæ* and *Belemnitidæ*) were respectively the first developed.

* Those genera are estimated as belonging to each System which occur in the strata *both above and below*, as well as those actually found in it.

† The *Palæoteuthis* of Bronn (not D'Orb.) appears to be a *fish-bone*, from the equivalent of the Old-red sandstone in the Eifel.

Amongst the *Brachiopoda* the hinge-less genera attained their maximum in the palæozoic age, and only three now survive, (*Lingula, Discina, Crania,*) —the representatives of as many distinct families. Of the genera with articulated valves, those provided with spiral arms appeared first and attained their maximum while the *Terebratulidæ* were still few in number. The subdivision with calcarious spires disappeared with the Liassic period, whereas the genus Rhynchonella still exists. Lastly, the typical group, *Terebratulidæ*, attained its maximum in the chalk period, and is scarcely yet on the decline. The number of sub-genera (as well as genera,) in each system, is stated in the preceding table, because this group shews a tendency to "polarity," or excessive development at the ends of the series.*

The genera of ordinary bivalves (*Conchifera*) are seven times more numerous in the newer tertiary than in the oldest geological system. The palæozoic formations contain numerous genera of all the families with an *open mantle; Cyprinidæ, Anatinidæ,* and the anomalous genus *Conocardium.* The mass of siphonated bivalves do not appear till the middle of the secondary age, and are only now at their maximum.

The *Gasteropoda* are represented in the palæozoic strata by several genera closely allied to the diminutive *Atlanta* and *Scissurella,* and by others perhaps related to *Ianthina.* The *Naticidæ,* and *Calyptræidæ* are plentiful, and there are several genera of elongated spiral shells referred to the *Pyramidellidæ.* In the secondary strata, *holostomatous* shells become plentiful; and in a few peculiar localities (especially Southern India) the genera of siphonated univalves make their appearance in strata of Cretaceous age. Fresh-water *Pulmonifera* of the recent genus *Physa* occur in the Purbeck strata, but the marine air-breathers and land-snails have not certainly been found in strata older than the Eocene tertiary.

Order of Succession of Groups of Shells.—It has been already pointed out that animals which are closely allied in structure and habits, rarely live together, but occupy distinct *areas,* and are termed "representative species." The same thing has been observed in the distribution of fossils; the species of successive strata are mostly representative.

At wider intervals of time and space, the representation is only generic, and the relative proportions of the larger groups are also changed.

The succession of forms is often so regular as to mislead a superficial observer; whilst it affords, if properly investigated, a valuable clue to the affinities of problematic fossils.

* See the anniversary address of Prof. E. Forbes to the Geological Society of London, Feb. 1854, p. 63. The hypothesis seems to have arisen out of an exclusive regard to the poverty of the Permian and Triassic strata in England, where they separate, like a desert, the palæozoic from the "neozoic" formations. The "Permian" should never have been esteemed more than a division of the Carboniferous system, and is poor in *species,* rather than in *types.* The Trias must be studied in Germany, or in the collection of Dr. Klipstein (in the Brit. Museum) to be properly appreciated.

It is now generally admitted that the earlier forms of life, strange as many of them seem to us, were really less metamorphosed—or departed less widely from their ideal archetypes—than those of later periods and of the present day.* The types first developed are most like the embryonic forms of their respective groups, and the progression observed is from these general types to forms more highly specialized. (*Owen.*)

Migration of Species and diffusion of Genera in Former Times.—Having adopted the doctrine of the continuity of specific and generic areas, it remains to be shewn that such groups as are now widely scattered *can* have been diffused from common centres, and that the barriers which now divide them have not always existed.

In the first place it will be noticed that the mass of the stratified rocks are of *marine origin*, a circumstance not to be wondered at, since the area of the sea is twice as great as the land, and probably has always been so; for the average depth of the sea is much greater than the general elevation of the land.†

The mineral changes in the strata may sometimes be accounted for by changes in the depth of the sea, or an altered direction of the currents. But in many instances the sea-bed has been elevated so as to become dry land, in the interval between the formation of two distinct marine strata; and these alterations are believed to occur (at least) *once in each formation.*

. If every part of what is now dry land has (on the average) been thirty times submerged, and has formed part of the sea-bed during two-thirds of all the past geological time;—there will be no difficulty in accounting for the migration of sea-shells, or the diffusion of marine genera.

On the other hand it may be inferred that every part of the present sea has been dry land many different times;—on an average not less than thirty times,—amounting to one-third of the whole interval since the Cambrian epoch.

The average duration of the marine species has been assumed at only one-third the length of a geological period, and this harmonises with the fact that so few (either living or extinct) have a world-wide distribution.

The life of the land-snails and of the fresh-water shells has been of longer

---

* Mr Darwin has pointed out that the *sessile* Cirripedes, which are more highly metamorphosed than the *Lepadidæ*, were the last to appear. The fossil mammalia afford, however, the most remarkable examples of this law. At the present day such an animal as the three-toed horse (*Hippotherium*) of the Miocene Tertiary would be deemed a *lusus naturæ*, but in truth the ordinary horse is far more wonderful. Unfortunately, a new " vulgar error " has arisen from the terms in which extinct animals have sometimes been described—as if they had been constructed upon *several distinct* types, and combined the character of several classes!

† The enormous thickness of the older rocks *in all parts of the world*, has been held to indicate the prevalence of deep water in the primæval seas.

average extent, enabling them to acquire a wide range, notwithstanding their tardy migrations.

But when we compare the estimated rate of change in physical geography with the duration of *genera* and *families* of shells, we not only find ample time for their diffusion by land or sea over large portions of the world, but we may perceive that such transferences of the scene of creation must have become inevitable.

Method of Geological Investigation.—In whatever way geological history is written, its original investigators have only one method of proceeding— from the known to the unknown—or backwards in the course of time.

The newest and most superficial deposits contain the remains of man and his works, and the animals he has introduced.

Those of pre-historic date, but still very modern, contain shells, &c., of recent species, but in proportions different from those which now prevail. (p. 384, 387). Some of the species may be extinct in the immediate neighbourhood of the deposits, but still living at a distance.

In the harbour of New Bedford are colonies of dead shells of the *Pholas costata,* a species living on the coast of the Southern States. At Bracklesham, Sussex, there is a raised sea-bed containing 35 species of sea-shells living on the same coast, and 2 no longer living there, viz.—*Pecten polymorphus,* a Mediterranean shell; and *Lutraria rugosa,* still found on the coasts of Portugal and Mogador.

Tertiary Age.—If any distinction is to be made between "Tertiary" and "Post-tertiary" strata, the former term should be restricted to those deposits which contain some *extinct* species. And the newest of these, in Britain, contain an assemblage of Northern shells. Prof. Forbes has published a list of 124 species of shells from these "Glacial beds," nearly all of which are now existing in British seas.*

In most of the localities for glacial shells, the species are all recent; but at Bridlington, Yorkshire, and in the Norwich *Crag,* a few extinct species are found. (e. g. *Nucula Cobboldiæ,* Pl. 17, f. 18.) At Chillesford, Suffolk, *Yoldia arctica* and *myalis* occur of large size and in excellent preservation, with numerous specimens of *Mya truncata,* erect as they lived, in the muddy sea-bed. *Trophon scalariforme, Admete viridula, Scalaria grœnlandica,* and *Natica grœnlandica,* also occur in the Norwich Crag; and *Astarte borealis,* with several arctic forms of *Tellina,* are amongst the commonest shells, and frequently occur in pairs, or with their *ligament* preserved; the deposit is extensively quarried for shell-sand.

Raised sea-beds with Arctic shells at Uddevalla in Sweden, have been repeatedly noticed ever since the time of Linnæus. Captain Bayfield disco-

* The species which have retired further north are marked (**) in the preceding Arctic List, p. 355.

vered similar beds near Quebec, 50—200 feet above the River St. Lawrence, containing an assemblage of shells entirely arctic in character; whereas in the present gulf he obtained an admixture of the American representatives of Lusitanian types, *Mesodesma, Periploma, Petricola, Crepidula.*

The glacial deposits of the northern hemisphere extend about 15° south of the line of "northern limit of trees;" but this comparatively recent extension of the Arctic ocean does not appear to have much influenced, if it ever invaded, the inland basin of the Aralo-Caspian, which contains only one species common to the White Sea, *Cardium edule,* var. *rusticum.**

The older pliocene period is represented in England by the *Coralline Crag,* a deposit containing 340 species of shells. Of these 73 are living British species, but (with two or three exceptions) they are such as range south of Britain. *(Forbes.)* The remainder are extinct, or living only to the south, especially in the Lusitanian province; e. g. *Fossarus sulcatus, Lucinopsis Lajonkairii, Chama gryphoides,* and species of *Cassidaria, Cleodora, Sigaretus, Terebra, Columbella* and *Pyramidella.* It also contains a few forms belonging to an earlier age,—a *Pholadomya,* a true *Pyrula,* a *Lingula,* and a large *Voluta,* resembling the Magellanic species.

The shells of the newer tertiaries are always identical, at least *generically,* with those of the nearest coasts. Thus, in Patagonia, are found species of *Trophon, Crepidula, Monoceros, Pseudoliva, Voluta, Oliva, Crassatella,* and *Solenella.* The tertiaries of the United States contain species of *Fulgur, Mercenaria* and *Gnathodon.* The miocene shells of St. Domingo appear at first sight to be all of recent species, but on comparison prove to be mostly distinct.

The proportion of extinct species in the *Pliocene* tertiary, varies from 1—50 per cent. If a deposit contains more than 50 per cent. of extinct species it is referred to the *Miocene* period; and this test is particularly valuable since the modern deposits are often isolated, and frequently no assistance can be derived from superposition, or even from identity of species.

In the *Eocene* tertiaries we perceive the "dawn" of the present order of things. All, or very nearly all, the species are different, but a large proportion of the genera are still existing, though not always in the seas nearest to the localities where they occur fossil.

Thus in the London clay are found—*Rostellaria, Oliva, Ancillaria,* and *Vulsella,* genera still living in the Red Sea; and many species of *Nautilus, Rimella, Seraphs, Conus, Mitra, Pyrula, Phorus, Liotia, Cardilia,*—genera characteristic of the Indian Ocean; *Cyprovula, Typhis* and *Volutilithes,* now

---

* Mr. Wm. Hopkins of Cambridge has investigated the causes which may have produced a temporary extension of the Arctic phenomena in Europe; and considers the most efficient and probable cause would be a diversion of the Gulf-stream, which he supposes to have flowed up what is now the valley of the Mississippi.—*(Geol. Journal.)*

living at the Cape; *Clavella*, at the Marquesas, and *Pseudoliva*, *Tröchita*, and species of *Murex*, whose recent analogues are found on the Western shores of S. America.

The freshwater shells of this period are Old-World forms; *Melanopsis*, *Potamides*, *Lampania*, *Melanatria* and *Nemutura* : whilst the land-shells form a group quite American in character; large species of *Glandina* and *Bulimus* (with reflected lip) *Megalomastoma* (*mumia*), a *Cyclotus* (with its operculum) like *C. Jamaicensis*, and the little *Helix labyrinthicus*.

Secondary Age.—In none of the older strata do we find indications of a warmer climate having prevailed, in the latitude of England, than that which marks the period of the London clay. And this is not more than can be accounted for by such a cause as the flow of an equatorial current from the direction of the Red Sea, until arrested by a continent to the south-west, as supposed by Mr. Prestwich, in the region of the Azores.

Some indications exist of a more moderate climate having obtained in the north polar regions; for remains of the *Ichthyosaurus* were found at Exmouth Id. the furthest point reached by Sir E. Belcher's expedition.

The peculiar physical conditions of the *Chalk period* are represented at the present day, not so much by the Coral-sea, as by the Ægean, where calcareous mud, derived from the waste of the *scaglia* regions, is being rapidly deposited in deep water. (*Forbes*).

The *Wealden period* was styled the "Age of Reptiles" by Dr. Mantell, who compared the state of England at that time with the present condition of the Galapagos Islands.

The *Oolitic period* finds its parallel in Australia, as long since pointed out by Prof. Phillips, and the comparison holds good to some extent, both for the Marine and Terrestrial Faunas.

The *Trias*, with its foot-prints of gigantic wingless birds, has been compared with the state of the Mascarene Islands only a few centuries ago, and with the New Zealand Fauna, where birds are still the highest aboriginal inhabitants.*

Palæozoic Age.—It has lately been suggested by Prof. Ramsay that signs of glacial action may be traced in some of the trappean conglomerates of the *Permian Period*; and Mr. Page has endeavoured to apply the same interpretation to phenomena of a much earlier date, in the old red sand-stone of Scotland.† Geologists generally have abandoned the notion, once very prevalent, of a universal high temperature in the earliest periods; a notion which

---

* In a paper read before the British Association, on the subject of the great extinct wingless birds of New Zealand, Prof. Owen suggested the notion of land having been propagated like a wave throughout the vast interval between Connecticut and New Zealand, since the Triassic period.

† See also the Rev. J. G. Cumming's "Isle of Man," (1849), p. 89.

they had derived from the occurrence of certain fossil plants, corals, and shells, in high latitudes.

❡ The absence of remains of mammalia in the palæozoic formations, is at present a remarkable fact, but it is completely paralleled in the great modern zoological province of the Pacific Islands.

Baron Humboldt has speculated on the possibility of some land being yet discovered, where gigantic lichens and arborescent mosses may be the princes of the vegetable kingdom*. If such exist, to shadow the Palæozoic age, its appropriate inhabitants would be like the cavern-haunting *Proteus*, and the *Silures* which find an asylum even in the craters of the Andes.

What then is it which has chiefly determined the character of the present Zoological provinces? What law, more powerful than climate, more influential than soil, and food, and shelter; nay, often seemingly producing results opposed to *a priori* probability, and at variance with the suitableness of conditions?†

The answer is, that each fauna bears, above all things, the impress of the age to which it belongs. Each has undergone a series of vicissitudes up to the time when its barriers became fixed, and after its isolation it has known no further change, but decline.

As regards the great types of terrestrial organization, their point of common origin seems to have been the centre of the Old World. Here they appear to have been formed in succession, and diffused outwards in all possible directions, to the ends of the earth; each wave of life developing in its progress special forms adapted to the circumstances of the times, and exemplifying the modifications of which each type was capable.‡

CHAPTER IV.
ON COLLECTING SHELLS.

The circumstances under which shells are found is a subject so intimately connected with the methods of collecting them, as to make it undesirable to treat of them separately.

Naturalists distinguish between the *habitats*, or geographical localities of species, and the stations or circumstances in which they are found: to the latter subject only slight allusion has been hitherto made. (p. 11).

Land-shells are most abundant on calcareous soils, (p. 37) and in warm and moist climates. The British species are collected with advantage in autumn, when full-grown, and showing themselves freely in the dews of morning and evening. Some species, like *bulimus acutus,* are found only near the sea;

---

* Views of Nature, p. 221. Bohn's ed.   † Burchell, in Darwin's Journal, p. 87.

‡ "The TIDE OF VEGETATION has, in the intertropical Pacific Islands, set in a direction contrary to the prevailing winds; namely, from the Asiatic, and not from the American shores." (Hooker, l. c. p. 211, note.)

Bulimus Lackhamensis ascends beech-trees on the Chalk downs and Cottes-woldes ; *Pupa Juniperi* and *Helix umbilicata* occur chiefly on rocks and stone walls. The moss-frequenting *Clausiliæ* may be obtained even in mild winter weather at the roots of trees; the small species of *Pupa* (or *Vertigo*) are sometimes taken abundantly when sweeping wet grass with an insect net ; *Acicula fusca* lives at the roots of grass ; *Cionella acicula* is found in old bones, (such as occur in Danish burial grounds !) and occasionally in moving garden-bulbs; *Helix aculeata* has been met with on the under sides of leaves (*e. g.* the sycamore), a few feet from the earth.

In tropical countries a large number of the land snails are *arboreal* in their habits. The West Indian Palms (such as *Oreodoxa regia*) are the chosen abode of many species of Helicidæ. Mr. Couthouy found *Bulimus auris leporis* on the orange and myrtle-trees near Rio, and *Partulæ* and *Helicinæ*, on the Dracænas and Bananas of the Polynesian Islands ; and the sailors of H.M.S. Rattlesnake, in Captain Owen Stanley's expedition, became expert in collecting *Geotrochi* in the trees of the Australian islands.

The great tropical *Bulimi* and *Achatinæ* will sometimes lay their eggs in captivity.*

The following are examples of the elevations at which land-snails have been found. (pp. 162, 166.)

> Helix pomatia, 5000 feet—Alps. (Jeffreys.)
> — rupestris, 1200—5000 ft.
> — bursatella, Gould, 2000—5000 ft. Taheiti.
> Bulimus vibex 7000 ft. India. (Benson.)
> — nivicola, and ornatus, 14,000 ft.   „
> — Lamarckianus, 8000 ft. New Granada.
> Achatina latebricola, 4—7000 ft. Landour.
> Pupa Halleriana, 1200—2500 ft. Alps.
> — tantilla, 2,000 ft. Taheiti.
> Clausilia Idæa, 5500 ft. Mr. Ida.
> Vitrina glacialis, Forbes, 8000 ft. Mte. Rosa.
> — annularis, 2000—3000 ft. Burgos. (M'Andrew.)
> — Teneriffæ, 2000—6210 ft. Madeira.
> Helicina occidentalis, Guilding, 2000 ft. St. Vincents.
> (Limnæa Hookeri, 18000 ft. Thibet.)

The land-snails of warm and dry regions remain dormant for long periods (p. 19), and require no attention for many months after being collected.†

Freshwater shells are collected with an insect-net or "landing-net" of strength suited to the work of raising masses of weed. The strongly-rooted

---

* Such giants require to be collected in a basket, while the small land shells of open and rocky countries may be put in a cotton bag, hung on a coat button.

† Land and freshwater snails may be killed instantaneously with boiling water, if a few are done at a time ; and cooled by removal to cold water. Every collector finds expedients for removing the animals more or less completely from their shells ; those which, like *Clausilia*, retire beyond the reach of a bent pin may be drowned in tepid water.

flags and rushes may be pulled up with a boat-hook; and *Cyclades* as well as univalves, may be obtained by shaking aquatic plants over the net. For getting up the Pearl-mussels, the most efficient instrument is a tin bowl, perforated like a sieve, and fitted on the end of a staff, or jointed rod. (*Pickering.*)

In some situations the freshwater shells are all much eroded, (p. 41, 273,) or coated with a ferruginous deposit. It may be desirable to find out the localities where the specimens are in best condition before collecting extensively. The *opercula* should always be preserved with the shells to which they belong; those of the *Cyclostomidæ*, and *Melaniadæ* are particularly interesting.

The *Auriculidæ* are especially met with in damp places by the sea; in mangrove-swamps, and creeks and river-banks where the water becomes brackish. *Amphibola* and *Assiminea* are found in salt-marshes, *Siphonaria* and *Peronia* on the shore, between tide-marks.

Collecting Sea-shells.—The following remarks are from the pen of an experienced conchologist, Mr. W. J. Broderip.—"When the tide is at the lowest, the collector should wade among the rocks and pools near the shore, and search under overhanging ledges of rock as far as his arms can reach. An iron rake, with long close-set teeth, will be a useful implement on such occasions. He should turn over all loose stones and growing sea-weeds, taking care to protect his hands with gloves, and his feet with shoes and stockings against the sharp spines of *echini*, the back-fins of sting-fishes, and the stings of *medusæ*. In detaching chitons and limpets which are all to be sought for on rocky coasts, the *spatula* or case knife will prove a valuable assistant. Those who have paid particular attention to preserving chitons have found it necessary to suffer them to die under pressure between two boards. Ormers (*Haliotides*) may be removed from the rocks to which they adhere by throwing a little warm water over them, and then giving them a sharp push with the foot sideways, when mere violence would be of no avail without injuring the shell. Rolled madrepores and loose fragments of rock should be turned over; Cowries and other shell-fish frequently harbour under them. Numbers of shell-fish are generally to be found about coral-reefs." In coral-regions the services of *natives*, should be obtained, as they may render much assistance by diving or wading.

Advantage may be taken of *spring-tides*, especially at the equinoxes, to examine lower tracts of sea-shore than are ordinarily accessible. Many *bivalves* bury in sand and mud at extreme low-water, and may be obtained alive by digging with a spade or fork; others may be found boring in piles and rocks and require the hammer and chisel for their extraction.*

---

* *Bivalves* may be boiled, and their soft parts removed when the shells gape. Care should be taken not to injure the ligament, or hinge, especially in the genera like the *Anatinid*) provided with an *ossicle*.

Mr. Joshua Alder remarks that, "in collecting among rocks the principal thing is to look close, particularly in crevices and under stones. Minute species inhabiting sea-weed are best obtained by gathering the weed and immersing it for some time in a basin of sea-water, when the little mollusks will generally creep out. If the shells only are wanted, the surer and more ready way is to plunge the weed into fresh-water, when the animals immediately fall to the bottom."

The *floating mollusca* of the open sea, especially in tropical latitudes, are comparatively little known. Good drawings, and descriptions made from the life, are most valuable. "Of the animal of the *Spirula*, entire specimens are greatly wanted. If captured alive, its movements should be watched in a vessel of sea-water, to see whether it has the power of rising and sinking at will; its mode of swimming, and position during these movements, and when at rest. The chambered shell should be opened under water, to ascertain if it contain a gas, the nature of which should if, possible, be made out. The pearly nautilus requires the same observations, which would be attended with more precision and facility from its larger size." (*Owen*.)[*]

The *towing-net* used by Mr. Mc Gillivray "consisted of a bag of *bunting* (used for flags) two feet deep, the mouth of which was sewn round a wooden hoop 14 inches in diameter; three pieces of cord, a foot and a half long, were secured to the hoop at equal intervals and had their ends tied together. When in use the net was towed astern, clear of the ship's wake, by a stout cord secured to one of the quarter boats, or held in the hand. The scope of the line required was regulated by the speed of the vessel at the time, and the amount of strain caused by the partially submerged net."[†]

Trawling.—Mr. John W. Woodall, of Scarbro', has kindly furnished the following sketches and particulars;—" Fig. 227, is intended to represent a *trawl-net*, at work on the bottom of the sea. The side frames are of iron, the upper beam of wood, and the lower edge of the net is kept down to the ground by means of a chain, which is wolded or wrapped round with old rope. The beam is generally from 40 to 50 feet in length, and about 8 inches square. The net is about 30 yards in depth, and has a couple of pockets inside. The end is untied when the net is hauled on board for the purpose of taking the fish out. These nets can only be worked where the bottom of the sea is free from rocks. They are used by boats of 35 to 60 tons, manned by crews of from 4 to 6 men, and 2 to 3 or four boys. In the vicinity of Scarbro' they fish between the shore-reefs and the off rock which is 4—10 miles from land; the bottom is sand or clay, with 4—15 fathom water on the land-side, and 17—25 fathoms on the off side." Immense quantities of Crustacea and shell-fish are taken with the trawl, as well as ground-fish.

---

[*] Admiralty, Manual of Scientific Inquiry. 8vo. Lond. 1849.
[†] Voyage of H. M. S. Rattlesnake, vol. I. p. 27.

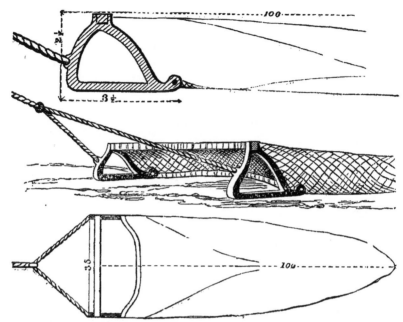

Fig. 227. A Trawl-net. A. Side view; C. Plan; B. Net in operation.

Kettle-nets.—On the flat, sandy coast of Kent and Sussex, the mackerel-fishery is pursued by setting up stakes 10 or 15 feet high, at distances of 10 feet apart, in lines running outwards from the shore at high-water, to low-water neap tides, where they are turned in the direction of the tide. To these stakes, nets are attached and leaded, which remain as long as the fish are on the coast. Cuttle-fish are frequently taken in these nets.

Deep-sea Fishery.—In North Britain an extensive ground-fishery is conducted by means of long lines,—often a mile in length—with hooks and baits every few yards. These lines are laid out at night, near the coast, and taken up the next morning. When used out at sea, the boats lay by for a few hours, and then take up the lines. The carnivorous whelks adhere to the baits (which have not been seized by fishes), and sometimes a bushel of them are taken in this way from a single line. *Rhynchonella psittacea, Panopæa Norvegica, Velutinæ,* and some of the scarce *Fusi,* have been obtained from these lines, the bivalves having been entangled accidentally by the hooks.

For trapping whelks on rocky ground a net may be made, such as is used for crabs and lobsters, by attaching a loose bag to an iron ring of a yard across. This is fastened to a rope by three equal strings, baited with dead fish, and let down from a vessel at anchor, or still better from a buoy. It is put down over-night, and hauled up gently in the morning.

Mr. D'Urban informs us that *Natica Alderi* and *monilifera* are fre-

U

quently found in the lobster-pots at Bognor, Sussex, which they enter to feed upon the bait.

Dredging.—The Dredges used in the Oyster and Whelk-fisheries are so rudely made as to injure the more delicate marine animals, and suffer all the minute things to escape. It is therefore necessary to have instruments specially adapted for the naturalist's work.

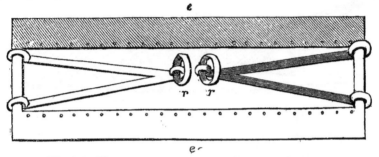

Fig. 228.   Plan of the Framework of a Dredge, reduced to ⅛.

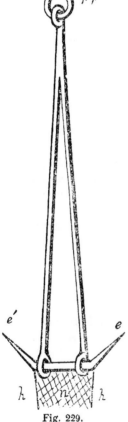

Fig. 229.

Fig. 228 is a plan, and Fig. 229 a side-view, of a small dredge, belonging to Mr. J. S. Bower-bank, and suited for such work as a private col-lector might do on the English coast. It is made of wrought iron, with moveable joints, so as to fold up and carry in the hand. The bag attached to the dredge is formed of two pieces of raw hide (*h*, *h*), connected at the ends and bottom by net (*n*) made of cod-line, to allow the water to escape; and is fastened to the frame with copper-wire, through the eyelet holes. The towing rope is attached to the rings (*r*, *r*), and when thrown overboard it scrapes with one or other of the cutting edges (*e*, *e'*). The opening is made narrow to prevent the admission of large and heavy stones.

Dredging should not be attempted in a *rowing-boat*, unless near shore, in smooth water, and with a depth not exceeding 5 or 10 fathoms. It may be managed in a light boat by two persons; one rowing, the other holding the rope of the dredge which is passed overboard near the stern.*

* "WEYMOUTH is pre-eminently the best place on the British coast for dredging. I can reckon 195 marine species of shells that I have collected within a range of five miles, and ten more species may safely be added. The dredging is also easy and safe. The cost of a suit-able boat and man is about 7s.6d. a day, *i. e.* from 10 or

The whelk and oyster-dredgers employ a decked sailing-vessel, and work several dredges simultaneously, each requiring a person to manage it. The dredges are put overboard on the weather-side, and the ropes made fast to a bulwark or thwart; each dredger holds the rope in his hand, after giving it a single turn round a thwart or "belaying-pin," to regulate the strain by means of the spare line. When a sufficient distance has been traversed, or the ropes strain with the weight of mud and stones, the vessel is brought to, and the dredges hauled up and emptied.*

The *length of line* required is about *double* the depth of the water. If the line is too short the dredge will only skim the bottom; if too long it will be in danger of getting fast. When the bottom is loose sand or soft mud, the line must be shortened, or the vessel have more way, or else the dredge will be apt to get buried.

The *strength of the line* ought to be sufficient to anchor the vessel in smooth water,—though not, of course, when there is much way on her,—so that if the dredge gets foul it is necessary to let out the spare line and relieve the strain, while the vessel is brought round. The dredge will then usually capsize, and may be hauled up.

If the bottom is at all rocky, a small strong dredge is best. The line must be shortened, and some additional precautions may be taken, such as fastening the rope to one ring of the dredge, and tieing the other with spun yarn, which will break under a sudden and dangerous strain, and release one end of the dredge.

In dredging on Coral-ground, Mr. Cuming employed a 3 inch hawser, and had a patent buoy attached to the dredge by a 1¼ inch rope. More than once the hawser parted, and the dredge was left down all night, but recovered the next day.

Mr. Mc Andrew's researches on the coast of Norway, were conducted in the " Naiad," a Yacht of 70 tons, and extended from the shore to 250 fathom water. The dredge employed was at least twice as strong and heavy as the one we have represented, and all forged in one piece, instead of folding up. The bag was fastened on the frame with thongs cut from the hide. Before using, it requires to be towed astern for a couple of hours to soften it. In three months work, only two cow-hides were used, and one of those was torn by accident on sharp rocks. Several spare dredges were on board (in case of emergency), but not used.

Dredging in deep water (50—300 fms.) can only be done in calm

11, a.m., to 4 or 5, p.m. Dredging can be carried on in Weymouth in almost any weather, the bay is so protected." (*R. Damon.*)

* The collector may go out with the fishermen, and superintend his own dredge, almost any time of the year, although oyster catching is illegal in the summer. The scallop-banks off Brighton are in 15 fms. water, and nearly out of sight of land. It is not always possible to work over them and return the same night.

weather, with a light breeze. The Yacht is brought to the wind (by putting up the helm), the foresheet hauled to windward, mainsail hauled up, and mizen taken in; the gaff topsail also hauled up; she then drifts to leeward, and the dredge is thrown overboard to windward, with the line made fast amidships; the spare line being coiled up so as to be given out readily. When the dredge is to be hauled in, the rope is passed through a moveable block, fixed to the shrouds, and the whole strength of the crew (15 hands) called into requisition if necessary. When the depth does not exceed 50 fathoms, the boat, with three men and the two dredgers, is used.

If the dredge gets fouled, the rope is passed into the boat, brought over the dredge, and hauled up. In very deep water (150 fm.) the line is carried forward and made fast to the bows, and the yacht itself hauled up till right over the dredge, which is then recovered without difficulty.

The contents of the dredge are washed, and sifted with two sieves, one " ¼ inch," the other very fine. They are made of *copper wire*, and one fits into the other. The dredge is emptied into the coarse sieve and washed in the sea from the boat, or if in the yacht, they are placed in an iron frame, over the side of the vessel, and buckets of water poured on. The sediment retained in the fine sieve may be dried and examined at leisure, for minute shells.

The following " dredging-papers," kept on the plan recommended by Prof. E. Forbes, have been selected by Mr. Barrett, to illustrate the kind of shells found at various zones of depth.

Note.—The shell-fish obtained by dredging should be at once boiled, and the animals removed, unless wanted for examination (p. 441). The bivalves gape, and require to be tied with cotton; the *opercula* of the univalves should be secured in their apertures with wool. The small univalves may be put up in spirit, or *glycerine*, to save time. In warm climates the flies and ants assist in removing any remains of the animals left in spiral shells, and *chloride of lime* may be necessary to deodorize them.

M. Petit de la Saussaye has given very full instructions for collecting and preserving shells, in the *Journal de Conchyliologie* for 1850, p. 215, and 1851, pp. 102, 226.

It is stated that both the form and colour of molluscous animals may be preserved in a saturated solution of *hydro-chlorate of ammonia* (10 parts) and corrosive sublimate (1 part—first dissolved in alcohol), but the preparation is expensive and dangerous.

Dredges and other apparatus, glazed boxes, and glass tubes for specimens, may be obtained of G. Sowerby, 70, Great Russell Street, Bloomsbury; and of R. Damon, Weymouth.

DREDGING PAPERS, AND RECORDS OF RESEARCHES ON THE COAST OF NORWAY.

By R. Mc Andrew, Esq. and Lucas Barrett, Esq. F.G.S.

I.

Date......................July 1st, 1855.
Locality ....Tromsoë (Nordland).
Depth ....Between tide marks.
Ground ....Rock and sand.

Species.	Number of living specimens.	Number of dead specimens	Observations.
Mya truncata	6	Many.	In sand.
Tellina incarnata	Many.	Many.	In sand.
Astarte compressa	1	0	On sand.
——— borealis	3	Many.'"*	On sand.
Cardium edule	Many.	Many.	In sand.
Crenella discors..	Many.	0	Covering the under sides of stones.
Acmæa testudinalis	Many.	0	On rock.
Margarita undulata	6	0	On weed.
——— helicina	8	0	On weed.
Litorina litorea	Many.	0	On rock.
——— rudis	Many.	0	On rock.
Lacuna vincta	2	0	On weed.
Natica pusilla	2	0	On sand.
——— clausa	Many.	0	On rock.
Purpura lapillus..	Many.	Many.	On rock.
Buccinum undatum	Many.	0	On rock and sand.
——— cyaneum	Many.	0	On rock.
Bela turricula	10	0	On rock.
Doris Johnstoni..	8	0	

(Note.) No specimens of *Trochus*, or *Patella vulgata* occurred.

II.

Date ....July 5th, 1855.
Locality ....Near Hammerfest (Finmarken).
Depth ....7 to 20 fathoms.
Distance from shore ....Close to shore
Ground.................Nullipore and sand.

Species.	Number of living specimens.	Number of dead specimens	Observations.
Saxicava arctica	4	0	Young.
Mya truncata	4	3	Young.
Thracia convexa..	4	0	In sand.
Tellina proxima	0	4'	
Mactra elliptica	1	0	
Venus ovata..	3	0	
——— striatula	Many.	0	

* The accented numbers in the column of "dead specimens" refer to disunited valves of *Conchifera* and *Brachiopoda*.

Species.	Number of living specimens.	Number of dead specimens.	Observations.
Cyprina Islandica	Many	Many.'	
Astarte compressa	Many.	0	
Cardium fasciatum	6	0	
Modiola modiolus	1	4'	
———— phaseolina	3	0	
Leda caudata	2	1	
Pecten Islandicus	0	2'	
Chiton asellus	2	0	
———— marmoreus	2	0	
Acmæa virginea	3	2	
———— testudinaria	0	1	
Patella pellucida	6	0	
Dentalium entale	4	2	
Trochus tumidus	Many.	Many.	
———— cinerarius	1	0	
Margarita helicina	12	0	
—————— undulata	Many.	Many.	
———— cinerea	6	2	
Velutina lævigata	0	1	
Buccinum undatum	0	3	
Trophon clathratus	1	0	
———— Gunneri	1	0	
Bela rufa	1	0	
—— turricula	0	4	
Mangelia nana	2	0	

III.

Date ....July 3rd, 1855.
Locality ....Island of Arnöe (Finmarken).
Depth ....7 to 22 fathoms.
Distance from shore ....Half a mile.
Ground.................Laminaria and red weed.

Species.	Living	Dead	Observations.
Saxicava arctica	3	Many.'	
Thracia convexa	1	0	
Venus ovata	1	3'	
Cyprina Islandica	2	Many.'	
Astarte crebricostata	Many.	Many.	
———— elliptica	12	Many.	
———— compressa	Many.	Many.	
Cardium fasciatum	Many.	Many.	
Cryptodon flexuosus	1	6'	
Modiola modiolus	1	Many.'	
Crenella decussata	Many.	Many.	
Leda pernula	Many.	Many.	
Pecten Islandicus	3	Fragments.	Young.
Anomia Ephippium	Many.	0	
————aculeata	Many.	0	
Chiton marmoreus	4	0	

Species.	Number of living specimens.	Number of dead specimens.	Observations.
Dentalium entale	4	Many.	
Trochus tumidus	Many.	Many.	
——— cinerarius	Many.	Many.	
Margarita cinerea	Many.	Many.	
——— undulata	Many.	Many.	
——— helicina..	Many.	Many.	
Lacuna vincta	Many.	Many.	
Litorina litoralis	3	0	
Rissoa parva	Many.	0	
Natica clausa	4	0	
——— pusilla	0	1	
Velutina lævigata	3	0	
——— flexilis	1	0	
Trichotropis borealis	3	0	
Nassa incrassata	1	0	
Mangelia nana..	8	0	
Bela turricula	Many.	0	
Trophon Gunneri	12	0	
——— clathratus	3	0	

IV.

Date ....July   , 1855.
Locality ....Vigten Island (N. Drontheim).
Distance from shore ....Quarter of a mile.
Depth ....30 fathoms.
Ground.................Coral-bank.

Species	Living	Dead
Arca nodulosa..	3	5
Leda caudata	2	0
Yoldia lucida	3	0
Astarte su cata	3	4'
Pecten Islandicus	0	2'
Lima excavata..	0	1'
Lucina Sarsii	0	1
Cryptodon flexuosus	2	0
Modiola phaseolina	10	0
Anomia ephippium	Many	0
Venus ovata	0	2
Terebratulina caput-serpentis ..	20	Many.
Chiton asellus..	4	0
Puncturella noachina	2	0
Emarginula fissura	1	2
——— crassa	0	1
Margarita cinerea	1	0
——— ai bastrum	1	0
Trophon barvicensis	1	0

V.

Date ....June 23rd, 1855.
Locality ....Omnaesöe (Nordland).
Depth ....30 to 50 fathoms.
Distance from shore ......Half a mile.
Ground....................Stones and sand.
No. of hauls.......... .....Four.

Species.	Number of living specimens.	Number of dead specimens.	Observations.
Saxicava arctica	6	2	
Tellina proxima	0	1	
Venus ovata	2	0	Small.
Cyprina Islandica	2	Many.	
Astarte elliptica	4	0	
—— compressa	6	0	
Cardium fasciatum	2	0	
—— suecicum	5	4	
Modiola phaseolina	200	Many.	Large.
Crenella nigra	0	1	Large.
Nucula nucleus	0	5	
—— tenuis	4	Many.	
Leda caudata	2	0	
Arca pectunculoides	12	10	Large.
Pecten striatus	2	0	
—— tigrinus	3	6	
—— similis	1	0	
—— islandicus	0	1	Large and Recent.
Terebratula cranium	80	10	
Terebratulina caput-serpentis ..	1	0	
Crania anomala	12	0	Many stones had on them the attached valve.
Chiton Hanleyi	3	0	
Lepeta cœca	4	0	
Acmæa virginea	10	6	
Pilidium fulvum	Many.	4	
Puncturella noachina	2	1	
Trochus millegranus	2	0	
Eulima polita	1	0	
Natica nitida	3	2	
—— helicoides	0	1	
—— pusilla	0	1	
Velutina lævigata	1	0	
Trichotropis borealis	5	3	Large.
Nassa incrassata	1	0	
Fusus antiquus	0	2	Carinated Var
Trophon clathratus	0	1	
Mangelia turricula	1	0	
Tornatella fasciata	0	2	
Buccinum undatum	6	0	Young.
Pleurotoma nivalis	10	15	

VI.

Date ....July 20th, 1855.
Locality ....North of Rolphsoe (Finmarken).
Depth ....130 to 180 fathoms.
Distance from shore ......Half a mile.
Ground....................Sand.
No. of hauls ....Two.

Species.	Number of living specimens.	Number of dead specimens.	Observations.
Cyprina Islandica..	0	3	
Neæra cuspidata	0	2′	
Leda caudata	0	3′	
Yoldia lucida	1	2′	
Pecten Islandicus	0	Many.	Small.
——— similis	0	1	
Arca pectunculoides	1	0	
Syndosmya prismatica..	0	1	
Cryptodon flexuosus	0	1	
Mactra elliptica	0	2″6′	
Cardium fasciatum	0	2	
——— suecicum..	0	3	
Astarte sulcata	1	0	
Anomia ephippium	Many.	0	
Crenella decussata..	2	Many.	
——— nigra	0	2′	
Terebratula cranium	3	0	
Rhynchonella psittacea	1	2	
Dentalium entale	Many.	Many.	
Puncturella noachina	Many.	0	
Lepeta cœca	2	0	
Pleurotoma nivalis	1	2	
Fusus? sp.	0	Fry.	
Buccinum Humphreysianum ..	0	1	
Bela turricula	2	0	
Margarita cinerea	3	4	
——— undulata	0	2	
——— alabastrum	0	1	

VII.

Date ....July 25th, 1855.
Locality..............Off the Island of Arnöe (Finmarken).
Depth ....200 fathoms.
Distance from shore ..Four miles.
Ground...............Mud.

Pecten similis..	0	2′	
Cryptodon flexuosus	4	0	
Neæra cuspidata	0	1	
Arca pectunculoides	1	3	

Species.	Number of living specimens.	Number of dead specimens.	Observations.
Nucula tenuis	2	0	
Yoldia lucida	4	6	
Modiola phaseolina	2	0	
Cardium suecicum..	2	0	
Crenella decussata..	1	0	
Astarte crebricostata	0	4'	
Terebratula cranium	0	2	
Dentalium entale	1	2	
——— sp...	1	8	
——— quinquangulare (Forbes).	1	0	
Eulima bilineata	2	2	
Eulimella Scillæ	0	3	
Mangelia trevelliana	0	1	
Bela rufa	0	1	
Philine quadrata	0	1	

DREDGING PAPERS, OR RECORDS OF RESEARCHES IN THE ÆGEAN SEA.

BY PROFESSOR E. FORBES

I.

Date.........................May 29th, 1841.
Locality ....Nousa Bay, Paros.
Distance from shore ....Within the Bay.
Depth ....Five to six fathoms.
Ground ....Mud and sandy mud.

Species	Living	Dead	Observations
Pinna squamosa	0	1	
Modiola tulipa..	1	0	In sandy mud.
Pecten polymorphus	4	6'	
——— hyalinus	1	0	
Nucula margaritacea	0	40'	In dark mud.
Cytherea chione	0	1	
——— venetiana	1	3-5'	
——— apicalis	1	2-12'	
Artemis lincta	0	1'	
Tapes virginea..	0	5'	
Venus verrucosa	0	5'	
Tellina donacina	0	1-3'	
——— balaustina	0	2'	
Syndosmya alba	0	2-10'	
Lucina lactea	0	2-28'	
——— squamosa	0	3'	
——— rotundata	0	4'	

Species.	Number of living specimens.	Number of living specimens.	Observations.
Cardium rusticum	0	1′	A strong valve.
———— exiguum	3	7′	
Cardita sulcata	0	1′	
Patella scutellaris	0	1	Washed in from shore.
Calyptræa Sinensis	0	2	
Bulla hydatis	0	1	
Turritella 3 plicata	0	1	
Trochus canaliculatus	0	4	
Cerithium lima	0	3	
———— vulgatum	12	8	
Murex fistulosus	1	0	In dark mud.
Aplysia depilans	1	0	
Ostræa plicatula	0	10′	

II.

Date..........................Sept. 14, 1842.
Locality ....Gulf of Smyrna.
Depth ....26 fathoms.
Distance from shore ....Two miles and a half.
Ground ....Fine brown mud.

Avicula Tarentina	3	3	Full grown, adhering to each other.
Saxicava arctica .. .. .., ..	4	0	

III.

Date...................August 5, 1841.
Locality ....Off northern extremity of Paros.
Depth ....40 fathoms.
Distance from shore ....Three miles and a half.
Ground ....Weedy

Pecten pusio	5	4′	
——— opercularis	0	1	Small.
Nucula margaritacea	0	2′	
Cytherea apicalis	0	1′	
Cardita squamosa	1	1′	
Cardium papillosum	0	2	
Fusus fasciolaroides	1	0	New.
Murex brandaris	0	3	
Vermetus gigas	0	1	
——— corneus	3	0	New.
Trochus exiguus	8	2	
Turbo rugosus	1	0	
Pleurobranchus sordidus	1	0	New.

Species.	Number of living specimens.	Number of dead specimens.	Observations.
Doris tenerrina	2		New.
———— gracilis	2		
———— coccinea	1		
Ascidium, four species			
Aplidium, two species			

IV.

Date.................Sept. 16th, 1841.
Locality ....Off Ananas Rocks.
Depth ....105 fathoms.
Ground ....Nullipore.
Distance from shore ..From Rocks three miles, from Milo ten miles·

Species.	Living	Dead	Observations.
Terebratula vitrea	0	2′	Dead and worn.
Megerlia truncata	30	100-20′	Of all ages.
Argiope decollata	100	400-6′	Of all ages.
———— seminulum	18	10.8′	
Morrisia anomioides	1	0	Adhering to T. vitrea. New.
Crania ringens	0	6′	
Lima elongata	0	5′	New
Pecten concentricus	0	1′	New.
———— fenestratus	0	2′	New.
Spondylus Gussoni	1	1′	
Arca lactea	1	7′	
———— scabra	0	2′	
Neæra cuspidata	0	1′	
———— attenuata	0	1′	New.
Fusus echinatus	0	2	
Pleurotoma crispata	0	2	Hitherto known only fossil.
———— maravignæ	0	2	New.
———— abyssicola..	0	4	New.
Mitra philippiana	0	4	New.
Cerithium lima	0	8	
Trochus tinei	0	6	
———— exiguus	1	9	
Turbo sanguineus	0	24	Hitherto known only fossil in the Mediterranean basin.
Rissoa reticulata	4	11	
Emarginula elongata	0	8	
Pileopsis Hungaricus	0	1	Small.
Acmæa unicolor	1	24	New.
Atlanta Peronii	0	2	Incrusted with nullipore, and thus rendered solid.
Hyalea gibbosa	0	1′	
Cleodora pyramidata	0	3	
Criseis clava	0	7	
———— spinifera	0	10	

V.

Date.........................Nov. 25th, 1841.
Locality ....S. extremity of Gulf of Macri.
Depth.......................230 fathoms.
Distance from shore ....One mile (shore steep).
Ground ....Fine yellowish mud.

Species.	Number of living specimens.	Number of dead specimens.	Observations.
Terebratula vitrea	0	2'	
Syndosmya profundissima	0	3'	
Arca imbricata..	1	1'	
Dentalium quinquangulare.. ..	1	0	
Hyalea gibbosa	0	1	
Cleodora pyramidata	0	8	
Criseis spinifera	0	5	

The Distribution of the Mollusca in Depth has been investigated by MM. Audouin and Milne-Edwards, M. Sars, and Prof. E. Forbes. By these observers the sea-bed is divided into four principal regions :—

1. The Litoral zone, or tract between tide marks.
2. The Laminarian zone, from low-water to 15 fms.
3. The Coralline zone, from 15—50 fms.
4. The deep-sea coral zone, 50—100 fms. or more.

1. *The Litoral zone* depends for its depth on the rise and fall of the tide, and for its extent on the form of the shore. The shells of this zone are more limited in their range than those which are protected from the vicissitudes of climate by living at some depth in the sea.* In Europe the characteristic genera of *rocky* shores are *Litorina, Patella* and *Purpura ;* of sandy beaches, *Cardium, Tellina, Solen;* gravelly shores, *Mytilus;* and on muddy shores *Lutraria* and *Pullastra.* On rocky coasts are also found many species of *Haliotis, Siphonaria, Fissurella,* and *Trochus;* they occur at various levels, some only at the high-water line, others in a middle zone, or at the verge of low-water. *Cypræa* and *Conus* shelter under coral-blocks, and *Cerithium, Terebra, Natica,* and *Pyramidella* bury in sand at low water, but may be found by tracing the marks of their long burrows. (*Macgillivray*).

2. *Laminarian zone.*—In this region, when rocky, the tangle (*Laminaria*) and other sea-weeds form miniature forests, the resort of the vegetable feeding mollusks—*Lacuna, Rissoa, Nacella, Trochus, Aplysia,* and various *Nudibranchiata.* On soft sea-beds bivalves abound and form the prey of *Bucci-*

---

* Some of the litoral shells, like *Purpura lapillus* and *Litorina rudis,* have no free-swimming larval condition, but commence life as crawlers, with a well-developed shell. Their habits are sluggish, and their diffusion by ordinary means must be exceedingly slow.

num, Nassa, and *Natica.* From low-water to the depth of one or two fathoms on muddy and sandy shores, there are often great *meadows* of grass-wrack (*Zostera*) which afford shelter to numerous shell-fish, and are the haunt of the cuttle-fish and calamary. In tropical seas, the reef-building corals often take the place of sea-weeds, and extend their operations to a depth of about 25 fathoms. They cover the bottom with living verdure, on which many of the carnivorous mollusks feed, while some, like *Ovulum* and *Purpura,* browse on the flexible *Gorgoniæ.* To this zone belong the oyster-banks of our seas, and the pearl-fisheries of the south ; it is richer than any other in animal life, and affords the most highly coloured shells.

3. *Coralline zone.* In northern seas the belt of sea-weed that fringes the coast is succeeded by a zone where horny zoophytes abound, and the chief vegetable growth consists of *Nullipore* which covers rocks and shells with its stony-looking incrustations. This zone extends from 15 or 25, to 35 or 50 fathoms, and is inhabited by many of the predacious genera—*Buccinum, Fusus, Pleurotoma, Natica, Aporrhais, Philine, Velutina ;* and by vegetable feeders, such as *Fissurella, Emarginula, Pileopsis, Eulima,* and *Chemnitzia.* The great banks of scallops belong to the shallower part of this region, and many bivalves of the genera *Lima, Arca, Nucula, Astarte, Venus, Artemis,* and *Corbula.*

4. *Deep-sea Coral-zone.* From 50 to 100 fathoms the *Nullipore* still abounds, and small branching corals to which the *Terebratulæ* adhere. In northern seas the largest corals (*Oculina* and *Primnoa*) are found in this zone, and shells are relatively more abundant, owing to the uniformity of temperature at these depths. These deep-water shells are mostly small and destitute of bright colours; but interesting from the circumstances under which they are found, their wide range, and high antiquity. Amongst the characteristic genera are *Crania, Thetis, Neæra, Cryptodon, Yoldia, Dentalium,* and *Scissurella.* In the mud brought up from deep water may be often found the shells of *Pteropoda,* and other mollusca which live at the surface of the sea. In the Ægean Sea there is deep-water within one or two miles of the coast; but in the British Channel the depth seldom amounts to more than 20—40 fathoms.

When registering the results of dredging-operations, it is important to distinguish between *dead and living shells,* as in the preceding Tables; for almost every species is met with, in the condition of *dead shells,* at depths far greater than those in which it actually lives. On precipitous coasts the litoral shells fall into deep water, and are mingled with the inhabitants of other zones; currents also may transport dead shells to some distance over the bed of the sea. But the principal agents by which so many decayed and broken shells are scattered over the bed of the deep sea, must be the mollusk-eating fishes. Of 140 species of boreal shells described by Dr. Gould (p. 358) more than half were obtained from the maws of fishes, in Boston market. Cod-fish do not swallow the large whelk-shells, but some idea of the number they consume may be derived from the fact that Mr. Warington has obtained the muscular foot and operculum of above 100 whelks, of large size, besides quantities of *crustacea*, from the maws of three cod-fish procured in the London market. Bivalve shells, like the Solens, and the rare *Panopæa Norvegica* are swallowed, and ejected again with eroded surfaces. The haddock swallows shells still more indiscriminately, and Mr. Mc Andrew has found great numbers of rare Pectens in them, but generally spoiled. The cat-fish and skate break up the strongest shell-fish with their teeth—accounting for the many angular fragments met with in the dredge, and in recent deposits.

The following are examples of shells obtained from great depths.

Norway. (Mc Andrew)

Living shells.

	Fathoms.
Cerithium metula	20—150
Margarita cinerea	10—130
Dentalium entale	200
Limea sarsii	120
Leda pygmæa	200
Yoldia limatula	120
Thetis koreni	40—100
Cryptodon flexuosus	200

Off the Cape. (Belcher.)

Buccinum? clathratum	136
Volutilithes abyssicola	132
Pectunculus Belcheri	120

Ægean. (Forbes.)

	Living.	Dead.
Terebratula vitrea	100	250
Argiope decollata	100	110
Crania ringens	90	150

Ægean. (Forbes.)

	Living.	Dead.
Murex vaginatus		150
Fusus muricatus	80—95	150
Nassa intermedia		45—185
Cerithium lima	3—80	140
Chemnitzia fasciata		110—150
Eulima distorta		69—140
Scalaria hellenica		110
Rissoa reticulata	55	185
Trochus exasperatus	10—105	165
Scissurella plicata		70—150
Acmea unicolor	60—105	150
Dentalium quinquangulare	150—230	
Bulla utriculus		40—140
Spondylus Gussonii	105	
Pecten Hoskynsii		185—200
Arca imbricata	90—230	
Neæra cuspidata	12—185	
Thetis anatinoides		40—150
Kellia abyssicola	70—180	200
Syndosmya profundissima		80—185

Preserving molluscous animals for examination.

When shell-fish are killed by sudden immersion in hot water or strong spirit, great and unequal contraction is caused, distorting the muscular parts and rupturing the membranes.

Experiments have yet to be made for the discovery of means whereby these and other marine animals may be paralysed and killed, without altering the ordinary condition of their organs.*

Glycerine is the best medium for preserving such objects as the univalve

---

* The brittle-stars (*Ophiocoma*) are killed by sudden immersion in fresh-water; and the *Actiniæ* may be stupified by adding fresh-water drop by drop, until they lose the power of retracting their tentacles. But the bivalves (such as *Pholas*) may be kept in stale water till their valves fall off with incipient decomposition, and yet the muscular siphons retain their irritability, and contract slowly and completely, when placed in spirit.

shell-fish, intended for the examination of their lingual teeth; for if put up in strong spirit they become so hard that it is almost impossible to make good preparations from them, and in weak spirit they will not keep for any length of time.

Alcohol.—The cheapest alcohol for preserving natural history objects, at home, is sold as "methylated spirit;" it contains ten per cent. of ordinary wood spirit, and being undrinkable, is free of duty. When many specimens are put up together the spirit becomes much diluted, and should be changed. The soft tissues of bivalves, and spiral bodies of the univalves soon decompose in weak spirit. But for permanent use, in Museums, proof spirit may be diluted with an equal bulk of water. Cotton wool may be put with the specimens in spirit, especially with cuttle-fish, to preserve them from distortion by pressure.

Goadby's solution is prepared by dissolving ½ lb. of bay salt, 20 grains of arsenious acid, or white oxide of arsenic, and 2 grains of corrosive sublimate, in 1 quart of boiling rain-water.

Burnet's solution (chloride of zinc), largely diluted, is now used at the British Museum for the preservation of fishes and other objects, in glass jars. It has several advantages over spirit; being undrinkable, and not inflammable, and the concentrated solution (sold by all druggists) is much less bulky.

Muriate of Ammonia is recommended, by Mr. Gaskoin, for removing any unpleasant odours which may arise from preparations when taken out of spirit for examination. (See p. 430.)

A solution of *Chloride of Calcium* has been employed by Gen. Totten, U.S. Engineers, for preserving the flexibility of the epidermis in various shells. The solution of this deliquescent salt (which any one can make by saturating hydrochloric acid with marble), keeps the object which has been steeped in it permanently moist, without injuring its colour or texture; while its antiseptic properties will aid in the preservation of matters liable to decay. (Prof. J. W. Bailey, in Silliman's Journal, July, 1854.)

Aquaria.

The establishment of fresh-water and marine *aquaria* by Mr. Mitchell, in the gardens of the Zoological Society, and the writings of Mr. Philip Gosse, have popularized the subject of aquatic animals, and shewn how easy and interesting it is to keep a few of them alive, and watch their habits even in the midst of London. Instead of the solitary gold-fish in its globe of glass, we may now have a variety of fishes in a little world of aquatic plants and water-insects and fresh-water shells. Salt-water may be brought from the sea, or manufactured at home; and a glass jar or tank of any size, may be tenanted with small sea-fish and soldier-crabs, sea-anemones, shrimps, and periwinkles.*

* All the materials for fresh-water and marine aquaria, including live plants and fishes, may be obtained of W. A. Lloyd, 164, St. John Street Road, London.

The woodcut (Fig. 228) represents a marine aquarium designed by Mr. Gosse, with a small fountain in the centre, which not only adds to its ornamental appearance, but serves to aërate the water, or mix with it a greater amount of the *fixed air* which gill-breathing creatures respire.* An

Fig. 228.

aquarium of this shape combines the advantage of a large surface *exposed to the air*, with the opportunity of watching its inhabitants through the glass sides. The form of *aquarium* best suited for aquatic animals, viz. a wide

* The use of the woodcut for this work, was kindly afforded at our request, by Mr. Gosse.

shallow pan, is the least convenient to keep; and therefore a large glass jar is usually adopted, or an oblong tank, made to fit the recess of a window, with slate ends and bottom, and plate-glass sides.

The most convenient form of tank, is that recommended by Mr. R. Warington; it is a four-sided vessel, having the back gradually sloping upwards from the bottom at an angle of 45 or 50 degrees, and the consequently extended top sloping slightly downwards and resting on the upper part of the back. The bottom is narrow, and the back may be covered with light rockwork, extending just above the water-line, to afford places of growth to the sea-weeds and fixed animals, and provide the litoral shell-fish with a feeding-ground close to the surface. The front and top of this aquarium are of glass, the rest of slate, fixed in a stout frame-work. (An. Nat. Hist. 14, p. 373.)

The aquarium should be covered, at least in towns, with a lid, or plate of glass, to check evaporation and exclude dust. If ventilation is necessary, the lid may be supported by small bent pieces of lead, hung on the rim of the tank.

The "balance of organic nature" is maintained in these aquaria by growing plants with the animals (p. 31, note). For fresh-water tanks, *Valisneria spiralis* is the best plant; but if there is space for the common flag (*Iris pseudacorus*) or water-plantain (*Alisma*) they will rise above the surface and blossom.* The *Anacharis alsinastrum* and *Hydrocharis* (like a miniature water-lily), may be grown at the surface. And if the tank is covered with a frame filling the window, some climbing plants may be trained in it, and the sides converted into a rockwork on which many ferns will thrive and expand their fronds in the moist air. (*Warington*).

For *marine aquaria* the green-weeds (*Ulva*, *Enteromorpha* and *Bryopsis*) are better oxygen-producers than the red sea-weeds, but the latter are so attractive as to be often tried.† The weed may become too luxuriant and require to be thinned in summer, but in the winter it dies down, and nearly disappears. Some of the threadlike weeds (diatomaceous algæ) are apt to gain admission, and in autumn break up spontaneously, filling the water with an opaque green cloud.

The surface of rockwork, in the aquarium is liable to be overspread, and the interior of the glass itself rendered opaque, by the early growth of *confervæ*. This may be in some degree prevented by keeping the water free from the grown plants, which are easily removed; and the green on the glass is kept in check by water-snails and periwinkles. These creatures occupy them-

* When small fishes are kept in an aquarium, however limited, in which the aquatic plants are grouped in the centre, they will swim round and round it in a little shoal

† Mr. Warington recommends the employment of glass tinted green for moderating the light when red sea-weeds are grown.

selves unceasingly in licking the glass (p. 161), and may be watched with a magnifier of moderate power.

Artificial salt-water.—The difficulty of obtaining sea-water has been obviated by the manufacture of salts for the formation of medicinal baths, by evaporating large quantities of the sea-water itself. This plan was suggested by Dr. E. Schweitzer, whose analysis of the water of the English Channel, taken off Brighton, shews the following salts in 100lbs. (or 10 gallons), stated in decimal fractions of the pound, and also in ounces and grains :—

Chloride of sodium	2.706	......	$43\frac{1}{4}$ ounces.
—— magnesium ..	0 367	......	6 ,,
—— potassium....	0.076	......	$1\frac{1}{4}$,,
Bromide of Magnesium	0.003	......	21 grains
Sulphate of magnesia..	0.230	......	$7\frac{1}{2}$ ounces ⎫
—— lime	0.140	......	$2\frac{3}{4}$,, ⎬ Crystals.
Carbonate of lime	0.003	......	21 grains ⎭

As the weight of the salts amounts to $60\frac{3}{4}$ ounces, the true proportion of water to be mixed with them will be 3 pints less then 10 gallons.*

The *temperature* of the aquarium should not range below 50° nor above 70°. The mean temperature of the sea is estimated to be about 56° Fahr. with a variation of about 12° throughout the year. In hot summer days a screen is necessary against strong sunlight.† (*Warington*).

Many little points, in the management of the *aquarium*, will be determined by experience; such as the number of living animals it is capable of maintaining, and the sorts which may be safely kept together. Everything dead or decaying should be removed as soon as detected. The loss by evaporation may be supplied occasionally by sprinkling with distilled water.‡

* These salts are manufactured by Messrs, Brew and Schweitzer, 71, East Street, Brighton ; the proportion ordered to be used is 6 oz. to the gallon of water, and stirred well until dissolved.

There are few inland towns without a fishmonger, through whom may be obtained live periwinkles (occasionally tenanted by the hermit-crab), and oyster-shells incrusted with *serpulæ* and sea-weed, some of which may be still living. The stickle-back is almost the only *fresh-water* fish capable of existing in a marine *aquarium*.

† In a sitting-room with a south aspect and good fire daily, the temperature of a thirty-gallon aquarium has been known to fall as low as 45° on several occasions, though screened at night by a blind. (Warington, An. Nat. Hist. 1855, p. 315.)

‡ Hand-book of the Marine Aquarium. P. H. Gosse. 12mo. Lond. 1855.

Chapter V.

SUPPLEMENTARY NOTES ON THE MOLLUSCA.

Class, I. Cephalopoda.

Development, (p. 54.)—" All that is at present known upon this subject, is contained in the very beautiful memoir by Kölliker, who gives an elaborate account of the development of *Sepia, Loligo,* and *Argonauta.**

" The process of yolk-division is partial, and the development of the embryo takes place within a distinct germinal area—whence a distinct yolk-sac is formed. This is proportionally very large in *Sepia* (Fig. 229) and *Loligo,* very small in *Argonauta* (Fig. 230) and therefore while the embryo is flattened and extended in the former genera, in the latter it more resembles the embryo of an ordinary Gasteropod.

" Development commences by the separation of the embryo into *mantle* and *body,* (foot). The part of the body in front of the mantle becomes the head ; that behind it becomes the branchio-anal surface.

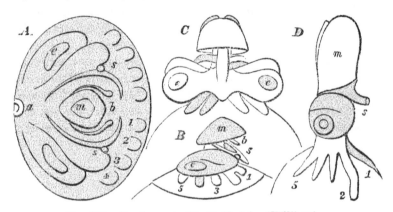

Fig, 229.   *Development of the Cuttle-fish.*   (Kölliker.)

A. Embryo two lines in diameter; *m.* mantle ; *b,* branchial processes ; *s,* siphonal processes; *a,* mouth ; *e,* eyes; 1—5 rudimentary arms.

B. Side view of the embryo, when more developed.

C. Front view, at a later period.

D. Young cuttle-fish, still attached to the yolk-sac, with the tentacular arms (2) longer than the rest.

" The latero-posterior margins of the body are produced into four or five processes on each side, which become the *arms.*

" On each side of the mantle, between it and the head and arms, a ridge

* Entwickelungs geschichte der Cephalopoden.   Zurich, 1844.

is formed upon the body. These ridges (*s. s.*) represent the *epipodium*; their anterior ends are continuous and attached, the posterior ends are at first free but eventually uniting, they form the funnel (D.*s.*). The rudimentary gills (*b.*) appear between the *epipodium* and mantle. The alimentary canal is at first straight; (the mouth being at *a*, the vent at *b*, in Fig. 229, A.)

"The embryo now grows faster in a vertical than in a longitudinal direction, so that it takes on the cephalopodic form. The intestine as a consequence, becomes bent upon itself; and the anterior pairs of arms grow over in front of the head and unite, so as eventually to throw the mouth nearly into the centre of the arms." (*Huxley.*)

At a later period of development (Fig. D.) the respiratory movements are performed by the alternate dilatation and contraction of the mantle; and the ink-bag is conspicuous by the colour of its contents, which are sufficient to blacken a considerable quantity of water. At the period of exclusion from the nidamental capsule, five layers of the shell of the young cuttle-fish have been formed; but except the nucleus, which is calcified, they are horny and transparent. The lateral fins are broader than in the mature animal.

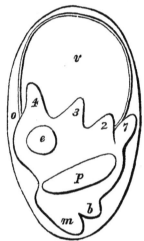

Fig. 230. *Argonaut,* embryo in the egg.

The observations of Madame Power respecting the young *Argonaut,* (quoted at p. 66), must have been made on the *Hectocotylus.* The embryo as described by Kölliker has simple, conical arms (1—4); and the elements of the funnel appear as a ridge (*p*) on each side of the body. In Fig. 230, *v,* is the yolk-sac; *o,* the position of the future mouth; *e,* the eye; *b,* the gill; *m,* mantle.

Octopoda, p. 65.—The account already given of the extraordinary condition of the male of the Argonaut and some other octopods has since been modified and extended by the observations of Dr. H. Müller[*] and M. Verany.[†]

According to Dr. Müller, the *Hectocotyle* of the Argonaut is an *arm irregularly metamorphosed,* spontaneously detached, (when the fluid formed in the true testis has been deposited in it,) enjoying an independent life, feeding on the female Argonaut, and fecundating by a true union.

The perfect male Argonauts are one inch in total length, and shell-less, (like the females of that size); their dorsal arms are pointed, not expanded. The testis is very large, and like that of the *Octopus* in structure

---

[*] Annales des Sciences Naturelles, t. 16, No. 3, and An. Nat. Hist., June, 1852.
[†] Moll. Medit. 4° Genes. 1851. An. Sc. Nat. t. 16, 1852.

and situation; it contains *spermatozoa* of different degrees of development, and the excretory duct probably debouches into the *Hectocotylus*. The Hectocotylus is developed in a coloured sac, *which occupies the place of the third arm of the left side;* the sac is cleft by the motions of the Hectocotylus, which extends itself, whilst the sac becomes inverted and forms the violet coloured capsule on its back. The sac never contains more than one Hectocotylus, which is attached by its base, whilst the rest of it is free and coiled up. It has no enlargement like the male *Tremoctopus* (Pl. I. f. 3); the filiform appendage proceeds from the smaller extremity, and sometimes remains entangled in the coloured cyst on the back of the Hectocotylus, near its base. It has a chain of nervous ganglia in its axis, (like that in the arms of ordinary cuttle-fish.)

M. Verany of Genoa, found the male of " Octopus carena " (*Tremoctopus granulosus*, Cuv.) with the *right arm of the third pair* more developed than the others, and bearing an oval globe at its free extremity (Fig. 231, C.) This abnormal arm, agreeing with the *Hectocotylus octopodis* of Cuvier, was found to be developed in a cyst (A.) like that of the male Argonaut.

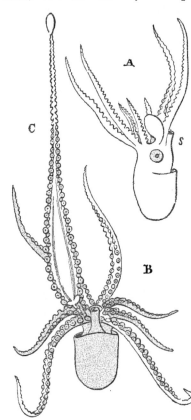

Fig. 231, *Octopus carena* ♂, Ver.
A. Side view, shewing cyst in place of third arm.
B. Ventral side of an individual more developed, with the Hectocotylus C.

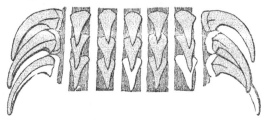

Fig. 232, *Lingual teeth of Seqia officinalis* (Cocken.)

The *Lingual dentition* of the cuttle-fish, as described by Lovèn, is most like that of the *Pteropoda* and *nucleobranchiata*. The central teeth are simple in *Sepia* and *Sepiola*, tricuspid in *Loligo*, and denticulated in *Eledone*. The lateral teeth, or *uncini*, are three on each side, and mostly simple and claw-like. There were 50 rows of teeth in the specimen of *Sepia* examined, the ribbon increasing in breadth backwards to twice its diameter in front.

Sciadephorus, Reinh. and Prosch; Bostrychoteuthis, Ag, = Cirroteuthis, p. 68.

Chondrosepia (loliginiformis) Leuckart, = Sepioteuthis p. 70.

Owenia, Prosch, = Cranchia megalops, N. Atlantic.

Leachia (cyclura) Les. 1821; Perotis, Esch. = Loligopsis, p. 71.

Belemnites.—Prof. Buckman of Cirencester possesses a *phragmocone* from the lias, containing the fossil ink-bag.

HELICERUS (Fugiensis) Dana, Sill. Journ. 1848. Shell like a Belemnite, half an inch in diameter; *guard* thick, subcylindrical, fibrous; *phragmocone* slender, terminating in a fusiform *spiral nucleus*. In slate-rock, Cape Horn.

Conoteuthis Dupinianus occurs in the *Gault* of Folkstone. (Mus. Bower bank.)

NAUTILUS.—The gas with which the air-chambers of the pearly nautilus are filled, consists chiefly of nitogen, without a trace of carbonic acid. (*Vrolik*, An. Nat. Hist. 12, 1843.)

Nautilus regalis, Sby, London Clay, Highgate. This species is distinguished by *serrated lines* on its external surface, nearly, but not quite concident with the lines of growth. (*Wetherell*, Lond. and Edin. Phil. Mag. IX. p. 462.)

ORTHOCERAS.—The species figured (Pl. II. 14) is *O. Ludense*, of the Ludlow-rock, Herefordshire. *O. giganteum* is an Actinoceras, with a large beaded siphuncle, differing in structure, however, from the Silurian species; the vascular tubes (or interspaces) connecting the inner siphuncle with the air-chambers exist in only one plane, on the ventral (?) side, whereas in *A. Bigsbyi* they radiate equally in all directions.

Hormoceras, Stokes.—The structure of the siphuncle in this fossil is essentially the same as in *Actinoceras*; the specimen (fig. 48, p. 88) is now in the British Museum.

DISCOSORUS (conoideus) Hall, 1852, Pal. New York, 99. This fossil appears to be a siphuncle similar to those figured by Dr. Bigsby in 1824 (Geol. Trans. Pl. 30. f. 6.) and which have been correctly referred by Quenstedt to the *Orthocerata*. It resembles a *pile of disks*, and is more or less curved, and conical, the *smaller* end being *upwards* or towards the last chamber!

Conoceras (angulosus) Bronn, 1830, was founded on *a figure* of a weathered fragment of *Gonioceras*, as pointed out by M. Saemann.

Thoracoceras (vestitum) Fischer, 1844 = Melia, Fischer (not L.) 1829, Carb. limestone of Moscow; the siphuncle is small and lateral. According to M. D'Orbigmy there are 20 species, ranging from the L. Silurian to the Carb. System, found in the U. States and Europe.

Apioceras (trochoides) Fischer, 1844. Dev.—Carb. Europe, Brit. (e. g. *O. fusiforme*) Aperture sub-circular, not much contracted.

Gyroceras, D'Orb. (not Meyer) has been employed for the principa᾽ fossils included in *Cyrtoceras* by Goldfuss. The name was originally given by Meyer, to *G. gracilis,* Bronn (*Spirula,* Goldf. MS. 1832, *Litnites,* Quenstedt) which is the *Goniatites compressus* of D'Arch. and Vern.

Trigonoceras (paradoxicum) Mc Coy, is a form of *Nautiloceras,* D'Orb. (Cyrt. ægocerus, Münst.) with a sub-spiral shell.

Discites (Mc Coy) is closely allied to the last, differing in the whirls being compact. It may be doubted whether any of the Palæozoic " Nautilidae" really belong to that family.

Ascoceras.—This curious fossil (which has been recently found at Ludlow, by Mr. Salter) only resembles *Ptychoceras* in appearance. It is slightly curved, and has a dorsal siphuncle, but the septa are bent and pro-longed forwards on the ventral side to such an extent as to give an appearance of the whole shell being doubled up.

Ammonites Jason, Reinecke (A. Gulielmi, Sby). The fossil figured, Pl. III 5, is *A. spinosus,* Sby. (=A. ornatus, Schl.) and is certainly distinct from the finely ribbed species which occurs with it, and to which the name *Jason* should be restricted.

CLASS II. GASTEROPODA.
Classification by lingual dentition.

The researches of Dr. Lovèn have been followed by many observations on genera not figured in his admirable memoir,* and by attempts to remodel the arrangement of the *Gasteropoda* by the aid of peculiarities in their dentition. Whatever improvements may be thus obtained, it does not appear desirable to introduce a new terminology for divisions long since well established, and already over-burdened with *classical* names.†

The patterns, or *types* of lingual dentition, are on the whole remarkably constant; but their *systematic value* is not uniform. It must be remembered that the teeth are essentially *epithelian cells,* and like other superficial organs liable to be modified in accordance with the wants and habits of the creatures. The instruments with which animals obtain their food are of all others most subject to these *adaptive* modifications, and can never form the *basis* of a philosophical system.‡

* Öfversigt af Kongl. Vetensk. Akad. Förhandl. 1847.

† The following names were proposed by Troschel (in Wiegman's *Handbuch der Zoologie,* 1848) and Gray (An. Nat. Hist.) for the principal types of lingual dentition.

   a. Tænioglossa, teeth 3. 1. 3 ; Litorina, Natica, Triton,

   b. Toxoglossa, teeth 1. 0. 1 ; Conus, Terebra?

   c. Hamiglossa, teeth 1. 1. 1 : Murex, Buccinum.

   d. Rachiglossa, teeth 0. 1. 0 ; Voluta. Mitra?

   e. Gymnoglossa, teeth 0 ; Pyramidella, Cancellaria, Solarium?

   f. Rhipidoglossa, teeth 00. 1. 00 ; Nerita, Trochus.

‡ The carnivorous opossums have teeth adapted for eating flesh, but are not on that account to be classified with the placental carnivora.

The lingual teeth, like the *operculum*, have usually a structure characteristic of the genera or sub-genera, and are sometimes uniform thoughout a whole family or group of families. They also exhibit minute differences in closely allied mollusks, and promise to be of great value in the discrimination of critical species. Mr. Wilton has ascertained that *Patella athletica* may be distinguished from the common limpet of our coast by its teeth; and a similar difference exists between two Cape species, *P. apicina* and *P. longicostata.*

In the account already given of the structure and use of the lingual teeth (p. 27 and 160), it has been pointed out that the *Carnivorous* families have a *retractile proboscis*; and it may now be added that in many instances the aperture of this organ is furnished with a prehensile spiny collar (fig. 239 and 260), apparently for the purpose of holding the prey whilst the lingual organ is employed in drilling or abrading it. The *spinose collar* coexists with a *lower* mandible in *Doris*; but appears not to be found in the genera provided with an upper jaw. The spiny buccal plates of *Natica* and *Lamellaria* are united above, like the lateral jaws of *Æolis*, of which they seem to be a modification. The *vegetable feeders* have a *rostrum*, or non-retractile muzzle, and frequently a horny upper *mandible* (fig. 260), which is sometimes divided, and forms two lateral jaws, articulated above. The *chemical composition* of the lingual teeth has not yet been examined by a competent observer. It is not improbable that the opaque brown teeth of *Chiton*, *Patella* and *Nerita*, are chitinous, like the mandibles and pen of the calamary.*

ORDER I. NUCLEOBRANCHIATA.

Lingual membrane plane, widening backwards; teeth 3. 1. 3. (p. 199.) Firola, Carinaria, Atlanta.

Fig. 233. *Carinaria cristata*, L. (Wilton.)

* The animal basis of shell is a peculiar organic substance, termed *conchioline*, insoluble in water, alcohol and ether, and resisting the long-continued action of acids; in caustic alkali it dissolves very slowly; its composition is—H. 5, 9; C. 50, 0; N. 17, 5, and O. 26, 6, (M. E. Fremy, Ann. de Chimie, 1855, p. 96.)

ORDER II. PROSOBRANCHIATA.

Section A. ZOOPHAGA, Lam. (*Proboscidea*, Troschel.)
[Family 1, *Strombidæ*: lingual teeth 3. 1. 3.] p. 104.

Fig. 234. *Strombus*. (Wilton).

Strombus (floridus) is described by Lovèn as having a non - retractile, produced *muzzle*, like *Aporrhais.* The teeth are 7 cusped; uncini—1 tridentate, 2 and 3

simple, claw-shaped. *S. gibberulus* is represented by Dr. Bergh with all the uncini denticulated.

The dentition of *Aporrhais* is most like *Strombus* and *Carinaria;* and quite unlike the *Cerithiadæ* with which it has been placed, in accordance with the views of Prof. Forbes. The animal is carnivorous. (p. 130.)

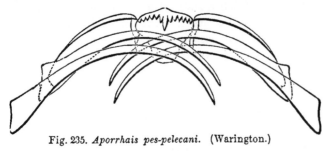

Fig. 235. *Aporrhais pes-pelecani*. (Warington.)

Fig. 236.
Operculum of
Struthiolaria.

[Family 2? *Cassidæ*: teeth 3. 1. 3.]

Cassis.	Dolium.	Ranella.
Cassidaria.	Pyrula.	Triton.

Fig 237. *Cassis saburon.* (Original.)

The spiny *buccal plates* of Cassis have been mistaken by Gray and Adams for the *teeth*, which in this genus, and also in *Triton*, are very minute and transparent.

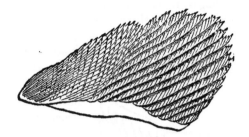

Fig. 238. Operc.
of *Cassis*.

Fig. 239. One of the buccal plates
of *Triton*, $\frac{40}{1}$. (Wilton.)

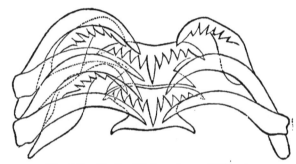

Fig. 240. *Teeth of Triton*, $\frac{240}{1}$. (Wilton.)

Fig. 241. *Dolium perdix.* (Original.)

Fam. **3.** *Muricidæ* (including *Buccinidæ*) : teeth 1. 1. 1. (p. 28).

Murex.	Fusus.	Buccinum.	Oliva
Trophon.	Fasciolaria.	Chrysodomus.	Ancillaria.
Purpura.	Turbinella.	Nassa.	Harpa.

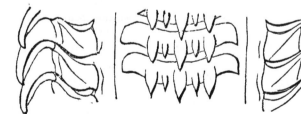

Fig. 242. *Murex tenuispina.* (Wilton.)

X 2

The lingual ribbon of *Murex, Purpura,* and most of the other members of the family, is very slender, and the teeth minute and glassy. It is quite certain that they drill holes in other shells to get at the animal; the process may be observed even in the confinement of a vivarium. The short, deeply notched canal of *Buccinum* and *Nassa* is related to their burrowing habits;

Mr. Warington has observed that when *Nassa reticulata* burrows, it maintains a communication with the surface by means of its long recurved siphon.

The teeth of *Fasciolaria* resemble those of *Fusus Islandicus.* In *Buccinum undatum,* the median tooth has 5, or rarely 6 denticles; and Mr. Wilton has observed that *B. limbosum,* ♂ has the teeth 7 cusped, whilst in the females they are 6 cusped.

Fig. 243. *Fasciolaria Tarentina.*
(Wilton.)

Fam. 4. *Conidæ:* teeth 1. 0. 1. (see p. 117).

Conus.   Pleurotoma.   Cithara.   Terebra?

Fam. 5. *Volutidæ:* teeth single, rachidian, or 1. 1. 1.

Voluta.   Cymba.   Melo.   Marginella?

?Mitra Grœnlandica, teeth 0. 1. 0. minute, voluta-like; in more than 120 rows.

„ Caffra   teeth 1. 1. 1. buccinoid.
„ episcopalis   „   1. 1. 1. resembling *Fasciolaria.*

Fam. 6. *Cypræidæ:* teeth 3. 1. 3. (p. 28).

Cypræa.   Ovulum.   Pedicularia.   Erato?

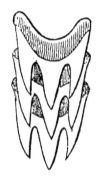

Fig. 244. *Voluta.*
(Wilton).

In *Ovulum* the teeth are 2. 1. 2. the outermost broad, with pectinated margins. Lovèn describes the Cypræidæ as having a short, non-retractile muzzle, and places them between the *Naticidæ* and *Lamellaria.*

Fam. 7. *Naticidæ:* teeth 3. 1. 3. or 1. 1. 1.

Natica   Sigaretus.   Lamellaria.   Velutina.

Fig. 245. *Natica monilifera.* (Wilton.)

The mouth of Natica is armed with buccal plates, shorter and broader in proportion than those of *Triton* (p. 239), and a similar armature exists in *Lamellaria.*

The dentition of *Lamellaria* is described and figured by Lovèn, and in the elaborate monograph by

Fig. 246. *Velutina lævigata.* (Warington.)

Dr. Rudolph Bergh. (Copenhagen, 1853.) It exhibits two modifications:—

1. *Lamellaria perspicua*, teeth 1. 1. 1., median with a bifid base, apex recurved, denticulated; laterals large, trapezoidal, hooked and serrulate.

2. *L. Grœnlandica* (Oncidiopsis, Beck.) and *L. prodita*, Lovèn, teeth 3. 1. 3., exactly as in *Velutina* (fig. 246). The dental canal is spiral.

Fam. 8. *Cancellariadæ*: teeth 3. 1. 3. or 0.

Cancellaria (teeth 0.)   Trichotropis.   Cerithiopsis?

Lovèn places *Trichotropis* in the same family with *Velutina*; Cancellaria is very closely allied, though it wants both teeth and operculum. Mr. Couthouy describes *Trichotropis cancellata* as having a muzzle like Litorina.

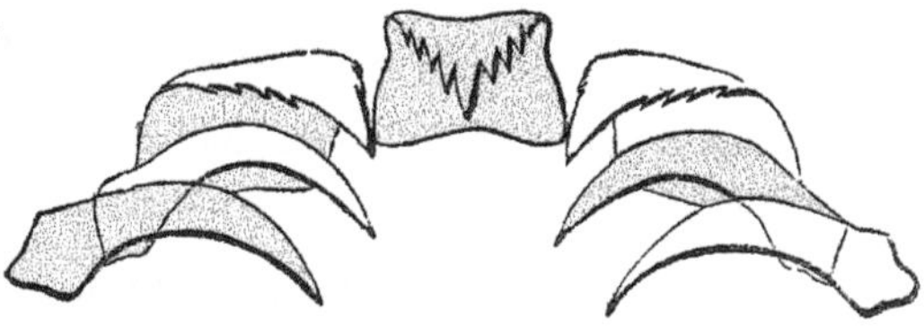

Fig. 247. *Trichotropis borealis.* (Warington.)

Fam. 9. *Pyramidellidæ*: teeth 0.

Pyramidella.   Aclis.   Eulima.   Monoptigma. (Lea.)

The *Pyramidellidæ* are related to *Tornatella*, which has numerous similar teeth (p. 180).

? Scalaria; uncini numerous, similar.

? Ianthina; dentition like *Testacella* (if the two halves were united by their outer margins), p. 148.

SECTION B.—PHYTOPHAGA, Lam. (*Rostrifera*, Lovèn).

Fam. 10. *Turritellidæ*: teeth 3. 1. 3. (p. 132).

Turritella.   Cæcum.   Vermetus.   Siliquaria.

Fam. 11. *Cerithiadæ*: teeth 3. 1. 3. (p. 127).

Cerithium.   Triforis.   Potamides.   Planaxis?

Mr. Wilton has examined the dentition of four *Cerithiadæ*; the teeth are broad, as in *Melaniadæ*, with incurved and dentated summits. In *Cerithidium* the median teeth are slender with minute hooks.

Fam. 12. *Melaniadæ*: teeth 3. 1. 3. (p. 130).

Melania.   Paludestrina.   Tanalia.   Melanopsis.   Pirena.

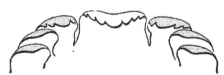

Fig. 248. *Pirena atra.* (Wilton.)

Fam. 13. *Paludinidæ:* lingual teeth 3. 1. 3. (p. 138).

Paludina. Paludomus? Ampullaria. Valvata.

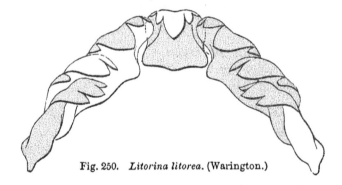

Fig. 249. *Ampullaria globosa.* (Wilton.)

The lingual uncini of *Paludina* and *Valvata* are denticulated; in *Am-pullaria* the first and second uncini are tricuspid.

Fam. 14. *Litorinidæ:* teeth 3. 1. 3. (p. 134).

Litorina.	Tectaria.	Modulus,	Risella.
Fossarus.	Narica.	Solarium?	Phorus?
Lacuna.	Litiopa.	Rissoa.	Truncatella.

The teeth of *Phorus* are like those of *Atlanta.* (Mörch.)

Fig. 250. *Litorina litorea.* (Warington.)

The lingual canal of the periwinkle passes from the back of the mouth under the œsophagus for a short distance, then turns up on the right side and terminates in a coil (like spare rope) resting on the plaited portion of the gullet. It is 2½ inches long, and contains about 600 rows of teeth; the part in use, arming the tongue, comprises about 24 rows. * The

* The opposite figure shows the manner in which a gasteropod may be laid out for examination, *under water*; the body requires to be fixed, and the cut edges of the mantle to be kept open with needle points. A convenient trough may be made of a plain earthenware *soap-dish*, by cutting a piece of sheet-cork (such as bootmakers use) to fit the bottom, and fixing it to a piece of sheet-lead of the same size with a couple of india rubber bands. The instruments required for dissecting are simply a pair of fine pointed scissors, a few broken needles, a penknife, or scalpel, and a pair of forceps with fine curved points.

dental ribbon of *Risella* is above 2 inches long, and coiled as in *Litorina*. (Wilton).

r, rostrum or muzzle.
k, buccal mass.
g, nervous ganglia
    (reproductive orifice,
    on the right side).
s, salivary gland.
œ, œsophagus.
l, lingual coil.
m, shell-muscle.
b, branchia or gill.
c, heart.
n, aorta.
e, stomach.
f, liver.
h, biliary canal.
i, intestine.
a, anus.
o, ovary.
d, oviduct.
u, nidament.
o', ovarian orifice.
x, renal organ.
y, mucus gland.

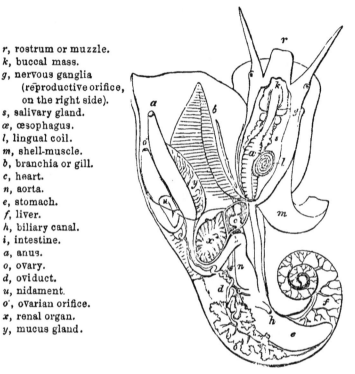

Fig. 251. *Litorina litoralis* ♀: (after Souleyet) Animal removed from its shell; branchial cavity and back laid open.

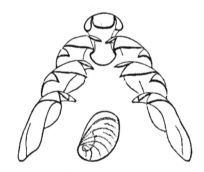

Fig. 252. Operculum and teeth of *Risella*. (Wilton.)

Fam. 15. *Calyptræidæ:* teeth 3. 1. 3. (p. 151.)

Calyptræa.   Pileopsis.   Hipponyx.   Metoptoma.

The rostrum is prominent and split, but non-retractile; the median tooth hooked and dentate; the first, or first and second laterals serrated, the third claw-shaped and simple. Lovèn places this family next to the *Velutinidæ*.

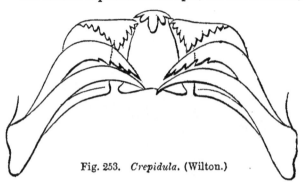

Fig. 253. *Crepidula*. (Wilton.)

(Section C. SCUTIBRANCHIATA, Cuv. *Rhipidoglossa*, Troschel.)

Fam. 16. *Turbinidæ*: lingual teeth 00. 5, 1, 5. 00 (pp. 28 and 142).

Fam. 17. *Haliotidæ*, p. 146. Fam. 18. *Fissurellidæ*, p. 149.

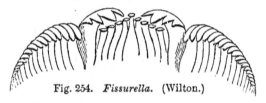

Fig. 254. *Fissurella*. (Wilton.)

Parmophorus differs from *Fissurella* in having a broad median tooth.

Fam. 19. *Neritidæ*: teeth 00. 3, 1, 3. 00 (p. 140).

Nerita.　　Neritopsis.　　Neritina.　　Navicella.　　Pileolus.

Fig. 255. *Navicella*. (Wilton).

Median tooth small; laterals 3,—1st·large, trapeziform, 2, 3, minute; uncini numerous,—1st large, strong and opaque, the rest slender, translucent, with denticulate hooks.

(*Cyclobranchiata*. Cuv.)

Fam. 20. *Patellidæ*: p. 153. Fam. 21. *Dentaliadæ*: p. 156.

Fam. 22. *Chitonidæ*.

Patella.　　Nacella.　　Acmæa.　　Gadinia.

Fig. 257. *Chitonellus*. Tasmania. (Wilton.)

The Cape limpets (e.g. *P. denticulata*) have a minute central tooth, which is wanting in any other species hitherto examined. (*Wilton.*)

ORDER III. PULMONIFERA.

Section A. In-operculata. Lingual teeth numerous, similar. (p. 160.)

Section B. Operculata. Lingual teeth 3. 1. 3. (p. 175.)

Glandina (*Algira*) has teeth like the Testacelle (p. 169, Raymond, Journ. Conch. 1853).

Fig. 256. *Patella vulgata.* (Original: Wilton.)

The anomalous genera *Siphonaria* and *Amphibola* have a dentition like the inoperculate land-snails. (Wilton). *Otina* (*Velutina*) *otis* has teeth similar to *Conovulus*. (Clark.)

The many points of agreement between the *Litorinidæ* and *Cyclostomidæ* have been already pointed out (pp. 32, 174).

ORDER IV. OPISTHOBRANCHIATA.

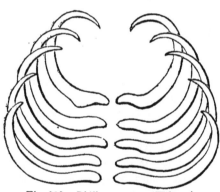

The lingual dentition is extremely varied in the *Bullidæ*. In Philine aperta there is no central tooth; and the laterals, which increase rapidly in size backwards, have a finely denticulated membranous inner edge.

In *Tornatella* and *Bulla* (physis) the rachis is unarmed, and the lateral teeth are numerous and similar; in *Acera, Cylichna,* and *Amphisphyra*, there is a minute central tooth.

Fig. 258. *Philine aperta.* (Wilton.)

ORDER V. NUDIBRANCHIATA.

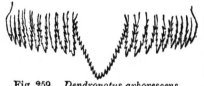

The *Dorididæ* are distinguished by having a short and wide lingual membrane with numerous similar teeth; the Æolids have a narow ribbon with a single series of larger teeth. In *Dendronotus* a large central tooth is

Fig. 259. *Dendronotus arborescens*

flanked by a few small denticulated teeth. (Alder and Hancock, Pl. II. fig. 8.)

The only Nudibranche with a solid upper jaw, is *Ægirus punctilucens* (A. and H. Pl. XVII. fig. 15). In other instances the two halves are arti-

X 3

culated and act as lateral jaws. In *Ægirus* the mouth is also furnished with membranous fringes (A. and H. Pl. XVII. fig. 14). *Ancula cristata* has a formidable spinous collar (Pl. XVII. fig. 7).

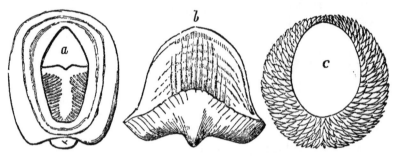

Fig. 260, *a.* Mouth of *Ægirus punctilucens.*
      b. Horny upper mandible detached.
      c. Prehensile collar of *Ancula.*
a, mandible; *x*, dental sac; *b*, insertion-plate of mandible; *c*, passage of mouth.

Note on the preparation of the Lingual Teeth as microscopic objects; by J. W. Wilton, Esq. The mollusk when taken from its shell must be pinned down in the dissecting trough, with needle-points passed through the sides of the muscular foot (fig. 251, and note). Water is then to be poured in till the animal is covered, and should be changed as often as the condition of the object renders it turbid. It is convenient to make these examinations under a simple lens, attached to an upright rod with a rack and screw, so that both hands may be free. A good light is necessary, and with lamp-light a bull's-eye condenser is useful. The lower point of the scissors should be passed into the mouth of the animal, and kept close to the upper side, which is to be cut open so as to expose the floor of the mouth, or *tongue*, with its teeth. When the cut edges have been pinned back, the whole length of the dental sack or canal may be carefully worked out with a lancet or other suitable instrument. Experience in this process may be gained by examining the periwinkle and whelk, or any others of which a number may be easily procured. The lingual ribbon, when detached, should be placed in a watch-glass of distilled water, and cleaned by repeated washings with a camel's-hair brush, and then placed in pure alcohol till wanted for mounting. If there is much difficulty in getting the membrane clean, it may be put for a time in *liquor potassæ*, care being taken to wash it in frequent change of water after-wards. Before mounting in balsam the preparation requires to be saturated with spirits of turpentine, which will more readily enter its structure if it be first soaked in *chloroform*. The slide is prepared by dropping a little Canada balsam on its centre, the quantity varying with the size and thickness of the object. The dental membrane is placed on the balsam with the side from which the teeth project upwards, and guided into the desired position; it is then covered with thin glass previously warmed over the flame of a spirit lamp.

Mr. Warington and Mr. Fisher Cocken recommend *glycerine* (which may be obtained at Price's, of Vauxhall) as the best medium for microscopic objects; the glass covers are cemented on with hatter's-varnish (shell-lac dissolved in spirits of wine), and painted over afterwards with asphalt dissolved in turpentine, such as the varnish-makers supply.

SUPPLEMENTARY NOTES ON THE GENERA.

DIBAPHUS, Phi. 1847. *Conohelix edentulus,* Sw. (*Strombidæ*? p. 104.) Subcylindrical, spire acute; aperture narrow, linear, edentulous, excised at the base; lip thickened, rectilinear, rounded and abbreviated below.

RHIZOCHILUS (antipathum) Stp. 1850. Founded on a sp. of *Purpura?* which lives on the *antipathes ericoides*. When adult they attach themselves, singly or in groups, to the branches of the coral, or to each other, by a solid extension of the lips of the shell. The aperture becomes closed, with the exception of the respiratory canal.

Planaxis, p. 114 (Cerithiadæ?). This genus was placed with the *Buccinidæ* on the statement of Mr. Gray, that the animal was like *Purpura.*

BORSONIA (prima) Bellardi, 1838. Is a *Pleurotoma* with the columella plaited like *Mitra*. *Miocene,* Turin. *Eocene,* Brit.

PACHYBATHRON (cassidiforme) Gaskoin. *Shell* small, oblong, striated with lines of growth; spire small, depressed, with channelled suture; aperture with callous, denticulated lips, like *Cypræa*. *Distr.* 3 sp.

Calpurnus, Montf. (name) = Ovulum verrucosum. p. 122.

Volva (Fleming) = Ovulum patulum, (*Calpurna*, Leach.)

Radius (Montf.) Schum. = Ovulum volva.

DESHAYESIA (Parisiensis), Raulin, 1844, (p. 123). *Miocene,* France. Some additional species have been found with a similar oblique aperture and corrugated inner lip. Baron Ryckholt has described a species (*D. Raulini*), from the *Devonian,* Belgium. The relation of the genus is uncertain.

NATICELLA (Munsteri, D'Orb.) Münster, 1841. This genus, abounding in the Trias of St. Cassian, has been referred to *Natica* by D'Orbigny. A characteristic species occurs in the Green-sand of Blackdown, and has been named *Natica carinata,* J. Sby. (Narica, D'Orb.) It is exactly intermediate between *Narica* (p. 124) and Fossarus (p. 135) and appears to form with them a little group nearly related to *Lacuna* (p. 136.)

Velutina inhabits the laminarian Zone, and ranges to 40 fms. *V. lævigata* is sometimes brought in on the fishermen's lines, (off Northumberland), generally adhering to *Alcyonium digitatum* (Alder). Dr. Gould obtained it from the stomach of fishes.

MONOPTIGMA (melanioides) Lea = M. striata, Gray (name only.)
Shell like *Chemnitzia*, rather fusiform, spirally grooved; columella slightly
folded, with a sinus at the base. *Distr.* 12 sp. Indo-Pacific, (p. 126).

Menestho, Möller, (Turbo albulus, Fabr. Greenland) v. Chemnitzia.

Aclis (p. 132) ascaris, Turt. (= A. supra-nitida, Wood) has the apex
sinistral, like the *Pyramidellidae.*

Vicarya (Verneuili) D'Arch. 1854. *Eocene.* Scinde. Shell like *Potamides;*
aperture with a broad callosity spreading over the body whirl, outer lip with
a deep narrow sinus like *Clionella.*

Holopella, Mc Coy, Turritella obsoleta, Sby. U. Silurian. Brit. Peristome
entire, not produced in front.

Scoliostoma (Dannenbergii) **Max.** Braun, 1838. *Syn.* Cochlearia, F.
Braun, 1841. Shell turreted, sometimes sinistral, whirls keeled or rounded,
aperture more or less twisted, trumpet shaped, sometimes with a widely
expanded outer peristome. *Fossil.* Devonian—Trias. Europe.

AMNICOLA, Gould and Haldemann, 1841 (p. 131) = Paludina porata,
Say, inhabits the fresh waters of New England, gregarious on stones and
submerged plants. The species are numerous.

PALUDOMUS, Sw. Shell turbinated, smooth or coronated; outer lip
crenulated; olivaceous, with dark brown spiral lines. *Distr.* 24 sp. Himalaya, Bombay, Ceylon, Seychelles. This genus was founded on *Melania conica*
and two other Indian species, having a concentric operculum, like *Paludina.*
In Reeve's monograph it was made to include, primarily, a group of Cingalese
shells for which Mr. Edgar Layard has revived Gray's MS. name TANALIA.
The description at p. 131. applies to this latter group.

PETALOCONCHUS (sculpturatus) Lea, 1843. Sub-genus of *Vermetus,*
p. 133. *Miocene,* U.S., St. Domingo, S. Europe. *Shell* with two internal
ridges, running spirally along the columella, becoming obsolete near the apex
and aperture.

Discohelix (calculiformis) Dunker, 1851. *Lias,* Gottingen. This name
was proposed for the depressed *Euomphali* of the Lower Oolites, of which
there are several species in Normandy and England. Shell usually sinistral,
flat or concave above, aperture quadrangular.

Platystoma (Suessi) Hörnes, 1855. *Trias,* Hallstadt. Shell discoidal,
sinistral? sculptured; peristome suddenly expanded, plain; aperture with an
inner rim, circular, and deflected (upwards,) at right angles to the plane of
the shell. Several examples have occurred.

BIFRONTIA, Desh. p. 135. *B. Zanclæa,* Phi. has been dredged alive off
Madeira, by Mr. Mc Andrew, the operculum is like *Torinia* (fig. 82) from
which the shell differs only in being more depressed.

PHILIPPIA (lutea) Gray, has a multi-spiral operculum, and the animal is
like *Trochus.* (Philippi.)

PALUDESTRINA (lapidum) D'Orb. part. Fresh-waters of S. America.

Shell conic, few-whirled, epidermis green; aperture oblique, peristome abruptly reflected; operculum claw-like. The typical species appear to be *Melaniadæ*, but some small shells like *Hydrobia* have been included in the genus.

VITRINELLA (valvatoides). C. B. Adams, 1850. *Shell* minute, hyaline, turbiniform, umbilicated; aperture large, orbicular. *Distr.* 18 sp. W. Indies (5), Panama.

SCISSURELLA (crispata). Animal like *Margarita*; tentacles long, pectinated, with the eyes at their base; foot with two pointed lappets and two long slender pectinated cirri on each side; operculum ovate, very thin, with an obscure sub-spiral nucleus.

Fig. 261. *Scissurella.* $\frac{8}{1}$.

No part of the animal was external to the shell. The only living example occurred at Hammerfest, in 40–80 fm. water; when placed in a glass of sea-water it crawled up the side and scraped the glass with its tongue. It was pale and translucent when living, but turned inky black after immersion in alcohol. (*Barrett*, An. N. H. 17, p. 206.)

Mr. Jeffreys found *S. elegans*, D'Orb. plentifully alive in sea-weed on the coast of Piedmont. It has a multispiral operculum, like *Margarita*. In this species, as noticed by Mr. G. Sowerby, the *slit* in the peristome of the young shell is converted into a *foramen* in the adult, as in the Jurassic *Trochotoma*.

Catantostoma (clathratum) Sandberger, 1842. *Shell* like *Pleurotomaria*; last whirl deflected, peristome incomplete, slightly varicose, irregular. *Fossil.* Devonian, Eifel.

RAPHISTOMA (angulata). Hall, (p. 147). L. Silurian. U. States. Canada. Shell depressed, out lip sinuated. In *R. compacta* (Salter) the spire is sunk and basin-shaped, the umbilical side flat, and the last whirl a little disunited.

Holopea (symmetrica). Hall. 1847. (Ianthinidæ?) Outer lip sinuated near the base. *L. Silurian*, New York.

BROWNIA (Candei) D'Orb. 1853. (Atlantidæ?) A minute discoidal shell, associated with *Helicophlegma* in the first instance, but distinguished by the serrated keels on its whirls, and lateral notches to the aperture. Cuba.

CALCARELLA (spinosa) Souleyet. 1850. (Atlantidæ?) Shell sub-globose, dextrally spiral, horny, pellucid, with three acutely serrated keels; aperture thickened, entire. Lat. 3 lines. South Seas. (= Echinospira, Krohn.)

RECLUZIA, Petit, 1853. R. Jehennei, Red Sea. R. Rollandiana, Atlantic, and Mazatlan. *Animal* pelagic, resembling Ianthina; one inch long. *Shell* paludiniform, thin, with a brown epidermis; whirls ventricose; aper-

ture ovate-oblique, slightly effused at the base, margins dis-united; inner lip
oblique, rather sinuated in the middle; outer lip acute, entire.

PATELLA, p. 154. The common limpet makes oval pits in timber as
well as in chalk. Small individuals sometimes roost, habitually, on larger
specimens, and make an oval furrow on the shell. The surface on which
limpets roost, and some space around it, is often covered with radiating striæ
not *parallel* like those *produced by their teeth on nullipore*. Mr. Gaskoin
has a limpet-shell incrusted with nullipore, which other limpets have rasped
all over. In M. D'Orbigny's collection of Cuban shells there is a group of
oysters (*O. cornucopiæ*), with a colony of the *Hipponyx mitrula* sheltered in
their interstices; these limpets have not only fed on the nullipore with which
the oysters are incrusted, but have extensively eroded the epidermal layer of
shell beneath.*

As to the *Calyptræidæ* generally, although furnished with lingual teeth
(fig. 248) like those of the animal-feeding *Velutina*, and themselves manifesting carnivorous propensities (p. 151), it is difficult to understand how
they can travel in quest of food.

The shape of some species of limpet is believed to vary with the nature
of the surface on which they habitually live. Thus the British *Nacella pellucida* is found on the *fronds* of the tangle, and assumes the form called
N. lævis, when it lives on their *stalks*. (Forbes.) The *Acmæa testudinalis*
becomes laterally compressed and is called *A. alvea* when it grows on the
blades of the *Zostera* (Gould); and *Patella miniata* of the Cape becomes a
new "genus" (*Cymba*, Adams, not Broderip) when it roosts on the round
stems of sea-weed, and takes the form called *P. compressa*. (Gray.)

TANYSTOMA (tubiferum) Benson, 1856. *Helicidæ*. Shell like *Anastoma*,
minute, umbilicated; aperture disengaged, trumpet-like, toothed. Banks of
the Irawadi, above Prome.

PFEIFFERIA (micans) Gray. *Helicidæ*. A *Nanina* without the mucus-
pore at the tail. Philippines.

SPIRAXIS, C. B. Adams, 1850. *Type*, Achatina anomala, Pfr. Shell
ovate-oblong, fusiform, or cylindrical; last whirl attenuated; aperture narrow, right margin usually inflected, columella more or less contorted, base
scarcely truncated, furnished with a deeply-entering callous lamina. *Distr.*
30 sp. W. Indies, Mexico, Juan Fernandez.

JANELLA, Gray, 1850 (not Grat. 1826). *Syn.* Athoracophorus (!) Gould.
Type. Limax bitentaculatus, Quoy. Elongate, limaciform, covered by a mantle with free margins; back grooved; tentacles 2, retractile, rising within
the edge of the mantle; respiratory orifice to the right of the dorsal groove,

---

* A similar circumstance has been noticed in the fresh-water *Paludinæ* and *Ampullaria*, by Dr. Bland and Mr. R. Swift; in the absence of other food they devour
the green vegetable matter incrusting one another's shells, and in doing this remove
the *epidermis*, or even make holes in the shell.

'reproductive orifice below it and beneath the mantle. *Distr.* New Zealand, on leaves.

TESTACELLA, p. 168. During winter and dry weather the Testacella forms a sort of cocoon in the ground by the exu-  dation of its mucus. If this cell is broken, the animal may be seen completely shrouded in its thin opaque white mantle, which rapidly contracts until it extends but a little way beyond the margin of the shell. Fig. 262 represents *T. Maugei* (lately found by Mr. Cunnington, in fields near Devizes), just disturbed from its sleep; *s*, the shell; *m*, the contracted mantle.

Fig. 262. *Testacella.*

LIMNÆIDÆ. Mr. R. Warington has observed that the fresh-water snails (and also *Neritina*) can lower themselves from aquatic plants by a mucous ·thread, *and reascend by the same;* a *Physa* could be lifted out of the water by its thread.

PLANORBULA, Haldemann, 1841. Planorbis armigerus, Say; aperture with 5 teeth, nearly closing the passage.

GUNDLACHIA (ancyliformis) Pfeiffer, 1850. Fresh-waters, Cuba. Shell thin, obliquely conic; apex inclined posteriorly; base closed for two-thirds by a flat, horizontal plate; aperture semicircular.

ADAMSIELLA (mirabilis) Pfeiffer, 1851 = Choanopoma, Pfr. (part) 1847. "Operculum thin, rather cartilaginous." *Distr.* 12 sp. Jamaica, Demarara. Named after the late Prof. C. B. Adams, of Amherst, Mass.

OPISTHOPORUS, Benson, 1855. O. biciliatus, Mouss. Shell like *Pterocyclus;* operculum double, margin grooved, interior concamerated. *Distr.* 4 sp. Singapore, Borneo, Java.

Aplysia (like Loligo, p. 69) has several shells when old.

Umbrella, p. 187, has a minute sinistral nucleus, like *Tylodina.*

STYLOCHILUS, Gould. Exped. shells. Aplysia longicauda Q. and G. Animal limaciform, cirrigerous, dilated at the sides, attenuated behind; neck distinct; tentacles 4, long, linear, papillose, far apart; lips dilated laterally into tentacular processes. *Distr.* 3 sp. New Guinea, on *Fuci.*

CHIORÆRA (leonina) Gould. Puget Sound. Appears to be a nudibranche resembling *Glaucus*, with oral cirri.

RHODOPE (Veranii) Kölliker, 1847. Animal minute, similar to *Limapontia ?* worm-shaped, rather convex above, flat beneath; without mantle, gills, or tentacles. Upon algæ, Messina.

BRACHIOPODA.

In the summer of 1855, Messrs. M'Andrew and Barrett obtained, on the coast of Norway, living examples of *Rhynchonella psittacea, Waldheimia cranium, Terebratulina caput-serpentis,* and *Crania anomala.* The two last projected their *cirri* beyond the margins of the opened valves, and moved them, as the Bryozoa move their oral tentacles; but in no instance were the *arms* extended. When the *Crania opened*, the upper valve turned upon its hinge-line. (Barrett, An. Nat. Hist.)

The anatomy of *Terebratula* and *Rhynchonella* has been further investigated by Dr. Gratiolet, Mr. Huxley, and Mr. A. Hancock.

The *pallial arteries* (mentioned p. 212, and figured p. 227, fig. 141) are regarded as "narrow bands from which the ovaria or testes are developed."

The nature of the organs previously described as *hearts* is rendered doubtful, as they appear to open *externally*, forming the "ovarian orifices" of Hancock; the plaited organs (*h, h,* fig. 165), described as *auricles*, are compared with nidamental glands.

Rhynchonella has two additional "hearts" above the others, one on each side of the liver. The peculiarity of the *ovarian spaces* in *Rhynchonella* and *Orthis* (described at p. 212, and represented in figs. 139, 140, 145, 147, letter *o*) is explained by the structure of the ovarian sinuses in the recent *Rhynchonella;* "the floor of this great sinus is marked out into meshes by the reticulated genital band, and from the centre of each mesh a flat band passes, uniting the two walls of the sinus, and breaking it up into irregular partial channels." The insertion of these *bands* produces the punctures in the shells represented in the figures above referred to. The membranes which support the alimentary canal are described, and explain the origin and nature of the *septa* in *Stringocephalus* and *Pentamerus*. The mode of termination of the alimentary canal is not yet satisfactorily made out.

Prof. Oscar Schmidt has observed the existence of flattened and radiated calcarious particles in the mantle, arms, and cirri of *Terebratulina caput-serpentis;* their occurrence appears to be very general in the Brachiopoda, and accounts for the frequent preservation of internal structures in fossil specimens.

Dr. Gratiolet has pointed out that the true function of the *cardinal muscles* of Terebratula was known to Prof. Quenstedt, and published by him in 1835. (Wiegm. Archiv. II. 220.)

SUESSIA (imbricata) Eugène Deslongchamps, 1855. (Dedicated to M. Suess.) Shell like *Spirifera;* texture fibrous; hinge area wide as the shell; foramen deltoid; large valve with two cardinal septa, and a prominent central septum, supporting a little plate; small valve with a tri-lobed cardinal process, and a broad 4-partite hinge plate, with processes from the outer angles of the dental sockets; crura of the spires united by a transverse band supporting a small process. *Fossil.* 2 sp. U. Lias, Normandy.

Davidsonia, p. 232. The upper valve sometimes exhibits markings derived from the surface on which the shell has grown.

ZELLANIA (Davidsoni) Moore, 1855. (*Etym.* Zella, a lady's name?) Shell minute, orthi-form; texture fibrous; hinge area short, foramen angular, encroaching on both valves; interior of dorsal valve as in *Thecidium*, with a single central septum and broad margin. *Fossil.* Lias—G. Oolite, 3 sp. Brit.

ANOPLOTHECA, (lamellosa) Fr. Sandberger, 1856. Dev. Rhine. = Atrypa.

MEGANTERIS, Suess, 1856. Terebratula Archiaci, Vern. *Devonian*, Asturias. Shell with a long, reflected, internal *loop*.

CONCHIFERA.

Development.—The observations of Dr. Lovèn on the development of *Cardium pygmæum* and *Crenella marmorata* (referrred to at p. 51, note) have been confirmed by M. M. Keber and Webb, who observed similar phenomena in the ova of the river-mussel (*Anodon*). The body described by Lovèn as the *nucleus of the germinal vesicle* is regarded by these later observers as a tubular orifice, analogous to the micropyle in the vegetable ovum, by which the spermatozoa penetrate the yolk.

In *Anodon* the embryonic mass divides, partially, into two halves, each having its own mouth and intestine; and its own distinct though simple heart; and it is by the approximation and ultimate fusion of the two ventricles that the common rectum of the originally distinct intestines is intercepted. (Quatrefages.—Lovèn.)

OSTREIDÆ, p. 253. The union of the *Ostreidæ* and *Pectinidæ*, as proposed by the authors of the " History of British Mollusca," has not proved satisfactory. The genus *Ostrea* stands quite alone, and distinct from all the *Pectinidæ* in the structure of its gills, which are like those of *Avicula*, and by resting on its *left* valve. The shell also is more nacreous than that of the scallops.

Dimya (Deshayesana) Rouault, 1859. Mém. Soc. Géol. b. III. 471. t. 15. fig. 3. *L. Eocene*, Paris. The figure is most like an oyster, and the " second adductor impression," on account of which it is named *Dimya*, is rather like the small anterior scar in *Pecten* (fig. 173, p. 249).

*Placuna** is essentially like *Anomia*, having the generative system attached to the right mantle-lobe, and the ventricle exposed. The mantle-margin is cirrated, and furnished with a *curtain*, as in Pecten; the foot is tubular and extensile, but has no distinct muscles except the small one, whose existence in *P. placenta* (Pl. XVI. fig. 6) we had predicated from examination of the shell (p. 256).† The small muscular impressions before and in the rear of the adductor are produced by suspensors of the gills.

Anomia. The description given at p. 255 requires correction; the lips

* Original figures and descriptions will be found in the An. Nat. Hist. 1855, p. 22.

† This organ appears to represent thê *byssal-sheath* of Anomia, rather than the foot, as there is no other opening for the passage of a byssus.

are extremely elongated and plain, the striated portion (or *palpi*) almost obsolete, whereas in *Placuna* the plicated surface is sufficiently extensive. The outer gill-laminæ, in both genera, are furnished with a broad reflected margin.

Plicatula, p. 259. The animal is like *Spondylus* in every essential respect, and only resembles *Ostrea* in the foot being nearly obsolete.

Streblo-pteria (lævigata) McCoy, 1856. *Carb.* Brit. (*Aviculidæ*).

Mytilidæ. Modiola pelagica (*Myrina*, Adams), p. 266, has the mantle open; the shell is peculiar from the large size of the anterior muscular impression; and the subcentral umbones distinguish it from *Modiolarca*.

Hoplomytilus (crassus) Sdbgr. *Devonian*, Nassau. *Shell* with a muscular plate in the umbo, like *Septifer* (p. 265). The *Mytilus squamosus*, Sby. Magnesian limestone, Brit. has a similar plate.

Arcadæ. Scaphula (celox) Benson, the fresh-water Ark, p. 268. A second species has been found in the R. Tenasserim, Birmah. The hinge is edentulous in the centre, and the posterior teeth are laminar and branched; the elements of the posterior muscular impression are distinct.

Limopsis, p. 268. *Syn.* Pectunculina, D'Orb. Mr. M'Andrew has dredged *L. pygmæa*, living, on the coast of Finmark; it is a fossil of the *Pliocene* of England, Belgium and Sicily.

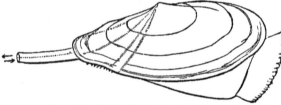

Nuculidæ, p. 270. The *Yoldia limatula* has been dredged, alive, by Mr. M'Andrew, on the coast of Finmark. It is also found in Portland Harbour, Mass.

Fig. 263. *Yoldia limatula* (after Barrett).

The animal is very active, and leaps to an astonishing height, exceeding in this faculty the scollop-shells. (Dr. Mighels.)

Unionidæ, p. 276. Mülleria; Fig. 246 represents the left, or attached valve, showing the single muscular impression, and projecting spur with the nucleus, consisting of *both* valves of the fry, united, and filled up with shell.*

Hippuritidæ, p. 279. The structure of these shells has been more fully described in the Quarterly Journal of the Geol. Soc. London. In all the genera the shell consists of *three layers*, but the outermost, which is thin and compact, is often destroyed by the weathering of the specimens. The principal layer in the lower valve of the *Hippurite* is not really very different from the upper valve in structure; the laminæ are corrugated, leaving irregular pores, or tubes, parallel with the long axis of the shell, and often visible on the rim. The umbo of the upper-valve of the *Radiolite* is marginal in the young shell. (Geol. Journ. vol. xi. p. 40.)

* M. D'Orbigny very liberally placed his suite of specimens of this remarkable genus in the British Museum. Oct. 1854.

Fig. 246. *Mulleria lobata*, Fér. (Original.)

Tridacnidæ, p. 289.   Animal of *Tridacna*, as seen on removing the left valve and part of the mantle within the pallial line.

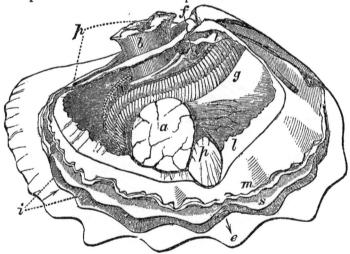

Fig. 265. *Tridacna crocea*, Lam. (Original).

a, the single adductor muscle; *p*, pedal muscle, and pedal opening in mantle; *f*, the small grooved foot; *b*, byssus; *t*, labial tentacles; *g*, gills; *l*, the broad pallial muscle; between *g* and *l* is the renal organ; *m*, the double mantle-margin; *s*, the siphonal border; *i*, inhalent orifice. *e*, valvular excurrent orifice. An. Nat. Hist. 1855, p. 190.

Lucinidæ, p. 294. Fig. 266, represents the animal of a species of *Diplo-*

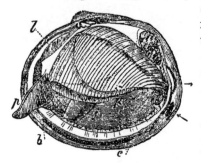

donta, from the Philippines, as seen on removing the left valve, and part of the mantle within the pallial line; *b-c*, the large pedal opening; the arrows indicate the small plain *incurrent* orifice, and the valvular *excurrent* orifice; *f*, the foot, contracted in spirit; *p, p*, the large striated palpi; *l*, the liver; the outer gill has a simple margin, the inner is grooved and conducts to the mouth. This genus has higher

Fig. 266. *Diplodonta*.

claims than *Kellia* to be regarded as the type of a family.

SCINTILLA (Cumingi) Desh. 1856. Small shells resembling *Lepton*, p. 296; minutely punctate; ligament internal, oblique; hinge-teeth 1. 2; posterior laterals 1. 2. *Distr.* 37 sp. (?) Philippines, N. Australia, Panama.

Family 12a. *Astartidæ*.

Astarte.　　　Opis.　　　Crassatella.　　　Circe?　　　Cardita.

Astarte (borealis); mantle-margins free, plain, slightly cirrated in the branchial region, united posteriorly by the branchial septum, forming a single, excurrent orifice; pedal muscles (*p. p'*), distinct from adductors; gills flat, finely striated, destitute of internal partitions; outer gill narrow, elliptical, with a simple margin; inner gill grooved, conducting to the mouth.

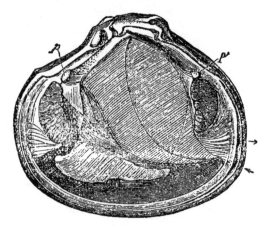

Fig. 267. *Astarte borealis*, var. semi-sulcata, Leach. $\frac{3}{2}$ Wellington Channel.

GOULDIA (Pacifica) C. B. Adams. *Shell* minute, triangular, furrowed; hinge like *Astarte*, with lateral teeth; pallial line simple. *Dist.* 4 sp. Panama, W. Indies.

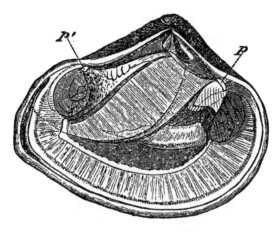

Fig. 268. *Crassatella pulchra.* Sandy Cape, J. B. Jukes.
Animal as seen on the removal of the *right* valve, and portion of the mantle.

Crassatella (pulchra) animal like *Astarte*; foot linguiform, slightly grooved; palpi short and broad, few-plaited; outer gill narrower in front.

Cypricardia rostrata, Lam. Philippines (p. 300). Animal with mantle-lobes united, and covered with wrinkled epidermis; siphonal orifices fringed; gills deeply plicated, anterior part of the outer gill *united to the inner;* dorsal border narrow, plaited; adductor muscles of two elements.

Fig. 269. *Cypricardia.*

Goniophora, Phillips, 1848. Cypricardia cymbæformis, Sby. *U. Silurian*, Brit. (*Mytilidæ* ?)

· *Redonia*, Rouault, Bull Soc. Geol. 8, 362. (= Pleurophorus ? p. 301.) Shell oval, tumid; hinge with cardinal and posterior teeth; anterior adductor bounded by a ridge. *Fossil*, L. Silurian, Brittany, Portugal. (Sharpe.)

Carbonicola, Mc Coy, 1856 = Anthracosia, p. 303.

Omalia, Ryck. 1856 = Pullastra bistriata, Portl. *Carb.* Belgium.

Verticordia, p. 304. Syn. Trigonulina (ornata) D'Orb. Jamaica. Hinge-teeth 2. 2; right valve with a long posterior tooth. Epidermis of large nucleated cells, as in *Trigoniadæ*, to which family it undoubtedly belongs. (Pl. XVII. f. 26.)

Lucinopsis, p. 306. The type of this genus having been erroneously placed in *Cyclina* by M. Deshayes, he has proposed a new genus (*Lajonkairia*) for the second species, *L. decussata*, Phi. a fossil of the English *Pliocene*, but still living in the Medit.

Glaucomya, p. 307.   See An. Nat. Hist. 1855, p. 23.

Sowerbya, p. 308. (Syn. Isodonta, Buv. p. 314).   The cavity described as a "cartilage-pit" receives a tooth of the opposite valve.

Tellinidæ, p. 311. Psammobia.

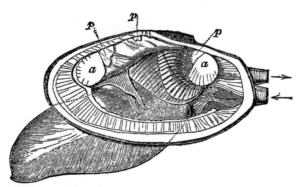

Fig. 270.   *Psammobia pallida,* Desh. Red Sea.   Left valve, part of the mantle, and retractor of the siphons removed.   Siphons much contracted; *a,a,* adductors; *p,p,* pedal muscles.

Solenidæ, Glycimeris, p. 320.   An. Nat. Hist. 1855, p. 99.

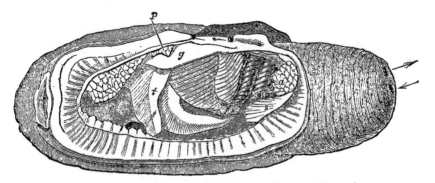

Fig. 271.   *Glycimeris siliqua,* Chemn.   Newfoundland.
a, a, adductor muscle; *p,* pedal muscle; *s,* siphonal muscle; *f,* foot; *t,* labial tentacles; *g,* gills, much contracted and crumpled.

RIBEIRIA (pholadiformis) Sharpe, 1853.   Ged. Journ.   Shell gaping at both ends; sub-ovate, rounded in front, elongated and rather attenuated behind; punctate-striate; casts of interior with a large umbonal impression (caused by a cartilage-plate, as in *Lyonsia?*) and a notch in front of it. *Fossil.* L. Silurian, Portugal.   (*Anatinidæ,* p. 320.)

Scaldia, Ryckholt, 1856.   Carb. Tournay.   Shell like *Edmondia* (p. 323,) with a single cardinal tooth in each valve.

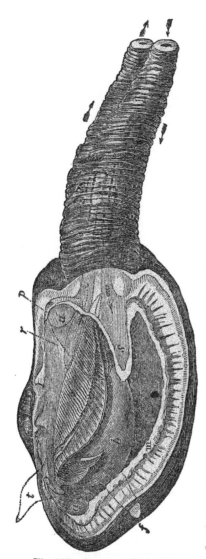

Fig. 272. *Panopæa glycimeris.*
$\frac{2}{7}$ The size of the original.
a, a', adductor muscles; *p*, posterior pedal
muscle; *r*, renal organ.

Myadæ. The description of the animal of *Panopæa*, at p. 319, was taken from the British species, *P. Norvegica*, which agrees both in the character of the shell and soft parts with *Saxicava*, and belongs to the *Gastrochænidæ.*

Fig. 272 represents the animal of the typical species of *Panopæa*, as seen on the removal of the left valve and thin part of the mantle. It was obtained on the coast of Sicily, and presented to the Gloucester Museum by Capt. Guise.

Mantle and siphons covered with thick, dark, wrinkled epidermis; siphons united, thick, contractile; pedal orifice small, in the middle of the anterior gape; foot small (*f*), body oval (*b*), with a prominent heel; pallial muscle (*m*) continuous, with a deep siphonal inflection (*s*); lips broad and plain, palpi triangular, deeply plaited (*t*); gills unequal, (much contracted in spirit), reaching the commencement of the siphons; inner gills prolonged between the palpi, plaits in pairs, each lamina being composed of vascular loops arranged side by side; margin grooved, dorsal border of inner lamina unattached; outer gills shorter and narrower, formed of a single series of branchial loops placed one behind the other, dorsal border wide and fixed.

Isoleda, Ryck. 1856 = Leda solenoides *and* Cucullella sp. p. 269.
Anòmianella, (proteus) Ryck. *Carb.* Tournay = Crania ?
Crenella (decussata) T. Br. 1827. p. 266 = Nuculocardia (divaricata) D'Orb. Cuba. = Myoparo, Lea (p. 269.) Brachydontes, Sw. p. 265, is more elongated; Lanistes (discors) Sw. nearly wants the crenulations.

ABBREVIATIONS OF AUTHOR'S NAMES.

C. B. Ad.	C. B. Adams, p. 375.	L.	Linnæus, 1787-
H. A. Ad.	H. and A. Adams.	Les.	Lesson, 1829.
Adans.	Adanson, p. 366.	Mant.	Mantell, 1822-54.
Ag.	Agassiz, p. 251.	Mart.	Martin, 1793.
Ant.	Anton, 1839.	Marti.	Martini, 1769-74.
A. & H.	Alder and Hancock.	Mtyn.	Martyn, 1784.
Bar.	Barrande, 1852.	Mc C.	Mc Coy, 1845—
Bl.	De Blainville, 1825.	Mke.	Menke, 1828.
Broc.	Brocchi, 1814.	Mid.	Middendorff, p. 354.
Brod.	Broderip, W. J.	Möl.	Möller, p. 355.
Bron.	Brongniart, 1835.	Mont.	Montagu, 1803-8.
Br.	Bronn, 1831—	Montf.	Montfort, 1799-1820.
T. Br.	T. Brown, 1827.	M. & L.	Morris and Lycett.
Buv.	Buvignier, 1852.	Mhl.	Muhlfeldt, 1811.
Charp.	Charpentier, 1837.	Müll.	Müller, O.F., 1773-6.
Chemn.	Chemnitz, 1780-95.	Münst.	Münster, 1826-43.
Chen.	Chenu, 1848—	Nils.	Nilsson, 1822-7.
Con.	Conrad, 1832—	Quenst.	Quenstedt, 1852.
Cuv.	Cuvier, 1799-1817.	Q. & G.	Quoy and Gaimard.
D'Arch.	D'Archiac.	Park.	Parkinson, 1804-11.
Defr.	Defrance, 1816-29.	Pen.	Pennant, 1776-7.
Dh.	Deshayes, 1825—	Pf.	Pfeiffer, 1848—
D'Orb.	D'Orbigny, 1835—	Phi.	Philippi, 1836—
Don.	Donovan, 1824-7.	Ph.	Phillips, 1829—
Drap.	Draparnaud, 1805.	Portl.	Portlock, 1843—
Eich.	Eichwald, 1828-30.	P. & M.	Potiez and Michaud.
F. Edw.	F. Edwards, 1850—	Ris.	Risso, 1826.
E. & S.	Eydoux and Sonleyet.	Rois.	Roissy, 1805.
Fabr.	O. Fabricius, 1780.	Röm.	Römer, F. A., 1836—
Fér.	Férussac, 1819.	Sdgr.	Sandberger, G. and F.
Flem.	Fleming, 1828.	Sav.	Savigny, 1816.
F. & H.	Forbes and Hanley.	Schl.	Schlotheim, 1813-23.
Gm.	Gmelin, 1788.	Sch.	Schumacher, 1816.
Gld.	Gould, 1841—	Sol.	Solander, 1765.
Gldf.	Goldfuss, 1826-44.	Sby.	Sowerby, 1812-30.
Hart.	Hartmann, 1840.	J. Sby.	J. Sowerby, 1830—
His.	Hisinger, 1837.	G. Sby.	Geo. Sowerby.
Johnst.	Johnston, G.	G. B. S.	G. B. Sowerby, 1843.
Kien.	Kiener, 1834—	Stp.	Steenstrup.
K. & D.	Koch and Dunker.	Sw.	Swainson, 1820-40.
Kon.	Koninck, 1837—	Turt.	Turton, 1822.
Küst.	Küster, 1837—	Vern.	Verneuil, 1845.
Lam.	Lamarck, 1799-1818.	Wahl.	Wahlenberg, 1821.

475

INDEX OF GENERA AND TECHNICAL TERMS.

The names of Genera commence with a capital letter, those *in italics* are Synonymes.

Y

ERRATA AND ADDENDA.

It is earnestly recommended that the *corrections* be made *with pen and ink* at the places indicated.

Page   5 The foot-prints referred to in the note, are now ascribed, by Prof. Owen to some unknown Crustaceous animal.

   23 second line from bottom *add* " but is more probably the seat of the olfactory sense."

   29 line 7 *for* " communicating " *read* " comminuting."

   32 lines 8 and 9 from bottom *erase* "in one family of tunicaries (*pelonæidæ*)."

   93 line 3 *Aganides*, D'Orb. (not Montf, = *Aturia zic zac.*)

   97 *Carinaria cymbium*, Desh. = C. cristata, L. sp. The same correction may be made at p. 200, and Pl. XIV., f. 19.

   108 line 4 from below *for* " Strombus " *read* " Velutina."

   109 line 20 *for* " Leiotomus " *read* " Leiostoma (bulbiformis)."

   115 line 5 *add* " U. States, S. Domingo."

   117 line 3 *Scaphula*, Sw. = Olivancillaria, D'Orb.

   121 *erase* lines 28—30.

   125 *erase* line 2, see p. 461.

   „ line 5 *erase* "like Velutina;" see p. 459.

   126   „   6 *for* " Gray " *read* " Lea, part," see p. 462.

   127   „   10 *add* " *Type* L. sinuata, U. Devonian, Petherwin."

   „   „   15 *add* " *Syn.* Polyphemopsis, Portlock."

   128   „   16 *for* " old world only ?" *read* " California."

   „   „   18 *for* " Vulsella," *read* " Ostrea."

   131   „   10 *for* " Eocene " *read* " Wealden."

   „   „   28 *for* " Anthony " *read* " G. & H.," see p. 462.

   137   „   10 *erase* " Paludestrina, D'Orb.," see p. 462.

   144   „   32 *for* " Otavia " *read* " Olivia."

   145   „   19 *for* " Eocene, Paris " *read* " *Type*, Euomphalus Serpula, Kon. *Carb.* Belgium."

   153   „   8 from bottom, *for* " jaws " *read* " upper jaw."

   154   „   16 *for* " tongue " *read* " dental canal."

   „   „   31 *for* " nocturnal " *read* " between tides."

   156   „   7 *add* " France."

   228   „   13 *for* " fig. 17 " *read* " fig. 21."

   184   „   15 *for* " BULLÆA, Lam." *read* " PHILINE, Ascanius. 1772 " and *erase* the foot-note.

   237 last line *for* " more like " *read* " setose, like."

   280 line 17 *erase* " Hippurite and."

Page 310 *for* "Diodonta" *read* "Gastrana," and *add* "*Syn.* Diodonta, F. & H. *not* Schum."

311 line 1 *for* "Greenland" *read* "Norway."

319 „ 8 from below, *for* "*Australis*" *read* "*Natalensis.*"

320 Saxicava belongs to the *Gastrochænidæ.*

321 line 6 *erase* "*Cochlodesma.*"

„ „ 30 *for* "without an" *read* "ossicle minute."

363 *Senegal; add* "Tellina lacunosa" *and* "Cymba olla."

364 line 13 *add* "*Typhis.*"

383 „ 9 from below *for* "Holsace" *read* "Holstein."

391 „ 6 from below *for* "all" *read* "nearly all."

419 „ 21 *for* "alterations" *read* "alternations."

450 lines 12 and 13 the terms *dorsal* and *ventral* are transposed.

457 in the figure of *Risella* the central tooth is worn round, it should be pointed as in *Litorina.*

Plate II. f. 14 O Ludense, Sby. ¼ Ludlow-rock.

III. f. 5 A. spinosus, Sby (ornatus, Schl. part.)

V. f. 12 *for* "W. America" *read* "Cape."

VI. f. 4 *for* "China" *read* "Cuba" D'Orb.

VIII. f. 23 *for* "Gray" *read* "Lam."

XI. f. 22 *for* "W. Indies" *read* "Cape."

XIV. f. 15 *for* "verrucosa, Gmel." *read* "scapula, Martyn."

„ f. 32 *for* "tridentata, Forsk." *read* "telemus, L."

„ *transpose* the numbers 46 and 47.

XIX. f. 1 *for* "China" *read* "W. Africa."

„ f. 22 *for* "Gray" *read* "Cailliaud."

XXI. f. 8 *for* "Diodonta" *read* "Gastrana."

„ f. 12 *for* "Bahamas" *read* "Peru."

„ f. 16 *for* "donacium, Lam." *read* "Chilensis, D'Orb."

XXII. f. 3 *for* "S. America" *read* "Penang."

XXIV. *for* "Bortyllidæ" *read* "Botryllidæ."

„ f. 1 *for* "tubulosa, Rathke" *read* "arenosa, A. & H.'

Alaria, Morris and Lycett, 1851. *Ex.* Rostellaria trifida, Ph. Shell like *Aporrhais* (p. 129) but having no channelled process of the lip extending up the spire. In most species the expanded lip is repeated, as in *Cerithium*, or produced periodically, as in *Ranella* and *Spinigera*. Fossil in the Oolites; the species are very numerous.

Amberlya (nodosa) M. & L. 1851. Gt. Oolite, Minchinhampton. Resembling *Tectaria* (p. 134) but slightly notched in front like *Purpurina.*

Anaulus (bombycinus) Pfr. 1855, Sarawak, Borneo. Shell like *Mega-*

lomastoma, with a small tubular orifice at the suture leading into the body-whirl at a little distance from the aperture. *A. Lorraini* is found at Penang.

Brachytrema (Buvignieri) M. & L. 1854. Gt. Oolite, Minchinhampton. Shell turbinated, whirls ornamented, columella twisted, canal short and oblique. *Fossil,* 10 sp. Oolites.

Ceritella (acuta) M. & L. 2851. Gt. Oolite, Minchinhampton. Shell turreted, acute, last whirl large, canal short. (= Rissoina, D'Orb. part.) *Fossil* 9 sp.

Coccoteuthis, (latipinnis) Owen, 1855, Geol. Journ. XI., pl. VII., p. 124 = *Geoteuthis,* part. Pen rather calcarious, rounded in front, lateral wings small. Kim. Clay and Oxford Clay, S. of England.

Corbicella (Bathonica) M. & L. 1855. Gt. Oolite, Minchinhampton, oval, smooth, posterior side elongated ; anterior lateral teeth wanting. *Fossil* 6 sp. Oolites.

Crossostoma (Prattii) M. & L. 1851. Gt. Oolite, Minchinhampton. *v. Liotia.* Columella toothed when young, concealed by callus in the adult.

Deslongchampsia (Eugenei) Mc Coy, MS. in M. & L. 1851, *Great Oolite,* Minchinhampton = Hemitoma, p. 151.

Diastoma, Desh. 1849 = Melania costellata, Lam.

Euspira, Ag. 1837. A subgenus of *Natica,* with angular whirls, *Fossil* Oolites.

Quenstedtia (oblita) M, &. L. 1855. Gt. Oolite, Minchinhampton. Like *Psammobia;* pallial sinus small; ligament in a narrow groove; cardinal teeth 0. 1.

Resania (lanceolata) Gray 1853, An. N. H. p. 43, (same shell as *Vanganella Taylori,* Gray, An. N. H. 1853, p. 475). New Zealand = Lutraria, subgenus, p. 309.

Fossil land-shells of Madeira, p. 387. Of the eleven species now common to Madeira and P. Santo, only two (*Helix paupercula,* and *H. compacta*) occur *fossil* in both islands. And of the species now peculiar to one island, two occur fossil in both, viz. *Helix sphærula* of P. Santo, and *Cyclostoma lucidum* of Madeira. (Wollaston).

LONDON : PRINTED BY W. OSTELL, HART STREET, BLOOMSBURY.

Pl.1.

wry.

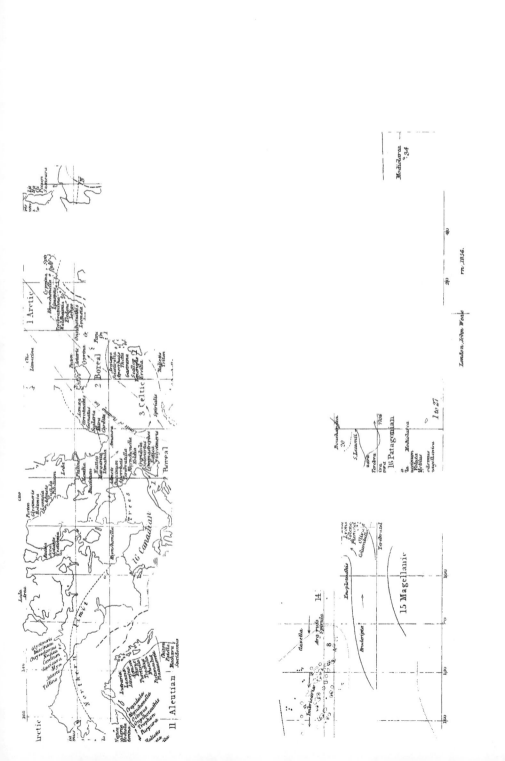

Pl.1.

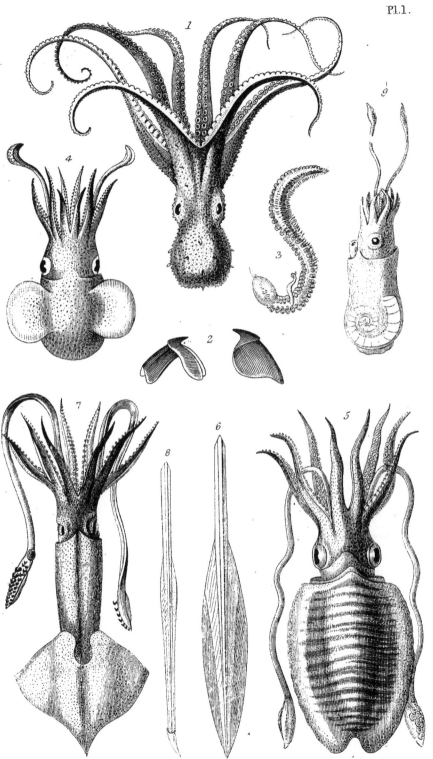

S.P. Woodward.   London, John Weale, 1851.   J.W. Lowry.

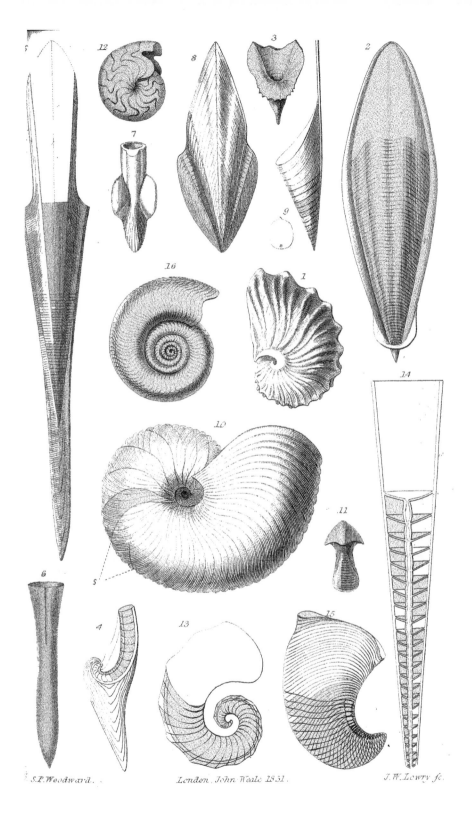

S.P. Woodward.

London, John Weale 1851.

J.W. Lowry sc.

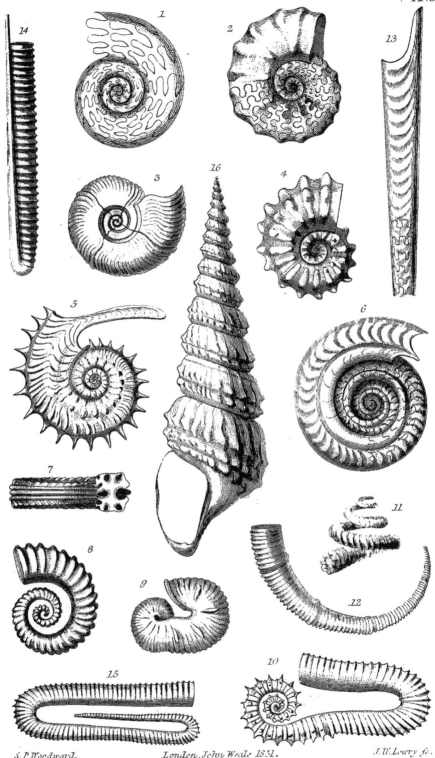

Pl. 3.

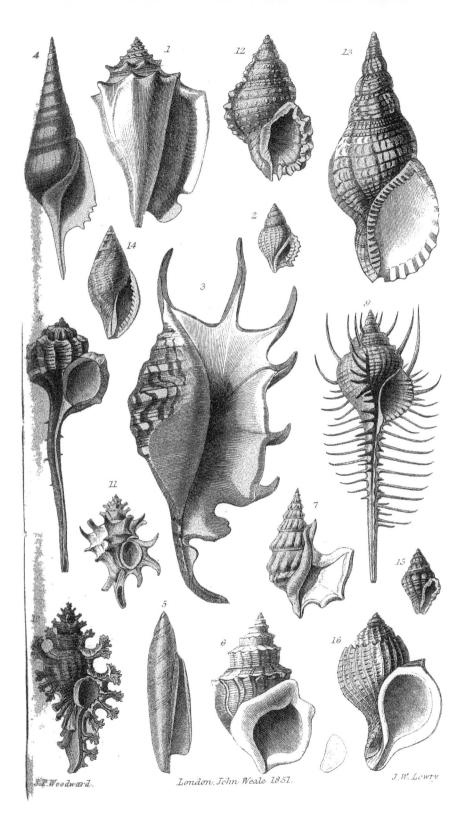

London, John Weale 1851.

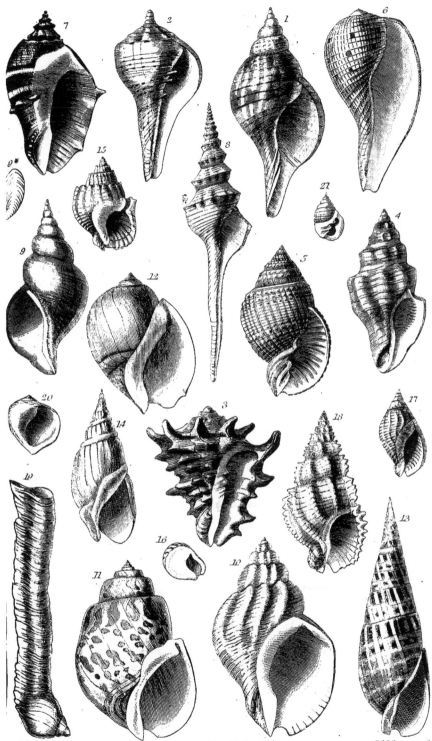

Pl.5.

S.P.Woodward.   London, John Weale 1851.   J.W.Lowry fc.

Pl. C.

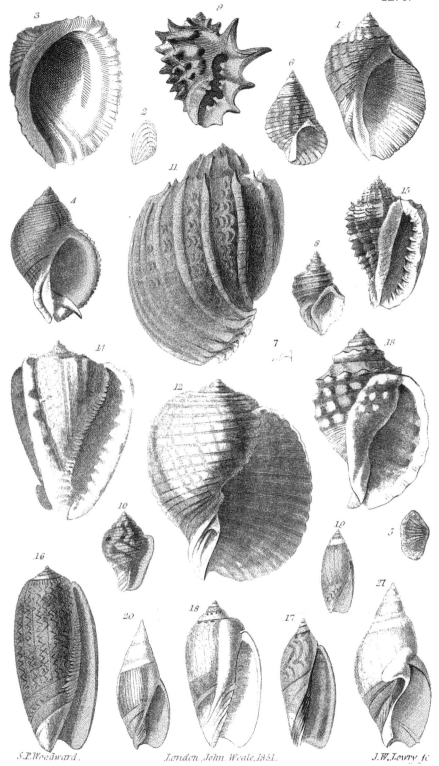

S.P.Woodward.   London, John Weale, 1851.   J.W.Lowry sc

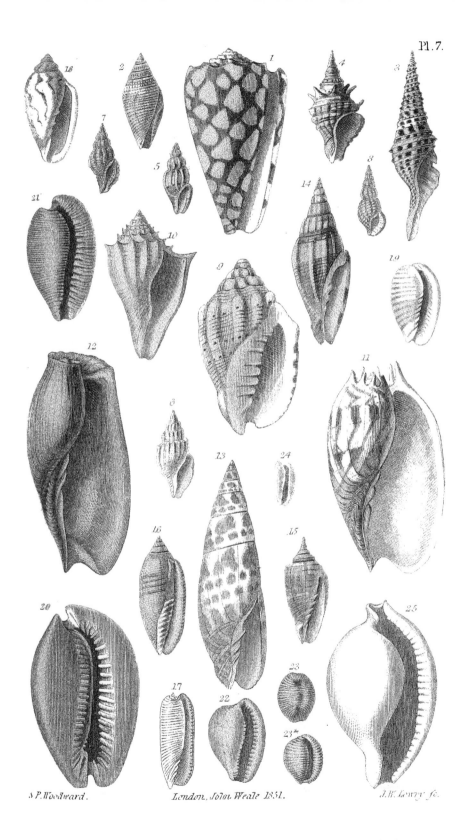

Pl. 7.

S.P. Woodward.　　　London. John Weale 1851.　　　J.W. Lowry sc.

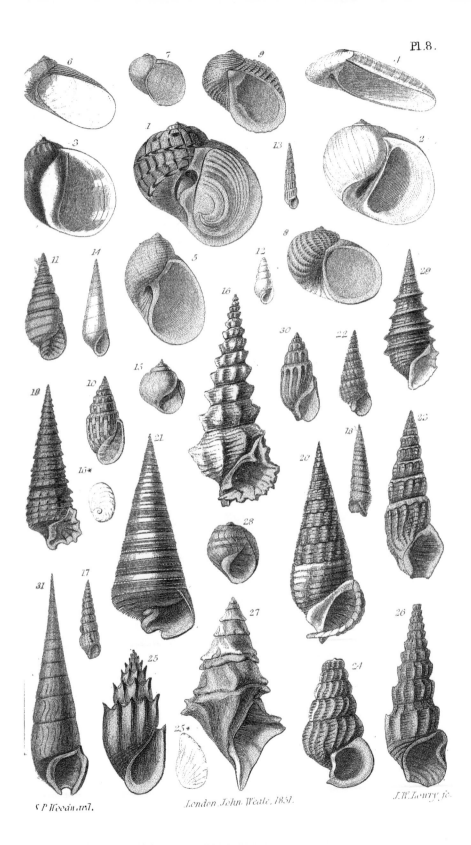

Pl.8.

S.P.Woodward.   London John Weale. 1851.   J.W.Lowry sc.

Pl.9.

Pl. 10.

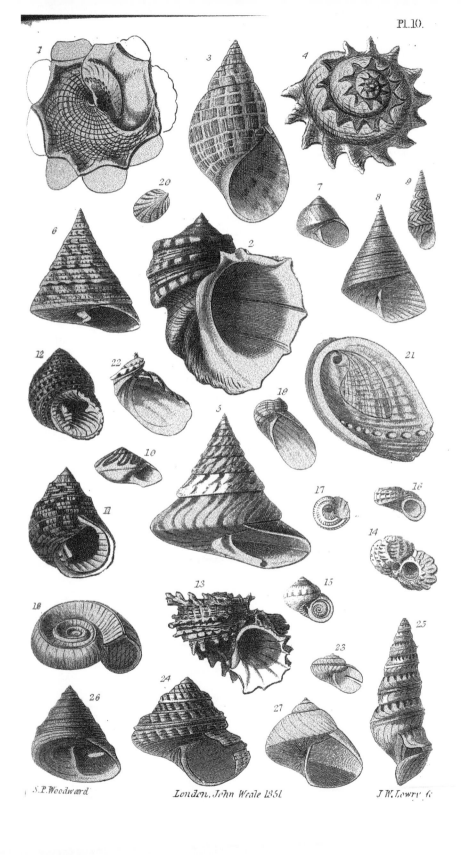

S.P.Woodward.   London, John Weale 1851.   J.W.Lowry sc.

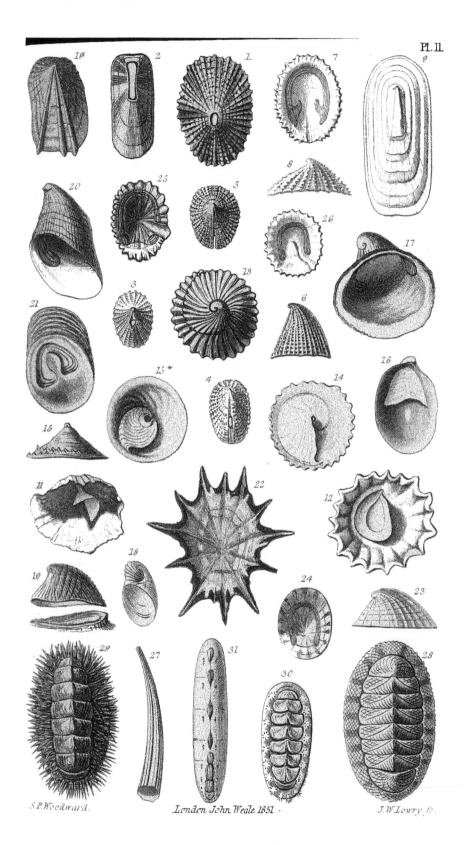

Pl. II.

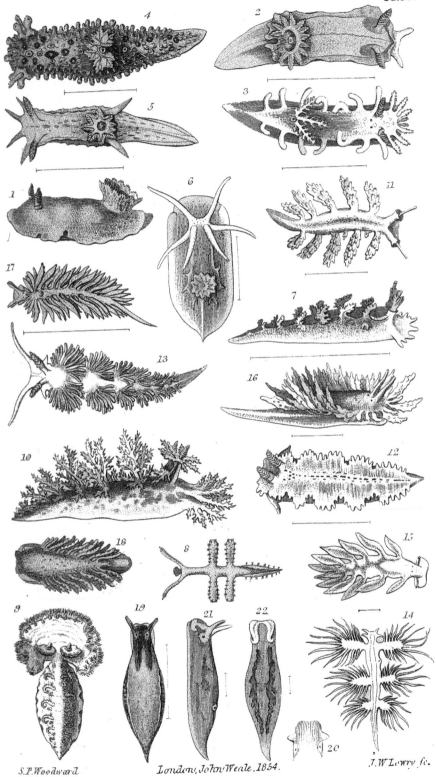

Pl. 13.

S.P. Woodward.

London, John Weale, 1854.

J.W. Lowry, sc.

Pl. 14.

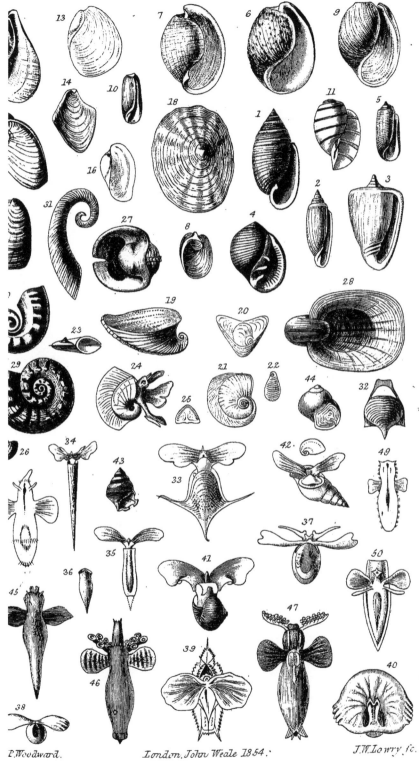

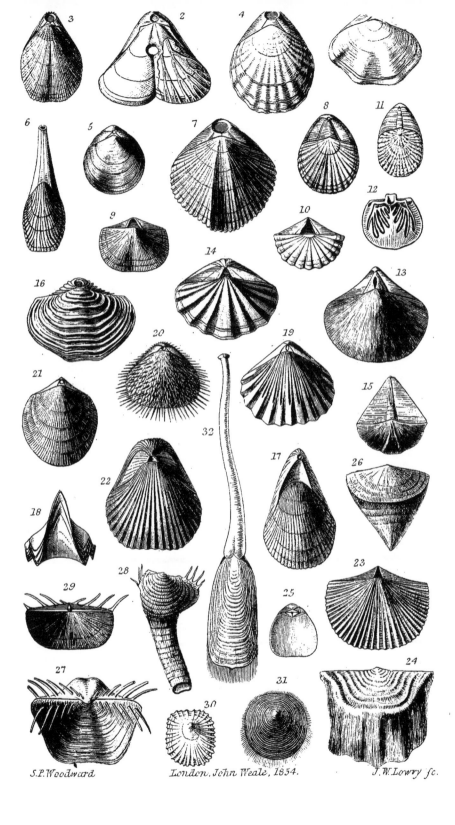

S.P.Woodward   London. John Weale, 1854.   J.W.Lowry sc.

Pl.16.

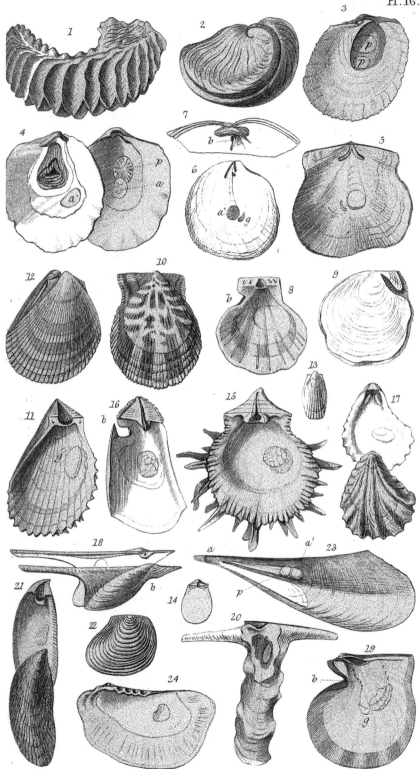

S.P.Woodward.   London, John Weale, 1854.   J.W.Lowry fc.

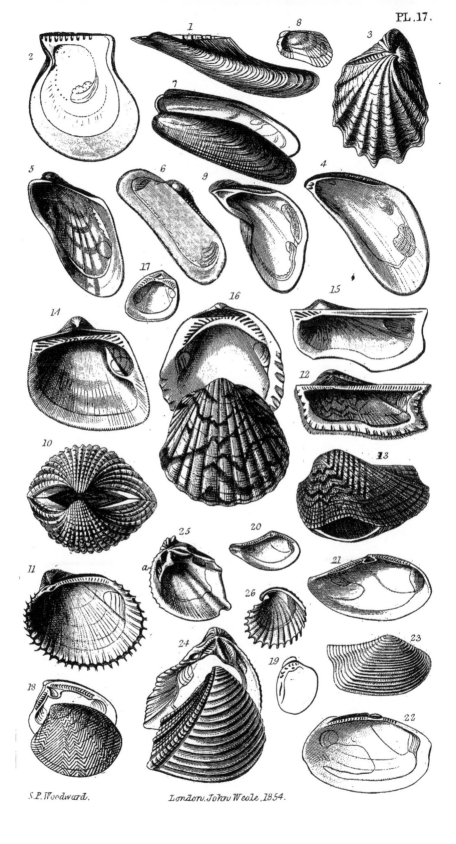

Pl. 18.

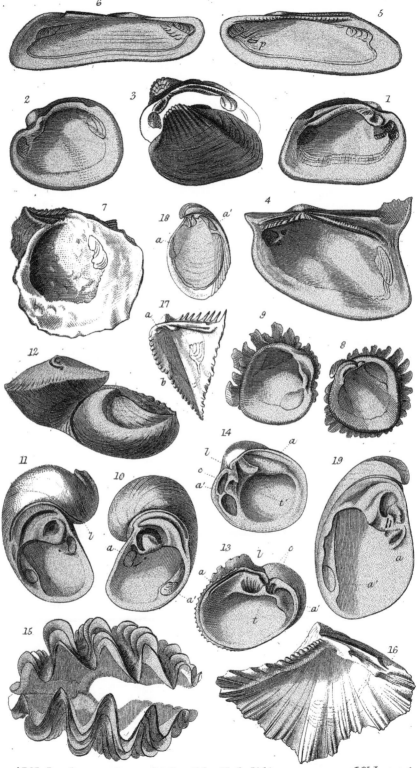

S. P. Woodward. London, John Weale 1854. J. W. Lowry fc.

S.P.Woodward. London, John Weale, 1854. J.W.Lowry sc.

Pl.20.

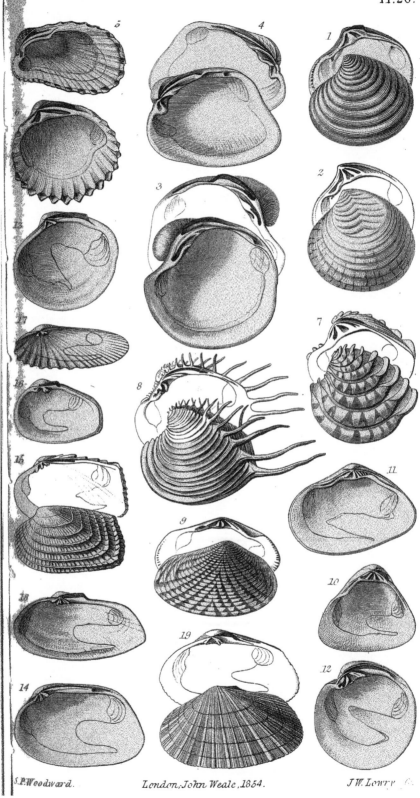

S.P.Woodward.   London, John Weale, 1854.   J.W.Lowry

Pl. 21.

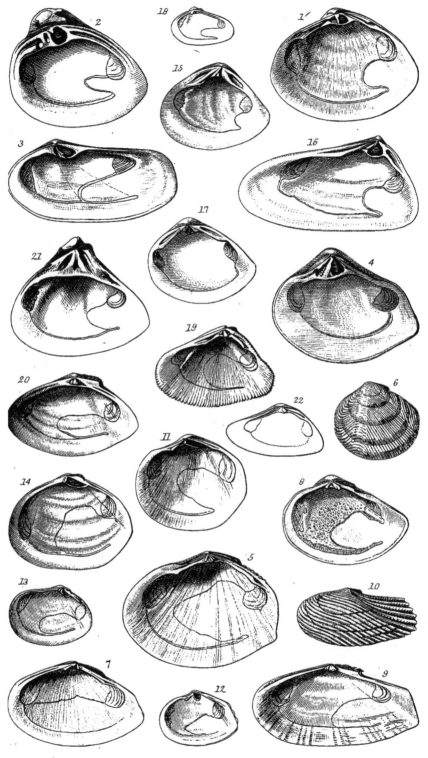

S.P.Woodward. London, John Weale 1854. J.W.Lowry fc.

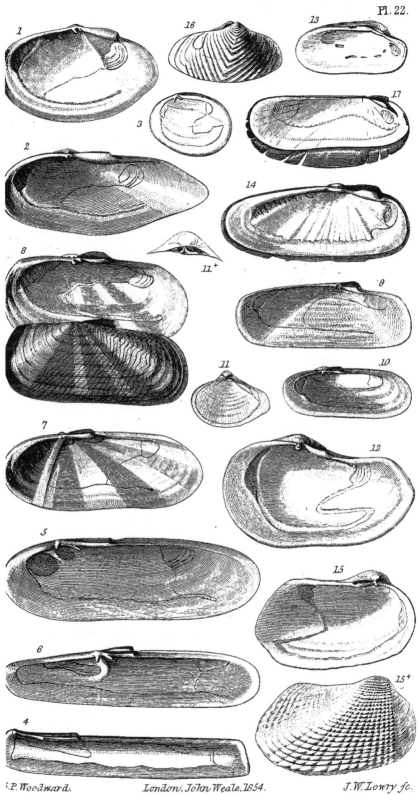

Pl. 22.

S.P. Woodward.   London, John Weale, 1854.   J.W. Lowry sc.

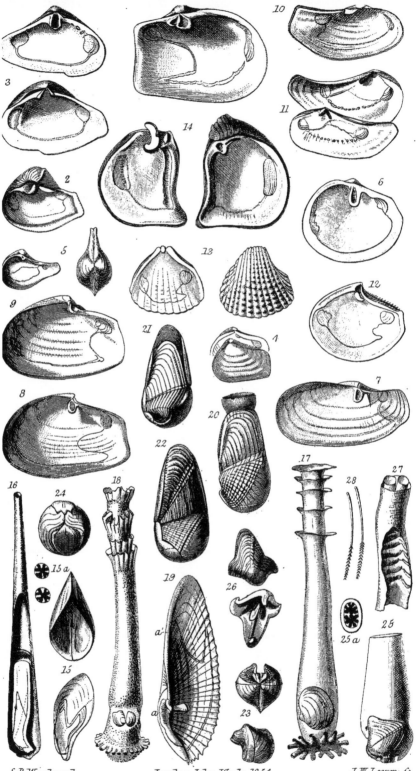

S.P. Woodward. London, John Weale, 1854. J.W. Lowry, sc.

Pl.24.

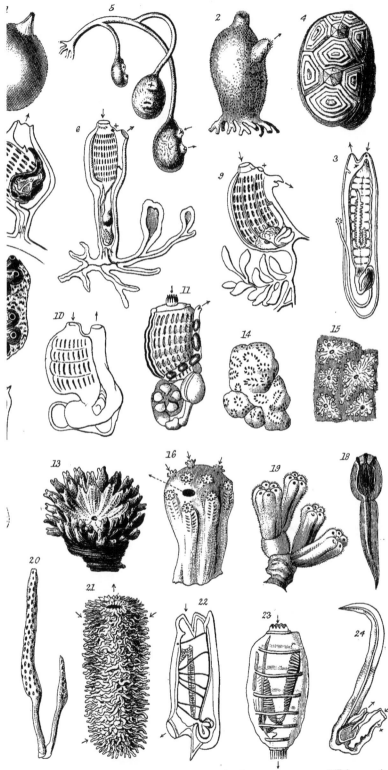

London, John Weale, 1854.

EXPLANATION OF THE PLATES.

The principal specimens figured were kindly communicated by Mrs. J. E. Gray, Mr. Hugh Cuming, Major W. E. Baker, Mr. Laidlay of Calcutta, Mr. Pickering, Sir Chas. Lyell, Mr. Sylvanus Hanley, Mr. James Tennant, and Mr. Lovell Reeve.

The fractions shew the number of times (or diameters) the figures are reduced, or magnified.

PLATE I.

PLATE II.

PLATE III.

Ammonitidæ.

PLATE IV.

Strombidæ.

Muricidæ.

PLATE V.

Muricidæ.

PLATE VI.

Buccinidæ.

PLATE VII.

Conidæ.

Volutidæ.

Cypræidæ.

PLATE VIII.

Naticidæ.

PLATE IX.

Turritellidæ.

PLATE X.

PLATE XI.

Fissurellidæ.

Calyptræidæ.

Patellidæ.

Dentaliadæ.

Chitonidæ.

PLATE XII.

Helicidæ.

1. Helix (*Acavus*) hæmastoma, L. $\frac{2}{3}$. Ceylon.
2. —— (*Polygyra*) polygyrata, Born. $\frac{1}{2}$. Brazil.
3. —— (*Carocolla*) lapicida, L. Britain.
4. —— (*Anastoma*) globulosa, Lam. Brazil.
5. —— (*Tridopsis*) hirsuta, Sby. U. States.
6. —— (*Streptaxis*) contusa, Fér. Brazil.
7. —— (*Sagda*) epistylium, Müll. Jamaica.
8. —— (*Helicella*) cellaria, Müll. Britain.
9. —— (*Stenopus*) lævipes, Müll. Malabar.
10. Bulimus oblongus, Müll. $\frac{1}{2}$. Guiana.
11, 12 —— decollatus, L. S. Europe.
13. —————— (*Partula*) faba, Martyn. Australian Islands.
14. —————— (*Zua*) lubricus, Müll. Britain.
15. —————— (*Azeca*) tridens, Pulteney. Britain.
16. Pupa uva, L. sp. Guadaloupe.
17. —— (*Vertigo*) Venetzii, Charp. $\frac{5}{4}$. Pliocene, Essex.
18. Megaspira elatior, Spix sp. $\frac{2}{3}$. Brazil.
19. Clausilia plicatula, Drap. Kent.
20. Cylindrella cylindrus, Chemn. sp. $\frac{2}{3}$. Jamaica. ⚮
21. Balæa perversa, L. sp. Britain.
22. Achatina variegata, Fab. Col. $\frac{1}{2}$. W. Africa.
23. Succinea putris, L. Britain.
24. —————— (*Omalonyx*) unguis, Orb. Paraguay.

Limacidæ.

25. Limax maximus, L. Britain.
26. Testacella haliotoides, Fèr. $\frac{2}{3}$. Britain.
27. Parmacella (*Cryptella*) calyculata, Sby. Canaries.
28. Vitrina Draparnaldi, Cuv. Britain.
29. —— (*Daudebardia*) brevipes, Drap. $\frac{2}{1}$. Austria.

Limneidæ.

30. Limnea stagnalis, L. sp. Britain.
31. —————— (*Amphipeplea*) glutinosa, Müll. Britain.
32. Physa fontinalis, Mont. sp. Britain.
33. Ancylus fluviatilis, Lister sp. Britain.
34. Planorbis corneus, L. sp. Britain.

Auriculidæ.

35. Auricula Judæ. L. $\frac{2}{3}$. India.
36. —— scarabæus, Gm. sp. Ceylon.
37. —— (*Conovulus*) coffea, L. W. Indies.
38. —— (*Alexia*) denticulata, Mont. sp. Britain.
39. Carychium minimum, Drap. sp. $\frac{5}{4}$. Britain.

Cyclostomidæ.

40. Cyclostoma elegans, Müll. sp. Britain.
41. Cyclophorus involvulus, Müll. sp. $\frac{2}{3}$. India.
42. Pupina bi-canaliculata, Sby. N. Australia.
43. Helicina Brownii, Sby. Philippines.
44. Acicula fusca, Walker, sp. $\frac{4}{1}$. Britain,

PLATE XIII.

The real size of each species is indicated by the accompanying line.

Dorididæ.

Tritoniadæ.

Æolididæ.

Elysiadæ.

PLATE XIV.

Opistho-branchiata. PAGE

1. Tornatella tornatilis, L. Brit. 180
2. Cylindrites acutus, Sby. Bath Oolite, Brit. 180
3. Acteonella Renauxiana, D'Orb. $\frac{1}{3}$. L. Chalk, France 180
4. Cinulia avellana, Brongn. U. Green-sand, Brit. 180
5. Tornatina voluta, Quoy sp. $\frac{3}{2}$. I. Guam, Australia 181
6. Bulla ampulla, L. $\frac{1}{2}$. India 182
7. —— (*Atys*) naucum, L. $\frac{1}{2}$. Philippines 182
8. Linteria viridis, Rang. Pitcairn's Id. 182
9. Acera bullata, Müll. Brit. 183
10. Cylichna cylindracea, Mont. Brit. 183
11. Aplustrum aplustre, L. sp. $\frac{1}{2}$. Mauritius 183
12. Scaphander lignarius, L. sp. $\frac{2}{5}$. Brit. 184
13. Bullæa aperta, L. sp. Brit. 184
14. Aplysia depilans, (hybrida, Sby.) Brit. 185
15. Dolabella verrucosa, Gmel. sp. $\frac{1}{3}$. Mauritius 186
16. Lobiger Philippii, Krohn. Sicily 186
17. Pleurobranchus membranaceus, Mont. $\frac{2}{5}$. Brit. 187
18. Umbrella umbellata, Dillw. $\frac{1}{4}$. Mauritius 187

Nucleobranchiata.

19. Carinaria cymbium, L. $\frac{1}{2}$. Medit. 200
20. Cardiapoda placenta, E. and S. $\frac{4}{1}$. Atlantic 200
21. Atlanta Peronii, Les. 22, operc. 23 fry. S. Atlantic 200
24. Oxygyrus Keraudrenii, Rang. 25, operc. S. Atlantic 201
26. Bellerophina minuta, Sby. Gault, Brit. 201
27. Bellerophon bi-carinatus, Lév. $\frac{1}{3}$. Carb. Limestone, Tournay ... 201
28. ———— expansus, Sby. $\frac{1}{4}$. U. Silurian, Brit. 201
29. Porcellia Puzosi, Lév. $\frac{1}{2}$. Carb. Limestone, Belgium 201
30. Cyrtolites ornatus, Conrad. (cast) $\frac{2}{3}$. L. Silurian, U. States 201
31. Ecculiomphalus Bucklandi, Portl. $\frac{1}{2}$. Silurian, Tyrone 201

Pteropoda.

32. Hyalea tridentata, Gmel. Atlantic — Medit 204
33. Cleodora pyramidata, L. Atlantic 205
34. Creseis aciculata, Rang. Atlantic 205
35. Cuvieria columnella, Rang. S. Atlantic 205
36. Vaginella depressa, Basterot. $\frac{3}{2}$. Miocene, Bordeaux 205
37. Eurybia Gaudichaudi, Souleyet. S. Pacific (*Huxley*) 206
38. Psyche globulosa, Rang. Newfoundland 206
39. Cymbulia proboscidea, Peron. Medit 206
40. Tiedemannia Neopolitana, Chiaje. Medit 206
41. Limacina antarctica (J. Hooker.) S. Polar Seas, 63°—46° 207
42. Spirialis bulimoides, D'Orb. sp. Atlantic 207
43. Cheletropis Huxleyi, Forbes. $\frac{5}{1}$. S. E. Australia 207
44. Macgillivraia pelagica, Forbes. $\frac{2}{1}$. C. Byron, E. Australia 207
45. Clio borealis, Brug. Arctic Seas 208
46. Pneumodermon violaceum, D'Orb. $\frac{3}{1}$. S. Atlantic 208
47. Spongio-branchæa australis, D'Orb. $\frac{3}{1}$. S. Atlantic, Falkland Ids. 209
48. Trichocyclus Dumerilii, Esch. $\frac{10}{1}$. South Seas 209
49. Pelagia alba, Q. and G. Amboina 209
50. Cymodocea diaphana, D'Orb. Atlantic 209

PLATE XV.

All, except those marked *, are *dorsal* views.

PLATE XVI.

Ostreidæ.

		PAGE
1.	Ostrea diluviana, Gmelin. $\frac{1}{3}$. Chalk-marl, Brit.	254
2.	—— (*Exogyra*) conica, Sby. $\frac{2}{3}$. U. Green-sand, Wilts	255
3.	Anomia Achæus, Gray. $\frac{2}{3}$. Kurachee, Scinde	255
4	Placunomia macroschisma, Desh. $\frac{1}{3}$. California	255
5.	Placuna sella, Gm. sp. $\frac{1}{4}$. China	256
6.	——— placenta, L. (young.) N. Australia	256
7.	Carolia placunoïdes, Cantr. (hinge.) Tertiary, Egypt	256
8.	Pecten plica, L. $\frac{2}{3}$. China	258
9.	—— (*Hemi-pecten*) Forbesianus, Ad. $\frac{2}{3}$. Sooloo Sea, 14 fms.	258
10.	—— (*Hinnites*) pusio, Pen. $\frac{2}{3}$. Brit.	258
11.	Lima squamosa, Lam. $\frac{1}{2}$. China	258
12.	—— (*Plagiostoma*) cardiiformis, Sby. Bath Oolite, Brit.	258
13.	—— (*Limatula*) sub-auriculata, Mont. Brit.	258
14.	—— (*Limæa*) strigilata, Brocchi sp. Pliocene, Italy	258
15.	Spondylus princeps, Gmel. $\frac{1}{2}$. Sooloo Sea	259
16.	———— (*Pedum*) spondyloïdes, Gml. $\frac{2}{5}$. Red Sea	259
17.	Plicatula cristata, *Lam.* $\frac{2}{3}$. W. Indies	259

Aviculidæ.

18.	Avicula hirundo, L. $\frac{1}{2}$. Medit.	260
19.	——— (*Meleagrina*) margaritifera, L. sp. $\frac{1}{4}$. Ceylon	260
20.	——— (*Malleus*) vulgaris, Lam. $\frac{1}{4}$. China	261
21.	——— (*Vulsella*) lingulata, Lam. $\frac{1}{3}$. Red Sea	261
22.	Posidonomya Becheri, Bronn. Carb. Hesse, Brit.	262
23.	Pinna squamosa, Lam. $\frac{1}{10}$. Medit.	263
24.	Crenatula viridis, Lam. $\frac{1}{3}$. Chinese Seas	263

---

a,a' adductor impressions.

p, pedal muscles.

g, suspensors of the gills.

b, byssal foramen or notch.

PLATE XVII.

* The figures marked are left valves ; (interiors).

Aviculidæ.

PLATE XVIII.

* The figures marked are *left* valves.

Unionidæ.

Chamidæ.

Hippuritidæ.

Tridacnidæ.

Cardiadæ (part).

Cyprinidæ (part).

† The animal of *Hyria* has two siphonal orifices.

PLATE XIX.

* The figures marked are *left* valves.

Cardiadæ.

PAGE

1. Cardium costatum, L. $\frac{1}{3}$. China 290
2. ——— lyratum, G. Sby. $\frac{1}{2}$. Madagascar 290
3. ——— hemicardium, L. $\frac{1}{2}$. China 291
4.*——— (*Adacna*) edentulum, Pallas. $\frac{2}{3}$. Caspian 291
5. Conocardium Hibernicum, Sby. $\frac{1}{2}$. Carb. limestone, Kildare ... 292

Lucinidæ.

6.*Lucina Pennsylvanica, L. $\frac{2}{3}$. W. Indies 292
7. ——— (*Cryptodon*) flexuosa, Mont. Brit.......................... 293
8. Corbis elegans, Desh. $\frac{1}{2}$. China 293
9. Diplodonta lupinus, Broc. sp. Miocene, Turin 294
10. Ungulina oblonga, Daud. $\frac{2}{3}$. W. Africa 294
11. Kellia Laperousii, Desh. California 295
12. ——— (*Poronia*) rubra, Mont. $\frac{4}{1}$. Brit. 294
13. Montacuta substriata, Mont. $\frac{3}{1}$. Brit. 295
14. Lepton squamosum, Mont. $\frac{4}{3}$. Brit................................ 296
15. Galeomma Turtoni, Sby. Brit. 296
16.*Cyamium antarcticum, Phil. $\frac{2}{1}$. Falkland Ids..................... 294

Cycladidæ.

17. Cyclas cornea, L. R. Thames... 297
18. ——— (*Pisidium*) amnica, Müll. $\frac{3}{2}$. R. Thames.................. 297
19. Cyrenoides Dupontii, Joan. $\frac{2}{3}$. 298
20. Cyrena cyprinoides, Quoy. $\frac{1}{2}$. Ceylon 297
21. ——— (*Corbicula*) consobrina, Gray. Alexandrian Canal 297

Cyprinidæ.

22. Cyprina Islandica, L. $\frac{1}{3}$. N. Brit. 298
23. Cardinia Listeri, Sby. sp. $\frac{1}{2}$. Lias, Cheltenham 302
24. Opis lunulata, Miller sp. Inf. Oolite, Bridport 302
25. Myoconcha crassa, Sby. $\frac{1}{4}$. Inf. Oolite, Dundry 303

PLATE XX.

(All the *interiors* are right valves.)

Cyprinidæ.

	PAGE
1. Astarte sulcata, Da Costa. Brit.	299
2. Circe corrugata, Chem. $\frac{4}{5}$. Red Sea	299
3. Isocardia cor. L. $\frac{1}{2}$. Brit.	300
4. Cypricardia obesa, Reeve. $\frac{1}{2}$. India	300
5. Cardita calyculata, L. S. Africa	303
6. ——— (*Venericardia*) ajar, Adans. W. Africa	303

Veneridæ.

7. Venus paphia, L. W. Indies	304
8. Cytherea dione, L. $\frac{2}{3}$. W. Indies	305
9. ——— (*Meröe*) picta, Schum. $\frac{2}{3}$. China	305
10. ——— (*Trigona*) tripla, L. $\frac{2}{3}$. W. Africa	305
11. ——— (*Grateloupia*) irregularis, Baster. $\frac{2}{3}$. Miocene, Bordeaux	326
12. Artemis exoleta, L. $\frac{1}{2}$. Brit.	326
13. Lucinopsis undata, Pennant sp. Brit.	326
14. Tapes pullastra, Wood. $\frac{2}{3}$. Brit.	326
15. Venerupis exotica, Lam. N. Australia	327
16. Petricola lithophaga, Retz. Medit.	327
17. ——— pholadiformis, Lam. $\frac{2}{3}$. New York	327
18. Glauçomya Chinensis, Gray. China	327

---

19. Capsula rugosa, Lam. sp. $\frac{1}{2}$. N. Australia	311

PLATE XXI.

(All the interiors are *right* valves.)

Mactridæ.

		PAGE
1.	Mactra stultorum, L. ⅔. Brit.	308
2.	Gnathodon cuneatus, Gray. ½. New Orleans	308
3.	Lutraria oblonga, Gmel. ½. Brit.	309

4.	Crassatella ponderosa, Gmel. sp. ½. Australia	299

Tellinidæ.

5.	Tellina lingua-felis, L. ½. Antilles	310
6.	—— carnaria, L. Antilles	310
7.	—— planissima, Anton. ⅔, India. (*T. rosea*, Sby.)	310
8.	Diodonta fragilis, L. ¾. Galway	310
9.	Psammobia Ferroënsis, Chemn. Brit.	311
10.	———— squamosa, Lam, Borneo	311
11.	Semele reticulata, Chemn. Antilles	312
12.	—— (*Cumingia*) lamellosa, G. Sby. Bahamas	312
13.	—— (*Syndosmya*) alba, Wood. Brit.	312
14.	Scrobicularia piperata, Gm. sp. ½. Brit.	312
15.	Mesodesma glabratum, Lam. Ceylon	313
16.	———— (*Donacilla*) donacium, Lam. ½. Peru	313
17.	———— (*Anapa*) Smithii, Gray. Tasmania	313
18.	Ervilia nitens, Mont. Antilles	313
19.	Donax denticulatus, L. Antilles	313
20.	—— (*Iphigenia*) Brasiliensis, Lam. ½. Antilles	314
21.	Galatea reclusa, Born sp. ½. R. Nile	314

22.	Tancredia extensa, Lycett. ½. Oolite, Brit.	293

PLATE XXII.

* The figures marked are *left* valves (interiors).

Tellinidæ.

		PAGE
1.	Sanguinolaria livida, Lam. $\frac{1}{2}$. N. Australia	311
2.	——— diphos, Chemn. $\frac{1}{3}$. India	311
3.	——— orbiculata, Wood. $\frac{1}{2}$. S. America	311

Solenidæ.

4.	Solen siliqua, L. $\frac{1}{3}$. Brit.	315
5.	Cultellus lacteus, Spengl. $\frac{1}{3}$. Tranquebar	315
6.	——— (*Cerati-solen*) legumen, L. $\frac{2}{3}$. Brit.	315
7.	——— (*Machæra*) politus Wood. $\frac{2}{3}$. India	316
8.	Solecurtus strigilatus, L. $\frac{2}{3}$. W. Africa	316
9.	——— Caribæus, Lam. $\frac{2}{3}$. U. States	316
10.	——— (*Novaculina*) Gangeticus, Bens. Calcutta	317

Myacidæ.

11.	Thetis hyalina, Sby. sp. $\frac{2}{3}$. China	319
12.	Panopæa Americana, Conrad. $\frac{1}{3}$. Miocene, Maryland	319
13.*	Saxicava rugosa, L. $\frac{2}{3}$. Brit. — Kamtschatka	320
14.	Glycimeris siliqua, Chemn. $\frac{2}{3}$. Arctic America	320

Anatinidæ.

15.*	Pholadomya candida, Sby. $\frac{1}{3}$. W. Indies	322
16.	Goniomya literata, Sby. $\frac{1}{2}$. Oolite, Brit.	322

17.	Solemya togata, Poli sp. $\frac{1}{2}$. Medit.	271

PLATE XXIII.

* The interiors marked are *left* valves.

Myacidæ.

	PAGE
1.*Mya truncata, L. ½. Brit.	317
2. Corbula sulcata, Lam. W. Africa	317
3. ——— (*Potamomya*) labiata, Maton sp. ⅗. Buenos Ayres	318
4. ——— (*Sphenia*) Binghami, Turt. Brit.	318
5. Neæra cuspidata, Olivi. Brit.	318
6. Anatinella candida, Chemn. ⅗. Ceylon	309

Anatinidæ.

7. Anatina subrostrata, Lam. ½. India	321
8. Cochlodesma prætenue, Mont. Brit.	321
9. Thracia pubescens, Pult. ⅓. Brit.	322
10.*Lyonsia Norvegica, Chemn. sp. ⅔. Brit.	323
11. Pandora rostrata, Lam ¾. Guernsey	324
12. Myodora brevis, Stutch. New South Wales	324
13. Myochama anomioides, Stutch. New South Wales	324
14. Chamostrea albida, Lam. sp. ½. New South Wales	325

Gastrochænidæ.

15. Gastrochæna modiolina, Lam. Galway	325
15a. ———— sp. siphonal orifices, in U. Green-sand, Haldon, Devonshire	326
16. ———— mumia, Spengl. ⅗. India	326
17. Clavagella bacillaris, Desh. ½. Pliocene, Sicily	32
18. Aspergillum vaginiferum, Lam. ½. Red Sea	327

Pholadidæ.

19. Pholas Bakeri, Desh. ½. India	328
20. ——— (*Pholadidea*) papyracea, Solr. ⅔. Brit.	329
21. ——— (*Martesia*) striata, L. W. Indies	329
22. ——— (*Parapholas*) bisulcata, Conrad. California	329
23. Xylophaga dorsalis, Turt. Brit.	329
24, 25. Teredina personata, Lam. London Clay, Bognor	330
25a. ———— siphonal orifice	
26.*Teredo Norvegica, Spengl. Brit.	329
27. ———————— siponal end of the tube, broken to show septa.	
28. ——— bi-pennata, Turt. (styles) Brit.	330

PLATE XXIV.

(Tunicated Mollusca described in the Supplement.)

Ascidiadæ.

1. Molgula tubulosa, Rathke.   N. Brit.
2. Cynthia papillosa, Brug. sp. $\frac{1}{3}$.   Medit.
3. Pelonæa glabra, Forbes, $\frac{2}{3}$.   N. Brit.
4. Chelyosoma Macleayanum, Brod: $\frac{1}{2}$.   Greenland.
5. Boltenia pedunculata, M. Edw. $\frac{1}{18}$.   New Zealand.

Clavellinidæ.

6. Clavellina lepadiformis, O. F. Müll.   North Sea.
7. Perophora Listeri, Wiegm. $\frac{2}{1}$.   Brit.

Bortyllidæ.

8. Botryllus violaceus, M. Edw. $\frac{2}{1}$.   France.
9.*Botrylloides rotifera, M. Edw.   France, N. Coast.
10.*Didemnium gelatinosum, M. Edw.   France.
11.*Eucœlium hospitiolum, Sav.   Medit.
12.*Distomus fuscus, M. Edw.   France.
13. Diazona violacea, Sav. $\frac{1}{4}$.   Ivica, Medit.
14. Aplidium lobatum, Sav. $\frac{1}{2}$.   Gulf of Suez.
15. Polyclinum constellatum, Sav.   Red Sea.
16. Parascidium flavum, M. Edw. $\frac{1}{3}$.   France.
17.*Amorœcium argus, M. Edw.   France.
18. ————— proliferum, M. Edw. (larva).   France.
19. Synœcium turgens, Phipps. $\frac{2}{3}$.   Spitzbergen.
20. Sigillina australis, Sav. $\frac{1}{3}$.   Australia.

Pyrosomidæ.

21. Pyrosoma giganteum, Lesueur, $\frac{1}{6}$.   Atlantic. Medit.

Salpidæ.

22. Salpa maxima, Forsk. $\frac{1}{4}$.   Medit. Atlantic.
23. Doliolum denticulatum, Q. and G. $\frac{4}{1}$.   New Zealand.
24. Appendicularia flabellum, Chamisso. $\frac{3}{1}$.   New Guinea.

* Magnified figures of zoïds separated from the common mass.

Lightning Source UK Ltd.
Milton Keynes UK
UKHW020921241118
332794UK00008B/584/P